MATHEMATICS
for Business Careers

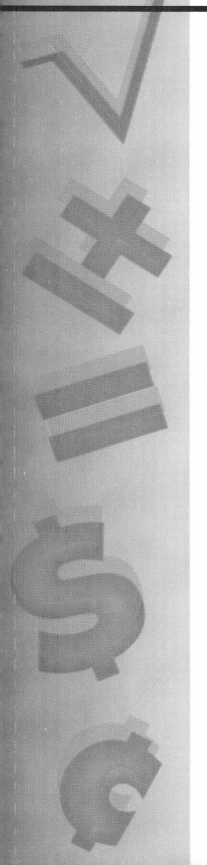

4th EDITION

MATHEMATICS
for Business Careers

JACK CAIN

ROBERT A. CARMAN

Prentice Hall

Upper Saddle River, New Jersey   *Columbus, Ohio*

Library of Congress Cataloging-in-Publication Data

    Mathematics for business careers / Jack Cain, Robert A. Carman.—
4th ed.
      p. cm.
    Includes index.
    ISBN 0-13-849258-1 (pbk.)
    1. Business mathematics. 2. Business mathematics—Problems,
exercises, etc.   I. Carman, Robert A.   II. Title.
HF5691.C26 1998
650′.01′513—dc21

                                                          96–49152
                                                            CIP

Cover art: Adam Niklewicz/SIS
Editor: Stephen Helba
Developmental Editor: Carol Hinklin Robison
Production Editor: Patricia S. Kelly
Design Coordinator: Julia Zonneveld Van Hook
Text Designer: Susan E. Frankenberry
Cover Designer: Brian Deep
Production Manager: Pamela D. Bennett
Illustrations: Steve Botts
Marketing Manager: Danny Hoyt

This book was set in Frutiger and Times by The Clarinda Company and was printed and bound by
Courier/Kendalville, Inc. The cover was printed by Courier/Kendalville, Inc.

Printed in the United States of America

10 9 8 7 6 5 4 3 2 1

ISBN: 0-13-849258-1

Prentice-Hall International (UK) Limited, *London*
Prentice-Hall of Australia Pty, Limited, *Sydney*
Prentice-Hall of Canada, Inc., *Toronto*
Prentice-Hall Hispanoamericana, S. A., *Mexico*
Prentice-Hall of India Private Limited, *New Delhi*
Prentice-Hall of Japan, Inc., *Tokyo*
Simon & Schuster Asia Pte. Ltd., *Singapore*
Editora Prentice-Hall do Brasil, Ltda., *Rio de Janeiro*

Preface

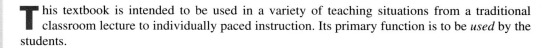

This textbook is intended to be used in a variety of teaching situations from a traditional classroom lecture to individually paced instruction. Its primary function is to be *used* by the students.

Learning business mathematics is much like learning tennis. One can watch an instructor demonstrate a topspin forehand or read about it, but to master the technique, one needs careful explanation and practice, practice, and more practice. This textbook provides students with clear detailed explanations of concepts followed by the necessary practice. Students participate actively in the course. A three-step procedure is followed:

Step 1 First, students read some explanatory material and are led through one or more **worked examples.**

Step 2 Second, students complete practice problems in the **Your Turn** feature.

Step 3 Finally, **immediate feedback** is provided using close-by answers and detailed solutions.

This three-step sequence facilitates the students' comprehension of various business math concepts by providing a means of "learning by doing."

Special features are included to make this text effective for students:

- Careful attention has been given to **readability** and the visual organization of the instructional material.

- All business math techniques are clearly outlined in a **Steps** sequence, giving step-by-step guidance.

- Each section ends with an extensive set of **practice problems** and real-life applications. Each student has abundant opportunity for practicing the business mathematics operations being learned. Answers to the odd-numbered section problems are at the end of the book.

- Each chapter ends with a **Self-Test** covering the work of that chapter.

- The **format** is designed for students; it is clear and easy to follow. This text respects the individual needs of readers so as to assure their understanding and continued attention.

- **Answers** to all the even problems in the text, along with the solution and answers to selected odd problems, are given in an Instructor's Manual.

- A light, lively **conversational style** of writing and a pleasant, easy-to-understand visual approach are used. **Humor** is used both to appeal to the student and to provide nonthreatening answers to the often-asked questions students hesitate to voice.

This book has been used in both classroom and individualized instruction settings, and it has been carefully field-tested with thousands of students. Students who have used this book tell us it is helpful, interesting, even fun to work through. More important, it works.

Flexibility of use was a major criterion in the design of the book, and field testing indicates that the book can be used successfully in a variety of course formats. It can be used as a textbook in traditional lecture-oriented courses. It is also very effective in situations where an instructor wishes to modify a traditional course by devoting a portion of class time to independent study. The book is especially useful in programs of individualized or self-paced instruction, whether in a learning lab situation, with tutors, or in totally independent study.

The fourth edition contains several new features:

- Many new up-to-date applications have been added;

- All problem sets, self-tests, exercises, and applications have been revised and updated;

- The sequencing of chapters has been changed so as to introduce basic percent calculations earlier;

- More calculator examples have been added, and emphasis has been placed on the use of business calculators rather than scientific calculators;

- The material on percent calculations has been revised;

- Minor changes have been introduced to make the text even more "user-friendly" and effective as a teaching tool.

ACKNOWLEDGMENTS

Many people have contributed to the development of this book. We are indebted to the following instructors who reviewed previous editions and offered many helpful criticisms and suggestions:

Christine Belles,	**Macomb Community College**
Ruth E. Branyan,	**Humphreys College**
John Brown,	**Savannah Area VoTech**
Janet Ciccarelli,	**Herkimer Community College**
Dick Clark,	**Portland Community College**
Joanne M. Collins,	**Kelley Business Institute**
Alton W. Evans,	**Tarrant County Junior College**
Anne Marie Gautier,	**DeVry Institute–Georgia**
Frank Goulard,	**Portland Community College**
Clo Hampton,	**West Valley Community College**
Donald Hollin,	**Mississippi County Community College**
David Lanning,	**Cortland Community College**
Donald Linner,	**Essex County College**
Jim McAnelly,	**Waubonsee Community College**
Michael R. Menaker,	**Montgomery College**

Eileen Norkunas,   **Hagerstown Business College**
Mary Lynn Pavone,   **ITT Technical Institute**
Nancy Regan,   **Pearl River Community College**
Elmer Shellenberger,   **Bethany Nazarene College**
Roger Waibel,   **Lockyear College**

We also wish to thank the reviewers of the fourth edition:

Dolores Foresman,   **Newport Business Institute**
Kevin W. Storatz,   **Denver Business College**
Elizabeth Garofalo,   **Central City Business Institute**

This book has benefited greatly from their excellence as teachers.

We are especially grateful to Marjorie K. Gross of Carteret Community College who carefully reviewed the manuscript, made many helpful suggestions, and contributed greatly to the new applications that are included.

Through the successive editions of this book many people have made important contributions. In particular, the staff of Prentice Hall have provided outstanding assistance at every step of the book's development and production. For this fourth edition we are especially grateful to developmental editor Carol Robison for her guidance and assistance during a difficult time and to production editor Patty Kelly for guiding the project through the publication process. A special thanks goes to copyeditor Lois Porter for her outstanding work, which went far beyond the call of duty, and for her unfailing good humor.

Brief Contents

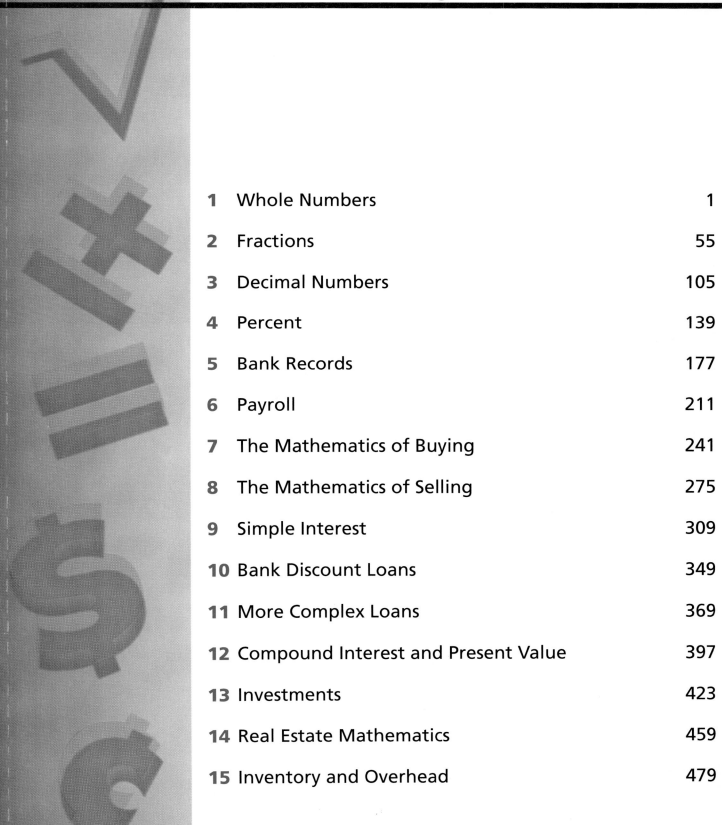

1	Whole Numbers	1
2	Fractions	55
3	Decimal Numbers	105
4	Percent	139
5	Bank Records	177
6	Payroll	211
7	The Mathematics of Buying	241
8	The Mathematics of Selling	275
9	Simple Interest	309
10	Bank Discount Loans	349
11	More Complex Loans	369
12	Compound Interest and Present Value	397
13	Investments	423
14	Real Estate Mathematics	459
15	Inventory and Overhead	479

16 Depreciation   509

17 Taxes   545

18 Insurance   581

19 Financial Statement Analysis   627

20 Statistics and Graphs   679

Appendix A: The Electronic Calculator   A-1

Appendix B: Metric System   B-1

Answers for Students   C-1

Contents

CHAPTER **1**

Whole Numbers   *1*

1.1   Writing Whole Numbers   2

Rounding Whole Numbers   5

1.2   Adding Whole Numbers   11

Adding   11 | How To Add Long Lists of Numbers   15 |
Checking Addition   17

1.3   Subtracting Whole Numbers   23

Subtracting   23 | Checking Subtraction   25 | Word Problems   26

1.4   Multiplying Whole Numbers   31

Multiplying   31 | Multiplication Shortcuts   35 |
Checking Multiplication   36

1.5   Dividing Whole Numbers   41

Dividing   41 | Checking Division   44 | Word Problems   45

Key Terms   51

Self-Test   53

CHAPTER **2**

Fractions   *55*

2.1   Renaming Fractions   56

Describing Fractions   56 | Changing an Improper Fraction to a Mixed Number
58 | Changing a Mixed Number to an Improper Fraction   59 | Raising a
Fraction to Higher Terms   60 | Reducing a Fraction to Lowest Terms   61

2.2   Multiplication of Fractions   67

2.3   Division of Fractions   75

Dividing Fractions   75 | Checking Division   77

2.4   Addition and Subtraction of Fractions   83

Adding Like Fractions   83 | Adding Mixed Numbers   84 | Adding Unlike Fractions   85 | Least Common Denominator   86 | Adding Mixed Numbers With Unlike Fractions   90 | Subtracting Fractions   92 | Subtracting Mixed Numbers   93

Key Terms   101
Self-Test   103

CHAPTER **3**

Decimal Numbers   *105*

3.1   Adding and Subtracting Decimals   106

Reading Decimal Numbers   106 | Adding Decimal Numbers   109 | Subtracting Decimal Numbers   110

3.2   Multiplying and Dividing Decimals   117

Multiplying Decimal Numbers   117 | Dividing Decimal Numbers   118 | Rounding   120

3.3   Converting Decimal Numbers and Fractions   127

Writing a Fraction as a Decimal Number   127 | Converting Mixed Numbers 128 | Writing Decimal Numbers as Fractions   129

Key Terms   135
Self-Test   137

CHAPTER **4**

Percent   *139*

4.1   Converting To and From Percent   140

Understanding Percent   140 | Changing a Decimal Number to a Percent   143 | Changing a Fraction to a Percent   144 | Changing a Percent to a Decimal 146 | Changing a Percent to a Fraction   147

4.2   Solving Percent Problems   153

General Guidelines   153 | Solving Percent Problems   153 | *P*-Type Problems   155 | %-Type Problems   159 | Calculating Percent Increase or Decrease   161 | *B*-Type Problems   162

Key Terms   173
Self-Test   175

CHAPTER **5**

Bank Records   *177*

5.1   Setting Up and Using a Checking Account   178

Opening an Account   178 | Making a Deposit   179 | Writing a Check 180 | Completing Check Stubs   182 | Using a Check Register   185

5.2   Reconciling Checking Accounts   195

Checking Your Reconciliation   199

Key Terms   205
Self-Test   207

C H A P T E R   **6**

Payroll   *211*

6.1   Paying Salaried and Hourly Employees   212

Computing Salary   212 | Calculating Hourly Pay   213 | Paying Overtime
214

6.2   Determining Piecework and Commission Pay   223

Calculating Piecework   223 | Computing Differential Piecework   224 |
Paying Commission   226 | Determining Sliding Scale Commissions   227 |
Calculating Salary Plus Commission   228

Key Terms   235
Self-Test   237

C H A P T E R   **7**

The Mathematics of Buying   *241*

7.1   Taking Advantage of Trade Discounts   242

Calculating Discounts   242 | Determining Trade Discounts   244 |
Calculating Chain Discounts   245 | Single Equivalent Discount Rate   250

7.2   Using Cash Discounts   257

Determining the Due Date   257 | Calculating Cash Discounts   258 |
EOM and Prox Dating   260 | ROG Dating   262 | Extra Dating   262 |
Crediting Partial Payments   263

Key Terms   269
Self-Test   271

C H A P T E R   **8**

The Mathematics of Selling   *275*

8.1   Calculating Markup Based on Cost   276

Comparing Cost, Markup, and Selling Price   276 | Finding the Markup When
Cost and Rate Are Known   277 | Finding the Rate When Markup and Cost Are
Known   278 | Finding the Cost When Markup and Rate are Known   279 |
Finding the Cost When the Selling Price and Rate Are Known   280

8.2   Calculating Markup Based on Selling Price   287

Finding the Markup When the Selling Price and Rate Are Known   287 |
Finding the Rate When the Markup and Selling Price Are Known   288 |

Finding the Selling Price When the Markup and Rate Are Known 289 |
Finding the Selling Price When the Cost and Rate Are Known 289

8.3 Setting Markdowns 297

Finding Markdown 297 | Calculating a Series of Markdowns 298 |
Determining Markdown Rate 300

Key Terms 305

Self-Test 307

C H A P T E R **9**

Simple Interest   *309*

9.1 Calculating Time 310

Using Exact Time 311 | Finding Exact Time by the Calendar Method 312 |
Calculating Ordinary or 30-Day-Month Time 314

9.2 Calculating Simple Interest 321

Determining Accurate Simple Interest 321 | Ordinary Simple Interest at
Ordinary or 30-Day-Month Time 325 | Calculating Ordinary Simple Interest
at Exact Time 327 | Recapping Simple Interest 329

9.3 Solving for Principal, Interest Rate, and Time 335

Solving for the Interest Variables 335 | Solving for Principal 336 |
Solving for Interest Rate 337 | Solving for Time 339 | Recapping Simple
Interest Problems 341

Key Terms 345

Self-Test 347

C H A P T E R **10**

Bank Discount Loans   *349*

**10.1 Promissory Notes and Bank Discount on Noninterest-Bearing
Notes 350**

Calculating Net Proceeds Under Bank Discount 351 | Determining Bank
Discount Using Banker's Interest 353

10.2 Bank Discount of an Interest-Bearing Note 357

Using Ordinary Interest to Calculate Discount 357 | Using Banker's Interest
to Calculate Discount 359

Key Terms 365

Self-Test 367

C H A P T E R **11**

More Complex Loans   *369*

11.1 Using Installment Loans 370

Determining Finance Charges 370 | Calculating the Annual Percentage Rate
371

11.2   Working With the Rule of 78   379

Finding the Sum of the Digits   379 | Calculating Refunds with the Rule of 78
380

11.3   Using Open-End Credit   385

Converting Between APR and Monthly Rates   385 | Finding Finance Charges
386 | Choosing a Payment Option   388 | Calculating Average Daily
Balance and New Balance   388

Key Terms   393
Self-Test   395

CHAPTER **12**

Compound Interest and Present Value   *397*

12.1   Compound Interest   398

Calculating Compound Interest   398 | Finding Maturity Values Using
Compound Interest   399 | Determining the Effective Rate   402 |
Calculating Interest Compounded Daily   404

12.2   Present Value   411

Finding Present Value   411

Key Terms   419
Self-Test   421

CHAPTER **13**

Investments   *423*

13.1   Annuities: Due and Ordinary   424

Calculating the Amount of an Annuity Due   424 | Determining the Value of an
Ordinary Annuity   427

13.2   Sinking Funds   433

Calculating the Payment for a Sinking Fund   433

13.3   Stocks and Bonds   439

Understanding Stocks   439 | Finding Dividends for Preferred Stock   439 |
Calculating Dividends for Cumulative Preferred Stock   440 | Determining
Dividends for Common Stock   441 | Calculating Stock Dividends   442 |
Paying for Stock   442 | Understanding Bonds   444 | Determining Current
Yield   447

Key Terms   453
Self-Test   457

CHAPTER **14**

Real Estate Mathematics   *459*

14.1   Mortgage Payments and Points   460

Paying Points   463 | Finding the Discount   464

14.2   Amortization   469

Determining Interest and Principal Payments   469

Key Terms   475
Self-Test   477

CHAPTER **15**

Inventory and Overhead   *479*

15.1   Inventory Valuation   480

Calculating Inventory Value Using the Specific Identification Method   481 |
Finding Inventory Value Using the Average Cost Method   483 | Determining
Inventory Value Using the FIFO Method   484 | Identifying Inventory Value
Using the LIFO Method   485 | Comparing Inventory Methods   487

15.2   Retail Method and Overhead   493

Estimating Inventory Value Using the Retail Method   493 | Allocating
Overhead by Sales Volume   495 | Allocating Overhead by Floor Space   497

Key Terms   503
Self-Test   507

CHAPTER **16**

Depreciation   *509*

16.1   Straight-Line and Units-of-Production Methods   510

Using the Straight-Line Method to Calculate Depreciation   511 | Constructing
a Depreciation Schedule   512 | Applying the Units-of-Production Method to
Depreciation   513

16.2   Declining-Balance and Sum-of-the-Years'-Digits Methods   521

Using the Declining-Balance Method of Depreciation   521 | Applying the
Sum-of-the-Years'-Digits Method of Depreciation   524

16.3   The ACRS and MACRS Methods   533

Using the ACRS Method of Depreciation   533 | Applying the MACRS Method
of Depreciation   535

Key Terms   541
Self-Test   543

CHAPTER **17**

Taxes   *545*

17.1   Federal Income Tax and FICA   546

Understanding Federal Income Tax   546 | Using the Wage Bracket Method
547 | Applying the Percentage Method   553 | Computing FICA   557

17.2   Unemployment Taxes   565

Calculating Federal Unemployment Tax   565 | Determining State
Unemployment Tax   566

17.3   Sales Tax, Excise Tax, and Property Tax   569

Calculating Sales Tax   569 | Determining Excise Tax   570 | Finding
Property Tax   571

Key Terms   575

Self-Test   577

C H A P T E R **18**

Insurance   *581*

18.1   Fire Insurance   582

Calculating Fire Insurance Premiums   582 | Determining Short-Term Policy
Premiums   584 | Canceling Property Insurance   586 | Determining Losses
Paid by Multiple Insurers   589 | Calculating Losses Paid by Coinsurance
Policies   590

18.2   Motor Vehicle Insurance   599

Determining Driver Classification   599 | Calculating Liability Insurance
Payouts   600 | Determining Liability Premiums on Motor Vehicle Insurance
601 | Finding Collision and Comprehensive Payments and Premiums   603 |
Calculating Medical Payments Premiums   605 | Total Motor Vehicle Insurance
Premiums   605

18.3   Life Insurance   611

Calculating Life Insurance Premiums   612 | Determining Surrender Options
613 | Selecting a Settlement Option   615

Key Terms   619

Self-Test   623

C H A P T E R **19**

Financial Statement Analysis   *627*

19.1   Balance Sheets   628

Reading Balance Sheets   628 | Performing a Horizontal Analysis   630 |
Performing a Vertical Analysis   634

19.2   Income Statements   647

Reading Income Statements   647 | Performing a Horizontal Analysis   649 |
Performing a Vertical Analysis   653

19.3   Financial Ratio Analysis   663

Calculating the Current Ratio   663 | Using the Acid-Test Ratio   664 |
Finding Return on Equity   665 | Determining Inventory Turnover   666

Key Terms   671

Self-Test   673

CHAPTER **20**

Statistics and Graphs   *679*

20.1   Business Statistics   680

Finding the Mean   680 | Finding the Mean for Grouped Data   682 |
Determining the Median   684 | Calculating the Mode   686 | Determining
the Range   686 | Reading Tables of Data   687

20.2   Business Graphs   697

Reading Bar Graphs   697 | Plotting Bar Graphs   699 | Reading Line
Graphs   702 | Plotting Line Graphs   703 | Reading Pie Charts   706 |
Constructing a Pie Chart   708

Key Terms   723
Self-Test   725

APPENDIX **A**

The Electronic Calculator   *A-1*

APPENDIX **B**

Metric System   *B-1*

Answers For Students   *C-1*

Index   *I-1*

Whole Numbers

CHAPTER OBJECTIVES

When you complete this chapter successfully, you will be able to:

1. Write whole numbers in both numerical form and word form.

2. Add, subtract, multiply, and divide whole numbers.

3. Use these operations to solve business problems.

■ During the first three months of the year, Micro Systems Support (MSS) reported the following income: January—$23,572; February—$22,716; and March—$24,247. What was the quarterly income for MSS?

■ On March 1, Mom and Pop's Grocery had $635 in its checking account. During March, Mom deposited checks of $352 and $114, and Pop wrote checks for $37, $216, and $106. How much was left in the account at the end of the month?

■ Your new office has 26 square yards of floor space. If carpeting costs $13 per square yard, what would it cost to carpet the room?

■ The annual insurance premium for the Daredevil Company is $3540. What are the monthly payments?

A s you can see from these examples, businesspeople frequently use the basic arithmetic skills of adding, subtracting, multiplying, and dividing. Calculators, of course, have made this work much faster. But to succeed in business, you must be able to perform simple calculations by hand and to check your calculator's result. It's easy to punch a wrong number on a calculator and get a wrong answer. For example, if you receive an order for 100 computers from each of 3 branch stores and punch in a 4 by mistake, you will have an extra 100 computers with no place to go (and maybe no job left for you). In this textbook, you'll learn the basic mathematical ideas and how to apply them to business problems. Then you'll get practice using your calculator.

Can You:

Convert a whole number in numerical form to word form?
Convert a whole number in word form to numerical form?
Round whole numbers.
. . . If not, you need this section.

Numeral Any way of expressing or naming a number, for example, 2 or II.

Digit The basic symbol used to construct a numeral; the decimal number system recognizes the digits 0, 1, 2, 3, 4, 5, 6, 7, 8, and 9.

Place Value System A characteristic of a number system in which the location of a digit indicates its value; places in the decimal system include ones, tens, hundreds, thousands, and so on.

We use **numerals** to name numbers. For example, the number of corners on a square is four, or 4, in our number system, IV in Roman numerals, and ⬧ in Chinese numerals.

In our decimal number system, we use the ten **digits** 0, 1, 2, 3, 4, 5, 6, 7, 8, and 9 to build numerals, just as we use the 26 letters of the alphabet to build words. Along with the digits, we use a **place value system** of naming numbers, where the value of any digit depends on the place in which it is located. The rightmost digit denotes the *ones* place. The next place to the left is called the *tens* place. For example, 49 represents 4 tens plus 9 ones, or forty-nine.

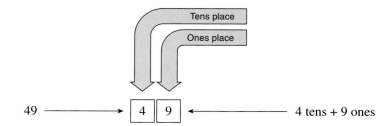

YOUR TURN

WORK THIS PROBLEM

The Question:

Try writing 68 in word form.

✔ YOUR WORK

The Solution:

68 = 6 tens plus 8 ones, so the answer is sixty-eight.

To the left of the tens place is the *hundreds* place, and the next place left is the *thousands* place. The number 2849 represents 9 ones, 4 tens, 8 hundreds, and 2 thousands, or two thousand, eight hundred forty-nine.

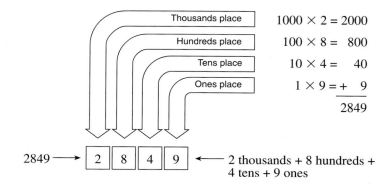

The *word form* of a number is very important in business. The dollar amount of a check is always written in word form.

| WORK THIS PROBLEM |

The Question:

Write out the following numbers in word form.

(a)   367 = _____

(b) 9712 = _____

(c) 3045 = _____

| ✔ YOUR WORK |

The Solution:

(a) 367 = three hundred sixty-seven.
(b) 9712 = nine thousand, seven hundred twelve.
(c) 3045 = three thousand, forty-five.

You can translate any large number into words by using the following diagram.

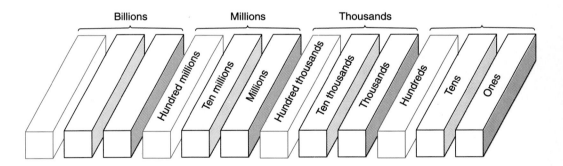

You would put the number 14,237 in this diagram as shown next. Read the number as "fourteen thousand, two hundred thirty-seven."

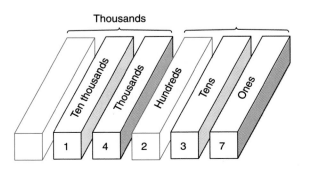

The number 47,653,290,866 becomes

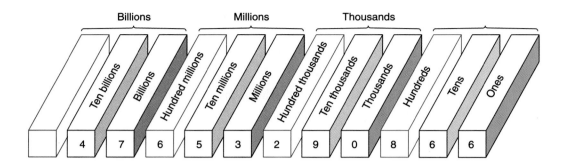

and is read "forty-seven billion, six hundred fifty-three million, two hundred ninety thousand, eight hundred sixty-six." In each block of three digits, read the digits in the normal way ("forty-seven," "six hundred fifty-three") and add the name of the block ("billion," "million").

YOUR TURN

WORK THIS PROBLEM

The Question:

Write the following numbers in word form.

(a) 6,109,276 = _____

(b) 156,768 = _____

(c) 538,251,740,459 = _____

✔ YOUR WORK

The Solution:

(a) 6,109,276 = six million, one hundred nine thousand, two hundred seventy-six.
(b) 156,768 = one hundred fifty-six thousand, seven hundred sixty-eight.
(c) 538,251,740,459 = five hundred thirty-eight billion, two hundred fifty-one million, seven hundred forty thousand, four hundred fifty-nine.

Rounding Whole Numbers

Rounding   The process of approximating a number.

While most business calculations require an accurate and precise answer, there are times when a quick approximation is useful. The approximation is often useful in checking your answer. The method used for approximation is **rounding.**

Steps: Rounding Whole Numbers

		Example
STEP 1.	*Determine* the place to which the number is to be rounded. Mark it with a ∧ to the left of that place value.	Round 24,147 to the nearest hundred. 24,1∧47
STEP 2.	If the digit to the right of the mark is less than 5, *replace* all digits to the right of the mark with zeros.	24,1∧47 becomes 24,100.
STEP 3.	If the digit to the right of the mark is equal to or larger than 5, *increase* the digit to the left of the mark by 1 and *replace* all digits to the right of the mark with zeros.	87,1∧72 becomes 87,200.

YOUR TURN

┌ WORK THIS PROBLEM ┐──────────────────────────────

The Question:

Round the following monthly sales figures to the nearest hundreds.

January	$56,818
February	73,264
March	62,549
April	73,682

┌ ✔ YOUR WORK ┐──────────────────────────────

The Solution:

January	$56,8∧18 is rounded to $56,800 because 1 is less than 5.
February	73,2∧64 is rounded to 73,300 because 6 is larger than 5.
March	62,5∧49 is rounded to 62,500 because 4 is less than 5.
April	73,6∧82 is rounded to 73,700 because 8 is larger than 5.

If you had difficulty with any of these problems, review this section. Otherwise, continue with Section Test 1.1 for practice problems on writing whole numbers and rounding.

WORKSPACE

SECTION TEST 1.1 Whole Numbers ━━━━━━

The following problems test your understanding of Section 1.1, Writing Whole Numbers. The answers to the odd-numbered problems are in the appendix.

A. Write the following in word form.

1. 28 _____

2. 53 _____

3. 243 _____

4. 910 _____

5. 767 _____

6. 421 _____

7. 1927 _____

8. 2135 _____

9. 17,925 _____

10. 63,785 _____

11. 645,003 _____

12. 902,009 _____

13. 67,000,532 _____

14. 1,333,333 _____

15. 973,574,810 _____

16. 81,999,646 _____

17. 72,901,375 _____

18. 19,070,702 _____

19. 1,255,987,123 _____

20. 59,831,121,111 _____

B. Write the following in numerical form.

1. Twenty-seven _____

2. Sixty-three _____

3. Three hundred eighty-seven _____

4. Five hundred twelve _____

5. Fifty-nine thousand, seven hundred sixty-five _____

6. Eleven thousand, six hundred ten _____

7. One hundred thousand, one hundred _____

8. Eighty-two thousand, five _____

9. Thirty-two million, four hundred sixteen thousand, seventy-five _____

10. One hundred million, four hundred thousand, three hundred twelve _____

11. Fifty-seven thousand, two hundred twelve _____

12. Thirty-seven million, four hundred seven thousand, thirteen _____

13. Sixteen billion, nine hundred eighty-nine million, three hundred thousand, fifty-two _____

_____

14. Seven hundred twelve thousand _____

15. Ten million _____

16. Nineteen billion, two hundred thirteen _____

17. Sixteen thousand, four hundred thirty-four _____

18. Sixty-seven million, five thousand, six hundred fourteen _____

19. Six hundred billion, two hundred twenty-seven thousand, five _____

_____

20. Six hundred fifty-five million, seven hundred twenty-eight thousand, forty-two _____

_____

C. Round the following numbers.

1. Round to the nearest hundred.
   (a) 327,596   (b) 512   (c) 64,251

   (d) 67,234   (e) 59,999   (f) 152,497

2. Round to the nearest thousand.
   (a) 874,592   (b) 4644   (c) 83,487

   (d) 45,555,555   (e) 41,714   (f) 759,601

3. Round to the nearest million.
   (a) 23,770,342   (b) 120,700,500   (c) 672,393

   (d) 12,007,999   (e) 79,900,999   (f) 2,496,867

D. Solve this problem.

For the following monthly sales:

January	$56,748
February	62,892
March	59,965
April	67,325
May	72,855
June	62,361
July	74,590
August	64,772
September	58,448
October	61,379
November	67,846
December	84,950

1. Round to the nearest hundreds.
2. Round to the nearest thousands.

WORKSPACE

SECTION 1.2: Adding Whole Numbers

Adding

Addition is the simplest arithmetic operation. For example,

$$\begin{matrix} \$ \ \$ \\ \$ \ \$ \end{matrix} + \begin{matrix} \$ \ \$ \\ \$ \end{matrix} = \begin{matrix} \$ \ \$ \ \$ \ \$ \\ \$ \ \$ \ \$ \end{matrix}$$

$$4 \ + \ 3 \ = \ \ 7$$

Addends Numbers to be added together.

Sum The result of adding two or more numbers together.

We add collections of objects by combining them into a single set and then counting and naming that new set. The numbers being added, 4 and 3 in this case, are called **addends** and 7 is the **sum** of the addition.

If you have not already memorized the addition of one-digit numbers, do so now.

YOUR TURN

[WORK THIS PROBLEM]

The Question:

Complete the following table by adding the number at the top to the number at the side and placing their sum in the proper square. We have added 1 + 2 = 3 and 4 + 3 = 7 for you.

Add	4	2	8	7	5	6	1	3	9
2									
4								7	
7									
5									
1		3							
9									
6									
8									
3									

✔ YOUR WORK

The Solution:

Add	4	2	8	7	5	6	1	3	9
2	6	4	10	9	7	8	3	5	11
4	8	6	12	11	9	10	5	7	13
7	11	9	15	14	12	13	8	10	16
5	9	7	13	12	10	11	6	8	14
1	5	3	9	8	6	7	2	4	10
9	13	11	17	16	14	15	10	12	18
6	10	8	14	13	11	12	7	9	15
8	12	10	16	15	13	14	9	11	17
3	7	5	11	10	8	9	4	6	12

Did you notice that changing the order in which you add numbers does not change their sum?

$$2 + 4 = 6 \quad \text{and} \quad 4 + 2 = 6 \quad \text{and so on}$$

The same facts apply when you add numbers with more than one digit.

WORK THIS PROBLEM

**YOUR
TURN**

The Question:

Add 35 + 42.

✔ YOUR WORK

The Solution:

77

Did you get this answer? If not, your error may stem from failing to line up the digits like this:

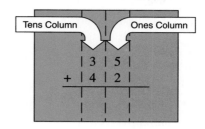

1. The numbers to be added are arranged vertically in columns.
2. The right end or ones digits are placed in the ones column, the tens digits are placed in the tens column, and so on.

Avoid the confusion of

$$
\begin{array}{r} 35 \\ + \ 42 \end{array} \quad \text{or} \quad \begin{array}{r} 35 \\ + \ 4\,2 \end{array}
$$

The most frequent cause of errors in arithmetic is carelessness, especially in simple procedures such as lining up the digits correctly. Carelessness in business may cost the company money and you a job.

Once the digits are lined up, the problem is easy.

$$
\begin{array}{r} 35 \\ +42 \\ \hline 77 \end{array}
$$

Sometimes the digits in a column will add up to more than nine. In such cases, follow this simple procedure:

Steps: Adding Whole Numbers

STEP 1. *Line up* the addends vertically.

STEP 2. *Add* each column, starting on the right, "carrying" when necessary.

For example, to add 27 + 48,

Step 1	*Step 2(a)*	*Step 2(b)*
Align the addends:	Add the ones column:	Add the tens column:
	$\overset{1}{}$	$\overset{1}{}$
27	27	27
+48	+48 7 + 8 = 15	+48 1 + 2 + 4 = 7
	5 Write 5.	75
	Carry 1 ten.	

The little 1 above the tens digit column is the "carry." The sum of the ones column is 7 + 8 = 15. Write down the 5 in the ones column in the sum and "carry" the 1 to the tens column.

YOUR TURN

| WORK THIS PROBLEM |

The Question:

Add the following pairs of numbers.

(a) 429 + 758   (b) $256 + 867
(c) 267 + 1135 + 2461   (d) $15,178 + 166 + 4415 + 27 + 13,001

| ✔ YOUR WORK |

The Solution:

After aligning the numbers *(Step 1),*

(a) **Step 2a**   **Step 2b**   **Step 2c**

$$
\begin{array}{r} 1 \\ 429 \\ +758 \\ \hline 7 \end{array}
$$
9 + 8 = 17
Write 7.
Carry 1 ten.

$$
\begin{array}{r} 1 \\ 429 \\ +758 \\ \hline 87 \end{array}
$$
1 + 2 + 5 = 8

$$
\begin{array}{r} 1 \\ 429 \\ +758 \\ \hline 1187 \end{array}
$$
4 + 7 = 11

(b) **Step 2a**   **Step 2b**   **Step 2c**

$$
\begin{array}{r} 1 \\ \$256 \\ +867 \\ \hline 3 \end{array}
$$
6 + 7 = 13
Write 3.
Carry 1 ten.

$$
\begin{array}{r} 1\ 1 \\ \$256 \\ +867 \\ \hline 23 \end{array}
$$
1 + 5 + 6 = 12
Write 2.
Carry 1 hundred.

$$
\begin{array}{r} 1\ 1 \\ \$256 \\ +867 \\ \hline \$1123 \end{array}
$$
1 + 2 + 8 = 11

(c) **Step 2a**   **Step 2b**

$$
\begin{array}{r} 1 \\ 267 \\ 1135 \\ +2461 \\ \hline 3 \end{array}
$$
7 + 5 + 1 = 13
Write 3.
Carry 1 ten.

$$
\begin{array}{r} 1\ 1 \\ 267 \\ 1135 \\ +2461 \\ \hline 63 \end{array}
$$
1 + 6 + 3 + 6 = 16
Write 6.
Carry 1 hundred.

Step 2c   **Step 2d**

$$
\begin{array}{r} 1\ 1 \\ 267 \\ 1135 \\ +2461 \\ \hline 863 \end{array}
$$
1 + 2 + 1 + 4 = 8

$$
\begin{array}{r} 1\ 1 \\ 267 \\ 1135 \\ +2461 \\ \hline 3863 \end{array}
$$
1 + 2 = 3

(d) **Step 2a**   **Step 2b**

$$
\begin{array}{r} 2 \\ \$15,178 \\ 166 \\ 4,415 \\ 27 \\ +13,001 \\ \hline 7 \end{array}
$$
8 + 6 + 5 + 7 + 1 = 27
Write 7.
Carry 2 tens.

$$
\begin{array}{r} 1\ 2 \\ \$15,178 \\ 166 \\ 4,415 \\ 27 \\ +13,001 \\ \hline 87 \end{array}
$$
2 + 7 + 6 + 1 + 2 + 0 = 18
Write 8.
Carry 1 hundred.

Step 2c

    1 2

$15,178

   166

 4,415

    27

+13,001 $1 + 1 + 1 + 4 + 0 = 7$

   787

Step 2d

 1   1 2

$15,178

   166

 4,415

    27

+13,001 $5 + 4 + 3 = 12$

2,787   Write 2.

            Carry 1 ten thousand.

Step 2e

 1   1 2

$15,178

   166

 4,415

    27

+13,001 $1 + 1 + 1 = 3$

$32,787

How To Add Long Lists of Numbers

Businesspeople often need to add long lists of numbers. A manager for an insurance company might need to add up all an agent's sales in order to figure out how much commission to pay that person. A manufacturing supervisor might need to add up all of the hours worked by people in a given department or all of the units of a certain good produced. Accounts for large conglomerates may involve adding the profits from hundreds of divisions in order to determine the firm's total profits.

In such cases, the best procedure is to break the problem down into a series of simpler additions. First add sets of two or three numbers, then add these sums to obtain the total. For example, to add 9, 3, 7, 6, 12, 4, 17, and 5,

```
   9
   3 ----- 12
   7
   6 ----- 13 ----- 25
  12
   4 ----- 16
  17
 + 5 ----- 22 ----- 38
                    63
```

You do a little more writing but carry fewer numbers in your head; the result is fewer mistakes. Better yet, keep your eye peeled for combinations that add to 10 or 15, and work with mental addition of three addends.

```
   9
   3 ⌐
   7 ⌐─ 10 ----- 19
   6
  12 ⌐─ 10
   4 ----- 22
  17
   5 ----- 22
                    63
```

YOUR TURN

WORK THIS PROBLEM

The Question:

Add the following columns of numbers.

(a)		(b)		(c)		(d)		(e)		(f)	
	8		7		3		11		3		13
	17		6		5		7		5		17
	3		8		7		2		12		11
	4		5		6		5		7		14
	11		9		5		6		6		15
	9		3		1		7		4		8
	16		7		3		13		1		9
	7		12		4		6		2		16
	11		8		2		5		18		12
	5		16		7		14		9		7
					3		16		7		8

✔ YOUR WORK

The Solution:

(a) 91   (b) 81   (c) 46   (d) 92   (e) 74   (f) 130

Very few business problems involve such small numbers. You can use the same techniques to solve problems with bigger numbers, though.

YOUR TURN

WORK THIS PROBLEM

The Question:

Robert, a salesman for the Jiffy Linoleum Company, earned the following monthly commissions: January—$1275; February—$1382; March—$987; April—$1129; May—$1264; June—$1383; July—$1492; August—$1537; September—$1487; October—$1572; November—$1731; December—$1645. What were his total earnings for the year?

✔ YOUR WORK

The Solution:

$ 1,275
1,382
987
1,129
1,264
1,383
1,492
1,537
1,487
1,572
1,731
+1,645
$16,884

Checking Addition

In business, accuracy is not only important, it's required. Most store clerks, bank tellers, or any businesspeople who deal with money must balance their receipts.

Businesses use many complicated methods of "checking" the accuracy of addition. The easiest method is by adding in the opposite direction. In addition, normally we start at the top and add downward. In "checking," we start at the bottom and add each column upward. For example, to add and check 456 + 862 + 57,

	Add		456	**Check:**		456
			862			862
Downward			+57	Add up		+57
			1375			1375

YOUR TURN

[WORK THIS PROBLEM]

The Question:

Mary had a savings account balance of $478 before she made deposits of $75 and $115. What is her current balance? Check your answer.

[✔ YOUR WORK]

The Solution:

	Add		$478	**Check:**		478
			75			75
Downward			+115	Add up		+115
			$668			$668

You should always check your answers even if you use a calculator. Although calculators rarely make errors, calculator operators often do.

If you had any trouble with the last problem, review this section. Otherwise, continue to Section Test 1.2.

WORKSPACE

Name

Date

Course/Section

SECTION TEST 1.2 Whole Numbers

The following problems test your understanding of Section 1.2, Adding Whole Numbers. The answers to the odd-numbered problems are in the appendix.

A. Add.

1. 28	2. 72	3. 53	4. 49	5. 58	6. 17
37	29	28	34	27	64

7. 87	8. 96	9. 48	10. 93	11. 75	12. 97
45	74	73	84	85	67

13. 256	14. 372	15. 394	16. 738	17. 357	18. 285
283	729	825	361	472	756

19. 752	20. 837	21. 576	22. 925	23. 419	24. 837
256	835	859	892	793	953

B. Add.

1. 7394	2. 3719	3. 8371	4. 7392	5. 1726
837	736	975	283	939

6. 6381	7. 1538	8. 8479	9. 9274	10. 8279
2575	9374	8462	5837	7368

11. 2734	12. 2574	13. 7318	14. 2597	15. 3947
7317	3857	8036	8374	4185

16. 2702	17. 2745	18. 8373	19. 8572	20. 7378
9374	8461	8563	9658	6722

C. Add.

1. 738	2. 739	3. 938	4. 495	5. 2635
234	375	736	362	127
752	864	379	857	481

6. 2748	7. 7395	8. 7268	9. 2608	10. 4829
863	571	473	572	738
135	738	823	717	513

11. 7261	12. 5724	13. 9488	14. 5727	15. 9027
927	824	1024	4782	9274
37	173	831	1836	932

16. 8263	17. 7382	18. 4826	19. 9346	20. 6246
836	1894	9273	9264	3364
48	383	825	846	632
878	3834	9363	5153	6133

D. Arrange vertically and add.

1. 737 + 78 + 737 + 42 + 7272 = _____

2. 813 + 246 + 84 + 839 + 12 = _____

3. 719 + 2780 + 7326 + 7378 + 43 = _____

4. 721 + 963 + 26 + 9641 + 2943 + 9 = _____

5. 7172 + 837 + 6589 + 73 + 726 + 1739 = _____

6. 839 + 8915 + 8 + 7837 + 2589 + 4793 = _____

7. 7261 + 7236 + 163 + 2582 + 27 + 7624 = _____

8. 8436 + 739 + 6281 + 1326 + 26 + 9626 + 271 = _____

9. 67 + 8362 + 273 + 8263 + 7 + 6738 + 3725 = _____

10. 7263 + 6378 + 12 + 74,734 + 278 + 63 + 8361 = _____

E. Solve these problems.

1. The Easy Street Diner needs 17 employees for the morning shift, 24 for the dinner shift, and 11 for the midnight shift. How many employees are needed for an entire day?

2. During the first three months of the year, Micro Systems Support, Inc., (MSS) reported the following income: January—$23,572; February—$22,716; and March—$24,247. What was the quarterly income for MSS?

3. Wired for Sound Electronics has three branches with the following daily sales. Calculate the total daily sales, the weekly total sales for each store, and the total sales for the week.

Branch	M	T	W	Th	F	S	Weekly Total
Downtown	$2574	$2148	$2014	$1957	$2683	$3189	
Suburb	2956	2751	2515	2283	3108	3573	
Mall3479	3216	3127	3065	3581	4157		
Totals							

4. Liki Valve, Inc., produced the following number of valves for the week: Monday—263; Tuesday—259; Wednesday—275; Thursday—252; and Friday—85. What was the total for the week?

5. The Gadget Company produced the following number of gadgets each day: Monday—28,672; Tuesday—29,736; Wednesday—31,736; Thursday—34,737; and Friday—29,638. Calculate the weekly total.

6. Calculate the total daily sales, weekly total sales for each salesperson, and the total sales for the week at The Write Place Bookstore.

Employee	M	T	W	Th	F	S	Weekly Total
C. Dickens	$1672	$1589	$1638	$1379	$1963	$2392	
H. Melville	1837	1525	1462	1783	1938	2308	
W. Irving	1452	1563	1783	2183	2619	2748	
G. Chaucer	2719	1893	2183	1974	2848	2892	
M. Twain	2748	2590	2863	2519	2581	2389	
Totals							

7. The local elevator music station KDOZ supplies music tapes to tall buildings. Herbie, the salesman, sold 26 tapes Monday, 24 tapes Tuesday, 19 tapes Wednesday, 18 tapes Thursday, and 12 tapes Friday. What were Herbie's total sales for the week?

8. Bob's Beeflike Burgers sells an amazing number of burgers. The following is a typical week of burger sales for the three Bob's stores. How many burgers did each store sell in the week? What is the total number of burgers sold by the stores in the week? What are the daily total burger sales for all three stores?

Store	Sun	M	T	W	Th	F	S	Weekly Total
B.B.B. #1	4513	6692	5436	5978	6123	7531	5942	
B.B.B. #2	4345	3344	2862	4323	4328	3978	3746	
B.B.B. #3	6232	5323	6453	4154	5463	4326	3910	
Totals								

9. Bits 'n' Bytes, a local computer store, sold the following computer system. The laser printer cost $879, the monitor cost $389, and the other components were packaged together for $1199. What was the total amount of the sale?

10. Hole-in-the-Wall Bagels has five delivery trucks. Truck 1 delivered 135 cases of bagels, truck 2 delivered 162 cases, truck 3 delivered 213 cases, truck 4 delivered 192 cases, and truck 5 delivered 167 cases. How many cases did Hole-in-the-Wall Bagels deliver?

11. Fit to Print, which specializes in computerized "desktop" publishing, is doing an inventory. It has 32 laser printers, 42 dot matrix printers, and 31 portable printers. How many printers does Fit to Print own?

12. The Zippy Delivery Service truck made five deliveries. The mileages for the deliveries were 87, 121, 213, 97, and 117. What was the truck's total delivery mileage?

13. The Butter-em-Up Bakery made $231 Monday, $361 Tuesday, $276 Wednesday, $307 Thursday, and $198 Friday. What were the total profits for the week?

14. Fly-By-Night Aviary sold 27 doves, 9 parrots, 31 parakeets, and 18 finches. How many birds did the Aviary sell?

15. Maria's Mannequins rented out 89 mannequins during January, 123 mannequins during February, and 116 mannequins during March. How many mannequins did Maria rent out during the first quarter?

Can You:

Find the difference between two whole numbers?
Check the accuracy of a subtraction solution?
Use subtraction to solve word problems?
. . . If not, you need this section.

Subtracting

In subtraction, we "take away" part or all of a preceding number. For example,

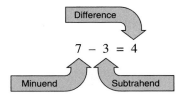

Minuend Generally, the larger of two numbers in a subtraction problem; it is the number from which the subtrahend is subtracted.

Subtrahend Generally the smaller of two numbers in a subtraction problem; it is the number subtracted from the minuend.

Difference The result of subtracting one number (the subtrahend) from another (the minuend).

To make it easier to talk about the numbers in a subtraction problem, we will refer to them by special names. The **minuend** is the larger of the two numbers in the problem. It is the number being decreased. The **subtrahend** is the number being subtracted from the minuend. The **difference** is the amount that must be added to the subtrahend to produce the minuend. It is the answer to the subtraction problem.

Most subtraction problems are written in this form, using the minus (−) sign. Sometimes, though, a subtraction problem should be set up to look more like an addition problem:

$$3 + \square = 7$$

Written this way, a subtraction problem asks the question, "How much must be added to a given number to total a required amount?"

To subtract numbers with multiple digits, you need only follow a few simple steps similar to those we used to solve addition problems:

Steps: Subtracting Whole Numbers

STEP 1. *Line up* the minuend and subtrahend vertically with the minuend above the subtrahend—larger number on top.

STEP 2. *Subtract* each column, starting on the right.

STEP 3. If a digit in the minuend is smaller than the number below it in the subtrahend, *borrow* one from the next column to the left in the minuend.

We'll return to step 3 shortly. First, let's see how steps 1 and 2 work.

WORK THIS PROBLEM

The Question:

Subtract 47 − 23.

✔ YOUR WORK

The Solution:

Step 1	*Step 2a*		*Step 2b*	
47	47		47	
−23	−23	Ones digits: 7 − 3 = 4.	−23	Tens digits: 4 − 2 = 2.
	4		24	

As we noted in the Steps, not all subtraction problems are this easy. In some cases, you need to take a third step and rewrite the minuend (larger) number using "borrowing." For example, to find the difference of 74 − 18, you would first follow step 1 and align the numbers. But in step 2, you cannot immediately subtract the 8 in the ones digits of the subtrahend from the 4 in the ones digits of the minuend. Instead, you must "borrow" from the next column to the left in the minuend, in this case a 10.

Step 3a

$$\overset{6\ 14}{7\cancel{4}}$$
−18   Borrow one 10, change the 7 in
  6   the tens place to 6, change 4 to 14,
       and subtract 14 − 8 = 6.

Step 3b

$$\overset{6\ 14}{7\cancel{4}}$$
−18   Tens digits: 6 − 1 = 5.
 56

WORK THIS PROBLEM

The Question:

Find the difference 64 − 37. (We know you have a calculator, but to check the process, work this one by hand.)

✔ YOUR WORK

The Solution:

Step 1	**Step 2a**		**Step 3a**		**Step 2b**	

64
−37

Note that the digits in the ones column cannot be immediately subtracted.

$\overset{5\ 14}{\cancel{6}\cancel{4}}$
−37
7

Borrow one 10, change the 6 in the tens place to 5, change 4 to 14, and subtract $14 − 7 = 7$.

$\overset{5\ 14}{64}$
−37 $5 − 3 = 2$
27

Checking Subtraction

Check subtraction problems by adding the answer and the subtrahend; their sum should equal the minuend.

$$
\begin{array}{ll}
37 & \text{Subtrahend} \\
+27 & \text{Difference} \quad \text{(answer)} \\
\hline
64 & \text{Minuend}
\end{array}
$$

YOUR TURN

WORK THIS PROBLEM

The Question:

Try these problems for practice. Be sure to check each problem.

(a)	71	(b)	$426	(c)	$36,467	(d)	902
	−39		−128		−19,269		−465

✔ YOUR WORK

The Solution:

(a) **Step 1**   **Step 3a**   **Check**

71          $\overset{6\ 11}{\cancel{7}\cancel{1}}$ ⎧ Borrow one 10,   39
−39   −39 ◄┤ change the 7 in the tens place to 6,   +32
                  32   │ change the 1 in the ones place to 11,   71
                              │ $11 − 9 = 2$, write 2,
                              ⎩ $6 − 3 = 3$, write 3.

(b) **Step 1**   **Step 3a**   **Step 3b**   **Check**

$426 $\overset{1\ 16}{\$4\cancel{2}\cancel{6}}$ $\overset{3\ 11\ 16}{\$\cancel{4}\cancel{2}\cancel{6}}$ ⎧ $16 − 8 = 8$, write 8. $128
−128   −128   −128 ◄┤ $11 − 2 = 9$, write 9.   +298
                              8          $298   ⎩ $3 − 1 = 2$, write 2. $426

Notice that in this case we must borrow twice. Borrow one 10 from the tens position to make 16, then borrow one 100 from the hundreds position to make 11 in the tens position.

(c)   **Step 1**   **Step 3a**   **Step 3b**   **Step 3c**

$$\begin{array}{c}
\text{\tiny 5 17} \\
\$36,467 \\
-19,269
\end{array}$$

$36,467	$36,467	$36,467	$36,467
−19,269	−19,269	−19,269	−19,269
	8	198	$17,198

17 − 9 = 8, write 8.
15 − 6 = 9, write 9.
3 − 2 = 1, write 1.
16 − 9 = 7, write 7.
2 − 1 = 1, write 1.

Check: $19,269
+17,198
$36,467

(d)   **Step 1**   **Step 3a**   **Step 3b**   **Check**

902	902	902	465
−465	−465	−465	+437
		437	902

12 − 5 = 7, write 7.
9 − 6 = 3, write 3.
8 − 4 = 4, write 4.

Notice that in this case we first borrow one 100 from the hundreds position to get a 10 in the tens place; then we borrow one 10 from the tens place to get a 12 in the ones place.

Word Problems

Businesspeople often need to use subtraction. Profit, after all, is a matter of subtracting all costs from all monies received.

It Computes, Inc., ordered $13,843 worth of computer parts in April. In May, It Computes shipped $4937 worth of defective parts back to the manufacturer. How much should It Computes' accounting department pay the manufacturer of the parts?

$13,843   **Check:** $ 4,937
−4,937   8,906
$ 8,906 $13,843

| WORK THIS PROBLEM |

The Question:

Jim originally had a bank balance of $1235, but he withdrew $477. What is his new balance?

| ✔ YOUR WORK |

The Solution:

$1235   **Check:** $ 477
−477   +758
$ 758 $1235

Check all subtraction problems, even if you are using a calculator.

SECTION TEST 1.3 Whole Numbers

The following problems test your understanding of Section 1.3, Subtracting Whole Numbers. The answers to the odd-numbered problems are in the appendix.

A. Subtract.

1. 36
  25
	2. 76
	  35
		3. 67
		  23
			4. 73
			  31
				5. 58
				  33
					6. 87
					  63

7. 78
  49
	8. 68
	  57
		9. 87
		  39
			10. 83
			  28
				11. 64
				  47
					12. 81
					  57

13. 265
  136
	14. 174
	  129
		15. 295
		  128
			16. 784
			  389
				17. 372
				  287
					18. 273
					  196

19. 653
  276
	20. 877
	  588
		21. 536
		  259
			22. 936
			  592
				23. 827
				  591
					24. 857
					  588

B. Subtract.

1. 838
  179
	2. 726
	  638
		3. 1268
		  975
			4. 2402
			  884
				5. 1762
				  939

6. 5361
  2576
	7. 4560
	  2374
		8. 8573
		  5667
			9. 9247
			  5858
				10. 5262
				  3368

11. 6743
  5357
	12. 5547
	  3859
		13. 7227
		  5136
			14. 8327
			  3379
				15. 7547
				  4182

16. 8705
  3377
	17. 5750
	  2861
		18. 8337
		  7563
			19. 3522
			  2658
				20. 7355
				  6766

21. 8370
  4479
	22. 8254
	  3778
		23. 8264
		  7277
			24. 6264
			  3665
				25. 7286
				  2778

C. Subtract.

1. 36,749
  17,929
	2. 72,636
	  63,827
		3. 11,373
		  9,777
			4. 42,305
			  18,838
				5. 42,167
				  39,392

6. 34,216
  12,558
	7. 63,670
	  37,374
		8. 37,633
		  17,667
			9. 39,158
			  25,859
				10. 56,126
				  23,358

11. 85,636	12. 74,681	13. 68,362	14. 47,457	15. 76,654
35,258	56,358	39,186	38,679	25,777

16. 68,670	17. 24,903	18. 48,272	19. 82,252	20. 47,546
23,877	12,867	37,693	52,668	27,627

21. 78,067	22. 38,462	23. 63,347	24. 45,353	25. 17,472
39,188	13,667	35,266	33,555	14,578

D. Solve these problems.

1. The Sparkle Cleaning Service has a limited amount of its secret cleaning fluid left. The service started the day with 128 gallons. The first cleaning job used 45 gallons. The next cleaning job used 34 gallons. How many gallons of the secret cleaning fluid are left?

2. Hole-in-the-Wall needs to deliver 797 cases of bagels. The delivery trucks have already delivered 392 cases. How many more cases of bagels remain to be delivered?

3. Maria's Mannequin Service provides mannequins for the local stores. She has already received orders for 127 mannequins for the week. She has 176 mannequins in stock. How many more can she rent out this week?

4. Fly-By-Night Aviary, a local bird store, has had very fertile birds for the last few months. Cages are getting crowded and the store needs to have a bird sale. Currently Fly-By-Night has 113 birds, but it also has a shipment of 50 more due tomorrow. How many birds must the store sell in the next few days to start next week with a maximum of 100 birds on hand?

5. Cindy's monthly salary is $1984. Every month she has the following deductions: federal income tax—$367; state tax—$124, FICA—$149; and insurance—$128. What is Cindy's take-home pay (pay after deductions)?

6. Zippy Delivery Service delivers packages incredibly fast in the local area. Today's deliveries have brought in $432. However, today's speeding fines total $175. How much is left over after the fines have been paid?

7. Nora's Collectibles originally bought an antique doll house for $1675. Its authenticity is in question, and the house is now worth $1205. How much has the value of the doll house decreased?

8. Harry's Mane and Tail Manager, a local horse grooming business, makes $75 for every horse tail braided. At the last horse show, Harry made $2250. However, Harry had to pay an unexpected $216 for his employees' dry cleaning bills. What was Harry's net profit after dry cleaning expenses?

9. Kate's Delicacies started with a business loan of $12,545. Kate has already paid $8345 on her loan. What is the current balance on her loan?

10. On March 1, Mom and Pop's Grocery had $635 in its checking account. During March, Mom deposited checks of $352 and $114, and Pop wrote checks for $37, $216, and $106. How much was left in the account at the end of the month?

11. The Write Stuff has 1067 cases of paper in stock. It needs to reduce stock to 450 cases. How many cases need to be sold?

12. Crystal's Diamonds is having a sale on tennis bracelets. Each bracelet is marked down $549. There are five bracelets on sale. The original prices are $3492, $5324, $4767, $4878, and $3989. What is the price of each bracelet after the markdown?

13. Bits 'n' Bytes sells the ultimate computer system for $5,782. Earlier this year the same system sold for $6,376. How much has the computer system been reduced?

14. An automobile originally costs $17,476. During the first year, it depreciated $3948. What is the value of the car at the end of the first year?

15. Nayle's Industrial Bolt Company is producing too many bolts. The company can produce 25,865 bolts daily, but can only package 23,520 daily. (a) How many bolts are not packaged daily? (b) How many bolts are not packaged after 5 days?

WORKSPACE

SECTION 1.4: Multiplying Whole Numbers

Can You:

Find the product of two whole numbers?
Check the accuracy of a multiplication solution?
Use multiplication to solve word problems?
. . . If not, you need this section.

Multiplying

Sam purchased six pens at $5 each. What was the total cost? You could count the dollars:

$$\overbrace{\$\$\$\$\$} \; \overbrace{\$\$\$\$\$} \; \overbrace{\$\$\$\$\$} \; \overbrace{\$\$\$\$\$} \; \overbrace{\$\$\$\$\$} \; \overbrace{\$\$\$\$\$} = \$30$$

Or you could add $5's:

$$\$5 + \$5 + \$5 + \$5 + \$5 + \$5 = \$30$$

Or you could use multiplication:

Multiplicand   The number to be multiplied.

Multiplier   The number of times the multiplicand is to be multiplied.

Product   The result of multiplying one number (the multiplicand) by another (the multiplier).

Factors   Numbers that, when multiplied, produce a given number; the multiplicands and multipliers that produce a given number.

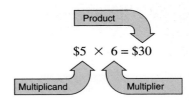

Multiplication is a shortcut method of counting or repeated addition. In a multiplication statement the **multiplicand** is the number to be multiplied, the **multiplier** is the number multiplying the multiplicand, and the **product** is the result of the multiplication. The multiplier and the multiplicand are called the **factors** of the product. In order to become skillful at multiplication, you *must* know the one-digit multiplication table from memory.

YOUR TURN

| WORK THIS PROBLEM |

The Question:

Complete the following table by multiplying the number at the top by the number at the side and placing their product in the proper square. We have multiplied $3 \times 4 = 12$ and $2 \times 5 = 10$ for you.

Multiply	2	5	8	1	3	6	9	7	4
1									
7									
5	10								
4					12				
9									
2									
6									
3									
8									

✔ YOUR WORK

The Solution:

Multiply	2	5	8	1	3	6	9	7	4
1	2	5	8	1	3	6	9	7	4
7	14	35	56	7	21	42	63	49	28
5	10	25	40	5	15	30	45	35	20
4	8	20	32	4	12	24	36	28	16
9	18	45	72	9	27	54	81	63	36
2	4	10	16	2	6	12	18	14	8
6	12	30	48	6	18	36	54	42	24
3	6	15	24	3	9	18	27	21	12
8	16	40	64	8	24	48	72	56	32

Not even the best of calculators can help you if you can't do one-digit multiplication. Practice until you are able to quickly perform these from memory.

Notice that the product of any number and 1 is that same number. For example,

$$1 \times 2 = 2$$
$$1 \times 6 = 6$$
$$1 \times 753 = 753$$

Zero has been omitted from the table because the product of any number and zero is zero. For example,

$$0 \times 2 = 0$$
$$0 \times 7 = 0$$
$$0 \times 395 = 0$$

Also notice that changing the order in which two numbers are multiplied does not change the answer. For example,

$$2 \times 6 = 6 \times 2 = 12$$
$$9 \times 5 = 5 \times 9 = 45$$

The Sexy Six

Here are the six most often missed one-digit multiplications:

"Inside"

$9 \times 8 = 72$
$9 \times 7 = 63$
$9 \times 6 = 54$
$8 \times 7 = 56$
$8 \times 6 = 48$
$7 \times 6 = 42$

It may help you to notice that in these multiplications the "inside" digits, such as 8 and 7, are consecutive and the digits of the answer add to nine: $7 + 2 = 9$. This is true for *all* one-digit numbers multiplied by 9.

Be certain you have these memorized.

(There is nothing very sexy about them, but we did get your attention, didn't we?)

The multiplication of larger numbers is based on the one-digit number multiplication table and follows a few simple steps:

Steps: Multiplying Whole Numbers

STEP 1. *Line up* the multiplicand and the multiplier vertically.

STEP 2. *Multiply* the digits in the multiplicand, starting at the right by the ones digit in the multiplier, "carrying" if necessary.

STEP 3. *Multiply* the digits in the multiplicand, starting at the right, by the next digits in the multiplier, beginning at the right. Place the first (rightmost) digit of each answer in the same column as the digit of the multiplier.

STEP 4. *Add* the partial products as aligned according to step 3.

Actually, these steps sound harder than they are. To show you how easy multiplication can be, let's start with a simple problem involving only steps 1 and 2: Find the product of 34×2.

Step 1	*Step 2a*		*Step 2b*	
34	34			34	
×2	×2	Multiply ones:		×2	Multiply 2 × 3 = 6,
	8	2 × 4 = 8; write 8.		68	write 6.

YOUR TURN

WORK THIS PROBLEM

The Question:

Find the product of $26,495 × 9. (Be careful, this problem requires "carrying.")

✔ YOUR WORK

The Solution:

Step 1	*Step 2a*		*Step 2b*	
		⁴		⁸ ⁴
$26,495	$26,495		$26,495	
×9	×9	9 × 5 = 45	×9	9 × 9 = 81
	5	Write 5, carry 4.	55	Add the carry, 81 + 4 = 85, write 5, carry 8.

Step 2c

⁴ ⁸⁴
$26,495
　×9　 9 × 4 = 36
　455　 Add the carry,
　　　　36 + 8 = 44,
　　　　write 4, carry 4.

Step 2d

⁵ ⁴ ⁸⁴
$26,495
　×9　 9 × 6 = 54
　8455　Add the carry, 54 + 4 = 58,
　　　　write 8, carry 5.

Step 2e

⁵ ⁴ ⁸ ⁴
$26,495
　　×9　 9 × 2 = 18
$238,455 Add the carry, 18 + 5 = 23,
　　　　　write 23.

Now that you've mastered steps 1 and 2, let's work a calculation involving a two-digit multiplier (and thus steps 3 and 4): 24 × 57.

Step 1

　57
×24

Step 2

　²
　57　 4 × 7 = 28. Write 8, carry 2.
×24　 4 × 5 = 20. Add the carry.
228　 20 + 2 = 22. Write 22.

Step 3

¹
²
　57
×24　 2 × 7 = 14. Write 4, carry 1.
228　 2 × 5 = 10. Add the carry.
114　 10 + 1 = 11. Write 11.

Step 4

¹
²
　57
×24
228 } Add.
114 }
1368

YOUR TURN

WORK THIS PROBLEM

The Question:

Multiply the following numbers.

(a) $64 × 37   (b) 327 × 145   (c) 342 × 102   (d) $5847 × 3256

✔ YOUR WORK

The Solution:

Note that we've omitted the steps here to save space. If your answers are different, trace the process step by step to find your problem.

(a)
$$\begin{array}{r} \$64 \\ \times 37 \\ \hline 448 \\ 192 \\ \hline \$2368 \end{array}$$
$7 × 64 = 448$
$3 × 64 = 192$

(b)
$$\begin{array}{r} 327 \\ \times 145 \\ \hline 1635 \\ 1308 \\ 327 \\ \hline 47{,}415 \end{array}$$
$5 × 327 = 1635$
$4 × 327 = 1308$
$1 × 327 = 327$

(c)
$$\begin{array}{r} 342 \\ \times 102 \\ \hline 684 \\ 000 \\ 342 \\ \hline 34{,}884 \end{array}$$
$2 × 342 = 684$
$0 × 342 = 000$
$1 × 342 = 342$

(d)
$$\begin{array}{r} \$5847 \\ \times 3256 \\ \hline 35082 \\ 29235 \\ 11694 \\ 17541 \\ \hline \$19{,}037{,}832 \end{array}$$
$6 × 5847 = 35{,}082$
$5 × 5847 = 29{,}235$
$2 × 5847 = 11{,}694$
$3 × 5847 = 17{,}541$

Multiplication Shortcuts

There are hundreds of quick ways to multiply various numbers. Most of them are quick only if you are already a math whiz and will confuse you more than help you. Here are a few that are easy to do and easy to remember.

1. To multiply by 10, attach a zero to the right end of the multiplicand. For example,

$$34 × 10 = 340$$
$$256 × 10 = 2560$$

Multiplying by 100 or 1000 is similar.

$$34 × 100 = 3400$$
$$256 × 1000 = 256{,}000$$

2. To multiply by a number ending in zeros, multiply the other numbers, then attach the zeros to the right. Two examples are $26 × 20$ and $34 × 2100$.

$$\begin{array}{r} 26 \\ \times 20 \end{array} \longrightarrow \begin{array}{r} 26 \\ \times 2\,|\,0 \\ \hline 52\,|\,0 \end{array} \qquad \begin{array}{r} 34 \\ \times 2100 \end{array} \longrightarrow \begin{array}{r} 34 \\ \times 2100 \\ \hline 34 \\ 68 \\ \hline 71{,}400 \end{array}$$

3. If both multiplier and multiplicand end in zeros, multiply the other numbers, then attach *all* zeros to the right. Two examples are 230×200 and 1000×100.

$$
\begin{array}{r}
230 \\
\times 200
\end{array}
\longrightarrow
\begin{array}{r}
23\,|\,0 \quad \longleftarrow\text{1 zero} \\
\times 2\,|\,00 \quad \longleftarrow\text{2 zeros} \\
\hline
46\,|\,000 \quad \longleftarrow\text{3 zeros}
\end{array}
\qquad
\begin{array}{r}
1000 \longleftarrow\text{3 zeros} \\
\times 100 \longleftarrow\text{2 zeros} \\
\hline
100{,}000 \longleftarrow\text{5 zeros}
\end{array}
$$

This sort of multiplication is mostly a matter of counting zeros.

Checking Multiplication

The easiest method for checking multiplication is to switch the multiplier and the multiplicand and then remultiply.

YOUR TURN

| WORK THIS PROBLEM |
_____

The Question:

Larry's Photo Shop sold 47 cameras for $198 each. What was the total income from these camera sales? Check your answer.

| ✔ YOUR WORK |
_____

The Solution:

$$
\begin{array}{r}
\$198 \\
\times 47 \\
\hline
1386 \\
792 \\
\hline
\$9306
\end{array}
\qquad
\textbf{Check:}
\qquad
\begin{array}{r}
47 \\
\times \$198 \\
\hline
376 \\
423 \\
47 \\
\hline
\$9306
\end{array}
$$

Always remember to check your work, even if you are using a calculator. It's extremely easy to hit the wrong keys.

SECTION TEST 1.4   Whole Numbers

The following problems test your understanding of Section 1.4, Multiplying Whole Numbers.

A. Multiply.

1. 36 __6_	2. 57 __5_	3. 82 __3_	4. 72 __9_	5. 39 __2_	6. 65 __8_
7. 87 __9_	8. 53 __7_	9. 79 __7_	10. 38 __2_	11. 46 __4_	12. 71 __5_
13. 26 _18_	14. 74 _12_	15. 95 _28_	16. 78 _39_	17. 72 _47_	18. 27 _67_
19. 53 _76_	20. 87 _58_	21. 56 _25_	22. 36 _59_	23. 82 _53_	24. 85 _86_

B. Multiply.

1. 238 __97_	2. 326 __63_	3. 278 __75_	4. 407 __84_	5. 726 __90_
6. 361 __67_	7. 560 __37_	8. 809 __67_	9. 272 __58_	10. 236 __36_
11. 673 __57_	12. 547 __38_	13. 729 __53_	14. 327 __37_	15. 537 __48_
16. 705 _127_	17. 750 _261_	18. 837 _253_	19. 352 _355_	20. 703 _764_

C. Multiply.

1. 370 _479_	2. 254 _378_	3. 826 _725_	4. 624 _365_	5. 726 _729_
6. 825 _206_	7. 582 _847_	8. 580 _306_	9. 836 _958_	10. 952 _937_
11. 3864 __862_	12. 9373 __782_	13. 9005 __803_	14. 9285 __870_	15. 8603 __458_
16. 8354 __723_	17. 8846 _2749_	18. 9362 _3758_	19. 9374 _6392_	20. 3692 _7362_

D. Solve these problems.

1. A copying machine rents for $635 per month. How much will it cost the company to rent the machine for one year?

2. Your new office has 26 square yards of floor space. If carpeting costs $13 per square yard, what would it cost to carpet the room?

3. Fit-to-Print's profits are $1,023 per week. What are its profits for one year (52 weeks)?

4. Kate's Delicacies can buy a new oven for $12,360 or the installment price of 24 monthly payments of $575. (a) What is the cost if paid in installments? (b) How much would be saved by paying cash?

5. Marcy's firm is restocking office supplies. Each subdivision will be given $5,550 for its supplies. There are 17 subdivisions. What is the total amount being budgeted for office supplies?

6. Bits 'n' Bytes, the constantly growing computer store, has just received a contract for 25 complete personal computer systems. (a) If each system sells for $1327, what is the total cost? (b) If Bits 'n' Bytes reduces the price of each system by $49, what would be the total cost?

7. Nayle's Industrial Bolt Company has an inventory of 1,320 cases. Each case contains 48 bolts. What is the total number of bolts?

8. The Sparkle Cleaning Service has 23 different cleaning jobs daily. Each job takes 13 gallons of the company's secret cleaning fluid. (a) How many gallons does the Sparkle Cleaning Service use daily? (b) How many gallons are used in a five-day workweek?

9. Maria's Mannequins needs to update some of the mannequins. A local company has offered a mannequin "facelift" at the low price of $32 per mannequin. Maria has 67 mannequins that need updated. What is her total cost?

10. Lew wants to change the decor at his trendy new restaurant. New tablecloths will cost $28 each, matching napkins are $9 each, and little matching hats for his waitpersons are $12 each. Lew needs 52 new tablecloths, 250 napkins, and 50 hats to make his color statement. How much will this cost Lew?

11. The Butter-em-Up Bakery has just received a rush order for 15 dozen croissants. Each croissant needs to be individually wrapped. How many wrappers are needed?

12. The town of Vistahue offers tourists a colorful brochure of the local area. The town uses 21 brochures each day that its information booth is open. The booth is open 25 days per month. How many brochures does Vistahue need per year?

13. Big Bob's Beeflike Burgers is expanding the menu to include a burger with a guacamole topping. Big Bob estimates that each of his three stores will need two dozen avocados daily. How many avocados will his Burger stores use in one week?

14. The Zippy Delivery Service needs to make extra deliveries to pay unexpected expenses. If each delivery returns a $28 profit, and each of the seven drivers can make 21 extra deliveries a day, how much extra profit can the Zippy people make in five days?

15. Ava's Air Tours charges $37 per hour per person for sightseeing tours. Biff and Jody want to go on a 5-hour tour of the coast. How much will Ava charge them?

16. Micro Systems Support sells several types of computers. The following business graph gives a visual picture of notebook computer sales for the first half of the year.

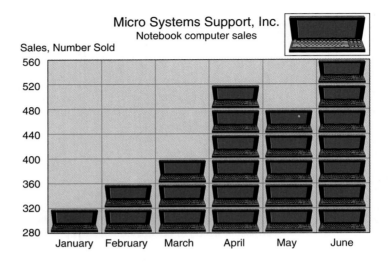

(a) What is the total sales for the first quarter (January to March)?

(b) What is the total sales for the second quarter (April to June)?

(c) What is the total sales for the first half of the year?

(d) If notebook computers sold for $2495 each in June, what was that month's total income from notebook sales?

(e) The price of notebook computers has been dropping. In January the price was $2789; February—$2745; March—$2698; April—$2624; May—$2595; and June—$2495. What was the total income from notebook computer sales for the first half of the year?

Can You:

Divide whole numbers?
Check the accuracy of division solutions?
Use division to solve word problems?
. . . If not, you need this section.

Dividing

Division enables you to separate a given quantity into equal parts. The mathematical phrase $12 \div 3$ asks you to separate a collection of 12 objects into three equal parts. The mathematical phrases

$$3\overline{)12} = 4, \quad \frac{12}{3} = 4, \quad \text{and} \quad 12/3 = 4$$

also represent division. You should read all of them as "twelve divided by three."

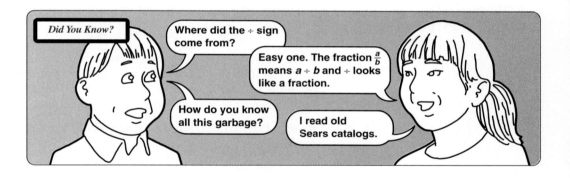

The most common way of writing a division problem is:

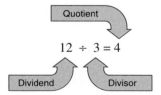

Dividend The number to be divided.

Divisor The number by which the dividend is to be divided.

Quotient The result of dividing one number (the dividend) by another (the divisor).

In division, the number being divided (in this case 12) is called the **dividend.** The number used to divide (3) is called the **divisor.** The result of the division (4) is called the **quotient,** from a Latin word meaning "how many times."

One way to perform division is to reverse the multiplication process.

$$24 \div 4 = \square \quad \text{means that} \quad 4 \times \square = 24$$

41

If the one-digit multiplication tables are firmly in your memory, you will recognize immediately that □ = 6. By placing a 6 in □, you make the statement 4 × □ = 24 true.

WORK THIS PROBLEM

YOUR TURN

The Question:

Perform the following divisions.

(a) 35 ÷ 7   (b) 45 ÷ 5   (c) 28 ÷ 4   (d) 70 ÷ 10

✔ YOUR WORK

The Solution:

(a) 35 ÷ 7 = 5   (b) 45 ÷ 5 = 9   (c) 28 ÷ 4 = 7   (d) 70 ÷ 10 = 7

How do we divide dividends that are larger than 9 × 9 and therefore not in the multiplication table? Obviously, we need a better procedure.

Steps: Dividing Whole Numbers

STEP 1.   *Arrange* the divisor and dividend horizontally.

STEP 2.   If the divisor can divide into the first (leftmost) digit of the dividend, *write* the number of times the division is possible above that digit. If not, move to the right, one place at a time, until you have a number large enough to be divided by the divisor; then write the appropriate dividing number as indicated.

STEP 3.   *Multiply* the divisor by the digit just entered over the dividend and subtract it from that digit.

STEP 4.   *Bring down* the next digit of the dividend and repeat steps 2 and 3 for the resulting number until all digits of the dividend have been addressed.

Again, these steps may sound hard but are easy when you see them in use. For example, let's find the quotient of 96 ÷ 8.

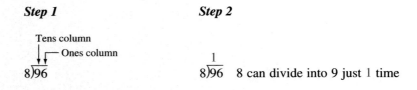

Step 1

Tens column
⌐ Ones column
8)96

Step 2

1
8)96   8 can divide into 9 just 1 time

Step 3

$$8 \overline{)\ 96} \quad \begin{cases} 8 \times 1 = 8 \\ 9 - 8 = 1 \end{cases}$$

with quotient 1, subtract -8, remainder 1

Step 4a

$$8 \overline{)\ 96} \quad \begin{cases} \text{Bring down the 6 to get 16} \\ 16 \div 8 = 2 \end{cases}$$

with quotient 12, -8, 16

Step 4b

$$8 \overline{)\ 96} \quad \begin{cases} 8 \times 2 = 16 \\ 16 - 16 = 0. \end{cases}$$

with quotient 12, -8, 16, -16, 0

So $96 \div 8$ equals 12.

WORK THIS PROBLEM

The Question:

Find the quotient of $976 \div 8$ without using a calculator.

✔ YOUR WORK

The Solution:

$$8 \overline{)\ 976} \quad \begin{array}{l} 122 \\ -8 \\ \overline{17} \\ -16 \\ \overline{16} \\ -16 \\ \overline{0} \end{array}$$

Step 1.	Line up divisor and dividend.
Step 2.	Divide 8 into 9? Write 1.
Step 3.	$8 \times 1 = 8$. Subtract $9 - 8 = 1$.
Step 4a.	Bring down 7. Divide 8 into 17, two times. Write 2.
Step 4b.	Multiply $8 \times 2 = 16$. Subtract $17 - 16 = 1$.
Step 4c.	Bring down 6. Divide 8 into 16, two times. Write 2.
Step 4d.	$8 \times 2 = 16$. Subtract $16 - 16 = 0$.

Sometimes a number "brought down" will not be divisible by the divisor. In these cases, you place a zero at the appropriate point in the quotient and bring down the next number. Repeat this process until the number "brought down" is divisible. For example,

$$5 \overline{)\ 2045} \quad \begin{array}{l} 409 \\ 20 \\ \overline{45} \\ -45 \end{array}$$

Remainder The amount "left over" when one number does not divide evenly into another.

In addition, not all division problems come out "even." For example, if you divide 50 by 6, the answer will be 8 with a **remainder** of 2.

YOUR TURN

WORK THIS PROBLEM

The Question:

Find the quotient of 3179 ÷ 6.

✔ YOUR WORK

The Solution:

$$\begin{array}{r} 529 \\ 6)\overline{3179} \\ -30 \\ \hline 17 \\ -12 \\ \hline 59 \\ -54 \\ \hline 5 \end{array}$$ ←——Remainder

Step 1.	Line up divisor and dividend.
Step 2.	Divide 6 into 3? No. Divide 6 into 31? 5 times. Write 5.
Step 3.	$6 \times 5 = 30$. Subtract $31 - 30 = 1$.
Step 4a.	Bring down 7. Divide 6 into 17, two times. Write 2.
Step 4b.	$6 \times 2 = 12$. Subtract $17 - 12 = 5$.
Step 4c.	Bring down 9. Divide 6 into 59, 9 times. Write 9.
Step 4d.	$6 \times 9 = 54$. Subtract $59 - 54 = 5$.
	5 is the remainder.

To solve a problem with a two-digit divisor, follow the same steps.

YOUR TURN

WORK THIS PROBLEM

The Question:

Find the quotient of 5096 ÷ 31.

✔ YOUR WORK

The Solution:

$$\begin{array}{r} 164 \\ 31)\overline{5096} \\ -31 \\ \hline 199 \\ -186 \\ \hline 136 \\ -124 \\ \hline 12 \end{array}$$ ←——Remainder

Step 1.	Line up divisor and dividend.
Step 2.	Divide 31 into 5? No. Divide 31 into 50? Write 1.
Step 3.	$31 \times 1 = 31$. Subtract $50 - 31 = 19$.
Step 4a.	Bring down 9. Divide 31 into 199, 6 times. Write 6.
Step 4b.	$31 \times 6 = 186$. Subtract $199 - 186 = 13$.
Step 4c.	Bring down 6. Divide 31 into 136, 4 times. Write 4.
Step 4d.	$31 \times 4 = 124$. Subtract $136 - 124 = 12$.
	Remainder = 12.

Checking Division

To check your answer, multiply the divisor by the quotient, which should yield the dividend. For example, the check for 96 ÷ 8 = 12 would be

$$8 \times 12 = 96$$

If there is a remainder, add it back in, as for 97 ÷ 8 = 12, remainder 1:

$$\begin{array}{r} 8 \times 12 = 96 \\ + 1 \\ \hline 97 \end{array}$$

Word Problems

Division—and *always* checking for accuracy—is a crucial business math skill. A sales manager wants to divide a special bonus among five salespeople. A personnel director wants to divide a yearly insurance premium into 12 monthly payments. A manufacturing supervisor wants to know a worker's average hourly rate of production.

YOUR TURN

| WORK THIS PROBLEM |

The Question:

The Giant Corporation purchased 849 giant makers for $646,938. What was the cost per giant maker? (Be sure to check your answer.)

| ✔ YOUR WORK |

The Solution:

$$
\begin{array}{r}
\$762 \\
849\overline{)\$646{,}938} \\
594\ 3 \\ \hline
52\ 63 \\
50\ 94 \\ \hline
1\ 698 \\
1\ 698 \\
\end{array}
$$

Check:

$$
\begin{array}{r}
762 \quad \leftarrow\text{Quotient} \\
\times 849 \quad \leftarrow\text{Divisor} \\ \hline
6858 \\
3048 \\
6096 \\ \hline
646{,}938 \quad \leftarrow\text{Dividend}
\end{array}
$$

WORKSPACE

SECTION TEST 1.5   Whole Numbers

The following problems test your understanding of Section 1.5, Dividing Whole Numbers.

A. Divide.

1. $84 \div 7 =$

2. $184 \div 8 =$

3. $135 \div 5 =$

4. $126 \div 2 =$

5. $222 \div 6 =$

6. $172 \div 4 =$

7. $468 \div 9 =$

8. $201 \div 3 =$

9. $657 \div 9 =$

10. $232 \div 8 =$

11. $510 \div 6 =$

12. $658 \div 7 =$

13. $\dfrac{549}{9} =$

14. $\dfrac{315}{5} =$

15. $\dfrac{354}{6} =$

16. $\dfrac{371}{7} =$

17. $\dfrac{316}{4} =$

18. $\dfrac{688}{8} =$

B. Divide.

1. $375 \div 15 =$

2. $864 \div 27 =$

3. $1081 \div 23 =$

4. $896 \div 32 =$

5. $247 \div 13 =$

6. $816 \div 34 =$

7. $361 \div 19 =$

8. $644 \div 28 =$

9. $1125 \div 25 =$

10. $\dfrac{805}{35} =$ 11. $\dfrac{4508}{49} =$ 12. $\dfrac{3741}{43} =$

13. $26\overline{)975}$ 14. $37\overline{)1085}$ 15. $41\overline{)1351}$

16. $24\overline{)427}$ 17. $52\overline{)1501}$ 18. $38\overline{)1373}$

C. Divide.

1. $7068 \div 76 =$ 2. $6375 \div 85 =$ 3. $4420 \div 68 =$

4. $4779 \div 76 =$ 5. $7636 \div 92 =$ 6. $4512 \div 47 =$

7. $49\overline{)4088}$ 8. $52\overline{)10,605}$ 9. $82\overline{)17,794}$

10. $97\overline{)31,719}$ 11. $81\overline{)22,075}$ 12. $93\overline{)33,371}$

13. $123\overline{)28,905}$ 14. $137\overline{)52,745}$ 15. $205\overline{)44,584}$

16. $362\overline{)100,738}$ 17. $473\overline{)151,164}$ 18. $793\overline{)503,582}$

D. Solve these problems.

1. The laser printer at your office prints 6 pages per minute. How long does it take to print a 174-page report?

2. The Butter-em-Up Bakery makes 216 croissants daily. Croissants are sold by the dozen. How many croissant boxes does the bakery use each day?

3. A Zippy Delivery Service driver made an overnight delivery. The round-trip took 13 hours for 702 miles. What was the average speed?

4. The Wonderful Widget Company makes a deluxe widget that sells for $27. How many deluxe widgets can Bob buy for $2160?

5. Lew wants the ultimate sound system for his restaurant. Wired for Sound Electronics will install it and finance it for one year. That total cost is $13,896. What are Lew's monthly payments?

6. The annual insurance premium for the Daredevil Company is $3540. What are the monthly payments?

7. Crystal's Diamonds wants its 17 managers to attend a local management workshop. The total cost to the company will be $5100. What is the cost per person?

8. Liki Valve, Inc., is upgrading its office equipment and has allocated $15,300 for computer equipment. If a standard personal computer system costs $1275, how many systems can Liki Valve purchase during its upgrade?

9. The Sparkle Cleaning Company obtained a new contract that will bring in $14,775 profit annually. How much will each of the three partners get if the profit is split equally?

10. The Pizza Palace has set aside $7280 for employee bonuses. If each of the 13 employees receives an equal bonus, how much is each bonus?

11. If your annual income is $24,180, what is your weekly income?

12. If your secretary types 65 words per minute, how long will it take him to type a 7800-word report?

13. Your new secretary types 75 words per minute. How long will it take for a 7800-word report to be typed?

14. If Nayle's Industrial Bolt Company produces 13,408 bolts and packages them 16 to a carton, how many cartons does it need?

15. Your cabinet business has just received a rush order for 68 cabinets. It takes two people 4 hours to build one cabinet. How many people will it take to complete the order in 8 hours?

16. Dynamite Deals, a new department store in town, rented all their mannequins from Maria's Mannequins. For the first three months, they rented 56 adult mannequins at $24 each per month and 27 child-sized mannequins at $14 each per month. After the three months, they renewed the rental for the next four months. What was the total rental amount that Dynamite Deals paid Maria for the seven-month period?

17. Sally, the buyer for the junior department of Dynamite Deals, purchased merchandise for sale with a total retail value of $225,392. One week after this purchase they still had $141,523 of this merchandise on hand. Sally's follow-up order of $253,904 arrived the next day and at the end of the second week they still had merchandise on the racks with a total retail value of $294,050. What was the total sales volume for this merchandise for these two weeks?

Key Terms

addends	factors	numeral	remainder
difference	minuend	place value system	rounding
digit	multiplicand	product	subtrahend
dividend	multiplier	quotient	sum
divisor			

Chapter 1 at a Glance

Page	Topic	Key Point	Example
2	Writing numbers	Use place value system.	4321 = four thousand three hundred two tens one ones
5	Rounding	Mark place to be rounded with a $_\wedge$. If the digit to the right of the mark is 4 or less, replace digits to the right with zeros. If the digit to the right is 5 or more, increase the digit to the left of the mark by 1 and replace digits to the right with zeros.	Round to tens place. 432$_\wedge$1 = 4320 678$_\wedge$9 = 6790
11	Adding	Add from top to bottom, "carrying" when the sum in a column exceeds the column's place value; check by adding up.	1 1 123 456 789 1,368 **Check:** 1,368 123 456 789
23	Subtracting	Subtract the subtrahend from the minuend, "borrowing" from the place to the left when the numeral in the minuend is too small; check by adding subtrahend and difference.	21111 4321 − 1234 3087 **Check:** 3087 + 1234 4321
31	Multiplying	Multiply the multiplicand by the multiplier, indenting to reflect place value of the multiplier, and add the answers; check by switching multiplier and multiplicand.	456 × 123 1368 912 456 56,088 **Check:** 123 × 456 738 615 492 56,088
41	Dividing	Divide the dividend by the divisor, placing numbers in the quotient over the appropriate place value in the dividend; check by multiplying quotient by divisor and adding any remainder.	37 R 16 123)4567 369 877 861 16 **Check:** 123 × 37 = 4551 + 16 4567

WORKSPACE

SELF-TEST   Whole Numbers ━━━━━━

The following problems test your understanding of whole numbers. The answers to the odd-numbered problems are in the appendix.

1. Write the number 34,756,105 in word form. _____

2. Write the number 83,250,615 in word form. _____

3. Write the following number in numerical form: one hundred seventy-five million, two thousand, six hundred forty-one. _____

4. Write the following number in numerical form: seven hundred two million, sixty-eight thousand, eight hundred thirty-five. _____

Perform the indicated operations.

5. $45,729 + 381,903 + 47 + 2679 + 25,148 =$ _____

6. $67 + 826,103 + 749 + 2951 + 72,918 =$ _____

7. $734,735 - 572,847 =$ _____

8. $4,382,735 - 3,191,966 =$ _____

9. $\$3,768 \times 473 =$ _____

10. $45,725 \times 794 =$ _____

11. $1,815,618 \div 387 =$ _____

12. $3,053,572 \div 652 =$ _____

13. Round the following monthly sales to the nearest hundreds.

January	$ 82,937
February	87,492
March	91,835
April	83,462
May	87,372
June	105,737
July	95,721
August	99,537
September	107,489
October	96,382
November	113,637
December	126,501

14. Round the monthly sales in problem 13 to the nearest thousands.

15. Micro Systems Support has three branches with the following daily sales. Calculate the total daily sales, the weekly sales for each store, and the total sales for the week.

Branch	M	T	W	Th	F	S	Weekly Total
Downtown	$8462	$8153	$7826	$7538	$9163	$9826	
Suburb	7592	7235	6836	7105	8472	8138	
Mall	7385	7104	6482	6242	8462	8936	
Totals							

16. The Big Mistakes Company produced the following number of erasers each day: Monday—238,624; Tuesday—247,472; Wednesday—234,258; Thursday—244,634; and Friday—201,337. Calculate the weekly total.

17. Trace's Surplus had $7325 in its checking account. After deposits of $952 and $1274, and writing checks for $737, $1264, and $2373, calculate the new account balance by adding the deposits and subtracting the checks.

18. The Nut Shop has 752 hexacentrics in stock. It needs to reduce its stock to 295. How many hexacentrics need to be sold?

19. A milling machine rents for $295 per month. How much will it cost the company to rent the machine for one year?

20. Chip 'n Dale's Cabinet Shop has just received an order for 257 cabinets. If each cabinet sells for $365, what is the total cost of the order?

21. The annual insurance premium for the Daredevil Company is $3312. What are the monthly payments?

22. If your annual income is $21,840, (a) what is your monthly income and (b) what is your weekly income?

Fractions

CHAPTER OBJECTIVES

When you complete this chapter successfully, you will be able to:

1. Write fractions as mixed numbers and improper fractions.

2. Reduce a fraction to lowest terms and raise a fraction to higher terms.

3. Multiply fractions.

4. Divide fractions.

5. Add and subtract fractions.

■ Each of David's office workers is required to have at least $66\frac{3}{4}$ square feet of office space. How much office space does David need for his 14 employees?

■ Stock from Wired for Sound is selling for $\$4\frac{3}{4}$ per share. You have $\$1900$ that you want to invest. How many shares of this stock can you buy?

■ The Bits 'n' Bytes computer store sold the following amounts of RS232 computer cable last week: $13\frac{1}{2}$ feet, $7\frac{7}{8}$ feet, and $34\frac{5}{6}$ feet. How much computer cable has the store sold?

■ Lots of Plots Realty originally owned $234\frac{1}{4}$ acres of land. After selling $145\frac{7}{8}$ acres, how many acres of land remain?

Fractions are very much a part of everyday life. Virtually every business involves some work with fractions. This "fraction" of the text is easy and fun, so let's get started.

Describing Fractions

Consider the following rectangular area. What happens when we break it into equal parts?

2 parts, each one-half of the whole

3 parts, each one-third of the whole

4 parts, each one-fourth of the whole

Notice that the four parts of "fourths" are equal in area. Any one of the four equal areas shown would be "one-fourth" of the entire area.

Fraction A way of expressing a number as the quotient of two whole numbers; usually (but not always) it is less than one.

A **fraction** is normally written as the division, or quotient, of two whole numbers:

$$\frac{2}{3}, \quad \frac{3}{4}, \quad \frac{26}{12}, \quad \text{or in the preceding example,} \quad \frac{1}{4}$$

Notice that the bottom number represents the whole and the top number represents the number of parts.

YOUR TURN

WORK THIS PROBLEM

The Question:

Write the fraction that describes the shaded portion of .

✔ YOUR WORK

The Solution:

$$\frac{3}{5} = \frac{3 \text{ shaded total}}{5 \text{ parts total}}$$

The fraction $\frac{3}{5}$ implies an area equal to three of the five original portions.

Let's look at this fraction more closely:

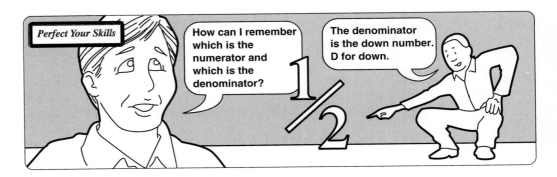

$$\frac{3}{5}$$

Numerator, the number of parts

denominator, the name of the part, "fifths"

Numerator The upper portion of a fraction; in the term $\frac{3}{4}$, 3 is the numerator.

The two numbers that form a fraction have special names that make it easier to talk about them. In the fraction $\frac{3}{5}$, the upper number (3) is called the **numerator** from the Latin *numero* meaning number. It is a count of the number of parts. The lower number is called the **denominator** from the Latin *nomen* or *name*. It tells us the name of the part being counted.

Denominator The lower portion of a fraction; in the term $\frac{3}{4}$, 4 is the denominator.

Perfect Your Skills

How can I remember which is the numerator and which is the denominator?

The denominator is the down number. D for down.

$1/2$

YOUR TURN

WORK THIS PROBLEM

The Question:

A calculator costs $6 and I have $5. What fraction of its cost do I have? Write the answer as a fraction and identify the numerator and the denominator.

✔ YOUR WORK

The Solution:

$\underline{\$\$\$\$\$}$ \$ \$5 is $\frac{5}{6}$ of the total cost

$\frac{5}{6}$ numerator = 5, denominator = 6

Proper Fraction A fraction with a value of less than 1; for example, $\frac{3}{4}$.

Fractions may be "proper" or "improper." A **proper fraction** is a number less than 1, as you would suppose a fraction should be. It represents a quantity less than the original whole that serves as the standard. For example, $\frac{1}{2}$, $\frac{2}{3}$, and $\frac{17}{20}$ are all proper fractions. Notice that for a proper fraction, the numerator is less than the denominator—the top number is less than the bottom number in the fraction.

Improper Fraction A fraction with a value equal to or more than 1; for example, $\frac{4}{3}$.

An **improper fraction** is a number greater than or equal to 1 and represents a quantity greater than the standard. Numbers such as $\frac{25}{16}$, $\frac{43}{20}$, $\frac{5}{2}$, and $\frac{84}{25}$ are all improper fractions. Improper fractions are hard to read though, so we use them only to multiply and divide fractions, as you will see later in this chapter.

YOUR TURN

WORK THIS PROBLEM

The Question:

Circle the proper fractions in the following list.

$$\frac{3}{2}, \ \frac{3}{4}, \ \frac{7}{8}, \ \frac{5}{4}, \ \frac{8}{8}, \ \frac{15}{12}, \ \frac{1}{16}, \ \frac{35}{32}, \ \frac{7}{54}, \ \frac{65}{64}, \ \frac{105}{100}$$

✔ YOUR WORK

The Solution:

The following are proper fractions (less than 1): $\dfrac{3}{4}, \ \dfrac{7}{8}, \ \dfrac{1}{16}, \ \dfrac{7}{54}$

Changing an Improper Fraction to a Mixed Number

Mixed Number A fraction written as the sum of a whole number and a proper fraction; for example, $1\frac{3}{4}$.

In business and everyday life, we need to convert improper fractions to **mixed numbers**—fractions written as the sum of a whole number and a proper fraction. For example, consider the problem of a hardware store owner who has received orders adding up to $\frac{7}{3}$ pounds of nails.

We can show the improper fraction $\frac{7}{3}$ graphically as follows:

Unit standard = ▭

with each part representing $\frac{1}{3}$ of the whole. Thus,

$$\frac{7}{3} = \boxed{}$$ (seven, count them)

We can convert this fraction to a mixed number by regrouping.

2 whole standard units + $\dfrac{1}{3}$ standard unit

We usually omit the + sign and write $2 + \frac{1}{3}$ as $2\frac{1}{3}$, and read it as "two and one-third."

$$\frac{7}{3} = 7 \div 3 = 2 \text{ remainder } 1 = 2\frac{1}{3}$$

The numbers $1\frac{1}{2}$, $2\frac{3}{5}$, and $16\frac{2}{3}$ are all mixed numbers.

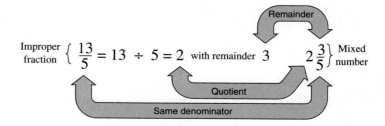

You can also convert improper fractions to mixed numbers without making a drawing.

Steps: *Converting an Improper Fraction to a Mixed Number*

STEP 1. *Divide* the numerator by the denominator.

STEP 2. *Place* any remainder over the denominator and place the resulting fraction to the right of the whole number found in step 1.

WORK THIS PROBLEM

YOUR TURN

The Question:

Convert the following improper fractions to mixed numbers.

(a) $\dfrac{23}{4}$ (b) $\dfrac{9}{5}$ (c) $\dfrac{41}{12}$

✔ YOUR WORK

The Solution:

(a) $\dfrac{23}{4} = 23 \div 4 = 5$ with remainder $3 \longrightarrow 5\dfrac{3}{4}$

(b) $\dfrac{9}{5} = 9 \div 5 = 1$ with remainder $4 \longrightarrow 1\dfrac{4}{5}$

(c) $\dfrac{41}{12} = 41 \div 12 = 3$ with remainder $5 \longrightarrow 3\dfrac{5}{12}$

Changing a Mixed Number to an Improper Fraction

The reverse process, rewriting a mixed number as an improper fraction, is equally simple.

Steps: *Converting a Mixed Number to an Improper Fraction*

STEP 1. *Multiply* the denominator times the whole number.

STEP 2. *Add* the numerator.

STEP 3. *Put* the sum over the original denominator.

Work in a clockwise direction,
first multiply $5 \times 2 = 10$
then add the numerator, $10 + 3 = 13$

$10 + 3 = 13$

$$2\dfrac{3}{5} = \dfrac{13}{5}$$

Same denominator

$5 \times 2 = 10$

WORK THIS PROBLEM

The Question:

Convert these mixed numbers to improper fractions.

(a) $3\frac{1}{6}$ (b) $4\frac{3}{5}$ (c) $1\frac{1}{2}$ (d) $8\frac{2}{3}$ (e) $15\frac{3}{8}$ (f) $9\frac{3}{4}$

✔ **YOUR WORK**

The Solution:

(a) $3\frac{1}{6} = \frac{19}{6}$ (b) $4\frac{3}{5} = \frac{23}{5}$ (c) $1\frac{1}{2} = \frac{3}{2}$

(d) $8\frac{2}{3} = \frac{26}{3}$ (e) $15\frac{3}{8} = \frac{123}{8}$ (f) $9\frac{3}{4} = \frac{39}{4}$

Raising a Fraction to Higher Terms

Two fractions are said to be *equivalent* if they are numerals or names for the same number. For example, $\frac{1}{2} = \frac{2}{4}$ since both fractions represent the same portion of some standard amount.

$$\frac{1}{2} = \frac{2}{4} = \frac{3}{6} = \frac{4}{8} = \frac{5}{10} = \cdots = \frac{48}{96} = \frac{61}{122} = \frac{1437}{2874} = \cdots$$

Each of the above fractions is a name for the same number, and we can use these fractions interchangeably.

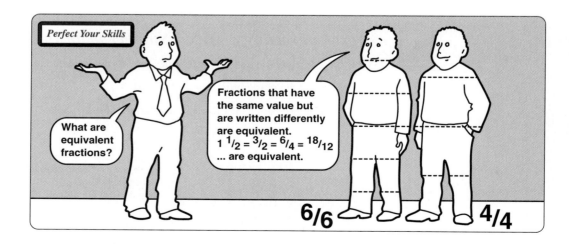

To obtain a fraction equivalent to any given fraction, simply multiply the original numerator and denominator by the same nonzero number. For example,

$$\frac{1}{2} = \frac{1 \times 3}{2 \times 3} = \frac{3}{6} \quad \text{and} \quad \frac{2}{3} = \frac{2 \times 5}{3 \times 5} = \frac{10}{15}$$

This method works because multiplying by $\frac{3}{3}$ or $\frac{5}{5}$ is the same as multiplying by 1 ($\frac{3}{3}$ and $\frac{5}{5}$ are both 1 standard unit). And, as we noted in Section 1.4, one times any number is equal to the number. The number value of the fraction has not changed; we have simply renamed it.

Using this method and your multiplication and division skills, you can find new numerators for partly converted fractions. For example,

$$\frac{3}{4} = \frac{?}{20}$$

$$\frac{3}{4} = \frac{3 \times ?}{4 \times ?} = \frac{3 \times 5}{4 \times 5} = \frac{15}{20}$$

$$4 \times ? = 20 \qquad \text{or} \qquad ? = 20 \div 4$$
$$? \text{ must be } 5 \qquad\qquad\qquad ? = 5$$

YOUR TURN

| WORK THIS PROBLEM |

The Question:

Complete the renaming of the following fractions.

(a) $\dfrac{5}{6} = \dfrac{?}{42}$ (b) $\dfrac{7}{16} = \dfrac{?}{48}$ (c) $\dfrac{3}{7} = \dfrac{?}{56}$ (d) $1\dfrac{2}{3} = \dfrac{?}{12}$

| ✔ YOUR WORK |

The Solution:

(a) $\dfrac{5}{6} = \dfrac{5 \times 7}{6 \times 7} = \dfrac{35}{42}$ (b) $\dfrac{7}{16} = \dfrac{7 \times 3}{16 \times 3} = \dfrac{21}{48}$

(c) $\dfrac{3}{7} = \dfrac{3 \times 8}{7 \times 8} = \dfrac{24}{56}$ (d) $1\dfrac{2}{3} = \dfrac{5}{3} = \dfrac{5 \times 4}{3 \times 4} = \dfrac{20}{12}$

Reducing a Fraction to Lowest Terms

Very often in working with fractions you will be asked to *reduce a fraction to lowest terms* or *simplify*. This means to replace it with the simplest fraction in its set of equivalent fractions. To reduce $\frac{15}{30}$ to its lowest terms means to replace it by $\frac{1}{2}$.

$$\frac{15}{30} = \frac{1 \times 15}{2 \times 15} = \frac{1}{2}$$

The fraction $\frac{1}{2}$ is the simplest equivalent fraction to $\frac{15}{30}$ because its numerator and denominator are the smallest whole numbers of any in the set $\frac{1}{2}, \frac{2}{4}, \frac{3}{6}, \frac{4}{8}, \ldots \frac{15}{30} \ldots$

How can you find the simplest equivalent fraction?

Steps: Reducing a Fraction

STEP 1. *Identify* a common factor of the numerator and denominator. A common factor is a number that evenly divides both the numerator and denominator.

STEP 2. *Divide* the numerator and denominator by that same factor.

STEP 3. *Continue dividing* by common factors until the fraction is reduced as far as possible.

For example, let's reduce $\frac{30}{42}$ to its lowest terms.

Step 1. Two is a factor of 30 and 42.

Step 2. $\dfrac{30}{42} = \dfrac{30 \div 2}{42 \div 2} = \dfrac{15}{21}$

Step 3. Three is a factor of 15 and 21.

$$\frac{15}{21} = \frac{15 \div 3}{21 \div 3} = \frac{5}{7}$$

$$\frac{30}{42} = \frac{15}{21} = \frac{5}{7}$$

If we had noticed that 6 is a factor of 30 and 42, we could have reduced $\frac{30}{42}$ in one step.

$$\frac{30}{42} = \frac{30 \div 6}{42 \div 6} = \frac{5}{7}$$

The important point is to continue reducing until there are no more common factors.

YOUR TURN

| WORK THIS PROBLEM |

The Question:

Reduce $\dfrac{90}{105}$ to lowest terms.

| ✔ YOUR WORK |

The Solution:

$$\frac{90}{105} = \frac{90 \div 5}{105 \div 5} = \frac{18}{21} = \frac{18 \div 3}{21 \div 3} = \frac{6}{7}$$

This process of dividing by, or eliminating, common factors is called *canceling*. Note that the order of canceling the factors doesn't affect the final answer.

Usually, we don't write out the division. Instead, we write the results above the numerator and below the denominator.

$$\frac{\overset{18\ \ \ \overset{6}{}}{\cancel{90}}}{\underset{21\ \ \ \underset{7}{}}{\cancel{105}}} = \frac{\overset{6}{\cancel{18}}}{\underset{7}{\cancel{21}}} = \frac{6}{7} \qquad \text{First, divide by 5. Second, divide by 3.}$$

YOUR TURN

WORK THIS PROBLEM

The Question:

Reduce the following fractions to lowest terms.

(a) $\dfrac{15}{84}$ (b) $\dfrac{21}{35}$ (c) $\dfrac{154}{1078}$ (d) $\dfrac{378}{405}$ (e) $\dfrac{256}{208}$

✔ YOUR WORK

The Solution:

(a) $\dfrac{15}{84} = \dfrac{5}{28}$ Divide by 3.   (b) $\dfrac{21}{35} = \dfrac{3}{5}$ Divide by 7.

(c) $\dfrac{154}{1078} = \dfrac{77}{539} = \dfrac{11}{77} = \dfrac{1}{7}$ Divide by 2. Divide by 7. Divide by 11.

(d) $\dfrac{378}{405} = \dfrac{126}{135} = \dfrac{42}{45} = \dfrac{14}{15}$ Divide by 3. Divide by 3. Divide by 3.

(e) $\dfrac{256}{208} = \dfrac{128}{104} = \dfrac{64}{52} = \dfrac{32}{26} = \dfrac{16}{13}$ Divide by 2. Divide by 2. Divide by 2. Divide by 2.

Remember, if you canceled the common factors in a different order, your final answer would be the same although the steps may be different.

Reducing an improper fraction may result in a whole number. For example,

$$\frac{6}{3} = \frac{6}{3} = \frac{2}{1} \quad \text{or simply} \quad 2$$

Any whole number may be written as a fraction by using a denominator equal to 1.

$$3 = \frac{3}{1} \qquad 4 = \frac{4}{1} \qquad \text{and so on}$$

WORKSPACE

SECTION TEST 2.1 Fractions

The following problems test your understanding of Section 2.1, Renaming Fractions. The answers to the odd-numbered problems are in the appendix.

A. Write as an improper fraction.

1. $4\frac{2}{3} =$ _____

2. $5\frac{3}{4} =$ _____

3. $3\frac{1}{2} =$ _____

4. $2\frac{3}{5} =$ _____

5. $7\frac{5}{6} =$ _____

6. $2\frac{5}{8} =$ _____

7. $7\frac{4}{9} =$ _____

8. $4\frac{3}{7} =$ _____

9. $8\frac{7}{11} =$ _____

10. $12\frac{7}{8} =$ _____

11. $15\frac{1}{4} =$ _____

12. $25\frac{3}{8} =$ _____

13. $12\frac{15}{16} =$ _____

14. $21\frac{7}{12} =$ _____

15. $20\frac{7}{15} =$ _____

16. $14\frac{7}{16} =$ _____

B. Write as a mixed number.

1. $\frac{14}{5} =$ _____

2. $\frac{23}{4} =$ _____

3. $\frac{23}{2} =$ _____

4. $\frac{33}{7} =$ _____

5. $\frac{27}{8} =$ _____

6. $\frac{47}{3} =$ _____

7. $\frac{75}{16} =$ _____

8. $\frac{39}{8} =$ _____

9. $\frac{41}{12} =$ _____

10. $\frac{51}{8} =$ _____

11. $\frac{51}{4} =$ _____

12. $\frac{53}{16} =$ _____

13. $\frac{47}{6} =$ _____

14. $\frac{67}{12} =$ _____

15. $\frac{101}{8} =$ _____

16. $\frac{165}{16} =$ _____

C. Reduce to lowest terms.

1. $\frac{12}{18} =$ _____

2. $\frac{28}{32} =$ _____

3. $\frac{24}{30} =$ _____

4. $\frac{45}{54} =$ _____

5. $\frac{45}{60} =$ _____

6. $\frac{42}{224} =$ _____

7. $\frac{48}{84} =$ _____

8. $\frac{63}{105} =$ _____

9. $\frac{72}{192} =$ _____

10. $\frac{112}{252} =$ _____

11. $\frac{105}{147} =$ _____

12. $\frac{225}{360} =$ _____

13. $\frac{120}{384} =$ _____

14. $\frac{96}{352} =$ _____

15. $\frac{252}{324} =$ _____

16. $\frac{630}{735} =$ _____

D. Complete.

1. $\frac{2}{5} = \frac{}{30}$

2. $\frac{5}{8} = \frac{}{24}$

3. $\frac{2}{3} = \frac{}{36}$

4. $\frac{3}{7} = \frac{}{35}$

5. $\frac{4}{9} = \frac{}{54}$

6. $\frac{5}{7} = \frac{}{56}$

7. $\frac{7}{8} = \frac{}{32}$

8. $\frac{7}{12} = \frac{}{96}$

9. $\frac{11}{16} = \frac{}{80}$

10. $\frac{7}{9} = \frac{}{72}$

11. $\frac{3}{4} = \frac{}{28}$

12. $\frac{5}{6} = \frac{}{66}$

13. $5\frac{3}{8} = \frac{}{40}$

14. $6\frac{1}{4} = \frac{}{12}$

15. $3\frac{5}{9} = \frac{}{36}$

16. $4\frac{7}{15} = \frac{}{45}$

WORKSPACE

Can You:

Multiply two or more given fractions?
. . . If not, you need this section.

The simplest arithmetic operation with fractions is multiplication and, happily, it is easy to show graphically:

$$3 \times \frac{1}{4} = \quad + \quad + \quad \text{or} \quad$$

We can express the same thing in numbers:

$$3 \times \frac{1}{4} = \frac{1}{4} + \frac{1}{4} + \frac{1}{4} = \frac{3}{4}$$

We can also show the product of two fractions graphically. Let's multiply $\frac{1}{2} \times \frac{1}{3}$.

$$\frac{1}{2} \times \frac{1}{3} \qquad \text{and}$$

The product is

$$\frac{1}{2} \times \frac{1}{3} = \frac{1}{6} = \frac{1 \text{ shaded area}}{6 \text{ equal areas in the } 1 \times 1 \text{ square}}$$

Thus multiplying fractions requires just two simple steps.

Steps: Multiplying Fractions

STEP 1. *Multiply* the numerators to find the new numerator.

STEP 2. *Multiply* the denominators to find the new denominator.

Do you see that $3 \times \frac{1}{4}$ is really the same sort of problem?

$$3 = \frac{3}{1}, \quad \text{so} \quad 3 \times \frac{1}{4} = \frac{3}{1} \times \frac{1}{4} = \frac{3 \times 1}{1 \times 4} = \frac{3}{4}$$

WORK THIS PROBLEM

The Question:

Find the product of $\dfrac{5}{6} \times \dfrac{2}{3}$ and reduce the answer to the lowest terms.

✔ YOUR WORK

The Solution:

$$\frac{5}{6} \times \frac{2}{3} = \frac{5 \times 2}{6 \times 3} = \frac{10}{18}$$

$$\frac{5}{6} \times \frac{2}{3} = \frac{\overset{5}{\cancel{10}}}{\underset{9}{\cancel{18}}} = \frac{5}{9}$$

It will save you time and effort if you eliminate common factors *before* you multiply. For example, in the previous problem, 6 and 2 are both evenly divisible by 2. Because one of these numbers is in the numerator and the other is in the denominator, we can reduce the fraction by dividing both 6 and 2 by 2.

$$\frac{5}{\underset{3}{\cancel{6}}} \times \frac{\overset{1}{\cancel{2}}}{3} = \frac{5 \times 1}{3 \times 3} = \frac{5}{9}$$

WORK THIS PROBLEM

The Question:

Multiply the following. (Hint: Change mixed numbers such as $1\frac{1}{2}$ and $3\frac{5}{6}$ to improper fractions, then multiply as usual. Multiply three fractions like two. Be sure to reduce all answers to lowest terms.)

(a) $\dfrac{8}{12} \times \dfrac{3}{16}$ (b) $\dfrac{3}{2} \times \dfrac{2}{3}$ (c) $1\dfrac{1}{2} \times \dfrac{2}{5}$

(d) $3\dfrac{5}{6} \times \dfrac{3}{10}$ (e) $1\dfrac{4}{5} \times \dfrac{2}{3} \times \dfrac{1}{4}$ (f) $2\dfrac{1}{4} \times \dfrac{1}{24} \times 2\dfrac{2}{3}$

✔ YOUR WORK

The Solution:

(a) $\dfrac{\overset{1}{\cancel{8}}}{\underset{4}{\cancel{12}}} \times \dfrac{\overset{1}{\cancel{3}}}{\underset{2}{\cancel{16}}} = \dfrac{1}{8}$ (b) $\dfrac{\overset{1}{\cancel{3}}}{\underset{1}{\cancel{2}}} \times \dfrac{\overset{1}{\cancel{2}}}{\underset{1}{\cancel{3}}} = 1$

(c) $1\dfrac{1}{2} \times \dfrac{2}{5} = \dfrac{3}{\underset{1}{\cancel{2}}} \times \dfrac{\overset{1}{\cancel{2}}}{5} = \dfrac{3}{5}$ (d) $3\dfrac{5}{6} \times \dfrac{3}{10} = \dfrac{23}{\underset{2}{\cancel{6}}} \times \dfrac{\overset{1}{\cancel{3}}}{10} = \dfrac{23}{20} = 1\dfrac{3}{20}$

(e) $1\dfrac{4}{5} \times \dfrac{2}{3} \times \dfrac{1}{4} = \dfrac{\overset{3}{\cancel{9}}}{5} \times \dfrac{\overset{1}{\cancel{2}}}{\underset{1}{\cancel{3}}} \times \dfrac{1}{\underset{2}{\cancel{4}}} = \dfrac{3}{10}$ (f) $2\dfrac{1}{4} \times \dfrac{1}{24} \times 2\dfrac{2}{3} = \dfrac{\overset{3}{\cancel{9}}}{4} \times \dfrac{1}{\underset{3}{\underset{1}{\cancel{24}}}} \times \dfrac{\overset{1}{\cancel{8}}}{\underset{1}{\cancel{3}}} = \dfrac{1}{4}$

Many business situations require the multiplication of fractions.

YOUR TURN

WORK THIS PROBLEM

The Question:

Leroy's office measures $14\frac{2}{3}$ feet by $20\frac{3}{4}$ feet. What is the area? (Area is length times width.)

✔ YOUR WORK

The Solution:

$$14\frac{2}{3} \times 20\frac{3}{4} = \frac{\overset{11}{44}}{3} \times \frac{83}{\underset{1}{4}} = \frac{913}{3} = 304\frac{1}{3} \text{ square feet}$$

WORKSPACE

SECTION TEST 2.2   Fractions

The following problems test your understanding of Section 2.2, Multiplication of Fractions.

A. Multiply and reduce the answer to lowest terms.

1. $\dfrac{3}{4} \times \dfrac{8}{9} =$ _____

2. $\dfrac{1}{2} \times \dfrac{4}{5} =$ _____

3. $\dfrac{3}{5} \times \dfrac{10}{21} =$ _____

4. $\dfrac{2}{5} \times \dfrac{3}{8} =$ _____

5. $\dfrac{5}{6} \times \dfrac{12}{35} =$ _____

6. $\dfrac{7}{9} \times \dfrac{27}{28} =$ _____

7. $\dfrac{9}{16} \times \dfrac{8}{27} =$ _____

8. $\dfrac{5}{12} \times \dfrac{9}{20} =$ _____

9. $\dfrac{7}{9} \times \dfrac{18}{49} =$ _____

10. $\dfrac{4}{15} \times \dfrac{9}{40} =$ _____

11. $\dfrac{5}{32} \times \dfrac{12}{15} =$ _____

12. $\dfrac{18}{35} \times \dfrac{14}{27} =$ _____

13. $12 \times \dfrac{3}{4} =$ _____

14. $\dfrac{7}{4} \times \dfrac{12}{35} =$ _____

15. $\dfrac{18}{5} \times \dfrac{15}{14} =$ _____

16. $\dfrac{12}{15} \times \dfrac{21}{20} =$ _____

17. $\dfrac{21}{16} \times \dfrac{24}{35} =$ _____

18. $\dfrac{12}{7} \times \dfrac{35}{16} \times \dfrac{2}{5} =$ _____

19. $\dfrac{15}{4} \times \dfrac{14}{25} \times \dfrac{15}{14} =$ _____

20. $\dfrac{7}{3} \times \dfrac{12}{35} \times \dfrac{15}{2} =$ _____

21. $\dfrac{4}{7} \times \dfrac{42}{5} \times \dfrac{35}{48} =$ _____

B. Multiply and reduce the answer to lowest terms.

1. $6\dfrac{2}{3} \times \dfrac{3}{10} =$ _____

2. $\dfrac{5}{8} \times 6\dfrac{2}{5} =$ _____

3. $3\dfrac{3}{4} \times \dfrac{8}{9} =$ _____

4. $\dfrac{14}{22} \times 2\dfrac{4}{7} =$ _____

5. $4\dfrac{1}{2} \times \dfrac{8}{27} =$ _____

6. $2\dfrac{4}{5} \times \dfrac{10}{21} =$ _____

7. $4\dfrac{4}{7} \times 5\dfrac{1}{4} =$ _____

8. $4\dfrac{1}{6} \times \dfrac{9}{10} =$ _____

9. $2\dfrac{2}{9} \times 2\dfrac{2}{5} =$ _____

10. $3\dfrac{3}{8} \times 1\dfrac{1}{9} =$ _____

11. $3\dfrac{1}{3} \times 4\dfrac{1}{2} =$ _____

12. $3\dfrac{3}{4} \times \dfrac{2}{5} =$ _____

13. $2\dfrac{1}{12} \times 1\dfrac{4}{5} =$ _____

14. $5\dfrac{2}{5} \times 3\dfrac{1}{3} =$ _____

15. $5\dfrac{5}{6} \times 3\dfrac{3}{7} =$ _____

16. $2\dfrac{4}{7} \times 1\dfrac{5}{9} =$ _____

17. $2\dfrac{11}{12} \times 4\dfrac{2}{7} =$ _____

18. $2\dfrac{2}{5} \times \dfrac{5}{6} \times 3\dfrac{1}{2} =$ _____

19. $5\dfrac{1}{3} \times \dfrac{5}{6} \times 2\dfrac{1}{4} =$ _____

20. $2\dfrac{4}{5} \times \dfrac{2}{7} \times 1\dfrac{7}{8} =$ _____

21. $10\dfrac{2}{3} \times \dfrac{5}{12} \times 2\dfrac{7}{10} =$ _____

C. Solve these problems.

1. Bits 'n' Bytes, the quickly expanding computer store, sells computer cable for $4 a foot. How much would $22\dfrac{5}{8}$ feet of cable cost?

2. Kate's Delicacies makes a very rich croissant. Each dozen of croissants requires $1\dfrac{1}{4}$ cups of butter. If Kate has an order for 27 dozen croissants, how much butter does she need?

3. Woody's Lumber Mill will custom cut lumber. Woody has a special order for 35 fence posts. Each fence post takes $\frac{1}{4}$ of an hour to cut. How many hours will it take Woody to cut the 35 posts?

4. Bob's Beeflike Burgers normally sells 2856 burgers daily. Each of these burgers requires $\frac{1}{4}$ pound of beef, $\frac{1}{16}$ pound of tomato, $\frac{1}{10}$ pound of lettuce, and $\frac{1}{20}$ pound of onion. How many pounds of beef, tomatoes, lettuce, and onions does Bob's use daily?

5. Maria's Mannequins are rented out for $46 a day. Lou wants to use a mannequin for $2\frac{3}{4}$ days. How much does Lou have to pay Maria?

6. Each of David's office workers is required to have at least $66\frac{3}{4}$ square feet of office space. How much office space does David need for his 14 employees?

7. Studies show that computer programmers spend at least $\frac{1}{3}$ of their 8-hour workday on personal activities. How many hours would 23 programmers spend on personal activities in a five-day workweek?

8. Crystal's Diamonds is having a sale. Her fine jewelry will be reduced by $\frac{3}{8}$ of the original price. Susan would like to purchase a ring that originally cost $346, a bracelet that was $412 originally, and a watch that was $178. How much money will Susan be saving at Crystal's sale on each item?

9. Pia's Pizza Palace sells gourmet pizza. She uses $\frac{3}{16}$ cup olive oil, $\frac{5}{8}$ pound mozzarella, $\frac{4}{7}$ pound tomatoes, and $1\frac{7}{8}$ ounces of her special spice mix for her Pale Pizza. If she sells 134 Pale Pizzas daily, how much olive oil, mozzarella, tomatoes, and special spice mix does she use in a day?

10. Jack's Jackhammer Service employees usually get $\frac{1}{16}$ of each hour as break for stabilization. How many minutes per 8-hour shift are Jack's employees on break?

11. The Fly-By-Night Aviary uses $7\frac{4}{5}$ pounds of birdseed daily. How many pounds of seed do its birds eat in a 30-day month?

12. Bricks by the Ton is doing a booming business. If each truckload of bricks weighs $1\frac{7}{8}$ ton and the company sells 10 a day, (a) how many tons of bricks does it sell daily? (b) How many tons of bricks does it supply in a five-day week?

13. Your company pays $1\frac{1}{2}$ times the base pay for overtime. (a) If your base pay is \$13 an hour, what is the overtime pay rate? (b) How much do you earn for 6 hours of overtime?

14. If you purchased 90 shares of Wired for Sound stock at \$$4\frac{3}{8}$ per share, what was the total cost?

15. The Zippy Delivery Service's best driver averages $50\frac{3}{4}$ miles per hour. He drives $7\frac{1}{4}$ hours per day. (a) How many miles does he cover daily? (b) How many miles does he cover in a five-day workweek?

WORKSPACE

SECTION 2.3:   Division of Fractions

Can You:

Divide two given fractions?
. . . If not, you need this section.

Dividing Fractions

Note that division is not a reversible operation. That is, 8 divided by 4 is *not* the same as 4 divided by 8. Because fractions represent a kind of division,

$8 \div 4$ is read "8 divided by 4" and written $4\overline{)8}$ and $\dfrac{8}{4}$

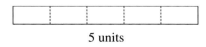

4 is the divisor

In the problem given, you are being asked to divide a set of 8 objects into sets of 4 objects. The divisor 4 is the denominator or bottom number of the fraction.

$8 \div 4$ or "8 divided by 4" is *not* equal to $\dfrac{4}{8}$ or $8\overline{)4}$.

YOUR TURN

| WORK THIS PROBLEM |

The Question:

In the division $5 \div \dfrac{1}{2}$, which number is the divisor?

| ✔ YOUR WORK |

The Solution:

The divisor is $\dfrac{1}{2}$.

The division $5 \div \frac{1}{2}$ is read "5 divided by $\frac{1}{2}$," and it asks how many $\frac{1}{2}$-unit lengths are included in a length of 5 units.

5 units

$\dfrac{1}{2}$ unit

If you look at this division problem as a form of multiplication, $5 \div \frac{1}{2} = \square$ asks that you find a number $\square$ such that $5 = \square \times \frac{1}{2}$.

| WORK THIS PROBLEM |

YOUR TURN

The Question:

Using the diagram in the preceding text, find the quotient for $5 \div \frac{1}{2}$.

| ✔ YOUR WORK |

The Solution:

$$5 \div \frac{1}{2} = 10$$

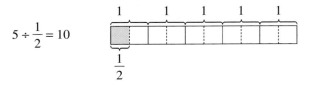

There are ten $\frac{1}{2}$-unit lengths contained in the 5-unit length.

Using a drawing of this sort to solve a division problem is difficult and clumsy. Fortunately, we can easily solve it numerically.

Steps: Dividing Fractions

STEP 1. *Invert* the divisor. (To invert a fraction simply means to switch top and bottom.)

STEP 2. *Multiply* the fractions normally.

For example, to divide $5 \div \frac{1}{2}$,

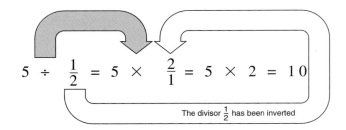

$$5 \div \frac{1}{2} = 5 \times \frac{2}{1} = 5 \times 2 = 10$$

The divisor $\frac{1}{2}$ has been inverted

| WORK THIS PROBLEM |

YOUR TURN

The Question:

Find the quotient: $\frac{3}{5} \div \frac{2}{3}$

✔ YOUR WORK

The Solution:

Step 1

$$\frac{3}{5} \div \frac{2}{3} = \frac{3}{5} \times \frac{3}{2}$$

Step 2

$$\frac{3}{5} \times \frac{3}{2} = \frac{3 \times 3}{5 \times 2} = \frac{9}{10}$$

We have converted a division problem that is difficult to picture into a simple multiplication.

Checking Division

The final, and very important, step in every division is checking the answer.

$$\text{If} \quad \frac{3}{5} \div \frac{2}{3} = \frac{9}{10} \quad \text{then} \quad \frac{3}{5} = \frac{2}{3} \times \frac{9}{10}.$$

Check: $\dfrac{\overset{1}{\cancel{2}}}{\underset{1}{\cancel{3}}} \times \dfrac{\overset{3}{\cancel{9}}}{\underset{5}{\cancel{10}}} = \dfrac{3}{5}$

WORK THIS PROBLEM

The Question:

Find the quotient of $\dfrac{7}{8} \div \dfrac{3}{2}$ and check your answer.

YOUR TURN

✔ YOUR WORK

The Solution:

$$\frac{7}{8} \div \frac{3}{2} = \frac{7}{\underset{4}{\cancel{8}}} \times \frac{\overset{1}{\cancel{2}}}{3} = \frac{7}{12} \qquad \textbf{Check:} \; \frac{\overset{1}{\cancel{3}}}{2} \times \frac{7}{\underset{4}{\cancel{12}}} = \frac{7}{8}$$

Whether they are using a calculator or working longhand, the chief source of confusion for many people in dividing fractions is deciding which fraction to invert. It will help if you

1. Put every division problem in the form (dividend) ÷ (divisor) first, then invert the *divisor,* and finally multiply to obtain the quotient.
2. Check your answer by multiplying. The product obtained by multiplying (divisor) × (quotient or answer) should equal the dividend.

YOUR TURN

┌─ WORK THIS PROBLEM ───

The Question:

Find quotients for the following and check your answers.

(a) $\dfrac{2}{5} \div \dfrac{3}{8}$ (b) $3\dfrac{3}{4} \div \dfrac{5}{2}$ (c) $4\dfrac{1}{5} \div 1\dfrac{4}{10}$

(d) $3\dfrac{2}{3} \div 3$ (e) Divide $\dfrac{3}{4}$ by $2\dfrac{5}{8}$ (f) Divide $1\dfrac{1}{4}$ by $1\dfrac{7}{8}$

┌─ ✔ YOUR WORK ──

The Solution:

(a) $\dfrac{2}{5} \div \dfrac{3}{8} = \dfrac{2}{5} \times \dfrac{8}{3} = \dfrac{16}{15} = 1\dfrac{1}{15}$ **Check:** $\dfrac{\overset{1}{\cancel{3}}}{\underset{1}{\cancel{8}}} \times \dfrac{\overset{2}{\cancel{16}}}{\underset{5}{\cancel{15}}} = \dfrac{2}{5}$

(b) $3\dfrac{3}{4} \div \dfrac{5}{2} = \dfrac{15}{4} \div \dfrac{5}{2} = \dfrac{\overset{3}{\cancel{15}}}{\underset{2}{\cancel{4}}} \times \dfrac{\overset{1}{\cancel{2}}}{\underset{1}{\cancel{5}}} = \dfrac{3}{2} = 1\dfrac{1}{2}$ **Check:** $\dfrac{5}{2} \times \dfrac{3}{2} = \dfrac{15}{4} = 3\dfrac{3}{4}$

(c) $4\dfrac{1}{5} \div 1\dfrac{4}{10} = \dfrac{21}{5} \div \dfrac{14}{10} = \dfrac{\overset{3}{\cancel{21}}}{\underset{1}{\cancel{5}}} \times \dfrac{\overset{\overset{1}{\cancel{2}}}{\cancel{10}}}{\underset{\underset{1}{\cancel{2}}}{\cancel{14}}} = 3$ **Check:** $1\dfrac{4}{10} \times 3 = \dfrac{14}{10} \times \dfrac{3}{1} = \dfrac{42}{10} = 4\dfrac{2}{10} = 4\dfrac{1}{5}$

(d) $3\dfrac{2}{3} \div 3 = \dfrac{11}{3} \div \dfrac{3}{1} = \dfrac{11}{3} \times \dfrac{1}{3} = \dfrac{11}{9} = 1\dfrac{2}{9}$ **Check:** $3 \times \dfrac{11}{9} = \dfrac{\overset{1}{\cancel{3}}}{1} \times \dfrac{11}{\underset{3}{\cancel{9}}} = \dfrac{11}{3} = 3\dfrac{2}{3}$

(e) $\dfrac{3}{4} \div 2\dfrac{5}{8} = \dfrac{3}{4} \div \dfrac{21}{8} = \dfrac{\overset{1}{\cancel{3}}}{\underset{1}{\cancel{4}}} \times \dfrac{\overset{2}{\cancel{8}}}{\underset{7}{\cancel{21}}} = \dfrac{2}{7}$ **Check:** $2\dfrac{5}{8} \times \dfrac{2}{7} = \dfrac{\overset{3}{\cancel{21}}}{\underset{4}{\cancel{8}}} \times \dfrac{\overset{1}{\cancel{2}}}{\underset{1}{\cancel{7}}} = \dfrac{3}{4}$

(f) $1\dfrac{1}{4} \div 1\dfrac{7}{8} = \dfrac{5}{4} \div \dfrac{15}{8} = \dfrac{\overset{1}{\cancel{5}}}{\underset{1}{\cancel{4}}} \times \dfrac{\overset{2}{\cancel{8}}}{\underset{3}{\cancel{15}}} = \dfrac{2}{3}$ **Check:** $1\dfrac{7}{8} \times \dfrac{2}{3} = \dfrac{\overset{5}{\cancel{15}}}{\underset{4}{\cancel{8}}} \times \dfrac{\overset{1}{\cancel{2}}}{\underset{1}{\cancel{3}}} = \dfrac{5}{4} = 1\dfrac{1}{4}$

Like all the math in this book, dividing fractions is necessary in the business world. Stock prices are usually written as mixed numbers.

YOUR TURN

┌─ WORK THIS PROBLEM ───

The Question:

Mary purchased some stock for \$556. If the cost per share was \$$17\dfrac{3}{8}$, how many shares did she purchase?

┌─ ✔ YOUR WORK ─┐

The Solution:

$$\$556 \div 17\frac{3}{8} = \frac{556}{1} \div \frac{139}{8} = \frac{\overset{4}{\cancel{556}}}{1} \times \frac{8}{\underset{1}{\cancel{139}}} = \frac{32}{1} = 32$$

Check: $32 \times \$17\frac{3}{8} = \frac{32}{1} \times \frac{139}{8} = \$556$

WORKSPACE

SECTION TEST 2.3 Fractions

Name

Date

Course/Section

The following problems test your understanding of Section 2.3, Division of Fractions.

A. Divide and reduce the answer to lowest terms.

1. $\dfrac{2}{3} \div \dfrac{8}{9} =$ _____

2. $\dfrac{4}{15} \div \dfrac{6}{25} =$ _____

3. $\dfrac{6}{7} \div \dfrac{8}{21} =$ _____

4. $\dfrac{3}{8} \div \dfrac{9}{32} =$ _____

5. $\dfrac{5}{7} \div \dfrac{25}{28} =$ _____

6. $\dfrac{5}{9} \div \dfrac{10}{27} =$ _____

7. $\dfrac{9}{32} \div \dfrac{3}{8} =$ _____

8. $\dfrac{5}{12} \div \dfrac{10}{21} =$ _____

9. $\dfrac{18}{49} \div \dfrac{9}{14} =$ _____

10. $\dfrac{9}{40} \div \dfrac{18}{25} =$ _____

11. $\dfrac{18}{35} \div \dfrac{24}{25} =$ _____

12. $\dfrac{14}{27} \div 8 =$ _____

13. $15 \div \dfrac{9}{10} =$ _____

14. $\dfrac{14}{15} \div \dfrac{21}{25} =$ _____

15. $\dfrac{5}{18} \div \dfrac{10}{27} =$ _____

16. $\dfrac{6}{25} \div \dfrac{9}{10} =$ _____

17. $\dfrac{8}{15} \div \dfrac{2}{5} =$ _____

18. $\dfrac{15}{16} \div \dfrac{5}{24} =$ _____

19. Divide 6 by $\dfrac{1}{2}$ _____

20. Divide $2\dfrac{1}{4}$ by $\dfrac{3}{8}$ _____

21. Divide $\dfrac{5}{8}$ by $\dfrac{2}{15}$ _____

B. Divide and reduce the answer to lowest terms.

1. $3\dfrac{1}{3} \div 4\dfrac{1}{6} =$ _____

2. $2\dfrac{1}{4} \div 1\dfrac{1}{2} =$ _____

3. $4\dfrac{1}{6} \div 3\dfrac{1}{3} =$ _____

4. $1\dfrac{3}{7} \div 1\dfrac{11}{14} =$ _____

5. $2\dfrac{2}{5} \div 2\dfrac{7}{10} =$ _____

6. $3\dfrac{3}{8} \div 11\dfrac{1}{4} =$ _____

7. $2\dfrac{1}{12} \div 5\dfrac{5}{6} =$ _____

8. $3\dfrac{3}{5} \div 2\dfrac{7}{10} =$ _____

9. $5\dfrac{1}{3} \div 2\dfrac{2}{9} =$ _____

10. $5\dfrac{1}{4} \div 5\dfrac{5}{6} =$ _____

11. $2\dfrac{2}{9} \div 4\dfrac{1}{6} =$ _____

12. $2\dfrac{2}{7} \div 1\dfrac{19}{21} =$ _____

13. $1\dfrac{7}{8} \div 6\dfrac{1}{4} =$ _____

14. $5\dfrac{3}{5} \div 3\dfrac{11}{15} =$ _____

15. $5\dfrac{3}{5} \div 2\dfrac{2}{15} =$ _____

16. $6\dfrac{3}{4} \div 6\dfrac{3}{10} =$ _____

17. $2\dfrac{1}{7} \div 1\dfrac{11}{14} =$ _____

18. $6\dfrac{2}{3} \div 6\dfrac{2}{9} =$ _____

19. $\dfrac{4}{\frac{1}{2}} =$ _____

20. $\dfrac{\frac{7}{8}}{2} =$ _____

21. $\dfrac{12}{\frac{3}{4}} =$ _____

C. Solve these problems.

1. Stock from Wired for Sound is selling for $\$4\frac{3}{4}$ per share. You have $1900 that you want to invest. How many shares of this stock can you buy?

2. Woody's Lumber Mill has received an order for 12-foot lumber that is to be custom cut into $1\frac{1}{2}$-foot sections. How many of these $1\frac{1}{2}$-foot sections can be cut from 25 pieces of 12-foot lumber?

3. Maria wants to make each of her mannequins a dust cover for days when they are not rented out. Each adult-sized mannequin requires $2\frac{1}{3}$ yards of material. Each child-sized mannequin requires half that amount. Maria has 70 yards of material for the adult mannequins and 28 yards for the child-sized mannequins. (a) How many adult mannequin covers can Maria make? (b) How many child mannequin covers can Maria make?

4. Lew's trendy eating establishment employs four waitpersons. One Saturday night the "tip pool" contained 124\frac{1}{2}$. If the tips are split equally, how much does each person receive?

5. If Kate of Kate's Delicacies has 8 pounds of butter left, and if a dozen of special chocolate croissants requires $1\frac{1}{3}$ pounds of butter, how many more chocolate croissants does she make with the remaining butter?

6. Bob's Beeflike Burgers has a new bigger and better burger that uses $\frac{7}{8}$ of a pound of beef. Bob just received a 140-pound shipment of beef; how many of his bigger beefier burgers can Bob make?

7. Bob's Beeflike Burgers also has a new vegetarian entry on its menu, the Totally Tofu Burger. Each Totally Tofu Burger requires $\frac{3}{8}$ of a cup of tofu. Bob just received a 25-gallon tofu shipment (1 gallon = 16 cups). How many Totally Tofu Burgers can he make?

8. The Sparkle Cleaning Company has 90 gallons of its secret cleaning potion in stock. Each cleaning job takes $1\frac{4}{5}$ gallons. How many cleaning jobs can the company complete?

9. Bits 'n' Bytes, your friendly neighborhood computer store, also connects personal computer systems. It takes $7\frac{1}{2}$ feet of cable to connect two computers. Currently the store has 165 feet of cable in inventory. How many computer connections can the store make before ordering new cable?

10. Big Ben's Industrial-Sized Bolts makes a big bolt that weighs $1\frac{3}{5}$ pounds. The cartons that are used for bolt packing can only hold 40 pounds. How many bolts can each carton hold?

11. The Fly-By-Night Aviary just received a 380-pound shipment of birdseed. Its birds consume $4\frac{3}{4}$ pounds of seed daily. How many days will this seed shipment feed the birds?

12. The Sparkle Cleaning Company has created a new condensed version of their secret cleaning fluid. It now takes $5\frac{1}{3}$ cups for a cleaning job. How many cleaning jobs can the Sparkle Cleaning Company handle with 40 gallons of the new condensed formula (1 gallon = 16 cups)?

13. Watt's Electrical Services installs phone lines in homes and businesses. A typical installation uses $12\frac{1}{2}$ feet of phone line. Phone line comes on 500-foot spools. How many phone lines can Watt install with one spool of phone line?

14. The Zippy Delivery Service truck gets $21\frac{1}{4}$ miles per gallon. How many gallons of gas are needed for a 170-mile trip?

15. How many shares of stock selling for 5\frac{3}{8}$ can be purchased for $2150?

Can You:

Add two or more fractions?
Subtract two fractions?
. . . If not, you need this section.

Adding Like Fractions

At heart, adding fractions is a matter of counting.

$$\frac{1}{5} + \frac{3}{5} = \frac{1+3}{5} = \frac{4}{5}$$

$\frac{1}{5}$ 1 fifth

+

$\frac{3}{5}$ + 3 fifths

=

$\frac{4}{5}$ = 4 fifths, count them

YOUR TURN

WORK THIS PROBLEM

The Question:

Add: $\dfrac{2}{7} + \dfrac{3}{7}$

✔ YOUR WORK

The Solution:

$$\frac{2}{7} + \frac{3}{7} = \frac{2+3}{7} = \frac{5}{7}$$

$\frac{2}{7}$ 2 sevenths

+

$\frac{3}{7}$ + 3 sevenths

=

$\frac{5}{7}$ = 5 sevenths or $\dfrac{5}{7}$

Like Fractions Fractions with the same denominator; for example, $\frac{3}{8}$ and $\frac{5}{8}$.

Fractions having the same denominator are called **like fractions.** In the problem given, $\frac{2}{7}$ and $\frac{3}{7}$ both have denominator 7 and are like fractions. Adding like fractions is easy.

Steps: Adding Like Fractions

STEP 1.   *Add* the numerators to find the numerator of the sum.

STEP 2.   *Use* the denominator the fractions have in common as the denominator of the sum.

For example, to add $\frac{2}{9} + \frac{5}{9}$,

$$\frac{2}{9} + \frac{5}{9} = \frac{2+5}{9} = \frac{7}{9} \quad \text{Add numerators.}$$
$$\text{Same denominator.}$$

Reduce to lowest terms, if possible, and write as a mixed number if the sum is an improper fraction. Adding three or more like fractions presents no special difficulty.

YOUR TURN

WORK THIS PROBLEM

The Question:

Add: $\dfrac{3}{12} + \dfrac{1}{12} + \dfrac{5}{12}$

✔ YOUR WORK

The Solution:

$$\frac{3}{12} + \frac{1}{12} + \frac{5}{12} = \frac{3+1+5}{12} = \frac{9}{12} = \frac{3}{4}$$

Notice that we reduce the sum to lowest terms.

Adding Mixed Numbers

To add together mixed numbers with the same denominators, just add the whole numbers together, add the fractions together, then simplify. For example,

$$2\frac{2}{5}$$
$$3\frac{4}{5}$$
$$7\frac{1}{5}$$

Convert to mixed number

$$12\frac{7}{5} = 12 + \frac{7}{5} = 12 + 1\frac{2}{5} = 13\frac{2}{5}$$

$$\frac{2}{5} + \frac{4}{5} + \frac{1}{5} = \frac{2+4+1}{5} = \frac{7}{5}$$

$$2 + 3 + 7 = 12$$

We could have solved this horizontally $(2\frac{2}{5} + 3\frac{4}{5} + 7\frac{1}{5})$, but lining up the mixed numbers vertically usually simplifies the process.

YOUR TURN

WORK THIS PROBLEM

The Question:

Add each of the following sets of fractions. Be sure to express your answer as a mixed number reduced to lowest terms.

(a) $2\frac{2}{7} + 5\frac{4}{7} + 3\frac{5}{7}$ (b) $3\frac{5}{8} + 4\frac{7}{8} + 1\frac{1}{8} + 12\frac{7}{8}$

✔ **YOUR WORK**

The Solution:

(a)
$$2\frac{2}{7}$$
$$5\frac{4}{7}$$
$$+3\frac{5}{7}$$
$$10\frac{11}{7} = 10 + \frac{11}{7} = 10 + 1\frac{4}{7} = 11\frac{4}{7}$$

(b)
$$3\frac{5}{8}$$
$$4\frac{7}{8}$$
$$1\frac{1}{8}$$
$$+12\frac{7}{8}$$
$$20\frac{20}{8} = 20 + \frac{20}{8} = 20 + 2 + \frac{4}{8} = 22 + \frac{4}{8} = 22 + \frac{1}{2} = 22\frac{1}{2}$$

Adding Unlike Fractions

How do we add fractions whose denominators are not the same?

$$\frac{2}{3} + \frac{3}{4} = ?$$

$$\frac{2}{3}\ \ \boxed{\ \ \ } \quad + \quad \frac{3}{4}\ \ \boxed{\ \ \ } \quad = \quad \boxed{\ \ \ }$$

The problem is to find a simple numeral that names this new number. One way to find it is to change these fractions to equivalent fractions with the same denominator. (Equivalent fractions are discussed in Section 2.1).

$$\frac{3}{4} = \frac{3 \times 3}{4 \times 3} = \frac{9}{12} \qquad \text{and} \qquad \frac{2}{3} = \frac{2 \times 4}{3 \times 4} = \frac{8}{12}$$

YOUR TURN

WORK THIS PROBLEM

The Question:

Add: $\dfrac{2}{3} + \dfrac{3}{4}$

✔ YOUR WORK

The Solution:

$$\frac{3}{4} + \frac{2}{3} = \frac{9}{12} + \frac{8}{12} = \frac{17}{12} = 1\frac{5}{12}$$

Least Common Denominator

Least Common Denominator (LCD)   The smallest number evenly divisible by the denominators of all the fractions involved in an addition or subtraction problem.

When you change original fractions to equivalent fractions with the same denominator, you want the new denominator to be the smallest number evenly divisible by the other denominators. That is, you want the new denominator to be the **least common denominator** or **LCD.**

How do you know what number to use as the LCD? Sometimes you can guess it or find it by trial and error. A better method is the **LCD finder.**

LCD Finder   A method of finding the least common denominator by using prime numbers.

The LCD finder uses **prime numbers**—numbers divisible only by themselves and one. Prime numbers include 2, 3, 5, 7, 11, 13, and so on. Numbers such as 4, 6, and 9 are not prime numbers because $4 = 2 \times 2$, $6 = 2 \times 3$, and $9 = 3 \times 3$.

Prime Number   Any number greater than one that is divisible only by itself and 1.

The following steps make up the LCD finder.

Steps: Finding the LCD

STEP 1.   *Arrange* the denominators in a row.

STEP 2.   *Divide* by any prime number that can be divided into one or more denominators. *Bring down* any denominators that are not divided.

STEP 3.   *Continue* until there are all 1s in the bottom row.

STEP 4.   *Multiply* all the divisors to find the LCD.

This method is really much easier than it sounds. An example will help clarify it, so let's find the LCD of the fractions $\frac{1}{4}$, $\frac{3}{7}$, and $\frac{5}{14}$.

Step 1.   Arrange the denominators.   4   7   14

Step 2.   Both 4 and 14 are divisible by 2. ──────→ 2 │ 4   7   14
              $4 \div 2 = 2$, $14 \div 2 = 7$
              Bring down the 7 to make ──────────→   2   7   7
              a new row.

Step 3a. There are now 2 sevens;   2 │ 4   7   14
              divide by 7. ────────────────────────
                                                       7 │ 2   7   7
              Bring down the 2 for a new row. ──────→   2   1   1

Step 3b. Divide the new row   2 │ 4   7   14
              by 2 to get all 1s
              in the last row. ────────   7 │ 2   7   7

                                                   2 │ 2   1   1

Step 4.   The LCD is $2 \times 7 \times 2 = 28$ ──────┘   1   1   1   ←── all 1s

Using the LCD finder may seem like a long procedure, but with a little practice, it's easy.

YOUR TURN

┌─────────────────┐
│ WORK THIS PROBLEM │
└─────────────────┘

The Question:

Find the LCDs of the following fractions.

(a) $\dfrac{3}{10}, \dfrac{1}{4}, \dfrac{7}{20}, \dfrac{5}{8}$ (b) $\dfrac{2}{3}, \dfrac{1}{8}, \dfrac{5}{12}, \dfrac{3}{4}$ (c) $\dfrac{7}{12}, \dfrac{5}{18}, \dfrac{1}{20}$

(d) $\dfrac{1}{10}, \dfrac{7}{8}, \dfrac{4}{15}, \dfrac{3}{4}, \dfrac{5}{6}$ (e) $\dfrac{2}{3}, \dfrac{3}{4}, \dfrac{5}{6}, \dfrac{7}{9}$

┌─────────────┐
│ ✔ YOUR WORK │
└─────────────┘

The Solution:

(a) The original denominators

Divide by 2 ──────→ 2 │ 10   4   20   8

Divide by 2 ──────→ 2 │ 5   2   10   4

Divide by 5 ──────→ 5 │ 5   1   5   2

Divide by 2 ──────→ 2 │ 1   1   1   2

                         1   1   1   1

LCD $= 2 \times 2 \times 5 \times 2 = 40$

(b)

2	3	8	12	4
2	3	4	6	2
3	3	2	3	1
2	1	2	1	1
	1	1	1	1

LCD $= 2 \times 2 \times 3 \times 2 = 24$

(c)

2	12	18	20
2	6	9	10
3	3	9	5
3	1	3	5
5	1	1	5
	1	1	1

LCD $= 2 \times 2 \times 3 \times 3 \times 5 = 180$

(d)

2	10	8	15	4	6
2	5	4	15	2	3
3	5	2	15	1	3
5	5	2	5	1	1
2	1	2	1	1	1
	1	1	1	1	1

LCD $= 2 \times 2 \times 3 \times 5 \times 2 = 120$

(e)

2	3	4	6	9
3	3	2	3	9
3	1	2	1	3
2	1	2	1	1
	1	1	1	1

LCD $= 2 \times 3 \times 3 \times 2 = 36$

Now that you understand the LCD, you can use it as a step toward solving addition problems with unlike fractions.

Steps: Adding Fractions

STEP 1. *Find* the LCD of the denominators of the fractions.

STEP 2. *Write* the fractions so that they are equivalent fractions with the LCD.

STEP 3. *Add* the numerators (the denominator is the LCD), convert any improper fractions to mixed numbers, and reduce to lowest terms.

For example, add $\frac{7}{15} + \frac{5}{6} + \frac{3}{10}$.

Step 1

2	15	6	10
3	15	3	5
5	5	1	5
	1	1	1

LCD $= 2 \times 3 \times 5 = 30$

Step 2

$$\frac{7}{15} = \frac{?}{30} = \frac{7 \times 2}{15 \times 2} = \frac{14}{30}$$

$$\frac{5}{6} = \frac{?}{30} = \frac{5 \times 5}{6 \times 5} = \frac{25}{30}$$

$$\frac{3}{10} = \frac{?}{30} = \frac{3 \times 3}{10 \times 3} = \frac{9}{30}$$

Step 3

$$\frac{7}{15} + \frac{5}{6} + \frac{3}{10} = \frac{14}{30} + \frac{25}{30} + \frac{9}{30} = \frac{14 + 25 + 9}{30} = \frac{48}{30} = 1\frac{18}{30} = 1\frac{3}{5}$$

If this way of adding fractions seems rather long and involved, here is why:

1. It is involved, but it is the only sure way to arrive at the answer unless you own a fancy calculator that works with fractions.
2. It is new to you and you will need lots of practice at it before it comes quickly. Take each problem step by step, work slowly at first, and gradually you will become very quick at adding fractions.

YOUR TURN

WORK THIS PROBLEM

The Question:

Add the following.

(a) $\dfrac{2}{15} + \dfrac{4}{9}$ (b) $\dfrac{5}{16} + \dfrac{7}{12} + \dfrac{1}{6}$ (c) $\dfrac{5}{12} + \dfrac{2}{9} + \dfrac{3}{8} + \dfrac{1}{6}$

✔ YOUR WORK

The Solution:

(a)

3	15	9
5	5	3
3	1	3
	1	1

$\text{LCD} = 3 \times 5 \times 3 = 45$

$$\frac{2}{15} = \frac{?}{45} = \frac{2 \times 3}{15 \times 3} = \frac{6}{45}$$

$$\frac{4}{9} = \frac{?}{45} = \frac{4 \times 5}{9 \times 5} = \frac{20}{45}$$

$$\frac{2}{15} + \frac{4}{9} = \frac{6}{45} + \frac{20}{45} = \frac{6 + 20}{45} = \frac{26}{45}$$

(b)

2	16	12	6
2	8	6	3
3	4	3	3
4	4	1	1
	1	1	1

$\text{LCD} = 2 \times 2 \times 3 \times 4 = 48$

$$\frac{5}{16} = \frac{?}{48} = \frac{5 \times 3}{16 \times 3} = \frac{15}{48}$$

$$\frac{7}{12} = \frac{?}{48} = \frac{7 \times 4}{12 \times 4} = \frac{28}{48}$$

$$\frac{1}{6} = \frac{?}{48} = \frac{1 \times 8}{6 \times 8} = \frac{8}{48}$$

$$\frac{5}{16} + \frac{7}{12} + \frac{1}{6} = \frac{15}{48} + \frac{28}{48} + \frac{8}{48} = \frac{15 + 28 + 8}{48} = \frac{51}{48} = 1\frac{3}{48} = 1\frac{1}{16}$$

(c)

2	12	9	8	6
2	6	9	4	3
3	3	9	2	3
3	1	3	2	1
2	1	1	2	1
	1	1	1	1

$\text{LCD} = 2 \times 2 \times 3 \times 3 \times 2 = 72$

$$\frac{5}{12} = \frac{?}{72} = \frac{5 \times 6}{12 \times 6} = \frac{30}{72}$$

$$\frac{2}{9} = \frac{?}{72} = \frac{2 \times 8}{9 \times 8} = \frac{16}{72}$$

$$\frac{3}{8} = \frac{?}{72} = \frac{3 \times 9}{8 \times 9} = \frac{27}{72}$$

$$\frac{1}{6} = \frac{?}{72} = \frac{1 \times 12}{6 \times 12} = \frac{12}{72}$$

$$\frac{5}{12} + \frac{2}{9} + \frac{3}{8} + \frac{1}{6} = \frac{30}{72} + \frac{16}{72} + \frac{27}{72} + \frac{12}{72} = \frac{30 + 16 + 27 + 12}{72} = \frac{85}{72} = 1\frac{13}{72}$$

Adding Mixed Numbers with Unlike Fractions

Adding mixed numbers together is similar to the steps for adding fractions.

Steps: Adding Mixed Numbers with Unlike Fractions

STEP 1.   *Find* the LCD of the denominators of the fractions.

STEP 2.   *Write* the fractions so they are equivalent fractions with the LCD.

STEP 3.   *Add* the whole numbers and add the fractions. Convert any improper fractions to mixed numbers, and reduce to lowest terms.

For example, let's add $3\frac{5}{6} + 17\frac{7}{12} + 9\frac{3}{10}$.

Step 1.

2	6	12	10
3	3	6	5
2	1	2	5
5	1	1	5
	1	1	1

$\text{LCD} = 2 \times 3 \times 2 \times 5 = 60$

Step 2.

$$\frac{5}{6} = \frac{?}{60} = \frac{5 \times 10}{6 \times 10} = \frac{50}{60}$$

$$\frac{7}{12} = \frac{?}{60} = \frac{7 \times 5}{12 \times 5} = \frac{35}{60}$$

$$\frac{3}{10} = \frac{?}{60} = \frac{3 \times 6}{10 \times 6} = \frac{18}{60}$$

Step 3.

$$3\frac{5}{6} = 3\frac{50}{60}$$

$$17\frac{7}{12} = 17\frac{35}{60}$$

$$9\frac{3}{10} = 9\frac{18}{60}$$

$$29\frac{103}{60} = 29 + 1\frac{43}{60} = 30\frac{43}{60}$$

YOUR TURN

WORK THIS PROBLEM

The Question:

Add the following mixed numbers.

(a) $4\dfrac{5}{6} + 12\dfrac{3}{7} + 8\dfrac{9}{14}$ (b) $13\dfrac{5}{12} + 2\dfrac{7}{8} + 5\dfrac{9}{16}$

✔ YOUR WORK

The Solution:

(a)

2	6	7	14
7	3	7	7
3	3	1	1
	1	1	1

$\text{LCD} = 2 \times 7 \times 3 = 42$

$$\dfrac{5}{6} = \dfrac{?}{42} = \dfrac{5 \times 7}{6 \times 7} = \dfrac{35}{42}$$

$$\dfrac{3}{7} = \dfrac{?}{42} = \dfrac{3 \times 6}{7 \times 6} = \dfrac{18}{42}$$

$$\dfrac{9}{14} = \dfrac{?}{42} = \dfrac{9 \times 3}{14 \times 3} = \dfrac{27}{42}$$

$$4\dfrac{5}{6} = 4\dfrac{35}{42}$$

$$12\dfrac{3}{7} = 12\dfrac{18}{42}$$

$$+8\dfrac{9}{14} = 8\dfrac{27}{42}$$

$$24\dfrac{80}{42} = 24 + 1\dfrac{38}{42} = 25\dfrac{38}{42} = 25\dfrac{19}{21}$$

(b)

2	12	8	16
2	6	4	8
2	3	2	4
3	3	1	2
2	1	1	2
	1	1	1

$\text{LCD} = 2 \times 2 \times 2 \times 3 \times 2 = 48$

$$\dfrac{5}{12} = \dfrac{?}{48} = \dfrac{5 \times 4}{12 \times 4} = \dfrac{20}{48}$$

$$\dfrac{7}{8} = \dfrac{?}{48} = \dfrac{7 \times 6}{8 \times 6} = \dfrac{42}{48}$$

$$\dfrac{9}{16} = \dfrac{?}{48} = \dfrac{9 \times 3}{16 \times 3} = \dfrac{27}{78}$$

$$13\dfrac{5}{12} = 13\dfrac{20}{48}$$

$$2\dfrac{7}{8} = 2\dfrac{42}{48}$$

$$+5\dfrac{9}{16} = 5\dfrac{27}{48}$$

$$20\dfrac{89}{48} = 20 + \dfrac{89}{48} = 20 + 1\dfrac{41}{48} = 21\dfrac{41}{48}$$

Subtracting Fractions

Once you have mastered the process of adding fractions, subtraction is very simple indeed. Just apply the same principles and your subtraction skills. If the denominators are the same, we subtract numerators and write this difference over the common denominator.

WORK THIS PROBLEM

The Question:

Calculate: $\dfrac{3}{8} - \dfrac{1}{8}$

✔ YOUR WORK

The Solution:

$$\frac{3}{8} - \frac{1}{8} = \frac{3-1}{8} = \frac{2}{8} = \frac{1}{4}$$

Subtracting fractions with different denominators is a bit harder, but again it's much like adding these fractions. When the two fractions have different denominators, we must change them to equivalent fractions with the LCD as denominator before subtracting. (If you have not yet learned how to find the least common denominator (LCD), review the Steps presented earlier in this section.

Steps: Subtracting Fractions

STEP 1.   *Find* the LCD of the denominators of the fractions.

STEP 2.   *Write* the fractions so that they are equivalent fractions with the LCD.

STEP 3.   *Subtract* the numerators (the denominator is the LCD) and reduce to lowest terms.

For example, to subtract $\frac{3}{4} - \frac{1}{5}$,

Step 1

The LCD of 4 and 5 is 20.

Step 2

$$\frac{3}{4} = \frac{?}{20} = \frac{3 \times 5}{4 \times 5} = \frac{15}{20} \qquad \frac{1}{5} = \frac{?}{20} = \frac{1 \times 4}{5 \times 4} = \frac{4}{20}$$

Step 3

$$\frac{3}{4} - \frac{1}{5} = \frac{15}{20} - \frac{4}{20} = \frac{11}{20}$$

WORK THIS PROBLEM

The Question:

Subtract the following.

(a) $\dfrac{5}{6} - \dfrac{1}{4}$ (b) $\dfrac{11}{15} - \dfrac{2}{9}$

✔ YOUR WORK

The Solution:

(a) The LCD of 6 and 4 is 12.

$$\frac{5}{6} = \frac{5 \times 2}{6 \times 2} = \frac{10}{12} \qquad \frac{1}{4} = \frac{1 \times 3}{4 \times 3} = \frac{3}{12}$$

$$\frac{5}{6} - \frac{1}{4} = \frac{10}{12} - \frac{3}{12} = \frac{10 - 3}{12} = \frac{7}{12}$$

(b) The LCD of 15 and 9 is 45.

$$\frac{11}{15} = \frac{11 \times 3}{15 \times 3} = \frac{33}{45} \qquad \frac{2}{9} = \frac{2 \times 5}{9 \times 5} = \frac{10}{45}$$

$$\frac{11}{15} - \frac{2}{9} = \frac{33}{45} - \frac{10}{45} = \frac{33 - 10}{45} = \frac{23}{45}$$

Subtracting Mixed Numbers

The easiest method of subtracting mixed numbers is to subtract the whole number parts and subtract the fractions. For example, let's subtract $17\frac{5}{6} - 12\frac{2}{9}$. The LCD of 6 and 9 is 18, so

$$
\begin{array}{r}
17\frac{5}{6} = 17\frac{15}{18} \\
-12\frac{2}{9} = -12\frac{4}{18} \\
\hline
5\frac{11}{18}
\end{array}
\qquad
\begin{array}{l}
\frac{5}{6} = \frac{5 \times 3}{6 \times 3} = \frac{15}{18} \\[6pt]
\frac{2}{9} = \frac{2 \times 2}{9 \times 2} = \frac{4}{18}
\end{array}
$$

Oftentimes you must use borrowing when subtracting mixed numbers, as when you subtract $28\frac{1}{6} - 15\frac{5}{8}$.

Steps: Subtracting Mixed Numbers and Borrowing

STEP 1. *Find* the LCD of the denominators of the fractions.

STEP 2. *Write* the fractions so that they are equivalent fractions with the LCD.

STEP 3. If necessary, *"borrow"* 1 from the whole number of the minuend, convert the 1 to the LCD, and add it to the fraction of the minuend.

STEP 4. *Subtract* the whole numbers, subtract the fractions, and reduce to lowest terms.

For example, to subtract $28\frac{1}{6} - 15\frac{5}{8}$,

Step 1

The LCD of 6 and 8 is 24.

Step 2

$$
\begin{array}{r}
28\frac{1}{6} = 28\frac{4}{24} \\
-15\frac{5}{8} = -15\frac{15}{24} \\
\hline
\end{array}
$$

Because $\frac{4}{24}$ is less than $\frac{15}{24}$, "borrow" $\frac{24}{24}$ or 1 from the whole number 28.

Step 3

$$28\frac{4}{24} = 27 + 1 + \frac{4}{24} = 27 + \frac{24}{24} + \frac{4}{24} = 27 + \frac{28}{24}$$

Step 4

$$\begin{array}{r} 27\dfrac{28}{24} \\ -15\dfrac{15}{24} \\ \hline 12\dfrac{13}{24} \end{array}$$

YOUR TURN

| WORK THIS PROBLEM |

The Question:

Subtract the following.

(a) $14\frac{2}{3} - 9\frac{1}{4}$ (b) $18\frac{1}{4} - 12\frac{9}{10}$ (c) $53\frac{5}{14} - 27\frac{3}{4}$ (d) $243\frac{1}{6} - 195\frac{5}{9}$

| ✔ YOUR WORK |

The Solution:

(a) The LCD is 12.

$$\begin{array}{r} 14\dfrac{2}{3} = 14\dfrac{8}{12} \\ -9\dfrac{1}{4} = -9\dfrac{3}{12} \\ \hline 5\dfrac{5}{12} \end{array}$$

(b) The LCD is 20.

$$18\frac{1}{4} = 18\frac{5}{20} = 18\frac{5}{20} = 17 + 1\frac{5}{20} = 17 + \frac{25}{20} = 17\frac{25}{20}$$
$$-12\frac{9}{10} = -12\frac{18}{20} = \qquad\qquad\qquad\qquad\qquad -12\frac{18}{20}$$
$$\overline{\qquad\qquad\qquad\qquad\qquad\qquad\qquad\qquad\qquad 5\frac{7}{20}}$$

(c) The LCD is 28.

$$53\frac{5}{14} = 53\frac{10}{28} = 53\frac{10}{28} = 52 + 1\frac{10}{28} = 52 + \frac{38}{28} = 52\frac{38}{28}$$
$$-27\frac{3}{4} = -27\frac{21}{28} = \qquad\qquad\qquad\qquad\qquad\qquad -27\frac{21}{28}$$
$$\overline{\qquad\qquad\qquad\qquad\qquad\qquad\qquad\qquad\qquad 25\frac{17}{28}}$$

(d) The LCD is 18.

$$243\frac{1}{6} = 243\frac{3}{18} = 243\frac{3}{18} = 242 + 1\frac{3}{18} = 242 + \frac{21}{18} = 242\frac{21}{18}$$
$$-195\frac{5}{9} = -195\frac{10}{18} = \qquad\qquad\qquad\qquad\qquad\qquad -195\frac{10}{18}$$
$$\overline{\qquad\qquad\qquad\qquad\qquad\qquad\qquad\qquad\qquad 47\frac{11}{18}}$$

Now use your fraction skills on this business application.

YOUR
TURN

WORK THIS PROBLEM

The Question:

The price of Micro Systems Support stock increased from 1\frac{7}{8}$ to 2\frac{1}{2}$ per share. What was the amount of the increase?

✔ YOUR WORK

The Solution:

Increase $= 2\frac{1}{2} - 1\frac{7}{8}$

$$2\frac{1}{2} = 2\frac{4}{8} = 1 + 1\frac{4}{8} = 1 + \frac{8}{8} + \frac{4}{8} = 1\frac{12}{8}$$

Therefore,

$$2\frac{1}{2} - 1\frac{7}{8} = 1\frac{12}{8} - 1\frac{7}{8} = \frac{5}{8}$$

The stock price for Micro Systems Support increased by $\$\frac{5}{8}$ per share.

WORKSPACE

SECTION TEST 2.4 Fractions

Name _____

Date _____

Course/Section _____

The following problems test your understanding of Section 2.4, Addition and Subtraction of Fractions.

A. Add or subtract as shown and reduce to lowest terms.

1. $\dfrac{3}{7} + \dfrac{5}{7} =$ _____

2. $\dfrac{5}{9} + \dfrac{2}{9} =$ _____

3. $\dfrac{5}{8} + \dfrac{7}{8} =$ _____

4. $\dfrac{9}{12} - \dfrac{5}{12} =$ _____

5. $\dfrac{7}{8} - \dfrac{3}{8} =$ _____

6. $\dfrac{8}{9} - \dfrac{5}{9} =$ _____

7. $\dfrac{9}{16} + \dfrac{11}{16} + \dfrac{5}{16} =$ _____

8. $\dfrac{9}{12} + \dfrac{7}{12} - \dfrac{11}{12} =$ _____

9. $\dfrac{7}{8} - \dfrac{5}{8} + \dfrac{3}{8} =$ _____

10. $\dfrac{1}{2} + \dfrac{3}{4} =$ _____

11. $\dfrac{5}{8} + \dfrac{1}{4} =$ _____

12. $\dfrac{5}{6} - \dfrac{7}{12} =$ _____

13. $\dfrac{1}{3} - \dfrac{1}{6} =$ _____

14. $\dfrac{7}{12} + \dfrac{1}{4} + \dfrac{2}{3} =$ _____

15. $\dfrac{5}{6} - \dfrac{5}{18} + \dfrac{4}{9} =$ _____

16. $\dfrac{1}{2} + \dfrac{2}{3} =$ _____

17. $\dfrac{3}{4} + \dfrac{1}{3} =$ _____

18. $\dfrac{4}{5} - \dfrac{1}{2} =$ _____

19. $\dfrac{2}{3} - \dfrac{2}{5} =$ _____

20. $\dfrac{4}{5} + \dfrac{1}{2} - \dfrac{2}{3} =$ _____

21. $\dfrac{1}{2} - \dfrac{1}{3} + \dfrac{3}{5} =$ _____

B. Add or subtract as shown and reduce to lowest terms.

1. $\dfrac{5}{6} + \dfrac{3}{4} =$ _____

2. $\dfrac{5}{9} + \dfrac{1}{6} =$ _____

3. $\dfrac{4}{9} + \dfrac{5}{12} =$ _____

4. $\dfrac{9}{10} - \dfrac{3}{4} =$ _____

5. $\dfrac{7}{8} - \dfrac{1}{6} =$ _____

6. $\dfrac{3}{4} - \dfrac{1}{6} =$ _____

7. $\dfrac{1}{4} + \dfrac{2}{3} + \dfrac{5}{6} =$ _____

8. $\dfrac{7}{10} + \dfrac{3}{5} - \dfrac{7}{8} =$ _____

9. $\dfrac{7}{8} - \dfrac{2}{3} + \dfrac{1}{6} =$ _____

10. $5\dfrac{2}{3} + 3\dfrac{1}{4} =$ _____

11. $12\dfrac{5}{8} + 8\dfrac{5}{6} =$ _____

12. $15\dfrac{5}{6} - 8\dfrac{1}{4} =$ _____

13. $28\dfrac{7}{9} - 15\dfrac{1}{6} =$ _____

14. $19\dfrac{7}{12} + 3\dfrac{9}{10} =$ _____

15. $38\dfrac{3}{15} + 18\dfrac{7}{10} =$ _____

16. $37\dfrac{2}{7} - 19\dfrac{2}{3} =$ _____

17. $25 - 17\dfrac{3}{4} =$ _____

18. $27\dfrac{5}{6} + 13\dfrac{4}{9} =$ _____

19. $6\dfrac{8}{15} + 9\dfrac{7}{9} =$ _____

20. $32\dfrac{3}{8} - 12\dfrac{11}{12} =$ _____

21. $84\dfrac{2}{9} - 58\dfrac{7}{12} =$ _____

C. Solve these problems.

1. Bob's Beeflike Burgers has a burger that weighs $\frac{7}{8}$ of a pound prior to cooking and $\frac{3}{4}$ of a pound after cooking. How much weight is lost in cooking?

2. J.R.'s Construction Company originally owned $23\frac{5}{6}$ acres of land. The company acquired another $17\frac{3}{8}$ acres of adjoining land. How many acres does the company own?

3. The Woody's Lumber Mill hires part-time workers. Marcia worked $2\frac{3}{4}$ hours Monday, $6\frac{5}{12}$ hours Wednesday, and $7\frac{1}{2}$ hours Friday. How many hours did she work during the week?

4. The Mouse Hole stocks cheeses that are sold by the wheel. Last week it sold $2\frac{3}{4}$ wheels of Cheddar, $5\frac{2}{3}$ wheels of Brie, and $7\frac{5}{6}$ wheels of lite Cheddar. How many wheels of cheese did it sell in all?

5. The Fly-By-Night Aviary had a high heating bill this winter. The Aviary used $122\frac{3}{4}$ gallons of oil in November, $187\frac{2}{3}$ gallons of oil in December, and $213\frac{5}{6}$ gallons in January. How many gallons of oil did the shop use in the three-month period?

6. Stock for Wired for Sound originally sold for $\$4\frac{1}{2}$ per share. It is now selling for $\$5\frac{1}{8}$ per share. How much has the stock increased?

7. The Bits 'n' Bytes computer store sold the following amounts of RS232 computer cable last week: $13\frac{1}{2}$ feet, $7\frac{7}{8}$ feet, and $34\frac{5}{6}$ feet. How much computer cable has the store sold?

8. The Zippy Delivery driver drove $6\frac{5}{12}$ hours Monday, $8\frac{3}{4}$ hours Tuesday, $5\frac{1}{2}$ hours Wednesday, $9\frac{1}{4}$ hours Thursday, and $8\frac{5}{6}$ hours Friday. How many hours did the Zippy driver drive during the week?

9. On Saturday, the Zippy Delivery Service driver had a very heavy day. In the morning, he made three deliveries that weighed $75\frac{3}{4}$ pounds each. During the afternoon he had to deliver 5 wedding cakes. Two of the cakes weighed $24\frac{2}{3}$ pounds each, and the other three cakes weighed $16\frac{1}{2}$ pounds each. How many pounds did he deliver on this particular day?

10. Your secretary has been working overtime to finish typing all of your dictation. He worked overtime $\frac{3}{4}$ of an hour Monday, $\frac{5}{6}$ of an hour Tuesday, $\frac{2}{3}$ of an hour Thursday, and $1\frac{1}{2}$ hours on Friday. What was his total overtime for the week?

11. Marco's Manilla Envelope Company sold $34\frac{1}{2}$ cases of manilla envelopes Monday, $45\frac{3}{8}$ cases Tuesday, $39\frac{1}{16}$ cases Wednesday, $41\frac{3}{4}$ cases Thursday, and $43\frac{5}{8}$ cases Friday. How many cases of manilla envelopes were sold?

12. Petulia's Petals sold seven flower arrangements. The arrangements used the following amounts of roses in each arrangement: $1\frac{1}{2}$ dozen, $\frac{3}{4}$ dozen, $\frac{1}{4}$ dozen, $1\frac{1}{4}$ dozen, $3\frac{5}{6}$ dozen, $2\frac{5}{12}$ dozen, and $1\frac{5}{6}$ dozen. How many dozens of roses were used in the seven arrangements?

13. Kate's Delicacies supplied beverages for a children's birthday party. The children consumed $\frac{3}{4}$ of a case of cherry pop, $\frac{7}{8}$ of a case of grape soda, $\frac{11}{12}$ of a case of Lively Lime soda, and $\frac{5}{6}$ of a case of Fruity Fizz. How many cases of soda did Kate sell?

14. Lots of Plots Realty originally owned $234\frac{1}{4}$ acres of land. After selling $145\frac{7}{8}$ acres, how many acres of land remain?

15. The four sides of a plot of land measured $125\frac{3}{4}$ feet, $98\frac{1}{2}$ feet, $132\frac{2}{3}$ feet, and $92\frac{5}{6}$ feet. What is the total distance around the edge of the lot?

16. On the stock exchanges, where stocks are bought and sold, prices are usually expressed as a fraction of a dollar. The following business graph traces the price of Micro Systems Support, Inc., for the first half of the year.

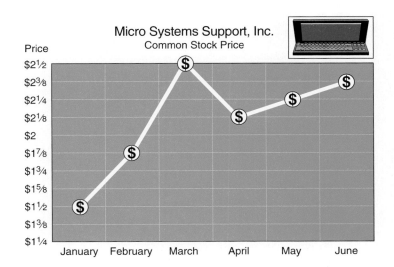

(a) What is the increase in stock price from January to February?
(b) What is the decrease in price from March to April?
(c) What is the increase in price from January to June?
(d) Which month had the largest price increase and what was the increase?
(e) What is the cost of 50 shares of stock in January?
(f) What is the cost of 50 shares of stock in June?
(g) What is the difference in cost of 50 shares of stock in January and June?
(h) How many shares of stock could you purchase in January for $285?
(i) How many shares of stock could you purchase in June for $285?

17. Carpets-R-Beautiful won the bid from Dynamite Deals to carpet the following areas:

Bridal and Formal Wear—$25\frac{1}{2}$ ft by $57\frac{1}{4}$ ft
Women's Wear—37 ft by $42\frac{3}{4}$ ft
Junior/Miss Department—$45\frac{3}{4}$ ft by $27\frac{1}{2}$ ft

If they plan to use the same type of carpet in each area, how many square feet of carpet are needed?

WORKSPACE

Key Terms

denominator
fraction
improper fraction
LCD finder
Least common denominator (LCD)

like fractions
mixed number
numerator
prime number
proper fraction

Chapter 2 at a Glance

Page	Topic	Key Point	Example
56	Parts of fractions	Numerator = upper number. Denominator = lower number.	$\dfrac{9}{10}$ = numerator = denominator
57	Types of fractions	Proper = less than 1. Improper = equal to or more than 1. Mixed number = whole number plus proper fraction.	$\dfrac{9}{10}$ = proper $\dfrac{11}{10}$ = improper, $\dfrac{10}{10}$ = 1 improper $1\dfrac{1}{10}$ = mixed number
58–60	Converting fractions	Improper to mixed: Divide numerator by denominator and put any remainder over the same denominator. Mixed to improper: Multiply denominator by whole number, add answer to numerator, and put result over denominator.	$\dfrac{987}{65} = 15\dfrac{12}{65}$ $12\dfrac{3}{4} = \dfrac{51}{4}$
60–61	Raising fractions to higher terms	Multiply numerator and denominator by the same number.	$\dfrac{3}{4} \times \dfrac{5}{5} = \dfrac{15}{20}$
61–63	Reducing fractions	Divide numerator and denominator by the same common divisor; repeat until no common divisors remain.	$\dfrac{90}{105} \div \dfrac{5}{5} = \dfrac{18}{21} \div \dfrac{3}{3} = \dfrac{6}{7}$
67–69	Multiplying fractions	Multiply the numerators; multiply the denominators; reduce to lowest terms	$\dfrac{2}{3} \times \dfrac{3}{4} = \dfrac{6}{12} = \dfrac{1}{2}$
75–79	Dividing fractions	Invert the divisor and multiply; reduce to lowest terms.	$\dfrac{7}{8} \div \dfrac{3}{2} = \dfrac{7}{8} \times \dfrac{2}{3} = \dfrac{14}{24} = \dfrac{7}{12}$
83	Adding like fractions	Add numerators; place result over denominator; reduce to lowest terms.	$\dfrac{3}{16} + \dfrac{5}{16} = \dfrac{8}{16} = \dfrac{1}{2}$
84	Adding mixed numbers	Add whole numbers; add fractions; reduce to lowest terms.	$2\dfrac{2}{5} + 3\dfrac{4}{5} = 5\dfrac{6}{5} = 6\dfrac{1}{5}$
86–88	Finding the least common denominator (LCD)	Arrange the denominators in a row and divide by any prime number that goes into one or more; bring down any denominators not divided; repeat until all 1s in bottom row; multiply all divisors.	To find the LCD of $\dfrac{1}{4}, \dfrac{3}{7}, \dfrac{5}{14}$ $\begin{array}{c\|ccc} 2 & 4 & 7 & 14 \\ \hline 7 & 2 & 7 & 7 \\ \hline 2 & 2 & 1 & 1 \\ \hline & 1 & 1 & 1 \end{array}$ LCD = $2 \times 7 \times 2 = 28$

Page	Topic	Key Point	Example
88–92	Adding unlike fractions	Find the LCD of the fractions; raise them to like fractions with the LCD; add the numerators; reduce to lowest terms.	To add: $\dfrac{7}{15} + \dfrac{5}{6} + \dfrac{3}{10}$ LCD = 30 $\dfrac{7}{15} = \dfrac{14}{30}; \dfrac{5}{6} = \dfrac{25}{30}; \dfrac{3}{10} = \dfrac{9}{30}$ $\dfrac{14}{30} + \dfrac{25}{30} + \dfrac{9}{30} = \dfrac{48}{30}$ $\dfrac{48}{30} = 1\dfrac{18}{30} = 1\dfrac{3}{5}$
92–93	Subtracting fractions	If fractions are like, subtract numerators and reduce to lowest terms. If fractions are unlike, find the LCD and raise them to fractions with the LCD; subtract the numerators and reduce to lowest terms.	$\dfrac{3}{8} - \dfrac{1}{8} = \dfrac{2}{8} = \dfrac{1}{4}$ To find: $\dfrac{5}{6} - \dfrac{1}{4}$ LCD = 12 $\dfrac{5}{6} = \dfrac{10}{12}; \dfrac{1}{4} = \dfrac{3}{12}$ $\dfrac{10}{12} - \dfrac{3}{12} = \dfrac{7}{12}$
93–95	Subtracting mixed numbers	Subtract fractions, if necessary "borrow" from the whole number in the minuend; subtract whole numbers; reduce to lowest terms.	$18\dfrac{1}{4} - 12\dfrac{9}{10}$ LCD = 20 $18\dfrac{1}{4} = \quad 18\dfrac{5}{20} = \quad 17\dfrac{25}{20}$ $-12\dfrac{9}{10} = -12\dfrac{18}{20} = -12\dfrac{18}{20}$ $\qquad\qquad\qquad\qquad\qquad 5\dfrac{7}{20}$

SELF-TEST Fractions

The following problems test your understanding of fractions. The answers to the odd-numbered problems are in the appendix.

1. Write $7\dfrac{5}{12}$ as an improper fraction. _____

2. Write $14\dfrac{4}{5}$ as an improper fraction. _____

3. Write $\dfrac{38}{7}$ as a mixed number. _____

4. Write $\dfrac{50}{9}$ as a mixed number. _____

5. Reduce to lowest terms $\dfrac{180}{225}$. _____

6. Reduce to lowest terms $\dfrac{135}{180}$. _____

7. $\dfrac{9}{16} = \dfrac{?}{80}$

8. $\dfrac{7}{12} = \dfrac{?}{96}$

9. $5\dfrac{1}{3} \times 4\dfrac{1}{6} \times 2\dfrac{1}{4} = $ _____

10. $2\dfrac{2}{5} \times \dfrac{4}{9} \times 3\dfrac{1}{8} = $ _____

11. $45\dfrac{5}{9} - 17\dfrac{5}{6} = $ _____

12. $21\dfrac{5}{12} - 8\dfrac{9}{10} = $ _____

13. $48\dfrac{9}{10} + 25\dfrac{8}{15} = $ _____

14. $83\dfrac{5}{6} + 27\dfrac{3}{4} = $ _____

15. $1\dfrac{5}{16} \div 1\dfrac{11}{24} = $ _____

16. $1\dfrac{3}{4} \div 2\dfrac{11}{12} = $ _____

17. Intense Tents, a local outdoor recreation store, sells climbing rope. The store used five pieces of rope for floor models, each $6\frac{2}{3}$ feet long, and sold two $165\frac{3}{4}$-foot lengths to climbers. (a) How much rope was used? (b) The ropes were cut from a 450-foot length of rope. How much rope is left?

18. The Butter-em-Up Bakery uses $17\frac{3}{4}$ pounds of flour daily. How much flour is used in seven days?

19. Pillory and Fetter, a local brokerage, sold 50 shares of Growing Fast Stock for $\$7\frac{3}{8}$ a share. It also sold 75 shares of Forever Steady Stock at $\$3\frac{5}{8}$ a share. (a) What is the value of each stock sold? (b) What is the total sale?

20. The Zippy Delivery Service's best driver averages $52\frac{3}{4}$ miles per hour. He drives $6\frac{1}{4}$ hours per day. (a) How many miles does he cover daily? (b) How many miles does he cover in a five-day workweek?

21. The four sides of a plot of land measured $135\frac{1}{2}$ feet, $115\frac{3}{4}$ feet, $142\frac{5}{6}$ feet, and $118\frac{2}{3}$ feet. What is the total distance around the edge of the lot?

22. Henry worked $7\frac{3}{4}$ hours Monday, $8\frac{2}{3}$ hours Tuesday, $7\frac{5}{6}$ hours Wednesday, $9\frac{1}{2}$ hours Thursday, and $8\frac{1}{3}$ hours Friday. How many hours did he work this week?

23. How many shares of stock selling for $\$6\frac{7}{8}$ per share can be purchased for $1100?

24. Lots of Plots Realty originally owned $1345\frac{3}{8}$ acres of land. After selling three portions of $548\frac{3}{4}$, $235\frac{7}{8}$, and $367\frac{1}{2}$ acres, how many acres of land remain?

CHAPTER 3

Decimal Numbers

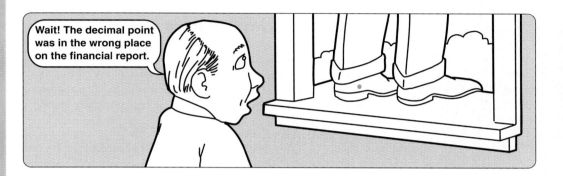

> Wait! The decimal point was in the wrong place on the financial report.

CHAPTER OBJECTIVES

When you complete this chapter successfully, you will be able to:

1. Add and subtract decimal numbers.

2. Multiply and divide decimal numbers.

3. Convert decimal numbers to fractions and fractions to decimal numbers.

■ The Moore Police Department is reequipping its police officers. Each officer gets a bulletproof vest ($379.72), kryptonite handcuffs ($25.95), and mirrored sunglasses ($34.27). How much does it cost to equip each police officer?

■ Sew What?, a fabric store, sells silk for $21.67 a yard. How much does it cost Taylor to buy 32.5 yards of silk?

■ The Pizza Palace has set aside $8412 for employee bonuses. If each of the 15 employees receives an equal bonus, how much is each bonus?

■ Frame-Up, the local picture shop, sells picture frame wood for $28.99 per yard. Calculate the cost of $56\frac{1}{3}$ yards of picture frame.

Every day, whether you know it or not, you use decimals. That little dot separating the digits of a number can make a huge difference. You'd probably love to buy a CD player for $25.99, but you'd be less enthusiastic about one priced at $2599.00.

In this chapter, you'll learn about how to use your arithmetic skills to work with decimals: adding, subtracting, multiplying, dividing, and solving problems involving these numbers. So sharpen your pencil and we'll get right to the "point"—the decimal point, that is.

105

SECTION 3.1:   Adding and Subtracting Decimals

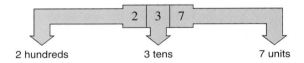

Can You:

Read decimal numbers?
Add decimal numbers?
Subtract decimal numbers?
. . . If not, you need this section.

Reading Decimal Numbers

Decimal Number   A way of expressing a fraction with a denominator of 10, 100, 1000, 10,000, and so on.

Look in your pockets or purse and see how much money is there, to the penny. If you wrote out your answer, you'd probably write a number like $7.23 or $84.42 or perhaps $0.02. All these numbers are decimal numbers. A **decimal number** is just another way of writing a mixed number whose fraction has a denominator of 10, 100, 1000, 10,000, and so on.

$$\$7.23 = \$7\frac{23}{100} \qquad \$84.42 = \$84\frac{42}{100} \qquad \$0.02 = \$\frac{2}{100}$$

In the case of U.S. money, the denominator in these fractions is always 100. But it's easy to convert any fraction or mixed number with a multiple-of-10 denominator to a decimal number.

In Chapter 1, you learned that whole numbers are written in a place value system. A number such as 237 means

| 2 | 3 | 7 |

2 hundreds   3 tens   7 units

We can also extend this way of writing numbers to decimal numbers. For example, consider the decimal number 3,254,935.4728:

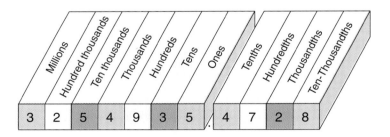

You could read this number as "three million, two hundred fifty-four thousand, nine hundred thirty-five, and four thousand seven hundred twenty-eight ten-thousandths." Notice that the decimal point is read "and."

More often, however, this number is read more simply as "three million, two hundred fifty-four thousand, nine hundred thirty-five, *point* four, seven, two, eight." This way of reading the number is easiest to write, to say, and to understand.

To write a fraction whose denominator is a multiple of 10 as a decimal number, put the last digit in the numerator of the fraction into the appropriate digit place. Fill in any spaces from the decimal point to the start of the numerator with zeros. For example,

$$\frac{6}{10} \qquad = 6 \text{ tenths} \qquad = 0.6$$

$$\frac{5}{100} \qquad = 5 \text{ hundredths} \qquad = 0.05$$

$$\frac{32}{100} \qquad = 32 \text{ hundredths} \qquad = 0.32$$

$$\frac{4}{1000} \qquad = 4 \text{ thousandths} \qquad = 0.004$$

$$\frac{267}{1000} \qquad = 267 \text{ thousandths} = 0.267$$

$$\uparrow \qquad\qquad\qquad\qquad\qquad \uparrow$$

Fraction form   Decimal form

The decimal number .6 is usually written 0.6. The zero added on the left is used to call attention to the decimal point. It is easy to mistake .6 for 6, but the decimal point on 0.6 cannot be overlooked.

WORK THIS PROBLEM

The Question:

(a) Write the decimal number 0.526 in fraction form.
(b) Write the fraction $\frac{92}{1000}$ in decimal form.

✔ YOUR WORK

The Solution:

(a) $0.526 = 526 \text{ thousandths} = \dfrac{526}{1000}$ (b) $\dfrac{92}{1000} = 0.092$

When a decimal number has both a whole number part and a fraction part, convert the fraction and add the whole number to it. For example,

$$324.576 = 324 + 0.576 = 324 + \frac{576}{1000} = 324\frac{576}{1000}$$

WORK THIS PROBLEM

The Question:

Write the following in fraction form.

(a) $86.42   (b) 43.607   (c) 14.5060   (d) 235.22267

✔ YOUR WORK

The Solution:

(a) $\$86.42 = \$86\dfrac{42}{100} = 86\dfrac{42}{100}$ dollars

(b) $43.607 = 43\dfrac{607}{1000}$

(c) $14.5060 = 14\dfrac{5060}{10000}$

(d) $235.22267 = 235\dfrac{22267}{100000}$

Decimal Digits All numbers to the right of the decimal point in a decimal number.

In the decimal number 86.423, the digits 4, 2, and 3 are called **decimal digits.** The number 43.6708 has four decimal digits. All digits to the right of the decimal point—those that name the fractional part of the number—are decimal digits.

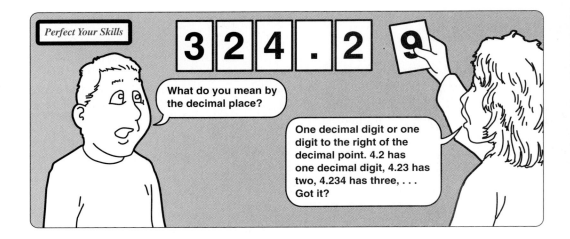

WORK THIS PROBLEM

YOUR TURN

The Question:

Lou Price was calculating compound interest and found the value 2.6915881 in the table. How many decimal digits are included in the number?

✔ YOUR WORK

The Solution:

The number 2.6915881 has seven decimal digits: 6, 9, 1, 5, 8, 8, and 1. These are the digits to the right of the decimal point.

Notice that the decimal point is simply a way of separating the whole number part from the fraction part; it is a place marker. In whole numbers the decimal point usually is not written, but its location should be clear to you.

$$2 = 2. \qquad \text{or} \qquad 324 = 324.$$

The decimal point   The decimal point

Very often additional zeros are annexed to the decimal number without changing its value. For example,

$$8.5 = 8.50 = 8.5000 \text{ and so on}$$

$$6 = 6. = 6.0 = 6.00 \text{ and so on}$$

The value of the number is not changed but the additional zeros may be useful, as you will see later in this chapter.

Adding Decimal Numbers

Because decimal numbers are fractions or mixed numbers, you could add them by converting them to fractions, adding, then converting the answer to a decimal number. For example, to add $0.2 + 0.5$,

$$0.2 + 0.5 = \frac{2}{10} + \frac{5}{10} = \frac{2+5}{10} = \frac{7}{10} = 0.7$$

This method works, but it is cumbersome. It's much easier just to arrange the digits in vertical columns and add directly. Be sure to put digits of the same place value in the same vertical column and to line up the decimal points vertically.

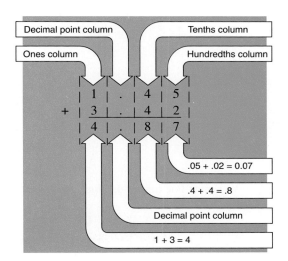

If one of the addends is written with fewer decimal digits than the other, annex as many zeros as needed to write both addends with the same number of decimal digits.

$$\begin{array}{r} 2.345 \\ +1.5 \\ \hline \end{array} \quad \text{becomes} \quad \begin{array}{r} 2.345 \\ +1.500 \\ \hline \end{array}$$

Except for the preliminary step of lining up decimal points, addition of decimal numbers is exactly the same process as addition of whole numbers.

YOUR TURN

___WORK THIS PROBLEM___

The Question:

Add the following decimal numbers by hand. Hold off on using your calculator here.

(a) $4.02 + $3.67   (b) 13.2 + 1.57   (c) 23.007 + 1.12

(d) 14.6 + 1.2 + 3.15   (e) 5.7 + 3.4   (f) 42.768 + 9.37

___✔ YOUR WORK___

The Solution:

               ─── Decimal points in line vertically.

(a) $4.02
     $3.67
     $7.69 ◄── 2 + 7 = 9   Add cents (hundredths).
             0 + 6 = 6   Add 10¢ units (tenths).
             4 + 3 = 7   Add dollars (units).

               ─── Decimal points in line.

(b) 13.20
    +1.57   Annex a zero to provide the same number of decimal
    14.77   digits as in the other addend.
             ─── Place answer decimal point in the same vertical line.

(c) 23.007   (d) 14.60
    +1.120   1.20
    24.127   +3.15
                                   18.95

      ¹   ¹¹ ¹
(e)   5.7   (f) 42.768
   +3.4   +9.370
    9.1   52.138
           7 + 4 = 11   Write 1.
                      Carry 1.
           1 + 5 + 3 = 9.

Be Careful!

You must line up the decimal points carefully to be certain of getting a correct answer.

Subtracting Decimal Numbers

Subtraction is equally simple if you are careful to line up decimal points and attach any needed zeros before you begin work.

For example, $437.56 − $41 is

        ³ ¹³
      $4̸3̸7.56   Decimal points in a vertical line.
    −$ 41.00 ◄── Attach zeros. (Remember that $41 is $41. or $41.00)
     $396.56
           ─── Answer decimal point in the same vertical line.

or, again, 19.452 − 7.3615

$$19.\overset{3\,15\,1\,10}{\cancel{4520}}$$ —— Decimal points in a vertical line.
$$-7.3615$$ —— Attach zeros.
$$12.0905$$

↑ —— Answer decimal point in the same vertical line.

WORK THIS PROBLEM

YOUR
TURN

The Question:

Subtract the following numbers and check your answers by comparing the sum of the difference (answer) and subtrahend (number subtracted) with the minuend.

(a) $37.66 − $14.57   (b) 248.3 − 135.921
(c) 6.4701 − 3.2   (d) 7.304 − 2.59

✔ **YOUR WORK**

The Solution:

Line up decimal points.

(a) $$\$37.\overset{5\,16}{\cancel{66}}$$
$$-14.57$$
$$\$23.09$$ **Check:** 14.57 + 23.09 = 37.66

(b) $$248.300$$ —— Line up decimal points.
$$-135.921$$ —— Attach zeros.
$$112.379$$

Check: 135.921 + 112.379 = 248.300

—— Answer decimal point in the same vertical line.

(c) $$6.4701$$
$$-3.2000$$
$$3.2701$$ **Check:** 3.2000 + 3.2701 = 6.4701

(d) $$7.304$$
$$-2.590$$
$$4.714$$ **Check:** 2.590 + 4.714 = 7.304

Perfect Your Skills

How do I do a problem like 90 − 25.4?

Easy. Write 90 as 90.0, then line up the decimal points

90.0
−25.4 and subtract

90.0
−25.4
64.6 Then check your answer:

25.4 + 64.6 = 90.0

Businesspeople often have to subtract decimal numbers. In fact, some businesses still have cash registers that do not let cashiers put in the amount received and tell them how much to give back in change. In such places, cashiers must be able to subtract decimals.

YOUR TURN

WORK THIS PROBLEM

The Question:

Jim had a savings account balance of $753.47 before withdrawing $165.19. What is his current balance?

✔ YOUR WORK

The Solution:

$753.47
−165.19
$588.28   **Check:** $165.19 + 588.28 = $753.47

By now, you should be ready for the set of practice problems on addition and subtraction of decimal numbers in Section Test 3.1.

SECTION TEST 3.1 Decimal Numbers

The following problems test your understanding of Section 3.1, Adding and Subtracting Decimal Numbers. The answers to the odd-numbered problems are in the appendix.

A. Add or subtract as shown.

1. $0.7 + 0.9 =$ _____

2. $0.7 + 0.3 =$ _____

3. $2.4 + 3.6 =$ _____

4. $4.6 + 7.5 =$ _____

5. $6.2 + 9.8 =$ _____

6. $5.7 + 4.7 =$ _____

7. $7.3 - 3.8 =$ _____

8. $3.6 - 0.8 =$ _____

9. $5.7 - 3.9 =$ _____

10. $8.1 - 3.4 =$ _____

11. $5.3 - 0.8 =$ _____

12. $7.4 - 2.5 =$ _____

13. $8.5 - 7.7 =$ _____

14. $5.4 - 4.5 =$ _____

15. $7.2 + 4.6 + 8.4 =$ _____

16. $6.2 + 9.3 + 7.9 =$ _____

17. $8.3 + 7.2 + 9.8 =$ _____

18. $8.2 + 4.7 + 8.4 =$ _____

19. $0.8 + 8.4 + 8.3 =$ _____

20. $4.6 + 0.3 + 3.5 =$ _____

21. $3.45 + 8.93 =$ _____

22. $6.83 + 3.05 =$ _____

23. $5.73 - 2.59 =$ _____

24. $8.38 - 3.49 =$ _____

B. Add or subtract as shown.

1. $23.6 + 3.9 =$ _____

2. $3.5 + 12.8 =$ _____

3. $\$6.73 + \$34.78 =$ _____

4. $\$56.83 + \$0.84 =$ _____

5. $17.64 + 3.5 =$ _____

6. $6.57 + 57.2 =$ _____

7. $\$35.83 - \$17.95 =$ _____

8. $\$34.78 - \$9.39 =$ _____

9. $84.3 - 29.45 =$ _____

10. $35.1 - 27.22 =$ _____

11. $4.273 + 0.47 =$ _____

12. $0.837 + 12.65 =$ _____

13. $12 + 5.836 =$ _____

14. $4.904 + 5.2 =$ _____

15. $3.623 - 0.654 =$ _____

16. $5.702 - 2.525 =$ _____

17. $3.75 - 1.495 =$ _____

18. $7.27 - 2.172 =$ _____

19. $2.47 + 5.284 + 6.5 =$ _____

20. $2.8 + 0.836 + 4.73 =$ _____

21. $735.2 + 4.264 + 3.74 =$ _____

22. $3.5 + 0.264 + 9.22 =$ _____

23. $35.46 - 7.837 =$ _____

24. $415.5 - 25.394 =$ _____

C. Calculate.

1. $0.3572 + 12.53 =$ _____

2. $8.7836 + 26.48 =$ _____

3. $\$50 - \$26.89 =$ _____

4. $\$20 - \$18.47 =$ _____

5. $34.2 + 9.37247 =$ _____

6. $0.7355 + 362.1 =$ _____

7. $6.7 - 0.7365 =$ _____

8. $2.1 - 0.83637 =$ _____

9. $26.735 + 1.7373 =$ _____

10. $246.3 + 2.84634 =$ _____

11. $47.2 - 24.7362 =$ _____

12. $35.1 - 27.7356 =$ _____

13. $1.2 - 0.72538 =$ _____

14. $1.5 - 0.73648 =$ _____

15. $0.2736 + 263.36 + 3.89463 + 54.2 =$ _____

16. $3.6 + 0.73952 + 836.24 + 34.73602 =$ _____

17. $9.3874 + 1.45 + 836 + 35.4 =$ _____

18. $84.281 + 27 + 0.83738 + 4.28 =$ _____

19. $48.834 + 2.72 - 27.7463 =$ _____

20. $83.2 + 3.736 - 81.8735 =$ _____

21. $84.72 + 2.7474 + 0.203 + 34 + 0.5 =$ _____

22. $3.58 + 72 + 28.93602 + 3.5 + 0.37 =$ _____

23. $9.02 + 56 + 0.786783 + 24.725 + 0.3 =$ _____

24. $3945 + 3.73 + 0.83683 + 38.3 + 0.1 =$ _____

D. Solve these problems.

1. During the first three months of the year, Micro Systems Support (MSS) reported the following income: January—$54,283.93; February—$49,234.53; and March—$43,859.82. What was the quarterly income for MSS?

2. Wired for Sound Electronics has three branches with the following daily sales. Calculate the total daily sales, the weekly sales for each store, and the total sales for the week.

Branch	M	T	W	Th	F	Weekly Total
Downtown	$2837.72	$3967.64	$3547.91	$2736.83	$5438.73	
Suburb	3524.82	5463.83	2635.83	3510.25	5362.53	
Mall	5362.74	5316.92	2527.54	5375.96	3251.85	
Totals						

3. Bits 'n' Bytes, a local computer store, sold the following computer system. The laser printer cost $827.73, the monitor cost $364.37, and the other components were packaged together for $1251.99. What was the total amount of the sale?

4. Maria's Mannequins rents out mannequins. During January, Maria made $5,723.82, during February she made $7,463.27, and during March she made $8,572.77. What was Maria's total mannequin rental income during the first quarter?

5. The Appliance Warehouse is having a sale. Each dryer is reduced by $49.98. What are the final prices for the following dryers?

Dryer	Original Price	Final Price
No-frills model	$212.75	
Middle of the road	275.42	
Even-folds model	417.67	

6. The Going Up Elevator Company has been making numerous maintenance visits. One of the repair people collected five accounts while making service calls this week: The amounts were $79.73, $181.68, $273.47, $82.91, and $281.03. What was the total amount collected?

7. The Moore Police Department is reequipping its police officers. Each officer gets a bulletproof vest ($379.72), kryptonite handcuffs ($25.95), and mirrored sunglasses ($34.27). How much does it cost to equip each police officer?

8. Bob's Beeflike Burgers has suffered a catastrophe—the cash register expired during lunch hour. You are the math wizard and have been chosen to stand in for the cash register. The following is a sample of some of your required calculations. Tax is included in each item's purchase price:

   (a) First order prices: $2.95, 0.99, 3.12, 1.99, 0.59
       Cash received: $10.75
       Change:

   (b) Second order prices: $3.99, 0.49, 0.59, 2.59, 0.99, 1.19, 3.27
       Cash received: $15.00
       Change:

   (c) Third order prices: $1.99, 0.29
       Cash received: $20.00
       Change:

   (d) Fourth order prices: $1.19, 2.99, 1.19, 0.59, 0.79, 1.19, 2.99, 0.49
       Cash received: $20.00
       Change:

9. The catastrophe has ended. H Adds Up, which sells and services cash registers, has Bob's register working again. The soda pop that was dumped into the keyboard was the culprit. The labor cost was $45.77, a new keyboard cost $89.95, and new connections cost $21.56. What was the total H Adds Up bill?

10. Bob's Beeflike Burgers has decided to take the repair costs from the previous problem out of clumsy Cliff's paycheck, since this was the second time he has spilled soda pop into the cash register. Cliff's paycheck after taxes was $312.13; what does he get after he pays Bob's for the H Adds Up bill?

11. J.R.'s Construction Company has expanded. Complete the following table to summarize the total quarterly profits for each division, the total monthly profits for all divisions, and the total profits for the quarter.

Month	North Division	South Division	East Division	West Division	Monthly Totals
April	$ 9,635.32	$ 7,345.19	$8,536.92	$10,534.83	
May	10,002.84	8,293.99	7,991.83	8,374.38	
June	8,928.47	10,475.29	9,010.10	16,732.89	
Quarter totals					

12. Plato has started a new business selling lumps of logic. These logic lumps come in graduated sizes, each one meant for differently sized situations. The four sizes are priced $13.99, $17.67, $19.91 and $24.59. Kay Oss is in dire need of all forms of logic, so she ordered one logic lump of each size. (a) What is Kay's total bill? (b) Kay paid Plato with the following currency: one fifty-dollar bill and two twenty-dollar bills. How much change does Plato need to give Kay?

13. Intense Outdoor Equipment is building a climbing wall. The original wall is 1.5 feet thick. The first layer is made of cement and is 0.75 feet thick. The second layer is 0.18 feet of small stones. There is an additional 0.085-foot coating of "Scrape No Knees," a specially made substance that helps prevent bloody knees and elbows. What is the total thickness of the climbing wall?

14. April has been a very damp month for the town of New Brighton. The weekly amounts of rainfall were 4.72 inches, 3.21 inches, 1.82 inches, and 5.69 inches. What was the total accumulation of rain for the month of April in New Brighton?

15. The Zippy Delivery Service Truck logged 124.7 delivery miles Monday, 209.1 miles Tuesday, 279.4 miles Wednesday, 185.4 miles Thursday, and 313.9 miles Friday. What was the total mileage for the week?

SECTION 3.2: Multiplying and Dividing Decimals

Can You:

Multiply decimal numbers?
Divide decimal numbers?
. . . If not, you need this section.

Multiplying Decimal Numbers

Since a decimal number is only a fraction, you could convert, multiply, and reconvert the numbers. For example, to multiply 0.5×0.3:

$$0.5 \times 0.3 = \frac{5}{10} \times \frac{3}{10} = \frac{5 \times 3}{10 \times 10} = \frac{15}{100} = 0.15$$

Of course it would be very, very clumsy and time-consuming to calculate every decimal multiplication in this way. Fortunately, there is a simpler method.

Steps: Multiplying Decimals

STEP 1. *Multiply* the two decimal numbers as if they were whole numbers. Pay no attention to the decimal points.

STEP 2. *Note* the sum of the decimal digits in both of the factors. It gives you the number of decimal digits in the product.

For example, let's multiply 3.2×0.41.

Step 1	*Step 2*	
32	3.2	= 1 decimal digit (2)
×41	0.41	= 2 decimal digits (4 and 1)
32		3 decimal digits total, so $3.2 \times 0.41 = 1.312$
128		
1312		

YOUR TURN

WORK THIS PROBLEM

The Question:

Try these simple decimal multiplications.

(a) 0.5×0.5 (b) 0.1×0.1 (c) 10×0.6
(d) 2×0.003 (e) 0.01×0.02 (f) 0.04×0.005

✔ YOUR WORK

The Solution:

(a) 0.25 ($5 \times 5 = 25$; there are two decimal digits total)
(b) 0.01 ($1 \times 1 = 1$; there are two decimal digits total)
(c) 6.0 ($10 \times 6 = 60$; there is one decimal digit total)

(d) 0.006 ($2 \times 3 = 6$; there are three decimal digits total)
(e) 0.0002 ($1 \times 2 = 2$; there are four decimal digits total)
(f) $0.00020 = 0.0002$ ($4 \times 5 = 20$; there are five digits total)

As you may have noticed in question (c), multiplication by 10 simply shifts the decimal point one digit to the right. Multiplying by 100 shifts the decimal point two digits to the right, by 1000 three digits to the right, and so forth. Similarly, multiplying a number by 0.1 shifts the decimal point one place to the *left,* by 0.01 two places to the left, and so forth. For example,

$$87 \times 0.1 = 8.7 \qquad 98 \times 0.01 = 0.98$$

Do not try to do *any* multiplication of decimal numbers mentally until you are *certain* you will not misplace zeros.

Multiplication of larger decimal numbers is performed in exactly the same manner.

YOUR TURN

WORK THIS PROBLEM

The Question:

Multiply the following.

(a) 4.302×12.05　　(b) 6.715×2.002　　(c) 3.144×0.00125

✔ YOUR WORK

The Solution:

(a) $51.83910 = 51.8391$ ($4302 \times 1205 = 5{,}183{,}910$; there are five decimal digits total)
(b) $13.443430 = 13.44343$ ($6715 \times 2002 = 13{,}443{,}430$; there are six decimal digits total)
(c) $0.00393000 = 0.00393$ ($3144 \times 125 = 393{,}000$; there are eight decimal digits total, so you must add two zeros to the left before the decimal point)

Dividing Decimal Numbers

Division of decimal numbers is very similar to division of whole numbers. For example,

$$6.8 \div 1.7 \quad \text{can be written} \quad \frac{6.8}{1.7}$$

and, if we multiply top and bottom of the fraction by 10,

$$\frac{6.8}{1.7} = \frac{6.8 \times 10}{1.7 \times 10} = \frac{68}{17}$$

The fraction $\frac{68}{17}$ is a normal whole number division.

$$\frac{68}{17} = 68 \div 17 = 4$$

Therefore,

$$6.8 \div 1.7 = 4 \qquad \textbf{Check:} \quad 1.7 \times 4 = 6.8$$

But it is easier to use a shortcut.

Steps: Dividing Decimals

Example: $6.8 \div 1.7$

STEP 1. *Write* the divisor and dividend in standard long division form.

$1.7 \overline{)6.8}$

STEP 2. *Shift* the decimal point in the divisor to the right so as to make the divisor a whole number.

$1.7 \overline{)}$

STEP 3. *Shift* the decimal point in the dividend *the same number of digits.* (Add zeros if necessary.)

$1.7 \overline{)6.8}$

STEP 4. *Place* the decimal point in the answer space directly above the new decimal position in the dividend.

$17.\overline{)68.}$

STEP 5. *Complete* the division exactly as you would with whole numbers. The decimal points in divisor and dividend may now be ignored.

$\begin{array}{r} 4. \\ 17.\overline{)68.} \\ \underline{68.} \end{array}$

Notice in steps 2 and 3 that we have simply multiplied both divisor and dividend by 10.

YOUR TURN

WORK THIS PROBLEM

The Question:

Find the quotient of $1.38 \div 2.3$ and check your answer.

✔ YOUR WORK

The Solution:

Step 1	Steps 2, 3	Step 4	Step 5
$2.3\overline{)1.38}$	$2.3\overline{)1.3\,8}$	$23.\overline{)13.8}$	$\begin{array}{r}.6\\23.\overline{)13.8}\\13\,8\end{array}$

$1.38 \div 2.3 = 0.6 \qquad \textbf{Check:} \quad 2.3 \times 0.6 = 1.38$

Some problems are a bit more complicated. For example, let's divide $2.6 \div 0.052$.

Steps 1, 2. $0.052\overline{)2.6}$

Step 3. $52.\overline{)2.600}$ To shift the decimal point three digits in the dividend, we must attach two zeros to its right.

Step 4. $52\overline{)2600.}$
                                               Now place the decimal point in the answer space above that in the dividend.

                              $50.$
Step 5. $52.\overline{)2600.}$
                              $\underline{260}$
                              0
                              $\underline{0}$

$2.6 \div 0.052 = 50$ **Check:** $0.052 \times 50 = 2.6$

Shifting the decimal point three digits and attaching zeros to the right of the decimal point in this way is equivalent to multiplying both divisor and dividend by 1000.

YOUR TURN

| WORK THIS PROBLEM |

The Question:

Divide the following:

(a) $3.5 \div 0.001$ (b) $\$9 \div 0.02$ (c) $0.365 \div 18.25$ (d) $\$8.80 \div 3.2$

| ✔ YOUR WORK |

The Solution:

(a) $0.001.\overline{)3.500.}$ $1.\overline{)3500.}$ with $3500.$

      $3.5 \div 0.001 = 3500$ **Check:** $0.001 \times 3500 = 3.5$

(b) $0.02.\overline{)\$9.00.}$ $2.\overline{)\$900.}$ with $\$450.$

      $\$9 \div 0.02 = \$450$ **Check:** $0.02 \times \$450 = \$9$

(c) $18.25.\overline{)0.36.5}$ $1825\overline{)36.50}$ with $.02$
                                      $36\ 50$

      $0.365 \div 18.25 = 0.02$ **Check:** $18.25 \times 0.02 = 0.365$

(d) $3.2.\overline{)\$8.8.0}$ $32.\overline{)88.00}$ with $\$2.75$

      $\$8.80 \div 3.2 = \$2.75$ **Check:** $3.2 \times \$2.75 = \$8.80$

Rounding

If the dividend is not exactly divisible by the divisor, we must either stop the process after some preset number of decimal places in the answer or we must *round* the answer. We do not generally indicate a remainder in decimal division.

To round decimal numbers, just use the same rules with which you rounded whole numbers (see Chapter 1).

Steps: Rounding Decimal Numbers

Example

STEP 1. *Determine* the decimal place to which the number is to be rounded. Mark it with a ∧ placed to the right of that decimal place.

Round 24.147 to the nearest tenth.
24.147 becomes 24.1∧47

STEP 2. If the digit to the right of the mark is *less* than 5, drop all digits to the right of the mark.

24.1∧or 24.1

STEP 3. If the digit to the right of the mark is *equal to* or *larger* than 5, *increase* the digit to the left of the mark by 1 and drop all digits to the right of the mark.

87.1∧72 becomes
87.2∧or 87.2

For example,

1.376521 = 1.377 rounded to three decimal digits
= 1.4 rounded to the nearest tenth
= 1 rounded to the nearest whole number

If you have a calculator with an eight-digit display, frequently your answer must be rounded. In business mathematics, you are often calculating money and the final answer must be rounded to the nearest cent or hundredths position. For example,

$24,847.847 = $24,847.85 rounded to the nearest cent or hundredths

YOUR TURN

WORK THIS PROBLEM

The Question:

When computing interest, Judy's calculator displayed 45.746753 as the final answer. Round her answer to two decimal places (cents).

✔ YOUR WORK

The Solution:

$45.746753 = $45.75 rounded to the nearest cents position

There are a few very specialized situations where our standard rounding rules are *not* used.

1. Engineers use a more complex rule when rounding a number that ends in 5.
2. In most retail stores, fractions of a cent are rounded up in determining selling price. Three items for 25¢ or $8\frac{1}{3}$¢ each are rounded to 9¢ each.

But our rules will be quite satisfactory for most of your work in business mathematics.

YOUR
TURN

WORK THIS PROBLEM

The Question:

Divide $6.84 by 32.7 and round your answer to two decimal places.

✔ YOUR WORK

The Solution:

```
                          $ 0.209
32.7.)$6.8.4          327.)$68.400
                          65 4↓↓
                           3 000
                           2 943
```

$6.84 ÷ 32.7 = $0.209 = $0.21 rounded to two decimal places.

Check: 32.7 × $0.21 = $6.867, which is approximately equal to $6.84. (The check will not be exact because we have rounded.)

Dividing decimal numbers is an everyday event at many companies. Just imagine a paint company that couldn't figure its cost per gallon when it paid $106.60 for a crate of 12 gallons!

YOUR
TURN

WORK THIS PROBLEM

The Question:

The Kozy Kitchen purchased 23.5 yards of carpet for $152.28. What was the cost per yard?

✔ YOUR WORK

The Solution:

```
                                              6.48
$152.28 ÷ 23.5      23.5.)152.2.8      235.)1522.80
                                            1410
                                            112 8
                                             94 0
                                             18 80
                                             18 80
```

$152.28 ÷ 23.5 = $6.48   **Check:** 23.5 × $6.48 = $152.28

The practice problems in Section Test 3.2 will help you make sure of your skills in multiplication and division of decimal numbers.

SECTION TEST 3.2   Decimal Numbers ━━━━

Name _____

Date _____

Course/Section

The following problems test your understanding of Section 3.2, Multiplying and Dividing Decimals.

A. Multiply.

1. $0.1 \times 0.01 =$ _____

2. $10 \times 0.001 =$ _____

3. $3.73 \times 10 =$ _____

4. $4.826 \times 100 =$ _____

5. $4.8 \times 0.6 =$ _____

6. $9.2 \times 0.09 =$ _____

7. $0.73 \times 2.6 =$ _____

8. $0.078 \times 8.3 =$ _____

9. $3.74 \times 0.037 =$ _____

10. $39.5 \times 0.0036 =$ _____

11. $0.836 \times 0.085 =$ _____

12. $0.0735 \times 0.29 =$ _____

13. $0.00783 \times 0.062 =$ _____

14. 0.00846×0.095 _____

15. $24.7 \times 0.00049 =$ _____

16. $8.46 \times 0.00075 =$ _____

17. $0.0374 \times 5.6 =$ _____

18. $0.00375 \times 0.39 =$ _____

19. $0.08461 \times 0.79 =$ _____

20. $0.09748 \times 0.0045 =$ _____

21. $0.8468 \times 0.0052 =$ _____

22. $0.8463 \times 0.0074 =$ _____

23. $0.0374 \times 0.0836 =$ _____

24. $0.0857 \times 0.0474 =$ _____

25. $5.836 \times 0.746 =$ _____

26. $94.72 \times 0.0743 =$ _____

27. $35.84 \times 0.8367 =$ _____

28. $285.6 \times 3.8552 =$ _____

29. $4.73 \times 9.5 \times 0.0084 =$ _____

30. $0.0372 \times 3.6 \times 0.74 =$ _____

31. $73.5 \times 0.36 \times 6.7 =$ _____

32. $3.73 \times 0.036 \times 0.86 =$ _____

B. Divide.

1. $9.2 \div 0.04 =$ _____

2. $8.5 \div 0.005 =$ _____

3. $0.0448 \div 0.007 =$ _____

4. $0.222 \div 0.06 =$ _____

5. $15.64 \div 0.023 =$ _____

6. $3.686 \div 0.38 =$ _____

7. $1.855 \div 0.070 =$ _____

8. $61.74 \div 0.90 =$ _____

9. $2.223 \div 0.95 =$ _____

10. $0.47025 \div 0.075 =$ _____

11. $1.38375 \div 0.0205 =$ _____

12. $0.442976 \div 0.0508 =$ _____

13. $4.4544 \div 0.0128 =$ _____   14. $121.545 \div 1.85 =$ _____

15. $1935.72 \div 2.83 =$ _____   16. $12.5944 \div 0.346 =$ _____

C. Divide and round to two decimal places.

1. $10 \div 6 =$ _____   2. $100 \div 0.3 =$ _____

3. $3 \div 0.7 =$ _____   4. $2.9 \div 0.09 =$ _____

5. $2.6 \div 0.06 =$ _____   6. $6.32 \div 1.3 =$ _____

7. $0.54 \div 0.039 =$ _____   8. $75 \div 0.076 =$ _____

9. $2.3 \div 0.15 =$ _____   10. $6.5 \div 0.27 =$ _____

Divide and round to the nearest thousandths.

11. $2.3 \div 12 =$ _____   12. $17 \div 0.06 =$ _____

13. $7.3 \div 0.29 =$ _____   14. $8.2 \div 0.47 =$ _____

15. $0.52 \div 6.3 =$ _____   16. $7.5 \div 0.026 =$ _____

17. $8.2 \div 0.072 =$ _____   18. $0.81 \div 6.8 =$ _____

19. $5.31 \div 0.0123 =$ _____   20. $2.73 \div 0.0237 =$ _____

D. Solve these problems.

1. Sew What?, a fabric store, sells silk for $21.67 a yard. How much does it cost Taylor to buy 32.5 yards of silk?

2. A laser printer rents for $87.35 per month. How much will it cost your company to rent the printer for one year?

3. Kate's Delicacies' income is $2,543.64 per week. What is Kate's income for one year (52 weeks)?

4. The Butter-em-Up Bakery can buy a new oven for $13,534.85 or the installment price of 24 monthly payments of $634.57. (a) What is the cost if paid in installments? (b) How much would be saved by paying cash?

5. The Zippy Delivery Service driver made an overnight delivery. The round-trip took 13.4 hours for 783.9 miles. What was the average speed?

6. The Wonderful Widget Company makes a deluxe widget that sells for $13.96. How many deluxe widgets can Harvy buy for $1717.08?

7. Lew wants the ultimate carpeting for his restaurant. On-the-Carpet will install it and finance it for one year. That total cost is $25,837.32. What are Lew's monthly payments?

8. The annual insurance premium for Risky Business, Inc., is $5324.88. What are the monthly payments?

9. Bits 'n' Bytes, the constantly growing computer store, has just received a contract for 25 complete personal computer systems. (a) If each system sells for $987.21, what is the total cost? (b) If Bits 'n' Bytes increases the price of each system $23.64, what would be the total cost?

10. Maria's Mannequins needs to update some of the mannequins. A local company has offered a mannequin "facelift" at the low price of $42.85 per mannequin. Maria has 67 mannequins that need updating. What is her total cost?

11. Bob's Beeflike Burgers is expanding the menu to include a burger with a guacamole topping. Big Bob estimates that each of his three stores will need 3 dozen avocados daily. (a) How many avocados will Big Bob's stores use in one week (seven days)? (b) Avocados typically cost 99 cents each. What will a week's supply of avocados cost?

12. The Zippy Delivery Service needs to make extra deliveries to pay unexpected expenses. If each delivery returns a $30.21 profit, and each of the 11 drivers can make 13 extra deliveries a day, how much extra profit can the Zippy people make in five days?

13. The Sparkle Cleaning Company obtained a new contract that will bring in $53,775.78 profit annually. How much will each of the six partners get if the profit is split equally?

14. The Pizza Palace has set aside $8412 for employee bonuses. If each of the 15 employees receives an equal bonus, how much is each bonus?

15. If your annual income is $34,345.48, what is your weekly income?

16. J.R.'s Construction Company has taken on a painting job. The paint will cost $12.67 a gallon. The job will require 39 gallons. (a) How much will the paint cost? (b) If the paint store owner gives J.R. a $1.19 discount on each gallon of paint, how much will the paint cost? (c) How much does J.R. save?

17. Now You See It Videos makes a $0.49 profit on each video rented. Average weekday rentals are 212 videos daily. On Saturdays and Sundays, the company rents an average of 372 videos each day. (a) What is the average profit for Monday through Friday? (b) What is the average Saturday–Sunday profit? (c) What is the average profit for an entire week?

18. The Major's chicken franchise is having a special: 8 pieces of extra crispy fried chicken for $3.99. (a) How much will 64 pieces of chicken cost Frank? (b) Frank pays for the chicken with two twenty-dollar bills. How much change does Frank get?

19.

Wing-Ding International

Dinging Wings Worldwide since 1896

2468 Dingaling Way
Paradise, CA 93121

INTERNET:http://www.wingding.com

Order No. 153775   July 4, 1998

Ricky Carman
222 Tractor Way
Rossdorf, CA 93222

INVOICE

Quantity	Description	Unit Cost	Amount
2	Double Wingies	49.95	
7	Single Dings	19.95	
1	Connector	4.95	
		Total	

Calculate the amounts for each item and the total amount.

Writing a Fraction as a Decimal Number

Fractions with 10-based denominators are easy to convert to decimal numbers. But many fractions have non-10-based denominators: $\frac{9}{23}$, $\frac{6}{577}$, $\frac{1}{9386}$, and so on. You can convert these fractions to decimals easily: Simply divide the numerator (top number) by the denominator (bottom number). For example, to convert $\frac{5}{8}$ to decimal form, divide 5 by 8.

$$
\begin{array}{r}
0.625 \\
8\overline{)5.000} \\
\underline{4\,8} \\
20 \\
\underline{16} \\
40 \\
\underline{40}
\end{array}
$$
$\longleftarrow$ Attach as many zeros as needed.

The conversion is $\frac{5}{8} = 0.625$.

Terminating Decimal A decimal number that results when one number divides evenly into another at some point.

If the division has no remainder, the decimal number is called a **terminating decimal.** The fraction $\frac{5}{8} = 0.625$ is a terminating decimal number. If a decimal does not terminate but does repeat, it is called a **repeating decimal.** For example, $\frac{1}{3} = 0.333333$ . . . , with the 3s going on forever.

Repeating Decimal A decimal number that results when one number *never* divides evenly into another, no matter how many decimal places the division is carried out to; as a result, the decimal digits repeat themselves endlessly.

You may round a repeating decimal to any desired number of decimal places. For example, let's convert $\frac{2}{13}$ to a decimal number with three decimal digits.

$$
\begin{array}{r}
0.1538 \\
13\overline{)2.0000} \\
\underline{1\,3} \\
70 \\
\underline{65} \\
50 \\
\underline{39} \\
110 \\
\underline{104} \\
6
\end{array}
$$

The fraction $\frac{2}{13} = 0.154$ rounded to three decimal digits. Notice that in order to round to three decimal digits, we must carry the division out to at least four decimal digits.

WORK THIS PROBLEM

YOUR TURN

The Question:

Convert the following fractions to decimal form and round to two decimal digits.

(a) $\dfrac{2}{3}$　　(b) $\dfrac{5}{6}$　　(c) $\dfrac{17}{7}$　　(d) $\dfrac{7}{16}$

✔ **YOUR WORK**

The Solution:

(a)
$$
\begin{array}{r}
0.666 \\
3\overline{)2.000} \\
\underline{1\,8} \\
20 \\
\underline{18} \\
20 \\
\underline{18} \\
2
\end{array}
$$

Carry the division to one more decimal digit than will be in the answer.

(b)
$$
\begin{array}{r}
0.833 \\
6\overline{)5.000} \\
\underline{4\,8} \\
20 \\
\underline{18} \\
20 \\
\underline{18} \\
2
\end{array}
$$

$\dfrac{2}{3} = 0.67$ rounded to two decimal digits

$\dfrac{5}{6} = 0.83$ rounded to two decimal digits

(c)
$$
\begin{array}{r}
2.428 \\
7\overline{)17.000} \\
\underline{14} \\
3\,0 \\
\underline{2\,8} \\
20 \\
\underline{14} \\
60 \\
\underline{56} \\
4
\end{array}
$$

(d)
$$
\begin{array}{r}
0.437 \\
16\overline{)7.000} \\
\underline{6\,4} \\
60 \\
\underline{48} \\
120 \\
\underline{112} \\
8
\end{array}
$$

$\dfrac{17}{7} = 2.43$ rounded to two decimal digits

$\dfrac{7}{16} = 0.44$ rounded to two decimal digits

Converting Mixed Numbers

Of course, not all fractions are quite so easily converted. Mixed numbers take a bit more work to convert, though it's not difficult.

Steps: Converting Mixed Numbers to Decimals

STEP 1. *Change* the fraction part to a decimal number.

STEP 2. *Add* the whole number part to the decimal.

For example, $37\frac{1}{4} = 37 + \frac{1}{4} = 37 + 0.25 = 37.25$.

WORK THIS PROBLEM

The Question:

Change the following mixed numbers to decimals. If the decimal does not terminate, round to two decimal digits.

(a) $29\frac{5}{9}$ square feet   (b) $186\frac{3}{8}$ yards

✔ YOUR WORK

The Solution:

(a)
$$
\begin{array}{r}
0.555 \quad = 0.56 \text{ rounded} \\
9\overline{)5.000} \\
\underline{4\,5} \\
50 \\
\underline{45} \\
50 \\
\underline{45} \\
5
\end{array}
$$

$29\frac{5}{9} = 29 + \frac{5}{9} = 29 + 0.56 = 29.56$ square feet

(b)
$$
\begin{array}{r}
0.375 \\
8\overline{)3.000} \\
\underline{2\,4} \\
60 \\
\underline{56} \\
40 \\
\underline{40}
\end{array}
$$

$186\frac{3}{8} = 186 + \frac{3}{8} = 186 + 0.375 = 186.375$ yards

Writing Decimal Numbers as Fractions

Converting decimal numbers to fractions is fairly easy. Just follow this procedure:

Steps: Converting Decimal Numbers to Fractions

STEP 1. *Write* the digits to the right of the decimal point as the numerator in the fraction.

STEP 2. In the denominator *write* 1 followed by as many zeros as there are decimal digits in the decimal number.

STEP 3. *Reduce* to lowest terms.

For example, let's convert 0.00325 to a fraction.

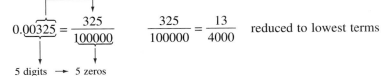

Step 1   *Step 2*   *Step 3*

$\dfrac{325}{?}$ $0.00325 = \dfrac{325}{100000}$ $\dfrac{325}{100000} = \dfrac{13}{4000}$ reduced to lowest terms

5 digits → 5 zeros

YOUR TURN

WORK THIS PROBLEM

The Question:

Write 0.036 as a fraction in lowest terms.

✔ YOUR WORK

The Solution:

$$0.\underbrace{036}_{} = \frac{36}{1\underbrace{000}_{}} = \frac{9}{250} \text{ reduced to lowest terms}$$

3 digits → 3 zeros

If the decimal number has a whole number portion, convert the decimal part to a fraction first and then add the whole number part. For example, let's convert 3.85 to a fraction.

$$3.85 = 3 + 0.85 \quad \text{and} \quad 0.85 = \frac{85}{100} = \frac{17}{20} \text{ reduced to lowest terms}$$

Therefore, $3.85 = 3 + \frac{17}{20} = 3\frac{17}{20}$.

YOUR TURN

WORK THIS PROBLEM

The Question:

Write the following decimal numbers as fractions in lowest terms.

(a) 0.0075   (b) 2.08   (c) 3.11

✔ YOUR WORK

The Solution:

(a) $0.0075 = \dfrac{75}{10,000} = \dfrac{3}{400}$ (b) $2.08 = 2 + \dfrac{8}{100} = 2 + \dfrac{2}{25} = 2\dfrac{2}{25}$

(c) $3.11 = 3 + \dfrac{11}{100} = 3\dfrac{11}{100}$

Now check your skills at converting decimal numbers and fractions in Section Test 3.3.

SECTION TEST 3.3 Decimal Numbers

Name

Date

Course/Section

The following problems test your understanding of Section 3.3, Converting Decimal Numbers and Fractions.

A. Write as decimal numbers.

1. $\dfrac{3}{4} =$ _____

2. $\dfrac{4}{5} =$ _____

3. $\dfrac{2}{5} =$ _____

4. $\dfrac{1}{4} =$ _____

5. $\dfrac{1}{2} =$ _____

6. $\dfrac{3}{8} =$ _____

7. $\dfrac{5}{16} =$ _____

8. $\dfrac{11}{16} =$ _____

9. $\dfrac{7}{8} =$ _____

10. $\dfrac{7}{20} =$ _____

11. $\dfrac{8}{25} =$ _____

12. $\dfrac{13}{16} =$ _____

13. $\dfrac{17}{8} =$ _____

14. $\dfrac{7}{4} =$ _____

15. $\dfrac{11}{5} =$ _____

16. $\dfrac{33}{20} =$ _____

17. $1\dfrac{9}{16} =$ _____

18. $2\dfrac{23}{32} =$ _____

19. $3\dfrac{11}{32} =$ _____

20. $4\dfrac{15}{16} =$ _____

B. Write as a decimal number and round to three decimal places.

1. $\dfrac{2}{3} =$ _____

2. $\dfrac{5}{6} =$ _____

3. $\dfrac{1}{7} =$ _____

4. $\dfrac{1}{3} =$ _____

5. $\dfrac{7}{12} =$ _____

6. $\dfrac{7}{11} =$ _____

7. $\dfrac{8}{9} =$ _____

8. $\dfrac{5}{12} =$ _____

9. $\dfrac{13}{15} =$ _____

10. $\dfrac{7}{9} =$ _____

11. $\dfrac{12}{17} =$ _____

12. $\dfrac{13}{19} =$ _____

13. $\dfrac{17}{6} =$ _____

14. $\dfrac{17}{12} =$ _____

15. $\dfrac{14}{3} =$ _____

16. $\dfrac{11}{7} =$ _____

17. $5\dfrac{2}{9} =$ _____

18. $6\dfrac{1}{12} =$ _____

19. $7\dfrac{1}{11} =$ _____

20. $8\dfrac{5}{7} =$ _____

C. Write as a fraction in lowest terms.

1. $0.7 =$ _____

2. $0.9 =$ _____

3. $0.4 =$ _____

4. $0.5 =$ _____

5. $0.2 =$ _____

6. $0.8 =$ _____

7. $0.27 =$ _____

8. $0.47 =$ _____

9. $0.45 =$ _____

10. $0.65 =$ _____

11. $0.48 =$ _____

12. $0.64 =$ _____

13. $0.56 =$ _____

14. $0.88 =$ _____

15. $3.24 =$ _____

16. $5.92 =$ _____

17. $0.125 =$ _____

18. $0.625 =$ _____

19. $1.875 =$ _____

20. $4.375 =$ _____

21. $0.0025 =$ _____

22. $0.0045 =$ _____

23. $7.6875 =$ _____

24. $2.5625 =$ _____

D. Solve these problems.

1. Bob's Beeflike Burgers pays its employees $9.72 per hour. What does the company owe Norm for working $6\frac{3}{4}$ hours?

2. The Sparkle Cleaning Service needs to purchase $3\frac{1}{4}$ gallons of cleaning fluid at $15.26 per gallon. How much does the cleaning fluid cost?

3. Lew is buying $25\frac{2}{3}$ square yards of trendy, slightly damaged carpet at $25.75 per yard for his new restaurant. What will the carpet cost?

4. Woody's Lumber Mill pays Sam $9.76 per hour for supervising its rough sawing. If he is paid $1\frac{1}{2}$ times his base pay for overtime, what is his overtime pay rate?

5. Sam's base pay is still $9.76 per hour. If he is paid $1\frac{3}{4}$ times his base pay for working overtime on Sunday, what is his Sunday overtime pay rate?

6. The Pizza Palace pays Olive $8.76 per hour for artistically slinging pizza dough into the air in front of the store window. What does The Pizza Palace owe Olive for working $12\frac{1}{3}$ hours?

7. Kate's Delicacies is having a croissant special. Kate is selling seven croissants for $2.00. What is the price of one croissant?

8. The Zippy Delivery Service driver is paid $8.46 per hour. Calculate the pay for $25\frac{5}{6}$ hours.

9. J.R.'s Construction Company is purchasing $4\frac{3}{5}$ gallons of solvent at $24.98 per gallon. How much does the solvent cost?

10. Bits 'n' Bytes, the local computer store, has grown drastically and is installing a new computer room. All of the flooring must be replaced. Calculate the cost of $573\frac{1}{2}$ square feet of flooring at $12.75 per square foot.

11. Food for Thought, the college's local sandwich shop, is selling sandwiches by the foot. How much will $48\frac{3}{4}$ feet of pastrami and cheese on rye at $5.89 per foot cost?

12. Frame-Up, the local picture shop, sells picture frame wood for $28.99 per yard. Calculate the cost of $56\frac{1}{3}$ yards of picture frame.

13. Crystal's Diamonds is having new counters installed. J.R.'s Construction Company's estimate was $65.50 per linear foot. What is the cost of $12\frac{3}{4}$ feet of cabinets?

14. The Butter-em-Up Bakery just hired a pastry assistant for $9.76 an hour. What will the bakery pay the new assistant for 1 hour and 50 minutes?

15. The amazing, new, miraculous Microwidgets are priced 11 for $3.00. What is the price of two amazing, new, miraculous Microwidgets?

16. If one hour is 60 minutes, find:

 (a) $\frac{1}{3}$ hr = _____ minutes (b) 0.6 hr = _____ minutes

 (c) 0.45 hr = _____ minutes (d) 42 minutes = _____ hr

 (e) 24 minutes = _____ hr (f) 2 hr 12 min = _____ hr

 (g) Calculate the wages that should be paid to a part-time sales clerk who worked the following: 13.7 hr at $12.50 per hr; 22 hr 40 min at $9.60 per hr; $9\frac{3}{4}$ hr at $10.40 per hr.

17. Dynamite Deals held a Grand Opening sale in all departments. Arnetta purchased 6 cereal bowls, normally priced at $3.95 each, for the sale price of $2.79 each. She also paid $7.74 for a matching turkey platter, which was usually $12.49. (a) How much did Arnetta pay for the bowls and platter? (b) How much did she save by purchasing these items on sale?

18. Claudia's Clip-Joint, a local beauty salon, purchased a 3-liter container of shampoo for use in the shop, paying a total of $17.85. (a) If the same shampoo is also available in a 0.5-liter size at $3.79 each, how much did Claudia save by buying in bulk? (b) What is the price per liter for the large container? (c) What is the price per liter for the smaller container?

WORKSPACE

Key Terms

decimal digits	repeating decimal
decimal number	terminating decimal

Chapter 3 at a Glance

Page	Topic	Key Point	Example
106	Reading decimals	Decimals are fractions expressed in another way.	$\dfrac{3}{100} = 0.03$; $7.23 = 7\dfrac{23}{100}$
109	Adding decimal numbers	Align decimal points and add, carrying if necessary.	$\begin{array}{r} 1.75 \\ +3.42 \\ \hline 5.17 \end{array}$
110	Subtracting decimal numbers	Align decimal points and subtract, borrowing if necessary.	$\begin{array}{r} 5.17 \\ -3.42 \\ \hline 1.75 \end{array}$
117	Multiplying decimal numbers	Multiply the numbers, ignoring the decimal points; the sum of the decimal digits in both factors is the number of decimal digits in the answer.	$\begin{array}{r} 3.2 = \underline{1}\text{ decimal digit} \\ \times 4.1 = \underline{1}\text{ decimal digit} \\ \hline 2\text{ decimal digits} \end{array}$ $\begin{array}{r} 32 \\ \times\ 41 \\ \hline 32 \\ 128 \\ \hline 1312 = 13.12 \text{ with two decimal digits} \end{array}$
118	Dividing decimal numbers	Shift the decimal points of both divisor and dividend an equal number of places to the right until the divisor has no decimal digits left; place a decimal point in the quotient over the new decimal point in the dividend; divide.	$2.3\,\overline{)1.3\,8} = 23\,\overline{)13.8}$ $\begin{array}{r} 0.6 \\ 23\overline{)13.8} \\ \underline{13\ 8} \end{array}$
120	Rounding decimal numbers	Determine the number of decimal places desired; round the desired last digit up if the next digit to the right is 5 or more	$1.3764 = 1.376$ to 3 decimal places $= 1.38$ to 2 decimal places $= 1.4$ to 1 decimal place $= 1$ to the units place
127	Converting fractions to decimal numbers	Divide the numerator by the denominator; round any repeating decimals.	$\dfrac{1}{3} = 3\,\overline{)1.0000}^{\,0.3333\ldots}$ $\dfrac{1}{3} = 0.333$ to 3 decimal places
128	Converting mixed numbers to decimal numbers	Convert the fraction to a decimal number; attach it to the whole number.	$37\dfrac{1}{4} = 37.25$
129	Converting decimal numbers to fractions	The decimal digits are the numerator; a 1 followed by as many zeros as there are decimal digits in the numerator is the denominator; reduce to lowest terms.	Convert 0.0325: $\dfrac{325}{10,000} = \dfrac{13}{400}$

WORKSPACE

SELF-TEST Decimal Numbers

The following problems test your understanding of decimal numbers. The answers to the odd-numbered problems are in the appendix.

1. Add: $4.724 + 75 + 6.3 + 0.0037 =$ _____

2. Add: $2.4 + 25.7253 + 12 + 0.02745 =$ _____

3. Subtract: $63.5 - 28.3865 =$ _____

4. Subtract: $26.7 - 18.8364 =$ _____

5. Multiply: $36.8 \times 0.295 =$ _____

6. Multiply: $5.82 \times 0.379 =$ _____

7. Divide 30.186 by $3.87 =$ _____

8. Divide 4.9726 by $0.529 =$ _____

9. Divide 45.73 by 6.7 and round to two decimal places. _____

10. Divide 6.723 by 7.3 and round to three decimal places. _____

11. Write 0.68 as a fraction and reduce to lowest terms. _____

12. Write 2.85 as a fraction and reduce to lowest terms. _____

13. Write $2\frac{3}{8}$ as a decimal number. _____

14. Write $3\frac{11}{16}$ as a decimal number. _____

15. During the last three months of the year, Micro Systems Support (MSS) reported the following income: October—$34,373.65; November—$38,836.97; and December—$42,361.45. What was the quarterly income for MSS?

16. Wired for Sound has three branches with the following daily sales. Calculate the total daily sales, the weekly sales for each store, and the total sales for the week.

Branch	M	T	W	Th	F	Weekly Total
Downtown	$3574.48	$3598.85	$3384.37	$2903.75	$4294.13	
Suburb	4537.47	4758.39	4201.97	4103.56	5103.38	
Mall	5635.75	5428.38	5117.24	4852.39	5583.31	
Totals						

17. Bob's Beeflike Burgers has suffered another serious catastrophe—the cash register expired during lunch hour again. You are the math wizard and have been chosen to stand in for the cash register. The following is a sample of some of your required calculations. Tax is included in each item's purchase price:

(a) First order prices: $1.59, 2.98, 1.99, 1.99, 0.75
Cash received: $20.00
Change:

(b) Second order prices: $1.99, 0.95, 3.49, 2.59, 0.95, 3.29, 3.48
Cash received: $20.00
Change:

(c) Third order prices: $2.98, 0.75
Cash received: $10.00
Change:

18. Kate's Delicacies had $5836.26 in its checking account. After writing checks for $2463.94, $56.98, $863.82, and $847.18, how much is left?

19. On-the-Carpet charges $27.98 per yard of installed carpeting. (a) What is the total cost of $36\frac{3}{4}$ square yards of carpet? (b) On-the-Carpet charges $6.99 a yard extra for the carpet padding. How much does the installed carpet cost with padding?

20. The pastry maker at The Butter-em-Up Bakery makes $8.37 per hour. Calculate the pay for $7\frac{1}{3}$ hours.

21. Bits 'n' Bytes sells six microblips for $2.00. What is the price of one microblip?

22. Pia is the ultimate pizza slinger at The Pizza Palace. Her base pay rate is $7.52 per hour. If she is paid $1\frac{1}{2}$ times her base pay for overtime, what is her overtime pay rate?

23. The annual insurance premium for Risky Business, Inc., is $1074.24. What are its monthly payments?

24. Bits 'n' Bytes offers a two-year installment package on the ultimate personal computer system. The total amount financed for two years is $6450. Calculate the monthly payments.

Percent

You can fool 100% of the people 57% of the time; you can fool 26% of the people 100% of the time; but you can't fool 100% of the people 100% of the time!

CHAPTER OBJECTIVES

When you complete this chapter successfully, you will be able to:

1. Write fractions and decimal numbers as percents and change percents to decimal and fraction form.

2. Solve problems involving percents.

■ Statistics show that 68% of students at Gigantic Urban and Rural University (GURU) change their majors at least once before graduating. What fraction of GURU students change their majors?

■ Shiring Brokerage's gross earnings last year were $87,500. This year, talented financial management has brought about an 8.5% increase. Calculate the amount of increase and the new earnings.

■ The Garden of Earthly Delights has discounted the price of installing a spring flower bed by 16%. If the amount of the discount is $15.68, calculate the original price and the discounted price.

As you can see, percents are an important part of everyday business. Businesspeople far removed from the accounting department—salespeople, manufacturing supervisors, and top executives—all require a knowledge of percents to talk about their operations and measure their success.

In this chapter, you will see how percents are just an extension of the fractions and decimals you have already mastered. The good news is that you probably already know more about percents than you think you do. The even better news is that, if you pay close attention to this chapter, you'll feel 100% confident about percents when you're done.

SECTION 4.1: Converting To and From Percent

Can You:

Define percent?
Convert a decimal number to a percent?
Convert a fraction to a percent?
Change percents to decimal form?
Change percents to fraction form?
. . . If not, you need this section.

Understanding Percent

Percent The relationship of one number (part) to a second number (base), expressed in hundredths.

Base The number against which another number is compared in calculating a percent.

The word **percent** comes from the Latin words *per centum* meaning "by the hundred" or "for every hundred." A number expressed as a percent is being compared with a second number called the standard or **base** by dividing the base into 100 equal parts and writing the comparison number as so many hundredths of the base.

For example, what part of the base is length *A*?

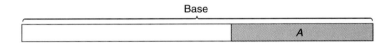

We could answer the question with a fraction or a decimal or a percent. First, divide the base into 100 equal parts. Then compare *A* with it.

The length of *A* is 40 parts out of the 100 parts that make up the base. We could write this result as

$$A = \frac{40}{100} \qquad \text{or} \qquad A = 0.40$$

To find an answer as a percent, write it as a fraction with a denominator equal to 100. Then delete the denominator and put a "%" (percent sign) after the numerator. So, in this example, *A* is 40% of the base.

Writing 40% means 40 parts in 100 or $\frac{40}{100}$. Fractions, decimals, and percents are alternative ways to talk about a comparison of two numbers.

WORK THIS PROBLEM
The Question:

What percent of this base is length *B*?

Base

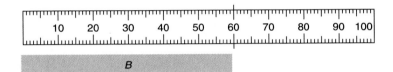

✔ YOUR WORK

The Solution:

B is $\frac{60}{100}$, which is 0.60 or 60%.

You might wonder why we bother with percents if they give the same information as decimals and fractions. Businesspeople often use percents because they are a more efficient way to express a comparison such as a discount or markup rate, interest rates, tax rates, or the like. Moreover, because our number system, and our money, is based on ten and multiples of ten, it is very handy to write comparisons in hundredths or percent.

WORK THIS PROBLEM
The Question:

What part of $1.00 is 50¢? Write your answer as a fraction, as a decimal, and as a percent.

✔ YOUR WORK
The Solution:

$$\frac{50¢}{100¢} = \frac{50}{100} = 0.50 = 50\%$$

Sometimes the compared number is larger than the base. An everyday example of this situation involves increased prices. For example, gasoline went from about $0.25 per gallon in 1970 to about $1.00 per gallon in 1990—a 300% increase.

Calculating such percents is easy. For example, what percent of the base is length *C*?

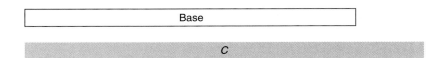

In this case, divide the base into 100 parts and extend it in length.

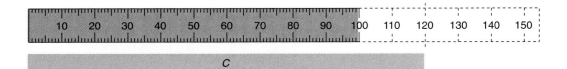

The length of *C* is 120 parts out of the 100 parts that make up the base, so *C* is $\frac{120}{100}$, or 120% of the base.

Although we always think of percent as dividing the base into 100 equal parts, that does not mean that the standard length is always 100. For example, suppose you have a 10 yard length of cloth and want to know what percent you are taking if you cut off 2 yards. First, write this relationship as a fraction: $\frac{2}{10}$. Next, you must convert this fraction to a fraction with a denominator of 100. (Remember: Percent means hundredths.) Since 10 goes into 100 ten times, multiply both the numerator and denominator by 10:

$$\frac{2}{10} = \frac{2 \times 10}{10 \times 10} = \frac{20}{100}$$

The answer, then, is 20%.

YOUR TURN

| WORK THIS PROBLEM |

The Question:

What percent of 20 is 4?

The Solution:

Because 20 goes into 100 five times, we multiply both the numerator and the denominator by 5:

$$\frac{4}{20} = \frac{4 \times 5}{20 \times 5} = \frac{20}{100} = 20\%$$

Changing a Decimal Number to a Percent

How do you change a decimal number to a percent? The procedure is simply to *multiply the decimal number by 100%*. For example,

$$
\begin{array}{r}
0.60 = \ 0.60 \\
\times 100\% \\
\hline
60.00\% = 60\%
\end{array}
$$

This procedure effectively *moves the decimal point two places to the right.*

Steps: Changing a Decimal Number to a Percent

STEP 1. *Move* the decimal point two places to the right.

STEP 2. *Attach* the percent sign on the right.

$$0.375 = 0.375 \times 100\% = 37.5\%$$
$$3.4 = 3.40 \times 100\% = 340\%$$
$$0.02 = 0.02 \times 100\% = 2\%$$

Notice that any number larger than 1 is more than 100%.

**YOUR
TURN**

WORK THIS PROBLEM

The Question:

Rewrite the following as percents.

(a) 0.70   (b) 1.25   (c) 0.001   (d) 3

✔ YOUR WORK

The Solution:

(a) $0.70 = 070.\% = 70\%$ (b) $1.25 = 125.\% = 125\%$

(c) $0.001 = 000.1\% = 0.1\%$ (d) $3 = 3.00 = 300.\% = 300\%$

Changing a Fraction to a Percent

Using this technique for changing a decimal number to a percent, we get a shortcut for writing a fraction as a percent.

_____

Steps: Changing a Fraction to a Percent

STEP 1. *Divide* the numerator of the fraction by its denominator. Use at least two decimal places.

STEP 2. *Move* the decimal point *two digits to the right* and attach the percent sign.

For example, to change $\frac{1}{2}$, $\frac{3}{4}$, $\frac{3}{20}$, and $1\frac{7}{20}$ to percents,

$$\frac{1}{2} = 2\overline{)1.00}^{\,0.50} = 0.50 = 50\% \qquad\qquad \frac{3}{4} = 4\overline{)3.00}^{\,0.75} = 0.75 = 75\%$$

$$\frac{3}{20} = 20\overline{)3.00}^{\,0.15} = 0.15 = 15\%$$
$$\begin{array}{r} 2\ 0 \\ \hline 1\ 00 \\ 1\ 00 \\ \hline \end{array}$$

$$1\frac{7}{20} = 20\overline{)27.00}^{\,1.35} = 1.35 = 135\%$$
$$\begin{array}{r} 20 \\ \hline 7\ 0 \\ 6\ 0 \\ \hline 1\ 00 \\ 1\ 00 \\ \hline \end{array}$$

WORK THIS PROBLEM
_____

The Question:

Calculators are currently on sale and have been reduced $\frac{5}{16}$ of their regular price. What is $\frac{5}{16}$ as a percent?

✔ YOUR WORK
_____

YOUR TURN

The Solution:

$$\frac{5}{16} = 16\overline{)5.0000}^{\,0.3125} = 0.3125 = 31.25\%$$
$$\begin{array}{r} 4\ 8 \\ \hline 20 \\ 16 \\ \hline 40 \\ 32 \\ \hline 80 \\ 80 \\ \hline \end{array}$$

Some fractions cannot be converted to exact decimals. For example,

$$\frac{1}{3} = 0.333 \ldots,$$ where the 3s continue to repeat endlessly.

We can round to get an approximate percent,

$$\frac{1}{3} = 0.333 = 33.3\% \text{ rounded}$$

or convert it to a fraction with 100 as denominator:

$$\frac{1}{3} = \frac{?}{100} \quad \text{gives} \quad \frac{1}{3} = \frac{1 \times 33\frac{1}{3}}{3 \times 33\frac{1}{3}} = \frac{33\frac{1}{3}}{100} = 33\frac{1}{3}\%$$

You may wonder why anyone would want to use percents containing a fraction. If such numbers are rounded, using them in arithmetic operations such as multiplication can give you inaccurate answers. Some interest rates, such as monthly credit card rates, use percents containing fractions. As you will see later in this chapter, percents written with fractions usually are more accurate than percents written with rounded decimals.

Aliquot Parts of 1

An *aliquot part* is any number that can be divided evenly into another number. In business, we frequently use aliquot parts of 1 in percent, decimal, and fraction form. The following is a short list of aliquot parts that you will find very helpful, if memorized.

Fraction	Decimal	Percent	Fraction	Decimal	Percent
$\frac{1}{2}$	0.50	50%	$\frac{1}{8}$	0.125	$12\frac{1}{2}\%$
$\frac{1}{3}$	0.333	$33\frac{1}{3}\%$	$\frac{3}{8}$	0.375	$37\frac{1}{2}\%$
$\frac{2}{3}$	0.667	$66\frac{2}{3}\%$	$\frac{5}{8}$	0.625	$62\frac{1}{2}\%$
$\frac{1}{4}$	0.25	25%	$\frac{7}{8}$	0.875	$87\frac{1}{2}\%$
$\frac{3}{4}$	0.75	75%	$\frac{1}{10}$	0.10	10%
$\frac{1}{5}$	0.20	20%	$\frac{3}{10}$	0.30	30%
$\frac{2}{5}$	0.40	40%	$\frac{7}{10}$	0.70	70%
$\frac{3}{5}$	0.60	60%	$\frac{9}{10}$	0.90	90%
$\frac{4}{5}$	0.80	80%	$\frac{10}{10}$	1.00	100%
$\frac{1}{6}$	0.1667	$16\frac{2}{3}\%$	$\frac{1}{12}$	0.0833	$8\frac{1}{3}\%$
$\frac{5}{6}$	0.833	$83\frac{1}{3}\%$	$\frac{1}{16}$	0.0625	$6\frac{1}{4}\%$
			$\frac{1}{20}$	0.05	5%

YOUR TURN

WORK THIS PROBLEM

The Question:

Rewrite the following fractions as percents.

(a) $\dfrac{7}{5}$ (b) $\dfrac{2}{3}$ (c) $3\dfrac{1}{8}$ (d) $\dfrac{5}{12}$

✔ YOUR WORK

The Solution:

(a) $\dfrac{7}{5} = 1.4 = 1.40 = 140\%$

(b) $\dfrac{2}{3} = \dfrac{?}{100}$ $\dfrac{2}{3} = \dfrac{2 \times 33\frac{1}{3}}{3 \times 33\frac{1}{3}} = \dfrac{66\frac{2}{3}}{100} = 66\frac{2}{3}\%$

(c) $3\dfrac{1}{8} = 3.125 = 312.5\%$

(d) $\dfrac{5}{12} = \dfrac{?}{100} = \dfrac{5 \times 8\frac{1}{3}}{12 \times 8\frac{1}{3}} = \dfrac{40\frac{5}{3}}{100} = \dfrac{41\frac{2}{3}}{100} = 41\frac{2}{3}\%$

Changing a Percent to a Decimal

In order to use percent in solving business problems, it is often necessary to change that percent to a decimal number. The procedure is to divide by 100%. For example,

$$50\% = \frac{50\%}{100\%} = \frac{50}{100} = 0.50 \qquad 100\overline{)\,50.0\,}^{\,0.5}$$

$$5\% = \frac{5\%}{100\%} = \frac{5}{100} = 0.05 \qquad 100\overline{)\,5.00\,}^{\,0.05}$$

$$0.2\% = \frac{0.2\%}{100\%} = \frac{0.2}{100} = 0.002 \qquad 100\overline{)\,0.200\,}^{\,0.002}$$

Notice that in each of these examples division by 100% has the effect of moving the decimal point two digits to the left and dropping the percent sign.

Steps: Changing a Percent to a Decimal

STEP 1. *Move* the decimal point two places to the left.

STEP 2. *Drop* the percent sign.

For example,

$$50\% = 50.\% = .50 = 0.50$$

$$5\% = 05.\% = .05 = 0.05$$

$$0.2\% = 00.2\% = .002 = 0.002$$

If a fraction is part of the percent number, it is easiest to change to a decimal number and round if necessary. If you round, be careful to include enough digits to avoid any error in your calculations—the eight digits in most calculators are plenty.

$$6\tfrac{1}{2}\% = 6.5\% = 06.5\% = 0.065$$

$$33\tfrac{1}{3}\% = 33.333\% = 0.333 \text{ rounded}$$

YOUR TURN

| WORK THIS PROBLEM |

The Question:

Write the following as decimal numbers. If necessary, round to three decimal digits.

(a) 4%   (b) 0.5%   (c) $16\tfrac{2}{3}\%$ (d) $79\tfrac{1}{4}\%$

| ✔ YOUR WORK |

The Solution:

(a) $4\% = 04.\% = 0.04$ (b) $0.5\% = 00.5\% = 0.005$

(c) $16\tfrac{2}{3}\% = 16.666\% = 0.167 \text{ rounded}$ (d) $79\tfrac{1}{4}\% = 79.25\% = 0.7925$

Changing a Percent to a Fraction

To change a percent to a fraction, just follow these simple steps.

Steps: Changing a Percent to a Fraction

STEP 1. *Divide* by 100%.

STEP 2. *Reduce* to lowest terms.

For example, convert (a) 36%   (b) $12\tfrac{1}{2}\%$ (c) 125%.

Step 1   **Step 2**

(a) $36\% = \dfrac{36\%}{100\%} = \dfrac{\overset{9}{\cancel{36}}}{\underset{25}{\cancel{100}}}$ $= \dfrac{9}{25}$

(b) $12\tfrac{1}{2}\% = \dfrac{12\tfrac{1}{2}\%}{100\%} = \dfrac{12\tfrac{1}{2}}{100} = \dfrac{\tfrac{25}{2}}{100} = \dfrac{25}{200}$ $= \dfrac{1}{8}$

(c) $125\% = \dfrac{125\%}{100\%} = \dfrac{125}{100} = \dfrac{5}{4}$ $= 1\tfrac{1}{4}$

YOUR TURN

WORK THIS PROBLEM

The Question:

Convert the following to fractions.

(a) 72%   (b) $16\frac{1}{2}\%$ (c) 240%   (d) $7\frac{1}{2}\%$

✔ YOUR WORK

The Solution:

(a) $72\% = \dfrac{72\%}{100\%} = \dfrac{\overset{18}{\cancel{72}}}{\underset{25}{\cancel{100}}} = \dfrac{18}{25}$

(b) $16\frac{1}{2}\% = \dfrac{16\frac{1}{2}\%}{100\%} = \dfrac{16\frac{1}{2}}{100} = \dfrac{\frac{33}{2}}{100} = \dfrac{33}{200}$

(c) $240\% = \dfrac{240\%}{100\%} = \dfrac{\overset{12}{\cancel{240}}}{\underset{5}{\cancel{100}}} = \dfrac{12}{5} = 2\frac{2}{5}$

(d) $7\frac{1}{2}\% = \dfrac{7\frac{1}{2}\%}{100\%} = \dfrac{7\frac{1}{2}}{100} = \dfrac{\frac{15}{2}}{100} = \dfrac{15}{200} = \dfrac{3}{40}$

Before going on, work the problems in Section Test 4.1 to be sure you can convert decimals, fractions, and percents.

SECTION TEST 4.1 Percent

The following problems test your understanding of Section 4.1, Converting To and From Percent.

A. Change each number to a percent.

1. $0.60 = $ _____   2. $0.20 = $ _____   3. $0.45 = $ _____

4. $0.72 = $ _____   5. $0.05 = $ _____   6. $0.09 = $ _____

7. $0.7 = $ _____   8. $0.2 = $ _____   9. $1.4 = $ _____

10. $1.7 = $ _____   11. $0.025 = $ _____   12. $0.042 = $ _____

13. $0.005 = $ _____   14. $0.008 = $ _____   15. $0.258 = $ _____

16. $7 = $ _____   17. $2 = $ _____   18. $5.01 = $ _____

19. $6.05 = $ _____   20. $4.875 = $ _____   21. $3.375 = $ _____

B. Change each fraction to a percent.

1. $\frac{1}{4} = $ _____   2. $\frac{1}{2} = $ _____

3. $\frac{7}{8} = $ _____   4. $\frac{1}{8} = $ _____

5. $\frac{7}{20} = $ _____   6. $\frac{5}{16} = $ _____

7. $\frac{13}{20} = $ _____   8. $\frac{11}{4} = $ _____

9. $\frac{25}{8} = $ _____   10. $\frac{21}{4} = $ _____

11. $2\frac{3}{8} = $ _____   12. $1\frac{19}{20} = $ _____

13. $4\frac{13}{16} = $ _____   14. $5\frac{3}{16} = $ _____

C. Change each percent to a decimal number.

1. $9\% = $ _____   2. $5\% = $ _____

3. $29\% = $ _____   4. $98\% = $ _____

5. $6.5\% = $ _____   6. $8.2\% = $ _____

7. $256\% = $ _____   8. $23.7\% = $ _____

9. $0.05\% = $ _____   10. $0.08\% = $ _____

11. $85\% = $ _____   12. $2.5\% = $ _____

13. $605\% = $ _____   14. $137.5\% = $ _____

D. Change each percent to a fraction in lowest terms.

1. 5% = _____
2. 8% = _____

3. 24% = _____
4. 75% = _____

5. 60% = _____
6. 20% = _____

7. 250% = _____
8. 0.5% = _____

9. 2.5% = _____
10. 1.2% = _____

11. 87.5% = _____
12. $7\frac{1}{2}$% = _____

13. $66\frac{2}{3}$% = _____
14. $33\frac{1}{3}$% = _____

E. Applied Problems.

1. Wreck-a-Mended Auto is able to restore to like-new condition $\frac{3}{4}$ of the damaged cars it receives. Express as a percent the fraction of its business this represents.

2. A surprising 32% of the buildings erected by J.R.'s Construction last year were medical offices. Express this percent as a decimal number.

3. Statistics show that 68% of students at Gigantic Urban and Rural University (GURU) change their majors at least once before graduating. What fraction of GURU students change their majors? (Be sure to express your answer in lowest terms.)

4. Watt's Electrical Service does $\frac{3}{5}$ of its work in industrial settings. Express as a percent the fraction of its work done for other businesses.

5. The Pizza Palace is proud of the fact that 75% of its customers rate the pizza as good to excellent. Express this percent as a decimal number.

6. According to Lew's latest figures, business at his trendy restaurant is off 25%. What fraction of his business has he lost? (Be sure to express your answer in lowest terms.)

7. Sparkle Cleaning Company has discovered a way to clean floors with only $\frac{11}{16}$ as much detergent as it formerly used. Express as a percent the fraction of the former amount Sparkle now uses.

8. Wired for Sound sold 125% more CD players in April than it did in May. Express this percent as a decimal number.

9. The cost to Crystal's Diamonds for unset diamonds has risen 130% in the last year. What fraction of the original cost does this represent? (Be sure to express your answer in lowest terms.)

10. Petal Pushers will have to raise its prices, because the cost of roses is $\frac{13}{8}$ what it was a year ago. Express the fraction of last year's price for roses to this year's as a percent.

11. The Fly-By-Night Aviary gets 45.3% of its birds from South America. Express this percent as a decimal number.

12. Big Ben's Bolts is planning to offer a new supermegabolt that is 0.05% longer than its old megabolt. What fraction longer than the old bolt will the new bolt be? (Be sure to express your answer in lowest terms.)

13. Bob's Beeflike Burgers recently switched to a roll that costs $1\frac{4}{5}$ times as much as its old roll but contains oat bran. The cost of the new roll is what percent of the cost of the old roll?

14. Female mannequins represent 58% of the stock at Maria's Mannequins. Express this percent as a decimal number.

15. Jack's Jackhammer Service was surprised to find that 37.5% of its customers want their driveways widened. What fraction of Jack's customers want wider driveways? (Be sure to express your answer in lowest terms.)

16. Bits 'n' Bytes offers a new low-price computer with $5\frac{3}{10}$ times the memory of its old low-price computer. The amount of memory of the new computer is what percent of the amount of memory of the old computer?

17. Only 5.1% of deliveries made by Zippy Delivery Service take more than 24 hours. Express this percent as a decimal number.

18. The I Scream Shop now offers an ice cream that is $7\frac{3}{4}$% butterfat. What fraction of this ice cream is butterfat? (Be sure to express your answer in lowest terms.)

19. Woody's Lumber Mill was shocked to find that logs now cost $12\frac{11}{32}$ times what they did a year ago. Express as a percent the rise in the price of logs.

20. Lots of Plots Realty made a tidy 287.5% profit on its sale of a plot to a condominium developer last month. Express this percent as a decimal number.

21. The Mouse Hole, a gourmet cheese shop, has done well in part because $83\frac{1}{3}$% of its customers buy 5 or more pounds per month. What fraction of the Mouse Hole's customers does this represent? (Be sure to express your answer in lowest terms.)

WORKSPACE

SECTION 4.2: Solving Percent Problems

Can You:

Solve a percent problem for the base?
Solve a percent problem for the percent?
Solve a percent problem for the part?
. . . If not, you need this section.

General Guidelines

Working with percents—whether in business, industry, science, or other areas—almost always involves one of three basic types of problems. All of these types of problems involve three quantities:

Base The number against which another number is compared in calculating a percent; in percent problems, it is sometimes called the *total* because it represents the whole.

1. The **base** (sometimes called the *total*)—the amount used for a comparison
2. The **part** (sometimes called the *percentage*)—the amount being compared with the base
3. The **percent** (sometimes called the *rate*)—the relationship of the part to the base

Each of the three basic percent problems involves finding one of these three quantities when the other two are known.

Part The amount being compared with the base; in percent problems, it is sometimes called the *percentage*.

Percent The relationship of one number (part) to a second number (base), expressed in hundredths; in percent problems, it is sometimes called the *rate*.

Solving Percent Problems

You can solve each of the three types of percent problems using five steps.

Steps: Solving Percent Problems

STEP 1. *Translate* the problem sentence into a math statement.

STEP 2. *Label* the quantities by type (percent, base, and part).

STEP 3. *Rearrange* the equation so that the unknown quantity is alone on the left of the equal sign and the other quantities are on the right. Use the Equation Finder on the next page.

STEP 4. *Solve* the problem by doing arithmetic.

STEP 5. *Check* your answer.

Before you try to use these steps, you should know a bit more about each.

Step 1. To *translate* the problem sentence, look for certain words and phrases that appear in most percent problems. They are signals alerting you to the mathematical operations to be done. The most common signal words in percent problems are

Signal Words	Translate as
is, is equal to, equals, will be	=
of	×

Use a □, letter of the alphabet, or ? for the unknown quantity you are asked to find. For example, the question

30% of what number is 15?   should be translated

30% ×   ?   = 15   or   30% × ? = 15

In this case, 30% is the *percent;* ? (the unknown quantity) is the *base;* and 15 is the part of the *total*.

Step 2. To *label* the parts, mark the base with a *B*. Identify the part with a *P*. Label the percent with a %. In our example, the result is

$$30\% \times ? = 15$$
$$\% \times B = P$$

Step 3. To *rearrange* the equation, use the following Equation Finder.

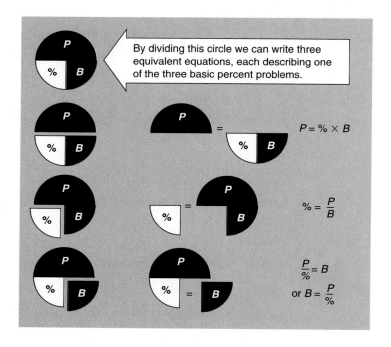

The three equations

$$P = \% \times B, \quad \% = \frac{P}{B}, \quad \text{and} \quad B = \frac{P}{\%} \quad \text{are all equivalent.}$$

In our example, the appropriate equation is $B = \dfrac{P}{\%}$. Therefore the equation $30\% \times B = 15$ becomes

$$B = \frac{P}{\%} = \frac{15}{30\%}$$

Step 4. To *solve* the problem, without the use of a calculator, you *must* rewrite all percents as fractions or decimals. Only then can you use them in a multiplication. In our example, $30\% = 0.30$, so

$$B = \frac{15}{0.30} = 50$$

Step 5. To *check* your answer, put the number you have found back into the original problem or equation to see if it makes sense. If possible, use the answer to calculate one of the other numbers in the equation as a check.

$$30\% \text{ of } 50 = 15$$
$$.30 \times 50 = 15$$

Most business calculators have a percent key that automatically converts a percent to a decimal number. If you have a calculator with a percent key, use the following sequence of keystrokes to solve this problem using a calculator.

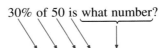

More instruction on the use of a calculator in percent calculations will be found in Appendix A.

Now let's look carefully at each type of problem. We'll explain each, give examples, show you how to solve them, and work through a few together.

P-Type Problems

P-Type Problems Percent problems that ask you to find the part, given the percent and the base.

Problems that ask you to find the part or percentage are called **P-type problems**. Such problems are usually stated in the form: "Find 30% of 50", or "What is 30% of 50?" The real question is

30% of 50 is what number?

Step 1. *Translate:* $30\% \times 50 = ?$

Step 2. *Label:* $\% \times B = P$

Step 3. *Rearrange:* $? = 30\% \times 50$

Remember from the Equation Finder that

Step 4. *Solve:* Before we give you the solution to this problem, you should try to solve it yourself. What answer did you get? The most common answers are 1500 and 15.

If you said 1500, you forgot our warning in step 4 and multiplied the percent without converting it to a decimal or fraction. Keep in mind that $30 \times 50 = 1500$, but *30% × 50* is *not* 1500. To find the correct answer, convert 30% to a decimal first.

$$30\% = 30.\% = 0.30$$
$$30\% \times 50 = 0.30 \times 50 = 15$$

Step 5. *Check:* $30\% \times 50 = 15$
$ 0.30 \times 50 = 15$

With a calculator, use the following sequence of keystrokes to solve this problem.

$50 \boxed{\times} 30 \boxed{\%} \longrightarrow \boxed{ 15.}$

Notice that when you use the percent key on a calculator, the base number is entered first $30\% \times 50$ is entered as $50 \times 30\%$.

YOUR TURN

WORK THIS PROBLEM

The Question:

Solve the following problems.

(a) Find $8\frac{1}{2}\%$ of 160.
(c) 35% of 20 = ?
(e) 120% of 15 is what number?

(b) Find 2% of 140.
(d) $7\frac{1}{4}\%$ of \$1000 = ?
(f) What is $5\frac{1}{3}\%$ of 3.3?

✔ YOUR WORK

The Solution:

(a) ***Step 1.*** *Translate:* $8\frac{1}{2}\% \times 160 = ?$

Step 2. *Label:* $\% \times B = P$

Step 3. *Rearrange:* $P = 8\frac{1}{2}\% \times 160$

Step 4. *Solve:* $8\frac{1}{2}\% = 8.5\% = 08.5\% = 0.085$

$ P = 0.085 \times 160 = 13.6$

Step 5. *Check:* $8\frac{1}{2}\% \times 160 = 13.6$
$ 0.085 \times 160 = 13.6$

(b) ***Step 1.*** $2\% \times 140 = ?$

Step 2. $\% \times B = P$

Step 3. $P = 2\% \times 140$

Step 4. $2\% = 02.\% = 0.02$

$ P = 0.02 \times 140 = 2.8$

Step 5. $2\% \times 140 = 2.8$
$0.02 \times 140 = 2.8$

(c) *Step 1.* $35\% \times 20 = ?$

Step 2. $\% \times B = P$

Step 3. $P = 35\% \times 20$

Step 4. $35\% = 35.\% = 0.35$
$P = 0.35 \times 20 = 7$

Step 5. $35\% \times 20 = 7$
$0.35 \times 20 = 7$

(d) *Step 1.* $7\frac{1}{4}\% \times \$1000 = ?$

Step 2. $\% \times B = P$

Step 3. $P = 7\frac{1}{4}\% \times \$1000$

Step 4. $7\frac{1}{4}\% = 7.25\% = 07.25\% = 0.0725$
$P = 0.0725 \times \$1000 = \$72.50$

Step 5. $7\frac{1}{4}\% \times \$1000 = \$72.50$
$0.0725 \times \$1000 = \$72.50$

(e) *Step 1.* $120\% \times 15 = ?$

Step 2. $\% \times B = P$

Step 3. $P = 120\% \times 15$

Step 4. $120\% = 120.\% = 1.20$
$P = 1.2 \times 15 = 18$

Step 5. $120\% \times 15 = 18$
$1.20 \times 15 = 18$

(f) *Step 1.* $5\frac{1}{3}\% \times 3.3 = ?$

Step 2. $\% \times B = P$

Step 3. $P = 5\frac{1}{3}\% \times 3.3$

Step 4.   Rounding $5\frac{1}{3}\%$ to a decimal number will result in an approximate answer.

$5\frac{1}{3}\% = 5.33\% = 05.33\% = 0.0533$ rounded

$P = 0.0533 \times 3.3 = 0.17589 = 0.176$ rounded

More decimal places in the rounded decimal number will result in greater accuracy. But to get the most accurate, we use fractions, not decimals.

$$5\frac{1}{3}\% = \frac{5\frac{1}{3}}{100} = \frac{\frac{16}{3}}{100} = \frac{16}{300}$$

$$P = \frac{16}{300} \times 3.3 = \frac{16}{300} \times \frac{33}{10} = \frac{528}{3000} = 0.176$$

Step 5. $5\frac{1}{3}\% \times 3.3 = 0.176$
$0.0533 \times 3.3 = 0.176$ rounded

How to Misuse Percent

1. In general you cannot add, subtract, multiply, or divide percent numbers. Percent helps you compare two numbers. You cannot use it in normal arithmetic operations.

   For example, if 60% of class 1 earned A grades and 50% of class 2 earned A grades, what was the total percent of A grades for the two classes?

   The answer is that you cannot tell unless you know the number of students in each class.

2. In advertisements designed to trap the unwary, you might hear that "children had 23% fewer cavities when they used . . ." or "50% more doctors use. . . ."

   Fewer than what? Fewer than the worst dental health group the advertiser could find? Fewer than the national average?

   More than what? More than a year ago? More than nurses? More than other adults? More than infants?

   There must be some reference or base given in order for the percent number to have any meaning at all.

   BEWARE of people who misuse percent!

A great many business problems are *P*-type problems.

YOUR TURN

WORK THIS PROBLEM

The Question:

A reasonable budget allows 25% of net income for housing. How much should be allowed for housing if the net monthly income is $1989.51? Round your answer to the nearest cent.

✔ YOUR WORK

The Solution:

Step 1. $25\% \times \$1989.51 = ?$

Step 2. $\% \times ? = P$

Step 3. $P = 25\% \times \$1989.51$

Step 4. $25\% = .25\% = 0.25$

   $P = 0.25 \times \$1989.51 = \$497.3775 = \$497.38$ rounded

Step 5. $25\% \times \$1989.51 = \$497.38$
   $0.25 \times \$1989.51 = \$497.38$ rounded

%-Type Problems

%-Type Problems Percent problems that ask you to find the percent, given the part and the base.

Problems that ask you to find the percent are called **%-type problems.** Such problems are usually stated in the form: "Find what percent 7 is of 16," or "7 is what percent of 16?" The real question is

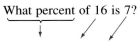

What percent of 16 is 7?

Step 1.   *Translate:* ?% × 16 = 7

Step 2.   *Label:* % × B = P   All of the problem statements are equivalent to this equation.

Step 3.   *Rearrange:* Remember the Equation Finder:

$$\frac{\boxed{\%}}{} = \frac{P}{B} \qquad \% = \frac{P}{B}$$

so ?% = $\dfrac{7}{16}$ 16 is the base and 7 is the part of the base being described.

Step 4.   *Solve:* ?% = $\dfrac{7}{16}$ = 16)7.0000 = 0.4375 = 43.75%

$$
\begin{array}{r}
0.4375 \\
16\overline{)7.0000} \\
\underline{6\,4} \\
60 \\
\underline{48} \\
120 \\
\underline{1\,12} \\
80 \\
\underline{80} \\
\end{array}
$$

Step 5.   *Check:* 43.75% × 16 = ?

0.4375 × 16 = 7

The solution to a %-type problem is always a fraction or decimal number that you must convert to a percent. If you had trouble converting $\frac{7}{16}$ to a percent, you should review this process in Section 4.1.

YOUR TURN

> **WORK THIS PROBLEM**

The Question:

Solve the following.

(a) What percent of 40 is 16?

(b) Find what percent of 25 is 65.

(c) $6.50 is what percent of $18.00?

(d) What percent of 2 is 3.5?

(e) $10\frac{2}{5}$ is what percent of 2.6?

✔ **YOUR WORK**

The Solution:

(a) ***Step 1.*** $?\% \times 40 = 16$

  Step 2. $\% \times B = P$

  Step 3. $\% = \dfrac{P}{B}$ (per the Equation Finder), so $?\% = \dfrac{16}{40}$

  Step 4. $?\% = \dfrac{16}{40} = 0.40 = 40\%$

  Step 5. $40\% \times 40 = 16$
      $0.40 \times 40 = 16$

(b) ***Step 1.*** $?\% \times 25 = 65$

  Step 2. $\% \times B = P$

  Step 3. $\% = \dfrac{P}{B}$, so $?\% = \dfrac{65}{25}$

  Step 4. $?\% = \dfrac{65}{25} = 2.60 = 260\%$

  Step 5. $260\% \times 25 = 65$
      $2.60 \times 25 = 65$

The most difficult part of this problem is in deciding whether the percent needed is found from $\frac{65}{25}$ or $\frac{25}{65}$. There is no magic to it. If you read the problem very carefully, you will see that it speaks of 65 as a part "of 25." The base is 25. The part is 65.

(c) ***Step 1.*** $\$6.50 = ?\% \times \$18.00$

  Step 2. $P = \% \times B$

  Step 3. $\% = \dfrac{P}{B}$, so $?\% = \dfrac{6.50}{18.00}$

  Step 4. There are two ways to change $\dfrac{6.50}{18.00}$ to a percent: the exact method and the rounding method.

     Exact: $\dfrac{6.50}{18.00} = \dfrac{6\frac{1}{2}}{18} = \dfrac{6\frac{1}{2} \times 5\frac{5}{9}}{18 \times 5\frac{5}{9}} = \dfrac{36\frac{1}{9}}{100} = 36\frac{1}{9}\%$ $?\% = 36\frac{1}{9}\%$

     Rounding: $\dfrac{6.50}{18.00} = 0.361 = 36.1\%$ rounded $?\% = 36.1\%$ rounded

  Step 5. $36.1\% \times \$18.00 = \$6.50$
      $0.361 \times \$18.00 = \$6.50$ rounded

(d) ***Step 1.*** $?\% \times 2 = 3.5$

  Step 2. $\% \times B = P$

  Step 3. $\% = \dfrac{P}{B}$, so $?\% = \dfrac{3.5}{2}$

  Step 4. $?\% = \dfrac{3.5}{2} = 1.75 = 175\%$

  Step 5. $175\% \times 2 = 3.5$
      $1.75 \times 2 = 3.5$

(e) ***Step 1.*** $10\frac{2}{5} = ?\% \times 2.6$

    Step 2. $P = \% \times B$

    Step 3. $\% = \dfrac{P}{B}$, so $?\% = \dfrac{10\frac{2}{5}}{2.6}$

    Step 4. $?\% = \dfrac{10\frac{2}{5}}{2.6} = \dfrac{10.4}{2.6} = \;\; 2.6\overline{)10.4} \quad = 4.00 = 400\%$
$$\underline{10\,4}$$

    Step 5. $400\% \times 2.6 = 10\frac{2}{5}$
$$4.00 \times 2.6 = 10.4$$

Calculating Percent Increase or Decrease

Businesspeople often have occasion to solve %-type problems. A common case is when they need to calculate percent increases or percent decreases. Such problems are not difficult if you follow these steps:

Steps: Calculating Percent Increases and Decreases

STEP 1. *Calculate* the amount of increase or decrease by subtraction.

STEP 2. *Use* the amount of increase or decrease and the *original* amount to calculate the percent increase or decrease.

WORK THIS PROBLEM

The Question:

If your original hourly pay rate was $8.75 per hour and you received an increase to $9.80, what would be your percent increase?

✔ YOUR WORK

The Solution:

1. Amount of increase = $9.80 − $8.75 = $1.05.
2. Calculate the percent increase: $1.05 is what percent of $8.75?

    Step 1. *Translate:* $\$1.05 = ?\% \times \$8.75$

    Step 2. *Label:* $P = \% \times B$

    Step 3. *Rearrange:* $\% = \dfrac{P}{B}$, so $?\% = \dfrac{\$1.05}{\$8.75}$

    Step 4. *Solve:* $?\% = \dfrac{1.05}{8.75} = 0.12 = 12\%$

    Step 5. *Check:* $12\% \times \$8.75 = \$1.05$
$$0.12 \times \$8.75 = \$1.05$$

B-Type Problems

B-Type Problems   Percent problems that ask you to find the base, given the part and the percent.

Problems that ask you to find the base or total are called ***B*-type problems.** Such problems are usually stated in the form: "Find a number of which 30% is 8.7," or "8.7 is 30% of what number?" The real question is

$$30\% \text{ of } \underbrace{\text{what number}} \text{ is equal to } 8.7?$$

Step 1.   *Translate:* 30% × ?   = 8.7

Step 2.   *Label:* % × B = P

Step 3.   *Rearrange:* Use the Equation Finder:

$$\frac{P}{\%} = B \qquad B = \frac{P}{\%}$$

$$\text{so } B = \frac{8.7}{30\%}$$

Step 4.   *Solve:* Remember to first convert the percent number to a decimal number. So

$$B = \frac{8.7}{30\%} = \frac{8.7}{0.30} = \quad 0.30\overline{)8.70} \quad \begin{array}{r} 29. = 29 \\ \hline \end{array}$$

$$\begin{array}{r} 6\ 0 \\ \hline 2\ 70 \\ 2\ 70 \\ \hline \end{array}$$

Step 5.   *Check:* 30% × 29 = ?
                      0.30 × 29 = 8.7

YOUR TURN

┌WORK THIS PROBLEM├─────────────────────────────────

The Question:

Solve the following.

(a) 16% of what number is equal to 5.76?
(b) 41 is 5% of what number?
(c) Find a number such that $12\frac{1}{2}\%$ of it is $26\frac{1}{4}$.
(d) 2 is 8% of a number. Find the number.
(e) 125% of what number is 35?

✔ YOUR WORK

The Solution:

(a) ***Step 1.*** $16\% \times ? = 5.76$

   Step 2. $\% \times B = P$

   Step 3. $B = \dfrac{P}{\%}$, so $B = \dfrac{5.76}{16\%}$

   Step 4. $16\% = 0.16$

$$B = \frac{5.76}{16\%} = \frac{5.76}{0.16} = \quad 0.16\overline{)5.76} = 36$$

$$\begin{array}{r} 36. \\ \hline 4\,8 \\ \hline 96 \\ \underline{96} \end{array}$$

   Step 5. $16\% \times 36 = 5.76$
   $0.16 \times 36 = 5.76$

(b) ***Step 1.*** $41 = 5\% \times ?$

   Step 2. $P = \% \times B$

   Step 3. $B = \dfrac{P}{\%}$, so $B = \dfrac{41}{5\%}$

   Step 4. $5\% = 0.05$

$$B = \frac{41}{5\%} = \frac{41}{0.05} = \quad 0.05\overline{)41.00} = 820$$

$$\begin{array}{r} 820. \\ \hline 40 \\ \hline 1\,0 \\ \underline{1\,0} \\ 00 \\ \underline{00} \end{array}$$

   Step 5. $5\% \times 820 = 41$
   $0.05 \times 820 = 41$

(c) ***Step 1.*** $26\frac{1}{4} = 12\frac{1}{2}\% \times ?$

   Step 2. $P = \% \times B$

   Step 3. $B = \dfrac{P}{\%}$, so $B = \dfrac{26\frac{1}{4}}{12\frac{1}{2}\%}$

   Step 4. $12\frac{1}{2}\% = 12.5\% = 0.125$

$$B = \frac{26\frac{1}{4}}{12\frac{1}{2}\%} = \frac{26.25}{0.125} = \quad 0.125\overline{)26.250} = 210$$

$$\begin{array}{r} 210. \\ \hline 25\,0 \\ \hline 1\,25 \\ \underline{1\,25} \\ 00 \\ \underline{00} \end{array}$$

   Step 5. $12\frac{1}{2}\% \times 210 = 26\frac{1}{4}$
   $0.125 \times 210 = 26.25$

(d) *Step 1.* $2 = 8\% \times ?$

    Step 2. $P = \% \times B$

    Step 3. $B = \dfrac{P}{\%}$, so $B = \dfrac{2}{8\%}$

    Step 4. $8\% = 0.08$

$$B = \frac{2}{8\%} = \frac{2}{0.08} = 0.08\overline{)2.00} = 25$$

$$\begin{array}{r} 25. \\ \hline 1\,6 \\ \hline 40 \\ \underline{40} \end{array}$$

    Step 5. $8\% \times 25 = 2$
              $0.08 \times 25 = 2$

(e) *Step 1.* $125\% \times ? = 35$

    Step 2. $\% \times B = P$

    Step 3. $B = \dfrac{P}{\%}$, so $B = \dfrac{35}{125\%}$

    Step 4. $125\% = 1.25$

$$B = \frac{35}{125\%} = \frac{35}{1.25} = 1.25\overline{)35.00} = 28$$

$$\begin{array}{r} 28. \\ \hline 25\,0 \\ \hline 10\,00 \\ \underline{10\,00} \end{array}$$

    Step 5. $125\% \times 28 = 35$
              $1.25 \times 28 = 35$

B-type problems are also a common aspect of business math, especially when figuring out sales taxes.

YOUR TURN

WORK THIS PROBLEM

The Question:

The sales tax on a coat was $4.71. If the tax rate was 6%, what was the price of the coat?

✔ YOUR WORK

The Solution:

6% of what is $4.71?

Step 1. $6\% \times ? = \$4.71$

Step 2. $\% \times B = P$

Step 3. $B = \dfrac{P}{\%}$, so $B = \dfrac{\$4.71}{6\%}$

Step 4. $B = \dfrac{\$4.71}{0.06} = \$78.50$

Step 5. $6\% \times \$78.50 = \$4.71$
$0.06 \times \$78.50 = \$4.71$

Are you ready for a bit of practice on the three basic kinds of percent problems? Wind your mind and turn to Section Test 4.2.

WORKSPACE

Name _____

Date _____

Course/Section _____

The following problems test your understanding of Section 4.2, Solving Percent Problems.

A. Complete the following by solving for the part, percent, or base.

	Base	Percent	Part
1.	20	35%	_____
2.	60	30%	_____
3.	75	80%	_____
4.	16	75%	_____
5.	_____	120%	180
6.	_____	160%	320
7.	_____	35%	42
8.	_____	65%	52
9.	260	_____	247
10.	750	_____	735
11.	900	_____	216
12.	860	_____	731
13.	125	24%	_____
14.	925	160%	_____
15.	370	_____	629
16.	150	_____	360
17.	_____	350%	980
18.	_____	3%	24
19.	_____	17%	935
20.	2400	23%	_____

B. Solve.

1. 32% of 25 is _____.

2. _____ is 85% of 80.

3. 52 is 80% of _____.

4. $33\frac{1}{3}$% of _____ is 4.

5. 10 is what percent of 15? _____

6. What percent of $27 is $675? _____

7. _____ is 5% of $260.

8. $5.88 is what percent of $4.90? _____

9. 124% of _____ is $9.30.

10. 105% of 72 is _____.

11. 303.75 is what percent of 135? _____

12. 232.8 is 97% of _____.

13. What percent of 85 is 66.3? _____

14. 52.2 is 87% of _____.

15. _____ is 350% of 2.5.

16. 14.85 is what percent of 8.25? _____

17. 35% of $12.80 is _____.

18. $10.62 is 36% of _____.

19. 25.6% of 6.5 is _____.

20. 68.5% of _____ is 2.466.

C. Applied problems.

1. Fantasy Realty, Inc., has bought a new office building for $127,500 with a down payment of 18%. Calculate the down payment.

2. Ava's Air Tours has a 4-hour tour that originally cost $275. The price is reduced by 24% for frequent flyers. Calculate (a) the discount and (b) the new price.

3. Pia, a pizza dough slinger at the Pizza Palace, was originally paid $9.20 per hour. However, Pia developed a two-handed, two-pizza sling that has made her the talk of the town. Her new pay rate is $10.58. Calculate (a) the amount of increase and (b) the percent increase.

4. Floating Holidays is buying a new sailboat for $92,000 with a down payment of $33,120. Calculate the percent of down payment.

5. Astute Accents has a classic designer belt on sale at 22% off. The original price was $72.50. Calculate (a) the discount and (b) the sale price.

6. The Zippy Delivery Service is buying a new delivery truck. The purchase price is $22,100. The down payment was 26% of the purchase price. Calculate the down payment.

7. Maria's Mannequins pays mannequin makeup artists $8.50 per hour. If makeup artists are given an 8% pay raise, calculate (a) the amount of the raise and (b) the new pay rate.

8. One dozen megabolts from Big Ben's Bolts sell for $12.50 plus 6% sales tax. Calculate (a) the sales tax and (b) the total cost.

9. Bits 'n' Bytes has a replacement keyboard that is sold for $125.60. The total price with tax is $131.88. Calculate (a) the amount of sales tax and (b) the sales tax rate.

10. The Fly-By-Night Aviary has extra large bird cages in stock. The original price for this bird cage was $246.80. However, the cage maker has raised the price to $259.14. Calculate (a) the increase and (b) the percent increase.

11. The sales manager at Wired for Sound had an original salary of $1875 per month. Due to the growth of the business, the sales manager is getting an 11% salary increase. Calculate (a) the amount of increase and (b) the new salary.

12. Bob's Beeflike Burgers purchased a new burger former for $8,700. The first year it lost value (depreciated) 28%. Calculate (a) the depreciation and (b) the new value.

13. The head chef at Lew's new gourmet restaurant originally had a weekly salary of $1550. Customers have raved about the food, so Lew has raised his chef's salary to $1612. Calculate (a) the amount of increase and (b) the percent increase.

14. The Butter-em-Up Bakery typically sells a dozen croissants for $12.50. This weekend the sale price will be $10.75. Calculate (a) the amount of decrease and (b) the percent decrease.

15. Kate's delicacies makes 240 Black Forest cakes per day. The cakes always sell out, so Kate has decided to increase production to 255 cakes. Calculate (a) the increase and (b) the percent increase.

16. Shiring Brokerage's earnings last year were $87,500. This year, talented financial management has brought about an 8.5% increase. Calculate (a) the amount of increase and (b) the new earnings.

17. J.R.'s Construction Company bought a new truck that originally cost $24,500. After one year the truck was worth $16,660. Calculate (a) the decrease in value and (b) the percent decrease.

18. Risky Business's earnings increased from $124,000 last year to $128,464 this year. Calculate (a) the amount of increase and (b) the percent increase.

19. Woody's Lumber Mill originally produced 450 curved stairway posts per week but now produces 567 such posts per week. Calculate (a) the increase and (b) the percent increase.

20. Rowdy's Music has been providing disc jockeys and continuous entertainment for $84.75 an hour, but wants to increase the hourly price by 24%. Calculate (a) the increase and (b) the new hourly price of an hour of Rowdy's Music.

21. The Cherches la Prix restaurant originally earned $15,200 per month while it was the place to be. Lew's trendy new restaurant has stolen the limelight, and Cherches' profits have decreased by 27%. Calculate (a) the decrease and (b) the new earnings.

22. Bull's Gutter Service earnings decreased 7% during the winter months. The original earnings were $7700. Calculate (a) the decrease and (b) the new monthly earnings.

23. Sew What? has a sewing machine that originally sold for $235.50. The reduced price is $202.53. Calculate (a) the amount of decrease and (b) the percent decrease.

24. The Bits 'n' Bytes firmware programmers typically produce 1550 lines of computer code per week. The expanding business needs to increase programmer productivity to 1953 lines per week. Calculate (a) the increase and (b) the percent increase.

25. Calculate the percent increase in revenues for the Web World Corporation from 1996 to 1997.

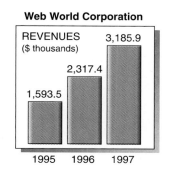

Web World Corporation

REVENUES ($ thousands)

26. Ken, the computer guru at the Bits 'n' Bytes Corporate office, found that his computer had 34,381,824 bytes of disk used on his hard drive and 5,157,274 bytes unused in file clusters. Calculate the percent of wasted disk space.

27. Phil was in charge of the produce department at the Everyman's Grocery. From past experience, Phil anticipates that 13% of the tomatoes that are purchased will spoil before they are sold. He recently purchased 162 pounds of tomatoes from a local farmer. How many pounds of tomatoes can Phil actually expect to sell?

28. The buyer for Dynamite Deals purchased $23,482 worth of novelty items for a spring promotion. When the order arrived, she discovered that $1,248 worth of items were damaged. She returned the damaged goods, and received a 23% discount on the part of the order that she kept. How much was the check that she sent for this order?

WORKSPACE

Key Terms

base

B-type problems

part

percent

%-type problems

P-type problems

Chapter 4 at a Glance

Page	Topic	Key Point	Example
143	Changing a decimal number to a percent	Multiply the decimal number by 100%.	$0.375 \times 100\% = 37.5\%$
144	Changing a fraction to a percent	Divide the numerator by the denominator to find a decimal number. To convert the decimal number, multiply by 100%.	$\dfrac{3}{4} = 4\overline{)3.00}\;\overset{0.75}{}$ $0.75 \times 100\% = 75\%$
146	Changing a percent to a decimal number	Divide by 100%.	$50\% = 100\overline{)50.0}\;\overset{0.5}{} = 0.5$
147	Changing a percent to a fraction	Divide by 100%; reduce to lowest terms.	$50\% = \dfrac{50}{100} = \dfrac{1}{2}$
155	Solving *P*-type problems	Use the formula: $P = \% \times B$	What is 30% of 50? $P = 30\% \times 50$ $\quad = 0.30 \times 50 = 15$
159	Solving %-type problems	Use the formula: $\% = \dfrac{P}{B}$ express as a percent	What percent of 40 is 16? $\% = \dfrac{16}{40} = 0.40 = 40\%$
162	Solving *B*-type problems	Use the formula: $B = \dfrac{P}{\%}$	125% of what number is 35? $B = \dfrac{35}{125\%} = \dfrac{35}{1.25} = 28$

WORKSPACE

Name _____

Date _____

Course/Section _____

The following problems test your understanding of percents.

1. Write $\frac{7}{8}$ as a percent. _____

2. Write $\frac{11}{16}$ as a percent. _____

3. Write 0.075 as a percent. _____

4. Write 2.4 as a percent. _____

5. Write 185% as a decimal number. _____

6. Write 8.2% as a decimal number. _____

7. Write 85% as a fraction and reduce to lowest terms. _____

8. Write 124% as a fraction and reduce to lowest terms. _____

9. What is 65% of 140? _____

10. 124% of 250 is what? _____

11. 14% of what is 1008? _____

12. 817 is 95% of what? _____

13. 468 is what percent of 720? _____

14. What percent of 450 is 621? _____

15. A Black Forest cake costs $17.50 plus 6% sales tax at Kate's Delicacies. Calculate (a) the tax and (b) the total price.

16. Maria's Mannequins is buying new mannequin wigs. The original cost of a lifelike wig was $58.60. There is a wig sale in progress and the price is discounted by 25%. Calculate (a) the discount and (b) the sale price.

17. Shiring's Brokerage has increased the apprentice pay rate from $7.60 to $8.74 per hour. Calculate (a) the increase and (b) the percent increase.

18. J.R.'s Construction company has increased the price of exterior door installation from $128.00 to $174.08. Calculate (a) the increase and (b) the percent increase.

19. The Garden of Earthly Delights has discounted the price of installing a spring flower bed by 16%. If the amount of the discount is $15.68, calculate (a) the original price and (b) the discounted price.

20. The music system in Lew's restaurant has depreciated $187.50, which is 15% of its original value. (a) What was the system's original value? (b) What is its new reduced value?

21. Bob's Beeflike Burgers needs to increase bun production from 5400 buns per day to 7290 buns per day. Calculate (a) the increase and (b) the percent increase.

22. Hiram's Employment Agency earned $342 interest on an account that paid 9% interest. Calculate the amount invested.

23. The Fly-By-Night Aviary has decreased the price of a singing canary from $24.50 to $18.13. Calculate (a) the decrease and (b) the percent decrease.

24. Ava's Air Tours made $6700 profit during the month of July. Profits increased by 17% during August. Calculate (a) the increase and (b) the August profit.

25. Harriet, a beautician, needs to increase the price of her permanents by 15% due to an increase in the cost of supplies. If the original price was $52.80, what is the new cost?

Bank Records

You mean all I have to do is write a check, and you pay my bills?

Yes, but you must have enough money in your account to cover the amount of the check.

CHAPTER OBJECTIVES

When you complete this chapter successfully, you will be able to:

1. Set up and use a checking account.

2. Reconcile your checking account records with those of your bank.

■ Bits 'n' Bytes deposits on September 3, 1998 included $472 in currency, $27.83 in coin, and checks for $473.95, $236.27, and $12.55.

■ The bank statement for Wired for Sound shows checks numbered 512, 513, 514, 515, 517, 519, and 520 and deposits on 2/7 and 2/13. There was a service charge of $6.75, and the bank balance was $623.12.

Virtually all businesses in the United States today deal with banks at least to the extent of having a checking account. In this chapter, you will learn how to use and monitor a checking account, a skill that will help you as an individual, even if you never use it in the business world.

It's really easy, so let's get started and check it out.

SECTION 5.1: Setting Up and Using a Checking Account

Can You:

Open a checking account?
Make a deposit to a checking account?
Write a check?
Fill out check stubs?
Keep track of checking account activities in a checkbook register?
. . . If not, you need this section.

Check A written order to a bank to remove money from an account and pay the indicated sum to the individual or organization indicated.

Cash is, as it says on a dollar bill, "legal tender [payment] for all debts" in the United States. Nevertheless, most businesses and many individuals in this country prefer to pay for their purchases with checks. A **check** is simply a written order to a bank to remove money from an account you have with them and pay the sum indicated. Provided you have put enough money into your checking account at the bank before you write the check, the bank will give the money you specified to the person to whom you made out the check.

Checks have increased in popularity because they offer a major advantage over cash: they act as a record of the transactions. After it cashes your check, the bank will return to you the canceled check or a record of the transaction. If the person you paid then claims not to have received the amount due, you have proof of payment. You can also use records of your checks to monitor your expenditures.

In this section, you will learn how to open a checking account at a bank and to complete the forms necessary to maintain such accounts.

Opening an Account

The first step in opening a checking account is to select a bank. One major factor in picking a bank is often the fees it charges on checking accounts. The following fees are the most common:

1. A *check printing charge*—a fee that covers the cost of printing the checks for the individual or business maintaining the account.
2. A *fixed charge per check*—a fee charged for each check written. This option is usually selected by people who write a small number of checks.
3. A *monthly service charge*—a flat fee charged for having the account, regardless of how much or little it is used. Such a flat fee usually means that there is no limit on the number of checks you may write each month.
4. A *variable monthly fee*—a fee based on the number of checks written and/or the amount of money a business or individual keeps in the checking account.

Signature Card A record kept by a bank for each account; it includes information about type of account, name of the account holder, references, and the signature(s) of those authorized to sign checks on the account.

Most banks do not charge any fees to individuals who maintain a minimum of $300 to $500 in their checking accounts at all times. However, nearly all banks charge some fees on business checking accounts, regardless of balance.

Once you have selected a bank, you will need to fill out a **signature card.** The signature card includes your name or company name, references, type of account, and the signature or signatures of the individual(s) authorized to sign the checks. Some businesses authorize more than one person to sign the company checks.

Making a Deposit

Deposit Money put into a bank account; it may take the form of cash or other checks.

To open a checking account, you must make an initial **deposit**—that is, you must put in some money. Information required on the deposit ticket accompanying the money includes

① Checking account number issued by the bank

② Company or individual's name

③ Address

④ Date

⑤ Amount of deposit, including currency, coin, and each check

⑥ Total deposit

The following is a deposit for Bits 'n' Bytes' account at the Friendly Bank. Karen M. Paisley, treasurer of Bits 'n' Bytes, is making an initial deposit of $1322.50, which includes $245 in currency and checks for $850.00 and $227.50.

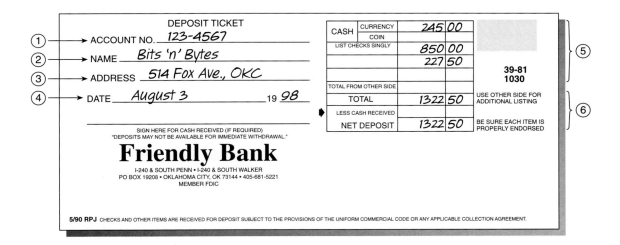

WORK THIS PROBLEM

The Question:

Bits 'n' Bytes' second deposit on August 10, 1998 included $892.00 in currency, $27.82 in coin, and checks for $829.98, $275.29, and $1589.12. Fill out the deposit ticket on the next page using the account number and company address from the previous deposit.

DEPOSIT TICKET

ACCOUNT NO. _____

NAME _____

ADDRESS _____

DATE _____ 19 _____

CASH	CURRENCY		
	COIN		
LIST CHECKS SINGLY			
TOTAL FROM OTHER SIDE			
TOTAL			
LESS CASH RECEIVED			
NET DEPOSIT			

39-81
1030

USE OTHER SIDE FOR ADDITIONAL LISTING

BE SURE EACH ITEM IS PROPERLY ENDORSED

SIGN HERE FOR CASH RECEIVED (IF REQUIRED)
"DEPOSITS MAY NOT BE AVAILABLE FOR IMMEDIATE WITHDRAWAL."

Friendly Bank

I-240 & SOUTH PENN • I-240 & SOUTH WALKER
PO BOX 19208 • OKLAHOMA CITY, OK 73144 • 405-681-5221
MEMBER FDIC

5/90 RPJ CHECKS AND OTHER ITEMS ARE RECEIVED FOR DEPOSIT SUBJECT TO THE PROVISIONS OF THE UNIFORM COMMERCIAL CODE OR ANY APPLICABLE COLLECTION AGREEMENT.

✔ **YOUR WORK**

The Solution:

DEPOSIT TICKET

ACCOUNT NO. *123-4567*

NAME *Bits 'n' Bytes*

ADDRESS *514 Fox Ave., OKC*

DATE *August 10* 19 *98*

CASH	CURRENCY	892	00
	COIN	27	82
LIST CHECKS SINGLY		829	98
		275	29
		1589	12
TOTAL FROM OTHER SIDE			
TOTAL		3614	21
LESS CASH RECEIVED			
NET DEPOSIT		3614	21

39-81
1030

USE OTHER SIDE FOR ADDITIONAL LISTING

BE SURE EACH ITEM IS PROPERLY ENDORSED

SIGN HERE FOR CASH RECEIVED (IF REQUIRED)
"DEPOSITS MAY NOT BE AVAILABLE FOR IMMEDIATE WITHDRAWAL."

Friendly Bank

I-240 & SOUTH PENN • I-240 & SOUTH WALKER
PO BOX 19208 • OKLAHOMA CITY, OK 73144 • 405-681-5221
MEMBER FDIC

5/90 RPJ CHECKS AND OTHER ITEMS ARE RECEIVED FOR DEPOSIT SUBJECT TO THE PROVISIONS OF THE UNIFORM COMMERCIAL CODE OR ANY APPLICABLE COLLECTION AGREEMENT.

Balance The amount in a bank account at a given time.

Service Charge A fee charged by a bank.

After you make an initial deposit, the bank will issue personalized checks with the name of the account holder (the person(s) or company), address, and account number. You may then write checks on the account, always being careful to keep the account **balance**—the amount still in the account—at or above $0. If your balance falls below $0, the account is *overdrawn* and the bank will not process any more checks until you make further deposits to bring the account balance over $0. Most banks also assess a **service charge** for being overdrawn.

Writing a Check

The following is an example of a typical check showing the six items you need to complete. To avoid alteration, *always* make out checks using nonerasable ink, not pencil.

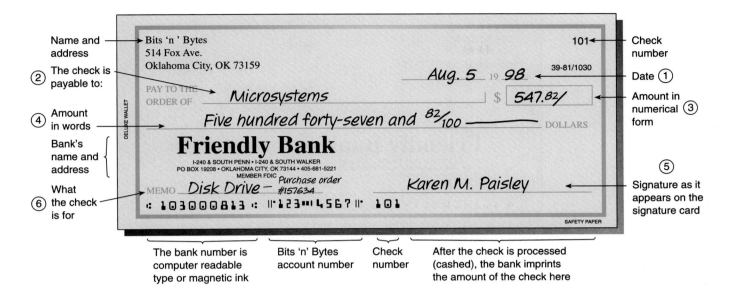

Name and address	Check number
② The check is payable to:	Date ①
④ Amount in words	Amount in numerical form ③
Bank's name and address	⑤
⑥ What the check is for	Signature as it appears on the signature card

The bank number is computer readable type or magnetic ink · Bits 'n' Bytes account number · Check number · After the check is processed (cashed), the bank imprints the amount of the check here

Information required on a typical check includes

Payee The person or business to whom the check is issued.

① The *date* of the check is entered in the upper right-hand corner.

② The **payee**—person or business to whom the check is issued—is entered in the blank denoted *"Pay to the order of."*

③ The amount of the check, in numerical form, is entered here. To avoid possible alteration, the amount should start immediately following the imprinted $. Do not write an amount as $547.82, for it can be easily altered to $547,820 by changing the period to a comma, and adding a zero to the end. Write the amount as $547.^{82} to avoid this problem.

④ The amount of the check in word form is entered in ④. The whole number portion of the amount is entered in word form and the decimal portion is entered as a fraction. To avoid alteration, the amount should start at the beginning of the space provided. After the amount, place a wavy line. The amount in ④ must be the same as in ③.

⑤ The individual's signature as it appears on the original signature card.

⑥ A space to note what the check is for is provided in the line marked "memo" or "for." Although completing this space is optional, it is a good idea to fill it out, as it can help you keep track of your purchases more closely.

YOUR TURN

⌐WORK THIS PROBLEM⌐

The Question:

Complete the following check from Bits 'n' Bytes to Paper Your Way for computer paper in the amount of $237.48, dated August 9, 1998.

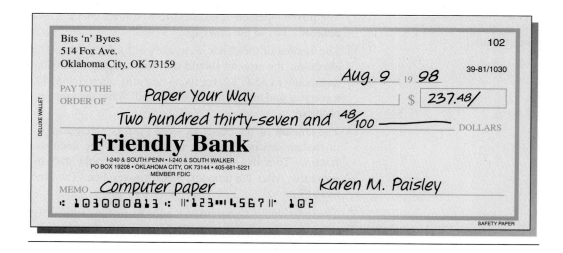

✔ **YOUR WORK**

The Solution:

Completing Check Stubs

Check Stub A record-keeping area to which a check is originally attached.

As you write out a check—and definitely before you hand it over—you need to record information about the check either on a check stub or in a check register. A **check stub** is attached to each check. On this stub, you enter the amount, date, to whom the check is issued, description of purchase, deposits, and the balance.

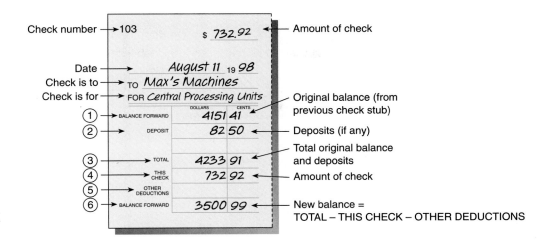

The information required at the top of the stub should be obvious to you at this point. The bottom of the stub is explained in the following steps:

Steps: Completing a Check Stub

STEP 1. The "balance forward" requires you to *take* the balance from the *bottom* of the previous check stub and enter it here.

STEP 2. Next *enter* any deposits you have made since writing the last check.

STEP 3. Now *add* together the numbers in 1 and 2 (balance forward plus deposits) to get the new total.

STEP 4. Enter the amount of the check.

STEP 5. *Enter* any other sums to be deducted, for example, if your bank charges you 25 cents for each check you write, write "25" in the cents column.

STEP 6. *Subtract* 4 and 5 from 3 (total – this check and other deductions) to get the new balance forward.

This process may sound a little complicated. Really, though, it's just a matter of adding any monies you've deposited to your account and subtracting any monies you are spending.

| WORK THIS PROBLEM |

The Question:

Complete the sequential check stubs on the following page:

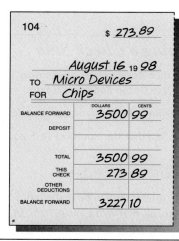

✔ YOUR WORK

The Solution:

Note that the new balance at the bottom of stub No. 104 is transferred as the beginning balance in stub No. 105.

Now that you've seen how easy it is to fill out a stub, see if you can fill out a check and a stub in the following situation.

WORK THIS PROBLEM

The Question:

On August 19, 1998, Bits 'n' Bytes had a balance of $2,874.85. On August 19, 1998, treasurer Karen M. Paisley made a deposit of $512.52 and wrote a check for $792.52 to Hardware Connection for printers. Complete the following check and check stub for these transactions.

✔ **YOUR WORK**

The Solution:

Using a Check Register

Check Register A record-keeping device separate from the checks but generally kept with the checks.

The second method of recording checks is in a **check register.** A check register has entries for the check number, date, to whom the check is issued, amount of check, deposits, and balance.

Notice that each deposit or check is recorded on a separate line. Each deposit is added to the previous balance to obtain the new balance. Each check amount is subtracted.

The √ column is used to record processed checks in a bank reconciliation. Bank reconciliation is discussed in Section 5.2.

RECORD ALL CHARGES OR CREDITS THAT AFFECT YOUR ACCOUNT

NUMBER	DATE	DESCRIPTION OF TRANSACTION	PAYMENT/DEBIT (−)	√ T	(IF ANY) (−) FEE	DEPOSIT/CREDIT (+)	BALANCE $ 0	
	8/3	Initial Deposit	$		$	$ 1322 50	1322	50
101	8/5	Microsystems For: Disk Drive	547 82				774	68

YOUR TURN

┌WORK THIS PROBLEM┐

The Question:

Complete the following check register by calculating the balance for each line.

RECORD ALL CHARGES OR CREDITS THAT AFFECT YOUR ACCOUNT

NUMBER	DATE	DESCRIPTION OF TRANSACTION	PAYMENT/DEBIT (−)	√ T	(IF ANY) (−) FEE	DEPOSIT/CREDIT (+)	BALANCE $ 0	
	8/3	Initial Deposit	$		$	$ 1322 50	1322	50
101	8/5	Microsystems Disk Drive	547 82				774	68
102	8/9	Paper Your Way Computer Paper	237 48					
	8/10	Deposit				3614 21		
103	8/11	Max's Machines Central Processing Units	732 92					
	8/11	Deposit				82 50		
104	8/16	Micro Devices Chips	273 89					
105	8/17	Maples Electronics Power Supplies	352 25					
	8/19	Deposit				512 52		
106	8/19	Hardware Connection Printers	792 52					

✔ YOUR WORK

The Solution:

RECORD ALL CHARGES OR CREDITS THAT AFFECT YOUR ACCOUNT

NUMBER	DATE	DESCRIPTION OF TRANSACTION	PAYMENT/DEBIT (−)		√ T	(IF ANY) (−) FEE	DEPOSIT/CREDIT (+)		BALANCE $ 0	
	8/3	Initial Deposit	$			$	$ 1322	50	1322	50
101	8/5	Microsystems Disk Drive	547	82					774	68
102	8/9	Paper Your Way Computer Paper	237	48					537	20
	8/10	Deposit					3614	21	4151	41
103	8/11	Max's Machines Central Processing Units	732	92					3418	49
	8/11	Deposit					82	50	3500	99
104	8/16	Micro Devices Chips	273	89					3227	10
105	8/17	Maples Electronics Power Supplies	352	25					2874	85
	8/19	Deposit					512	52	3387	37
106	8/19	Hardware Connection Printers	792	52					2594	85

WORKSPACE

SECTION TEST 5.1   Bank Records

The following problems test your understanding of Section 5.1, Setting Up and Using a Checking Account.

1. Bits 'n' Bytes' deposit on September 3, 1998 included $472 in currency, $27.83 in coin, and checks for $473.95, $236.27, and $12.55. Fill out the deposit ticket using the account number and company address from this section.

DEPOSIT TICKET

ACCOUNT NO. _____

NAME _____

ADDRESS _____

DATE _____ 19 ____

SIGN HERE FOR CASH RECEIVED (IF REQUIRED)
"DEPOSITS MAY NOT BE AVAILABLE FOR IMMEDIATE WITHDRAWAL."

CASH	CURRENCY		
	COIN		
LIST CHECKS SINGLY			
TOTAL FROM OTHER SIDE			
TOTAL			
LESS CASH RECEIVED			
NET DEPOSIT			

39-81
1030

USE OTHER SIDE FOR
ADDITIONAL LISTING

BE SURE EACH ITEM IS
PROPERLY ENDORSED

Friendly Bank

I-240 & SOUTH PENN • I-240 & SOUTH WALKER
PO BOX 19208 • OKLAHOMA CITY, OK 73144 • 405-681-5221
MEMBER FDIC

5/90 RPJ CHECKS AND OTHER ITEMS ARE RECEIVED FOR DEPOSIT SUBJECT TO THE PROVISIONS OF THE UNIFORM COMMERCIAL CODE OR ANY APPLICABLE COLLECTION AGREEMENT.

2. Bits 'n' Bytes' deposit on September 15, 1998 included $542 in currency, $37.87 in coin, and checks for $935.75, $28.12, and $563.29. Fill out the deposit ticket using the account number and company address from this section.

DEPOSIT TICKET

ACCOUNT NO. _____

NAME _____

ADDRESS _____

DATE _____ 19 ____

SIGN HERE FOR CASH RECEIVED (IF REQUIRED)
"DEPOSITS MAY NOT BE AVAILABLE FOR IMMEDIATE WITHDRAWAL."

CASH	CURRENCY		
	COIN		
LIST CHECKS SINGLY			
TOTAL FROM OTHER SIDE			
TOTAL			
LESS CASH RECEIVED			
NET DEPOSIT			

39-81
1030

USE OTHER SIDE FOR
ADDITIONAL LISTING

BE SURE EACH ITEM IS
PROPERLY ENDORSED

Friendly Bank

I-240 & SOUTH PENN • I-240 & SOUTH WALKER
PO BOX 19208 • OKLAHOMA CITY, OK 73144 • 405-681-5221
MEMBER FDIC

5/90 RPJ CHECKS AND OTHER ITEMS ARE RECEIVED FOR DEPOSIT SUBJECT TO THE PROVISIONS OF THE UNIFORM COMMERCIAL CODE OR ANY APPLICABLE COLLECTION AGREEMENT.

3. Complete the following two checks and check stubs from Bits 'n' Bytes. The "BALANCE FORWARD" is $2763.14. Check No. 152 is to Maples Electronics for surge suppressors in the amount of $96.82, dated September 12, 1998. Check No. 153 is to Under-Ware Electronics for monitors in the amount of $1267.95, dated September 14, 1998.

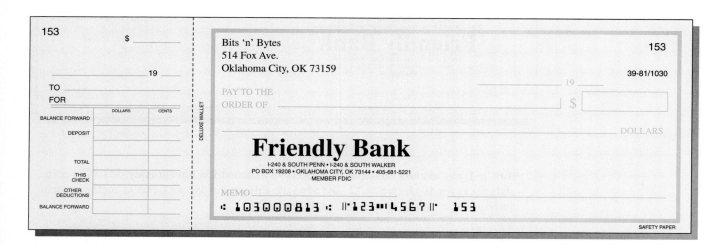

Check 152 stub:

152
$ _____

_____ 19 __

TO _____
FOR _____

	DOLLARS	CENTS
BALANCE FORWARD		
DEPOSIT		
TOTAL		
THIS CHECK		
OTHER DEDUCTIONS		
BALANCE FORWARD		

DELUXE WALLET

Check 152:

Bits 'n' Bytes
514 Fox Ave.
Oklahoma City, OK 73159

152

39-81/1030

_____ 19 ____

PAY TO THE
ORDER OF _____ $ _____

_____ DOLLARS

Friendly Bank

I-240 & SOUTH PENN • I-240 & SOUTH WALKER
PO BOX 19208 • OKLAHOMA CITY, OK 73144 • 405-681-5221
MEMBER FDIC

MEMO _____

⑈ 103000813 ⑈ ⑉123⑈⑈4567⑈⑉ 152

SAFETY PAPER

Check 153 stub:

153
$ _____

_____ 19 __

TO _____
FOR _____

	DOLLARS	CENTS
BALANCE FORWARD		
DEPOSIT		
TOTAL		
THIS CHECK		
OTHER DEDUCTIONS		
BALANCE FORWARD		

DELUXE WALLET

Check 153:

Bits 'n' Bytes
514 Fox Ave.
Oklahoma City, OK 73159

153

39-81/1030

_____ 19 ____

PAY TO THE
ORDER OF _____ $ _____

_____ DOLLARS

Friendly Bank

I-240 & SOUTH PENN • I-240 & SOUTH WALKER
PO BOX 19208 • OKLAHOMA CITY, OK 73144 • 405-681-5221
MEMBER FDIC

MEMO _____

⑈ 103000813 ⑈ ⑉123⑈⑈4567⑈⑉ 153

SAFETY PAPER

---

4. Complete the following two checks and check stubs from Bits 'n' Bytes. The "BALANCE FORWARD" is $4371.25. Check No. 156 is to Technologic Systems for printers in the amount of $2859.17, dated October 2, 1998. Check No. 157 is to Micro-Ware for buffers in the amount of $1355.34, dated October 4, 1998.

Check 156 stub:

156
$ _____

_____ 19 __

TO _____
FOR _____

	DOLLARS	CENTS
BALANCE FORWARD		
DEPOSIT		
TOTAL		
THIS CHECK		
OTHER DEDUCTIONS		
BALANCE FORWARD		

DELUXE WALLET

Check 156:

Bits 'n' Bytes
514 Fox Ave.
Oklahoma City, OK 73159

156

39-81/1030

_____ 19 ____

PAY TO THE
ORDER OF _____ $ _____

_____ DOLLARS

Friendly Bank

I-240 & SOUTH PENN • I-240 & SOUTH WALKER
PO BOX 19208 • OKLAHOMA CITY, OK 73144 • 405-681-5221
MEMBER FDIC

MEMO _____

⑈ 103000813 ⑈ ⑉123⑈⑈4567⑈⑉ 156

SAFETY PAPER

157			$ _____
			_____ 19
TO	_____		
FOR	_____		
		DOLLARS	CENTS
BALANCE FORWARD			
DEPOSIT			
TOTAL			
THIS CHECK			
OTHER DEDUCTIONS			
BALANCE FORWARD			

Bits 'n' Bytes
514 Fox Ave.
Oklahoma City, OK 73159

157

39-81/1030

_____ 19 _____

PAY TO THE
ORDER OF _____ $ [   ]

_____ DOLLARS

Friendly Bank

I-240 & SOUTH PENN • I-240 & SOUTH WALKER
PO BOX 19208 • OKLAHOMA CITY, OK 73144 • 405-681-5221
MEMBER FDIC

MEMO _____

⑆ 103000813 ⑆ ⑈ 123 ⑊ 4567 ⑈ 157

SAFETY PAPER

5. Complete the following check register by calculating the balance for each line.

RECORD ALL CHARGES OR CREDITS THAT AFFECT YOUR ACCOUNT

NUMBER	DATE	DESCRIPTION OF TRANSACTION	PAYMENT/DEBIT (−)		√ T	(IF ANY) (−) FEE	DEPOSIT/CREDIT (+)		BALANCE $ 2783 26	
836	11/2	Quang Van Tran Payroll	$ 1429	57		$	$			
837	11/5	P D Q Company Supplies	372	95						
838	11/7	Sparkle Cleaning Cleaning services	85	50						
	11/10	Deposit					2410	75		
839	11/10	Acme Widget Company Widgets	932	74						
840	11/15	Microsystems Computer equipment	1285	90						
841	11/16	DeTrop Surplus Misc.	207	28						
842	11/19	Bargin Universe Office supplies	152	12						
843	11/22	Instant Art Art work	78	32						
844	11/25	Surety Insurance Insurance	42	50						
845	11/25	Paper Cutters Computer Paper	217	21						
	11/30	Deposit					562	07		

191

6. Complete the following check register by calculating the balance for each line.

RECORD ALL CHARGES OR CREDITS THAT AFFECT YOUR ACCOUNT

NUMBER	DATE	DESCRIPTION OF TRANSACTION	PAYMENT/DEBIT (−)		√ T	(IF ANY) (−) FEE	DEPOSIT/CREDIT (+)		$	BALANCE 1926	53
1253	12/4	Radio Hut Computer equipment	$ 832	15		$	$				
1254	12/6	Acme Widget Co. Widgets	617	26							
1255	12/7	Paper Cutters Computer paper	451	50							
	12/8	Deposit					955	75			
1256	12/9	The Productive Office Office supplies	751	68							
1257	12/14	Hardware Connection RS 232 connecting	27	95							
1258	12/14	Microsystems Service	83	52							
	12/15	Deposit					992	56			
1259	12/18	Sparkle Cleaning Cleaning	842	97							
1260	12/18	Surety Insurance Insurance	12	57							
1261	12/20	De Trop Surplus Misc.	236	19							
	12/29	Deposit					827	35			

7. (a) Using the given form, write a check for $32.95 to Sally's Shoes, dated Sept. 19, 1998.

182

39-81/1030

19 _____

PAY TO THE
ORDER OF _____ $ [   ]

_____ DOLLARS

FIRST BANK

MEMO _____

⑆ 103000813 ⑆ ⑆123⑈4567 ⑈ 182

SAFETY PAPER

(b) Enter the balance in the check register after the check written in part (a). Also, record check number 183 for $57.14 to Milly's Grocery dated Sept. 19, 1998 and find the ending balance.

#	DATE	DESCRIPTION OF TRANSACTION	CHECK (−)	✓T	DEPOSIT (+)	BALANCE
						98 95
176	8/22	CP & L	35 27			63 68
	8/30	Deposit			827 46	891 14
177	8/31	Ace Finance	250 00			641 14
178	9/1	Mary Brown (rent)	200 00			441 14
179	9/9	Joe's Deli	82 90			358 24
180	9/10	First Bank	179 87			178 37
181	9/15	Telephone	35 01			143 36
182	9/19	Sally's Shoes				

(c) Suppose you deposit the following checks on 9/28/98:

$827.46 $93.07 $7.10

Complete the deposit slip below and then record this deposit in the check register of part (b). Next, record check number 184 for $12.29 to Short's Drugs dated 10/1/98. What is the final balance?

WRITE FIRMLY - PRESENT ALL COPIES OF DEPOSIT SLIP TO TELLER	CASH		
	CHECKS		
DATE_____ 19 _____			
	TOTAL		
→	LESS CASH REC'D		
SIGN FOR LESS CASH REC'D	NET DEPOSIT		

WORKSPACE

SECTION 5.2: Reconciling Checking Accounts

Can You:

Reconcile your check stubs or register with a bank statement?
Find and correct an error in balancing a checking account?
. . . If not, you need this section.

Canceled Checks Checks processed and paid out by a bank.

Statement A summary of all transactions on a bank account for a given period.

Balancing The process of reconciling the bank statement for an account with the account holder's check stubs or check register to make sure both agree.

Once a month, the bank sends back to you all your **canceled checks**—the checks that it has processed on your account that month. An account summary **(statement)** sent with the canceled checks lists all checks processed, deposits recorded, and any service fees. Some banks do not return canceled checks, but send only a summary statement. This enables them to charge a reduced service fee. The checks are stored and available for a small fee.

As soon as you receive this package, you should check the summary against your check stubs or check register. This procedure is called reconciling (or more commonly **balancing**) your account. A reconciliation form is usually included on the back of the summary. The reconciliation procedure is easy; just follow the steps included on the form.

Steps: Reconciling a Checking Account

STEP 1. In the check register, *check off* all checks shown as processed and deposits shown as received on the bank statement. Also enter in your register any monthly service fees shown on the statement and note the new balance.

STEP 2. *Note* on the reconciliation form any checks or deposits not checked off in your register. Then add the unlisted deposits to and subtract the unlisted checks from the balance shown *on the bank statement.*

STEP 3. *Compare* the result of step 2 with the current balance in your check register. If they do not agree, something—your math, your records, or, in *very* rare cases, the bank's records—is wrong.

Consider, for example, the monthly bank statement for Bits 'n' Bytes that follows. Using the bank statement, Bits 'n' Bytes' checkbook register, and the reconciliation form, let's balance the account.

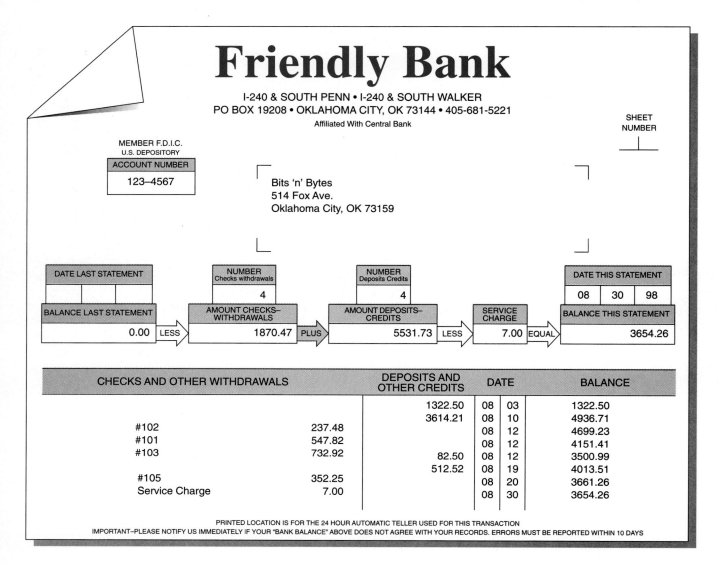

Friendly Bank

I-240 & SOUTH PENN • I-240 & SOUTH WALKER
PO BOX 19208 • OKLAHOMA CITY, OK 73144 • 405-681-5221

Affiliated With Central Bank

SHEET
NUMBER

MEMBER F.D.I.C.
U.S. DEPOSITORY

ACCOUNT NUMBER

123–4567

Bits 'n' Bytes
514 Fox Ave.
Oklahoma City, OK 73159

DATE LAST STATEMENT	NUMBER Checks withdrawals	NUMBER Deposits Credits	DATE THIS STATEMENT		
	4	4	08	30	98

BALANCE LAST STATEMENT	AMOUNT CHECKS– WITHDRAWALS	AMOUNT DEPOSITS– CREDITS	SERVICE CHARGE	BALANCE THIS STATEMENT
0.00 LESS	1870.47 PLUS	5531.73 LESS	7.00 EQUAL	3654.26

CHECKS AND OTHER WITHDRAWALS		DEPOSITS AND OTHER CREDITS	DATE		BALANCE
		1322.50	08	03	1322.50
		3614.21	08	10	4936.71
#102	237.48		08	12	4699.23
#101	547.82		08	12	4151.41
#103	732.92	82.50	08	12	3500.99
		512.52	08	19	4013.51
#105	352.25		08	20	3661.26
Service Charge	7.00		08	30	3654.26

PRINTED LOCATION IS FOR THE 24 HOUR AUTOMATIC TELLER USED FOR THIS TRANSACTION
IMPORTANT–PLEASE NOTIFY US IMMEDIATELY IF YOUR "BANK BALANCE" ABOVE DOES NOT AGREE WITH YOUR RECORDS. ERRORS MUST BE REPORTED WITHIN 10 DAYS

Step 1. In the check register, place a checkmark in the "√" column beside each item that was processed, according to the bank statement and enter any service charges to the register. Only checks numbered 104 and 106 and the deposit made 8/30 do not appear on the statement. Bits 'n' Bytes had a service charge of $7.00.

RECORD ALL CHARGES OR CREDITS THAT AFFECT YOUR ACCOUNT

NUMBER	DATE	DESCRIPTION OF TRANSACTION	PAYMENT/DEBIT (−)		√ T	(IF ANY) (−) FEE	DEPOSIT/CREDIT (+)		BALANCE $ 0	
	8/3	Initial Deposit	$		√	$	$ 1322	50	1322	50
101	8/5	Microsystems Disk Drive	547	82	√				774	68
102	8/9	Paper Your Way Computer Paper	237	48	√				537	20
	8/10	Deposit			√		3614	21	4151	41
103	8/11	Max's Machines Central Processing Units	732	92	√				3418	49
	8/11	Deposit			√		82	50	3500	99
104	8/16	Micro Devices Chips	273	89					3227	10
105	8/17	Maples Electronics Power Supplies	352	25	√				2874	85
	8/19	Deposit			√		512	52	3387	37
106	8/19	Hardware Connection Printers	792	52					2594	85
	8/30	Deposit					391	18	2986	03
		Service Charge from Bank Statement	7	00	√				2979	03

Outstanding Checks
Any checks written by an account holder but not yet processed by the bank.

Step 2. On the reconciliation form, enter the balance shown by the bank on the top line to the right. Note the unaccounted for deposit of 8/30 on the second line down on the right. List the amounts of the **outstanding checks**—those not processed (numbers 104 and 106) on the left. Perform the necessary additions and subtractions.

OUTSTANDING CHECKS WRITTEN TO WHOM	$ AMOUNT
104 Micro Devices	273.89
106 Hardware Connection	792.52
TOTAL OF CHECKS OUTSTANDING $	1066.41

BANK BALANCE (Last amount shown on this statement) $ 3654.26

ADD + Deposits not shown on this statement (if any) $ 391.18

TOTAL $ 4045.44

◄──── SUBTRACT ────► $ 1066.41

BALANCE $ 2979.03

Step 3. The amount in the last line on the right-hand side of the reconciliation form agrees with the balance in the register, so the account balances.

YOUR TURN

WORK THIS PROBLEM

The Question:

The September bank statement for Bits 'n' Bytes shows that checks numbered 104 and 106 and the deposit from 8/30 were processed in September. Checks numbered 107, 108, and 110 and the deposit on 9/5 were also processed. There was a service charge of $6.00, and the bank balance was $5995.33. Complete the check register and the bank reconciliation form.

RECORD ALL CHARGES OR CREDITS THAT AFFECT YOUR ACCOUNT

NUMBER	DATE	DESCRIPTION OF TRANSACTION	PAYMENT/DEBIT (–)		√ T	(IF ANY) (–) FEE	DEPOSIT/CREDIT (+)		BALANCE $ 2979 03	
107	9/2	Michael Brown Payroll	$1682	93		$	$			
108	9/2	Micro Devices Chips	172	85						
	9/5	Deposit					5293	27		
109	9/15	Guttenburg & Sons Printing	935	22						
110	9/20	Microsystems Hard drives	415	19						
	9/30	Deposit					1274	78		

OUTSTANDING CHECKS WRITTEN TO WHOM	$ AMOUNT		
		BANK BALANCE (Last amount shown on this statement)	$ _____
		ADD + Deposits not shown on this statement (if any)	$ _____
		TOTAL	$ _____
TOTAL OF CHECKS OUTSTANDING $		← SUBTRACT →	$ _____
		BALANCE	$ _____

✔ YOUR WORK

The Solution:

RECORD ALL CHARGES OR CREDITS THAT AFFECT YOUR ACCOUNT

NUMBER	DATE	DESCRIPTION OF TRANSACTION	PAYMENT/DEBIT (−)		√ T	(IF ANY) (−) FEE	DEPOSIT/CREDIT (+)		BALANCE $	
									2979	03
107	9/2	Michael Brown Payroll	$1682	93	√	$	$		1296	10
108	9/2	Micro Devices Chips	172	85	√				1123	25
	9/5	Deposit			√		5293	27	6416	52
109	9/15	Guttenburg & Sons Printing	935	22					5481	30
110	9/20	Microsystems Hard drives	415	19	√				5066	11
	9/30	Deposit					1274	78	6340	89
		Service charge	6	00	√				6334	89

OUTSTANDING CHECKS WRITTEN TO WHOM	$ AMOUNT
#109 Guttenburg & Sons	935.22
TOTAL OF CHECKS OUTSTANDING $	935.22

BANK BALANCE $ _5995.33_
(Last amount shown on this statement)

ADD +
Deposits not shown on this statement (if any) $ _1274.78_

TOTAL $ _7270.11_

← SUBTRACT → $ _935.22_

BALANCE $ _6334.89_

The balance at the bottom of the reconciliation form, $6334.89, is the same as the checkbook balance (after you subtracted the service charge). It balances.

Checking Your Reconciliation

Sometimes, your first attempt to balance your statement with your check register may not work. What do you do when your reconciliation statement does not balance? Follow these simple steps:

Steps: Checking Your Reconciliation

STEP 1.   *Check* your arithmetic on the reconciliation form.

STEP 2.   *Check* your check register or check stubs for an arithmetic error.

STEP 3.   *Make sure* that each check is recorded and that the amount is correct.

STEP 4.   *Check* that each deposit is recorded and that the amount is correct.

STEP 5.   *Check* that each service charge is recorded and that the amount is accurate.

STEP 6.   *Make sure* you haven't forgotten any outstanding checks from a previous month.

STEP 7.   If your last month's statement balanced and you have checked the steps 1–6, *check* with the bank to see if there was a bank error. Few errors are made by the bank.

Section Test 5.2 contains additional examples to test your skills.

SECTION TEST 5.2 Bank Records

The following problems test your understanding of Section 5.2, Reconciling Checking Accounts.

1. The bank statement for Wired for Sound shows checks numbered 512, 513, 514, 515, 517, 519, and 520 and the deposits on 2/7 and 2/13. There was a service charge of $6.75, and the bank balance was $623.12. Complete the check register (don't forget to enter the service charge), and complete the bank reconciliation.

RECORD ALL CHARGES OR CREDITS THAT AFFECT YOUR ACCOUNT

NUMBER	DATE	DESCRIPTION OF TRANSACTION	PAYMENT/DEBIT (−)		√ T	(IF ANY) (−) FEE	DEPOSIT/CREDIT (+)		BALANCE	
									$ 952	06
512	2/6	Wreck-A-Mended Garage Truck repair	$ 517	89		$	$			
513	2/6	Copy! Copy! Copy! Copy service	425	62						
	2/7	Deposit					531	50		
514	2/8	Midwest Labs Alignment	322	36						
515	2/12	Aladdin Lock & Key Alarm system	208	95						
	2/13	Deposit					1232	22		
516	2/15	Surety Insurance Insurance	57	91						
517	2/20	Maples Electronics Cabinets	128	85						
518	2/20	Bargin Universe Supplies	562	80						
519	2/22	Microsystems Support service	419	74						
520	2/24	Sparkle Cleaning Cleaning	62	50						
	2/27	Deposit					1012	25		

OUTSTANDING CHECKS WRITTEN TO WHOM	$ AMOUNT
TOTAL OF CHECKS OUTSTANDING $	

BANK BALANCE
(Last amount shown on this statement) $ _____

ADD +
Deposits not shown on this statement (if any) $ _____

TOTAL $ _____

← — — — SUBTRACT — — — → $ _____

BALANCE $ _____

2. The bank statement for Kate's Delicacies shows checks numbered 712, 713, 714, 715, 716, 718, 720, and 721 and the deposit on 6/5. There was a service charge of $7.25, and the bank balance was $1176.52. Complete the check register (don't forget to enter the service charge), and complete the bank reconciliation.

RECORD ALL CHARGES OR CREDITS THAT AFFECT YOUR ACCOUNT

NUMBER	DATE	DESCRIPTION OF TRANSACTION	PAYMENT/DEBIT (−)		√ T	(IF ANY) (−) FEE	DEPOSIT/CREDIT (+)		BALANCE $ 1452 25	
712	6/2	Sparkle Cleaning Service Yearly cleaning contract	$ 872	32		$	$			
713	6/4	Cool Air Air conditioner repair	516	97						
	6/5	Deposit					2310	52		
714	6/6	Paper Cutters Envelopes	123	65						
715	6/12	Zippy Delivery Service Special delivery	29	98						
716	6/12	Master Systems Monitor	455	74						
717	6/15	Ulysses Travel Agency Conference travel	842	17						
718	6/20	Petal Pushers Secretary's day	8	21						
719	6/20	On Alert Security guard	235	83						
720	6/22	Panes Unlimited Front window replacement	137	46						
721	6/23	J. R.'s Remodeling Front entry way	434	67						
	6/28	Deposit					650	25		

OUTSTANDING CHECKS WRITTEN TO WHOM	$ AMOUNT
TOTAL OF CHECKS OUTSTANDING $	

BANK BALANCE (Last amount shown on this statement) $ _____

ADD + Deposits not shown on this statement (if any) $ _____

TOTAL $ _____

← SUBTRACT → $ _____

BALANCE $ _____

3. Use the checkbook register in problem 5 for Section 5.1 to complete the following bank reconciliation. The bank statement shows checks numbered 836, 837, 838, 839, 840, 842, 844, and 845 and the deposit on 11/10. There was a service charge of $7.50, and the bank balance was $668.02. Enter the service charge in the check register, and complete the bank reconciliation.

OUTSTANDING CHECKS WRITTEN TO WHOM	$ AMOUNT
TOTAL OF CHECKS OUTSTANDING $	

BANK BALANCE (Last amount shown on this statement) $ _____

ADD + Deposits not shown on this statement (if any) $ _____

TOTAL $ _____

◄——————— SUBTRACT ———————► $ _____

BALANCE $ _____

4. Use the checkbook register in problem 6 for Section 5.1 to complete the following bank reconciliation. The bank statement shows checks numbered 1253, 1254, 1255, 1256, 1259, and 1261 and the deposits on 12/8 and 12/15. There was a service charge of $8.25, and the bank balance was $134.84. Enter the service charge in the check register, and complete the bank reconciliation.

OUTSTANDING CHECKS WRITTEN TO WHOM	$ AMOUNT
TOTAL OF CHECKS OUTSTANDING $	

BANK BALANCE (Last amount shown on this statement) $ _____

ADD + Deposits not shown on this statement (if any) $ _____

TOTAL $ _____

◄——————— SUBTRACT ———————► $ _____

BALANCE $ _____

WORKSPACE

Key Terms

balance	deposit
balancing	outstanding checks
canceled checks	payee
check	service charge
check register	signature card
check stub	statement

Chapter 5 at a Glance

Page	Topic	Key Point	Example
179	Making a deposit	Total the cash and checks deposited.	See the illustration on page 179.
180	Writing a check	Be sure the amount in numbers agrees with the amount in letters.	See the illustration on page 181.
182	Completing a check stub	Find the new balance forward.	See the illustration on page 183.
185	Using a check register	Find the new balance.	See the illustration on page 186.
195	Reconciling a bank account	1. Check off in register all processed checks and deposits on statement. 2. Subtract service fees from register balance. 3. Add unlisted deposits to and subtract outstanding checks from the statement balance. Register and statement should now agree.	1. See the illustration on page 197. 2. See the illustration on page 197. 3. See the illustration on page 197.

WORKSPACE

Name

Date

Course/Section

The following problems test your understanding of bank records.

1. Bits 'n' Bytes' deposit on December 2, 1998 included currency of $519, coin of $32.64, and checks for $836.93, $284.72, and $89.25. Fill out the deposit ticket using the account number and company address from this chapter.

DEPOSIT TICKET

ACCOUNT NO. _____

NAME _____

ADDRESS _____

DATE _____ 19 _____

SIGN HERE FOR CASH RECEIVED (IF REQUIRED)
"DEPOSITS MAY NOT BE AVAILABLE FOR IMMEDIATE WITHDRAWAL."

Friendly Bank

I-240 & SOUTH PENN • I-240 & SOUTH WALKER
PO BOX 19208 • OKLAHOMA CITY, OK 73144 • 405-681-5221
MEMBER FDIC

CASH	CURRENCY		
	COIN		
LIST CHECKS SINGLY			
TOTAL FROM OTHER SIDE			
TOTAL			
LESS CASH RECEIVED			
NET DEPOSIT			

39-81
1030

USE OTHER SIDE FOR
ADDITIONAL LISTING

BE SURE EACH ITEM IS
PROPERLY ENDORSED

5/90 **RPJ** CHECKS AND OTHER ITEMS ARE RECEIVED FOR DEPOSIT SUBJECT TO THE PROVISIONS OF THE UNIFORM COMMERCIAL CODE OR ANY APPLICABLE COLLECTION AGREEMENT.

2. Bits 'n' Bytes' deposit on December 12, 1998 included currency of $693, coin of $43.28, and checks for $1283.32, $384.56, and $73.35. Fill out the deposit ticket using the account number and company address from this chapter.

DEPOSIT TICKET

ACCOUNT NO. _____

NAME _____

ADDRESS _____

DATE _____ 19 _____

SIGN HERE FOR CASH RECEIVED (IF REQUIRED)
"DEPOSITS MAY NOT BE AVAILABLE FOR IMMEDIATE WITHDRAWAL."

Friendly Bank

I-240 & SOUTH PENN • I-240 & SOUTH WALKER
PO BOX 19208 • OKLAHOMA CITY, OK 73144 • 405-681-5221
MEMBER FDIC

CASH	CURRENCY		
	COIN		
LIST CHECKS SINGLY			
TOTAL FROM OTHER SIDE			
TOTAL			
LESS CASH RECEIVED			
NET DEPOSIT			

39-81
1030

USE OTHER SIDE FOR
ADDITIONAL LISTING

BE SURE EACH ITEM IS
PROPERLY ENDORSED

5/90**RPJ** CHECKS AND OTHER ITEMS ARE RECEIVED FOR DEPOSIT SUBJECT TO THE PROVISIONS OF THE UNIFORM COMMERCIAL CODE OR ANY APPLICABLE COLLECTION AGREEMENT.

3. Complete the following check and check stub for Bits 'n' Bytes. The "BALANCE FORWARD" is $1236.24. Check No. 183 is to Top Brand for paper in the amount of $265.74, dated December 16, 1998.

4. Complete the following check and check stub for Bits 'n' Bytes. The "BALANCE FORWARD" is $836.38. Check 185 is to Office Products for office supplies in the amount of $345.81, dated December 21,1998.

5. The bank statement for Big Ben's Bolts shows checks numbered 592, 593, 594, 595, 596, 598, 599, and 601 and the deposit on 2/8 were processed. There was a service charge of $5.75, and the bank balance was $300.91. Complete the check register (don't forget to enter the service charge), and complete the bank reconciliation.

RECORD ALL CHARGES OR CREDITS THAT AFFECT YOUR ACCOUNT

NUMBER	DATE	DESCRIPTION OF TRANSACTION	PAYMENT/DEBIT (−)		√ T	(IF ANY) (−) FEE	DEPOSIT/CREDIT (+)		BALANCE $ 650 12	
592	2/2	Bargin Universe Office supplies	$ 89	27		$	$			
593	2/4	Under-Ware Electronics Connectors	42	95						
594	2/7	Paper Cutters Computer paper	473	18						
	2/8	Deposit					1650	25		
595	2/10	Spiffy Insurance Co. Insurance	119	55						
596	2/10	Instant Art Company Artwork	238	21						
597	2/19	J. R.'s Remodeling Front office	251	79						
598	2/20	Microsystems Computer equipment	892	13						
599	2/25	Biff's Trash Service Trash pickup	18	57						
600	2/25	Kate's Delicacies Breakfast catering	23	95						
601	2/26	Sparkle Cleaning Co. Cleaning	119	85						
	2/28	Deposit					532	24		

OUTSTANDING CHECKS WRITTEN TO WHOM	$ AMOUNT			
		BANK BALANCE (Last amount shown on this statement)	$ _____	
		ADD + Deposits not shown on this statement (if any)	$ _____	
		TOTAL	$ _____	
TOTAL OF CHECKS OUTSTANDING $		← SUBTRACT →	$ _____	
		BALANCE	$ _____	

6. The following bank statement for J.R.'s Construction Company shows checks numbered 672, 673, 675, 676, 677, 678, and 680 and the deposits on 5/4 and 5/16 were processed. There was a service charge of $6.15, and the bank balance was $523.06. Complete the check register (don't forget to enter the service charge), and complete the bank reconciliation.

RECORD ALL CHARGES OR CREDITS THAT AFFECT YOUR ACCOUNT

NUMBER	DATE	DESCRIPTION OF TRANSACTION	PAYMENT/DEBIT (−)		√ T	(IF ANY) (−) FEE	DEPOSIT/CREDIT (+)		BALANCE $ 1275 51	
672	5/2	Office Products Supplies	$ 582	19		$	$			
673	5/4	Microsystems Printer	655	25						
	5/4	Deposit					751	95		
674	5/7	Preferred Insurance Co. Truck insur.	213	19						
675	5/12	Air-Care Company Air conditioning service	175	25						
676	5/15	Maples Electronics Chips	325	72						
	5/16	Deposit					625	42		
677	5/18	Acme Widget Company Widgets	25	72						
678	5/20	Sparkle Cleaning Co. Cleaning	39	95						
679	5/20	Hottaire Inc. Gas	245	21						
680	5/21	Learned Books Reference books	319	59						
	5/28	Deposit					617	49		

OUTSTANDING CHECKS WRITTEN TO WHOM	$ AMOUNT
TOTAL OF CHECKS OUTSTANDING $	

BANK BALANCE (Last amount shown on this statement) $ _____

ADD + Deposits not shown on this statement (if any) $ _____

TOTAL $ _____

←———— SUBTRACT ————→ $ _____

BALANCE $ _____

Payroll

I'm going to pay you with cash, so I don't have to hassle with all that paperwork!

CHAPTER OBJECTIVES

When you complete this chapter successfully, you will be able to:

1. Calculate salary and hourly wages.

2. Use salary, hourly wages, piecework, differential piecework, straight commission, sliding scale commission, and salary plus commission to calculate payroll.

■ Tom earns $6.57 per hour painting fences. What is his gross pay for a week in which he worked $38\frac{1}{4}$ hours?

■ Eleanora earns a 5% commission on her weekly sales of dance equipment. What was her commission for a week in which she sold $5958.76 in merchandise?

When you hear people talk about the resources that go into producing something, you probably think about raw materials like wire and wood, plastic, and peanuts. But no business can operate without another type of resource: people. Just as businesses must pay for the other resources they use, so they must pay for the human resources they use. (It's a sad fact of business life that unpaid employees seldom stick around.)

In this chapter, you will learn about a variety of ways employers have developed for paying their workers. A little concentration, you'll find, will really pay off here.

> *Can You:*
>
> Compute the salary per pay period?
> Calculate the regular wages of hourly employees?
> Compute overtime pay?
> . . . If not, you need this section.

Computing Salary

Salary Payment to individuals for their labor based on a fixed amount regardless of the hours worked.

The most common method of paying white-collar and management personnel is by salary. **Salary** is a fixed amount of money paid to an employee for certain assigned duties. The number of hours worked or the productivity of the employee does not affect the salary paid. Of course, they *do* affect the salaried employees' chances of keeping their jobs, being promoted, and getting raises. (Paying an employee with a combination of salary and commission is covered in Section 6.2 of this chapter.)

Salaries are often stated as an amount per year. Few people are paid only once a year, however. Most annual salaries are converted to more frequent pay periods. *Common pay periods* are

Weekly	52 times per year
Biweekly	Every other week—26 times per year
Semimonthly	Twice a month—24 times per year
Monthly	12 times per year

For example, if Keith's annual salary is $22,365 per year, what is his monthly salary? To find Keith's monthly salary, simply divide the annual salary by the number of months in a year: 12.

$$\frac{\$22,365}{12} = \$1863.75$$

WORK THIS PROBLEM

The Question:

Judy's annual salary is $24,960. Compute her salary payments if she is paid

(a) weekly   (b) biweekly   (c) semimonthly   (d) monthly

✔ YOUR WORK

The Solution:

(a) Weekly: $\dfrac{\$24{,}960}{52} = \$480$

(b) Biweekly: $\dfrac{\$24{,}960}{26} = \$960$

(c) Semimonthly: $\dfrac{\$24{,}960}{24} = \$1040$

(d) Monthly: $\dfrac{\$24{,}960}{12} = \$2080$

Calculating Hourly Pay

Hourly Wages Payment to individuals for their labor based on a rate per hour worked.

Gross Pay Pay before taxes.

The most common method of paying wages is based on the number of hours worked. Straight **hourly wages** are very easy to calculate. The **gross pay** is the number of hours worked times the pay rate per hour.

Gross pay = hours worked × rate per hour

For example, if Patti worked 38 hours last week and her pay rate is $6.78 per hour, what is her gross pay?

Hours		Hourly rate

Gross pay = 38 × $6.78 = $257.64

YOUR TURN

WORK THIS PROBLEM

The Question:

Tom earns $6.57 per hour painting fences. What is his gross pay for a week in which he worked $38\frac{1}{4}$ hours?

✔ YOUR WORK

The Solution:

$$38\tfrac{1}{4} \times \$6.57 = 38.25 \times \$6.57 = \$251.3025 = \$251.30 \text{ rounded}$$

The easiest way to work this problem is to first change $38\frac{1}{4}$ to a decimal form, $38\frac{1}{4} = 38.25$, then multiply. If you need a review of changing fractions to decimal numbers, see Section 3.3.

Most businesses use a payroll sheet to compute gross pay. The payroll sheet for straight hourly wages will include the employee's name, Social Security number, number of hours worked per day, total hours for the week, rate per hour, gross pay, and total gross pay.

The following is a sample payroll sheet. We can complete this payroll sheet by calculating each employee's total hours per week, multiplying it by that worker's hourly wage to find the gross pay for each employee, and then totaling the gross pay column vertically.

┌─ multiply ─┐

Name	Hours					Total Hours	Rate per Hour	Gross Pay
	M	T	W	T	F			
Eyre, Jane	6	8	5	8	8	35	6.50	227.50
Marner, Silus	8	8	8	8	8	40	5.25	210.00
Sawyer, Thomas	7	7	6	7	7	34	6.84	232.56
Total								670.06

add

YOUR TURN

WORK THIS PROBLEM

The Question:

Complete the following payroll sheet.

Name	Hours					Total Hours	Rate per Hour	Gross Pay
	M	T	W	T	F			
Ewing, Patrick	7	8	6	7	8		6.00	
Joyner, Florence	8	8	8	8	7		6.29	
Louganis, Gregory	8	8	8	8	8		7.58	
Valenzuela, Fernando	8	6	8	8	7		7.18	
Witt, Katerina	6	8	7	7	8		5.94	
Total								

✔ YOUR WORK

The Solution:

Name	Hours					Total Hours	Rate per Hour	Gross Pay
	M	T	W	T	F			
Ewing, Patrick	7	8	6	7	8	36	6.00	216.00
Joyner, Florence	8	8	8	8	7	39	6.29	245.31
Louganis, Gregory	8	8	8	8	8	40	7.58	303.20
Valenzuela, Fernando	8	6	8	8	7	37	7.18	265.66
Witt, Katerina	6	8	7	7	8	36	5.94	213.84
Total								1244.01

Paying Overtime

By law, nonexempt employees must be paid at a higher rate per hour for overtime. Exempt employees are not covered and salaried employees are compensated on a different basis. For

Overtime In the United States, anything above 40 hours worked; individuals paid hourly wages must receive $1\frac{1}{2}$ times the regular rate for all these hours.

most hourly employees, **overtime** is the time worked over 40 hours per week. The usual overtime rate is:

$$\text{Overtime rate} = 1\tfrac{1}{2} \times \text{regular rate}$$

For example, if the regular rate is $5.58, the overtime rate would be:

$$1\tfrac{1}{2} \times \$5.58 = 1.5 \times \$5.58 = \$8.37$$

Note that you need to write $1\frac{1}{2}$ as 1.5 before you multiply.

For employees working over 40 hours per week, four steps are necessary to compute their gross pay.

Steps: Calculating Hourly Pay with Overtime

STEP 1. *Compute* the regular pay. (Note that the regular hours cannot exceed 40 for a week.)

$$\text{Regular pay} = \text{regular hours} \times \text{regular rate per hour}$$

STEP 2. *Compute* the overtime rate.

$$\text{Overtime rate} = 1\tfrac{1}{2} \times \text{regular rate}$$

STEP 3. *Compute* the overtime pay.

$$\text{Overtime pay} = \text{overtime hours} \times \text{overtime rate}$$

STEP 4. *Compute* the total wages or gross pay.

$$\text{Total wages} = \text{regular pay} + \text{overtime pay}$$

It's really quite easy. For example, Pauline earns $5.86 per hour. Last week she worked 46 hours. What were her regular pay, overtime rate, overtime pay, and total wages?

Step 1. Regular pay = $40 \times \$5.86 = \$234.40$

Step 2. Overtime rate = $1\tfrac{1}{2} \times \$5.86 = 1.5 \times \$5.86 = \$8.79$

Step 3. Overtime pay = $6 \times \$8.79 = \$52.74$

Step 4. Total wages = $\$234.40 + 52.74 = \$287.14$

YOUR TURN

| WORK THIS PROBLEM |

The Question:

Dave worked 45 hours last week. His regular pay rate is $5.88 per hour. What is his total wage?

| ✔ YOUR WORK |

The Solution:

$$\text{Regular pay} = 40 \times \$5.88 = \$235.20$$
$$\text{Overtime rate} = 1\tfrac{1}{2} \times \$5.88 = \$8.82$$
$$\text{Overtime rate} = 5 \times \$8.82 = \$44.10$$
$$\text{Total wage} = \$235.20 + 44.10 = \$279.30$$

When computing the overtime rate, you can end up with more than two decimal places. Carry all the decimal places and round only after you compute the overtime pay.

WORK THIS PROBLEM

The Question:

Mary worked 47 hours last week. Her regular pay rate is $5.19 per hour. What is her regular pay, overtime rate, overtime pay, and gross pay?

✔ YOUR WORK

The Solution:

$$\text{Regular pay} = 40 \times \$5.19 = \$207.60$$
$$\text{Overtime rate} = 1\tfrac{1}{2} \times \$5.19 = \$7.785 \longleftarrow \text{Don't round here.}$$
$$\text{Overtime pay} = 7 \times \$7.785 = \$54.495 = \$54.50 \quad \text{Rounded.}$$
$$\text{Gross pay} = \$207.60 + 54.50 = \$262.10$$

Of course, few businesspeople (if any) solve such problems by hand, preferring to use a calculator.

Finding Overtime with a Calculator

A calculator can be very handy when solving overtime pay problems with extra decimals. Begin by finding the regular pay. For the problem in the example, press

40 $\boxed{\times}$ 5.19 $\boxed{=}$ $\longrightarrow$ 〔 207.6 〕

If your calculator can store a number in memory, store this answer. If not, jot it down on a piece of paper.

Next, find the overtime pay by multiplying 1.5 by the regular rate by the number of overtime hours. Then add in the regular pay, either by hand or from the computer memory. The following shows how to complete this process:

1.5 $\boxed{\times}$ 5.19 $\boxed{\times}$ 7 $\boxed{=}$ $\longrightarrow$ 〔 54.495 〕 $\boxed{+}$ 207.6 $\boxed{=}$ $\longrightarrow$ 〔 262.095 〕

Rounding this answer gives the same $262.10.

We can easily incorporate the overtime calculations in the payroll sheet by adding columns for overtime hours, overtime rate, overtime pay, and gross pay.

WORK THIS PROBLEM

The Question:

Complete the following payroll sheet. Each employee's calculations should be made as in the previous problem. Add columns to find the totals for regular pay, overtime pay, and gross pay.

Name	Hours M	T	W	T	F	Total Hours	Regular Hours	Regular Rate	O.T. Hours	O.T. Rate	Regular Pay	O.T. Pay	Gross Pay
Lauper, Cynthia	8	9	7	9	9			$6.56					
O'Connor, Sinead	10	8	8	9	9			5.98					
Palmer, Robert	10	9	10	9	8			7.79					
Smith, William	8	8	7	8	8			6.42					
Van Halen, Edward	8	9	8	10	8			6.16					
Totals													

✔ **YOUR WORK**

The Solution:

Name	Hours M	T	W	T	F	Total Hours	Regular Hours	Regular Rate	O.T. Hours	O.T. Rate	Regular Pay	O.T. Pay	Gross Pay
Lauper, Cynthia	8	9	7	9	9	42	40	$6.56	2	$ 9.84	$262.40	$ 19.68	$ 282.08
O'Connor, Sinead	10	8	8	9	9	44	40	5.98	4	8.97	239.20	35.88	275.08
Palmer, Robert	10	9	10	9	8	46	40	7.79	6	11.685	311.60	70.11	381.71
Smith, William	8	8	7	8	8	39	39	6.42	0	—	250.38	0	250.38
Van Halen, Edward	8	9	8	10	8	43	40	6.16	3	9.24	246.40	27.72	274.12
Totals											$1309.98	$153.39	$1463.37

Be careful with the third employee above. The overtime rate is $1\frac{1}{2} \times \$7.79 = \$11.685$. Do not round the overtime rate. Check your addition by adding the total regular pay and total overtime pay. This should equal the total gross pay: $1309.98 + 153.39 = 1463.37$.

Accuracy is essential! Whether you compute the payroll by hand or with a calculator, you should always check your answers. An incorrect answer will cost either the business or an employee money. It may possibly cost a job.

Now check your understanding of payroll calculations by completing the practice problems in Section Test 6.1.

WORKSPACE

SECTION TEST 6.1 Payroll

The following questions test your understanding of Section 6.1, Paying Salaried and Hourly Employees.

A. Convert the following annual salaries to the required pay period.

1. $22,380 = _____ monthly   2. $23,640 = _____ semimonthly

3. $31,800 = _____ monthly   4. $25,870 = _____ biweekly

5. $14,144 = _____ weekly   6. $29,640 = _____ semimonthly

7. $22,776 = _____ biweekly   8. $23,660 = _____ weekly

9. $41,100 = _____ monthly   10. $25,800 = _____ semimonthly

11. $23,270 = _____ biweekly 12. $34,580 = _____ weekly

B. Convert the following annual salaries to the various pay periods.

Annual	Monthly	Semimonthly	Biweekly	Weekly
1. $31,200				
2. 20,280				
3. 21,840				
4. 29,640				
5. 34,320				
6. 56,160				
7. 46,800				
8. 15,600				
9. 39,000				
10. 26,520				
11. 54,600				
12. 62,400				
13. 17,160				
14. 48,360				
15. 23,400				
16. 24,960				

C. Complete the following weekly payroll sheets.

1.

Name	Hours					Total Hours	Rate per Hour	Gross Pay
	M	T	W	T	F			
1. Bender, Keith	8	8	8	8	8		$ 8.12	
2. Chao, I-Na	8	7	6	5	4		7.55	
3. Greenwood, Robert	6	8	5	8	7		6.38	
4. Luliak, John	7	8	8	8	7		9.83	
5. Pitt, David	8	6	0	8	6		10.27	
6. Rhine, Alan	8	0	8	0	8		9.42	
7. Tanner, Christie	8	7	7	7	8		6.83	
8. Vasquez, Jose	8	6	5	6	0		7.98	
Total								

2.

Name	Hours					Total Hours	Regular Hours	Regular Rate	O.T. Hours	O.T. Rate	Regular Pay	O.T. Pay	Gross Pay
	M	T	W	T	F								
1. Czarnecki, Randall	8	9	9	10	9			$ 7.56					
2. Fazlian, Mohsen	9	9	9	9	9			8.24					
3. Hudson, Teresa	9	8	7	10	8			8.98					
4. Lawson, Dennis	9	9	10	9	7			9.48					
5. Martinez, Marie	10	10	10	10	10			10.52					
6. Ngvyen, Lam	9	8	7	6	5			9.46					
7. Querry, Frank	9	8	9	8	7			8.72					
8. Titus, Patrick	8	8	9	8	8			10.10					
Totals													

3.

Name	Hours					Total Hours	Regular Hours	Regular Rate	O.T. Hours	O.T. Rate	Regular Pay	O.T. Pay	Gross Pay
	M	T	W	T	F								
1. Cramer, Marvin	8	9	10	9	8			$ 9.27					
2. Etheredge, Hugh	8	8	9	9	9			9.45					
3. Lampron, George	9	8	7	6	5			8.52					
4. McCahe, Lowell	10	10	10	10	9			8.99					
5. Ngvyen, Tri	9	9	10	7	9			7.82					
6. Reed, Harold	9	8	9	8	9			7.95					
7. Tivis, Jerome	8	9	8	9	8			8.25					
8. Wood, Regena	9	8	7	9	10			9.77					
Totals													

WORKSPACE

SECTION 6.2: Determining Piecework and Commission Pay

Can You:

Calculate piecework?
Compute differential piecework pay?
Pay commissions?
Determine a sliding scale commission?
Calculate salary plus commission?
. . . If not, you need this section.

Calculating Piecework

Piecework   Payment to
individuals for their labor
based on a rate per item
produced.

Some businesses pay their employees based on actual production. One such method is by **piece-work,** where each employee's pay is based on the number of pieces completed during a shift or work period. The gross pay is the product of the number of pieces and the rate per piece.

> Gross pay = number of pieces × rate per piece

For example, Brad works for the Big Ben Bolt Company, making steel bolts. If he earns $0.23 per bolt and completed 237 in a day, what is his pay?

Gross pay = 237 × $0.23 = $54.51

Piecework calculations can be easily incorporated in a payroll sheet. This payroll sheet will include the employee's name, number of pieces completed per day, total pieces completed per week, rate per piece, and gross pay.

WORK THIS PROBLEM

The Question:

Complete the following payroll sheet by finding the total pieces and gross pay. Don't forget to total the gross pay column.

Name	Pieces Completed					Total Pieces	Rate	Gross Pay
	M	**T**	**W**	**T**	**F**			
Fillmore, Millie	40	38	35	42	41		$1.94	
Harding, Warren	35	33	38	40	36		2.15	
Polk, James	26	27	22	24	25		2.21	
Taylor, Zachary	56	56	57	58	55		1.54	
Van Buren, Marsha	96	87	85	89	88		1.35	
Total								

The Solution:

Name	Pieces Completed					Total Pieces	Rate	Gross Pay
	M	**T**	**W**	**T**	**F**			
Fillmore, Millie	40	38	35	42	41	196	$1.94	$ 380.24
Harding, Warren	35	33	38	40	36	182	2.15	391.30
Polk, James	26	27	22	24	25	124	2.21	274.04
Taylor, Zachary	56	56	57	58	55	282	1.54	434.28
Van Buren, Marsha	96	87	85	89	88	445	1.35	600.75
Total								$2080.61

Computing Differential Piecework

Differential Piecework
Payment to individuals for their labor based on a series of rates per piece produced that increases as the number of pieces produced increases.

Some businesses use **differential piecework** as an incentive to increase production. This method uses a scale in which the rate per piece increases as the number of pieces completed increases. For example, the Ali Box Works pays its employees for each box constructed according to the following schedule:

1–100 boxes	@	$0.29 each
101–150 boxes	@	$0.31 each
151 and up boxes	@	$0.34 each

If Ray constructed 162 boxes in a day, we can compute his gross pay as follows:

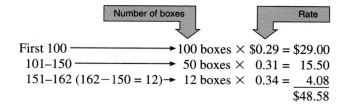

	Number of boxes		Rate	
First 100	→	100 boxes ×	$0.29 =	$29.00
101–150	→	50 boxes ×	0.31 =	15.50
151–162 (162−150 = 12)→		12 boxes ×	0.34 =	4.08
				$48.58

Note: Ray is not paid at the $0.34 rate for all 162 boxes, only for those he constructed after he has completed 150 boxes.

YOUR TURN

The Question:

Signs of the Times pays its employees for each medium-sized sign by the following daily schedule:

1–12 signs	@	$2.55 each
13–16 signs	@	2.65 each
17 and up signs	@	2.78 each

Calculate Mona's gross wages or total pay if she painted 18 signs in one day.

✔ YOUR WORK

The Solution:

$$
\begin{array}{llll}
\text{First 12 signs} & \longrightarrow & 12 \times \$2.55 = & \$30.60 \\
\text{Signs 13–16} & \longrightarrow & 4 \times 2.65 = & 10.60 \\
\text{Signs 17–18} & \longrightarrow & 2 \times 2.78 = & \underline{5.56} \\
\text{Gross wages} & \longrightarrow & & \$46.76
\end{array}
$$

In a differential piecework payroll sheet, the rate column is deleted, since the rate varies.

YOUR TURN

WORK THIS PROBLEM

The Question:

Steele's Cabinet Shop pays its employees for each cabinet made per week according to the following schedule:

$$
\begin{array}{lll}
\text{1–25 cabinets} & @ & \$12.45 \text{ each} \\
\text{26–30 cabinets} & @ & 12.67 \text{ each} \\
\text{31 and up cabinets} & @ & 12.98 \text{ each}
\end{array}
$$

Complete the following payroll sheet for Steele's Cabinet Shop.

Name	Cabinets per Day					Total Cabinets	Gross Wages
	M	**T**	**W**	**T**	**F**		
Baer, Max	6	6	5	6	5		
Douglas, Donna	6	5	5	5	4		
Ebsen, Buddy	6	7	7	6	7		
Ryan, Irene	4	4	5	5	4		
Total							

✔ YOUR WORK

The Solution:

Baer, Max
$$
\begin{array}{l}
25 \times \$12.45 = \$311.25 \\
3 \times 12.67 = \underline{38.01} \\
\$349.26
\end{array}
$$

Douglas, Donna
$$
25 \times \$12.45 = \$311.25
$$

Ebsen, Buddy
$$
\begin{array}{l}
25 \times \$12.45 = \$311.25 \\
5 \times 12.67 = 63.35 \\
3 \times 12.98 = \underline{38.94} \\
\$413.54
\end{array}
$$

Ryan, Irene
$$
22 \times \$12.45 = \$273.90
$$

Name	Cabinets per Day					Total Cabinets	Gross Wages
	M	**T**	**W**	**T**	**F**		
Baer, Max	6	6	5	6	5	28	$ 349.26
Douglas, Donna	6	5	5	5	4	25	311.25
Ebsen, Buddy	6	7	7	6	7	33	413.54
Ryan, Irene	4	4	5	5	4	22	273.90
Total							$1347.95

If you are having trouble with these arithmetic calculations, return to Chapter 3 to review the addition and multiplication of decimal numbers.

Paying Commission

Commission Payment to individuals for their labor based on a percentage of their total sales.

Straight Commission Payment to individuals for their labor based *solely* on a fixed percentage of their total sales.

Another procedure used for calculating wages is commission. A **commission** is usually paid to people selling merchandise. There are several different commission plans. The easiest to calculate is straight commission. In **straight commission** the commission is a percent of the total sales.

> Commission = rate of commission × sales

For example, Richard earns a 6% commission on his weekly sales. What was his commission for a week in which he sold $4847.65 in merchandise?

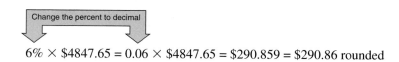
Change the percent to decimal

$$6\% \times \$4847.65 = 0.06 \times \$4847.65 = \$290.859 = \$290.86 \text{ rounded}$$

Remember that you must always change a percent to a decimal number before multiplying. If you are having difficulty changing 6% to a decimal, turn to Section 4.1.

YOUR TURN

| WORK THIS PROBLEM |

The Question:

Eleanora earns a 5% commission on her weekly sales of dance equipment. What was her commission for a week in which she sold $5958.76 in merchandise?

| ✔ YOUR WORK |

The Solution:

$$5\% \times \$5958.76 = 0.05 \times \$5958.76 = \$297.938 = \$297.94 \text{ rounded}$$

Determining Sliding Scale Commissions

Sliding Scale Commission
Payment to individuals for their labor based *solely* on a percent of their total sales, with the percent increasing as sales increase.

A second method of calculating commissions uses a sliding scale. Much like differential piecework, **sliding scale commission** rewards employees for increased production. The commission rate increases as sales increase. For example, the Wilton Sporting Goods Supply Company pays its employees on the following weekly schedule:

> 4% on sales up to $3000
> 5% on sales from over $3000 to $5000
> 6% on sales over $5000

What were Bo's gross wages for a week in which he had a sales total of $5473.86?

First $3000	4% × $3000 = 0.04 × $3000 =	$120.00
$3000 to $5000	5% × 2000 = 0.05 × 2000 =	100.00
Over $5000	6% × 473.86 = 0.06 × 473.86 =	28.43 rounded
		$248.43

If you have trouble changing fractional percents such as $7\frac{1}{2}\%$ and $8\frac{1}{4}\%$ to decimal numbers, review Section 4.1.

YOUR TURN

【WORK THIS PROBLEM】─────────────────────────

The Question:

Yucks Novelty Company pays its salespeople on the following weekly schedule:

> 7% on sales up to $8000
> $7\frac{1}{2}\%$ on sales from over $8000 to $10,000
> $8\frac{1}{4}\%$ on sales over $10,000

Complete the following payroll sheet.

Name	Sales	Gross Wages
Barr, Roseanne	$ 9567.42	
Leno, Jay	7526.98	
Williams, Robin	10,846.29	
Total		

【✔ YOUR WORK】─────────────────────────

The Solution:

Barr, Roseanne
7% × $8000.00 = 0.07 × $8000.00 = $560.00
$7\frac{1}{2}\%$ × 1567.42 = 0.075 × 1567.42 = 117.56 rounded
Total = $677.56

Leno, Jay
7% × $7526.98 = 0.07 × $7526.98 = $526.89 rounded

Williams, Robin

$$7\% \times \$8000.00 = 0.07 \times \$8000.00 = \$560.00$$
$$7\tfrac{1}{2}\% \times 2000.00 = 0.075 \times 2000.00 = 150.00$$
$$8\tfrac{1}{4}\% \times 846.29 = 0.0825 \times 846.29 = \underline{69.82}\quad \text{rounded}$$
$$\$779.82$$

Name	Sales	Gross Wages
Barr, Roseanne	$ 9567.42	$677.56
Leno, Jay	7526.98	526.89
Williams, Robin	10,846.29	779.82
Total		$1984.27

Calculating Salary Plus Commission

Salary Plus Commission
Payment to individuals for their labor based on a combination of salary and either a straight or a sliding scale commission.

A third method of calculating commissions is **salary plus commission.** The salary assures employees of a regular income and the commission acts as an incentive plan. With this method, you calculate the commission and add it to the salary to compute the gross pay.

For example, Marissa earns a weekly salary of $235.50 plus a commission of $4\tfrac{1}{2}\%$ on her total sales. Let's calculate Marissa's gross pay for a week with total sales of $1295.98.

$$\text{Commission} = 4.5\% \times \$1295.98 = 0.045 \times \$1295.98 = \$58.32 \quad \text{rounded}$$
$$\text{Salary} = \underline{+ \$235.50}$$
$$\text{Gross Pay} = \$293.82$$

Never multiply by a percent. *Always* change the percent to a decimal number first or use a calculator with a percent key.

WORK THIS PROBLEM

The Question:

The Chic Boutique pays each salesperson a salary plus a commission of 5% on his or her total sales for the week. Complete the following payroll sheet.

Name	Weekly Salary	Sales	Commission	Gross Pay
Armani, Giorgio	$225.50	$8475.60		
Beene, Geoffrey	217.75	5024.50		
Claiborne, Liz	232.25	6754.90		
Karan, Donna	221.40	5174.90		
Total				

✔ YOUR WORK

The Solution:

Name	Weekly Salary	Sales	Commission	Gross Pay
Armani, Giorgio	$225.50	$8475.60	$423.78	$ 649.28
Beene, Geoffrey	217.75	5024.50	251.23	468.98
Claiborne, Liz	232.25	6754.90	337.75	570.00
Karan, Donna	221.40	5174.90	258.75	480.15
Total				$2168.41

Once again, businesspeople generally use calculators to perform these calculations.

Determining Salary Plus Commission with a Calculator

You can use a calculator to find salary plus commission in one smooth flow. For example, in the case of Mr. Armani,

8475.6 ×̄ .05 =̄ ⟶ `423.78` +̄ 225.5 =̄ ⟶ `649.28`

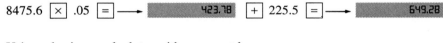

Using a business calculator with a percent key,

8475.6 ×̄ 5 %̄ ⟶ `423.78` +̄ 225.5 =̄ ⟶ `649.28`

Commission   Gross pay

Be sure to check your understanding by working through the practice problems in Section Test 6.2.

WORKSPACE

Name

Date

Course/Section

The following problems test your understanding of Section 6.2, Determining Piecework and Commission Pay.

A. Complete the following piecework payroll sheet.

Name	Pieces Completed					Total Pieces	Rate	Gross Pay
	M	T	W	T	F			
1. Adams, Charles	40	42	38	36	35		$2.25	
2. Bilbery, Dianna	42	45	39	40	41		2.30	
3. Dooley, Donald	37	36	38	35	34		2.13	
4. Herrmann, Kay	41	40	45	38	40		2.27	
5. McGee, George	35	34	34	36	35		1.98	
6. Nelson, Christina	45	44	44	46	45		2.42	
7. Ta, Long	42	42	41	41	41		2.31	
8. Vanbeber, Jerry	39	42	37	39	40		2.19	

B. Complete the following payroll sheet.

The Pelota Bearing Company pays its employees on a piecework basis using the following schedule:

$$
\begin{array}{lll}
1\text{--}200 & @ & \$1.50 \\
201\text{--}250 & @ & 1.75 \\
251 \text{ and up} & @ & 1.95
\end{array}
$$

Name	Bearings per Day					Total	Gross Wages
	M	T	W	T	F		
1. Ashley, Glenda	52	55	50	49	52		
2. Custer, Bradley	40	41	39	38	40		
3. Hill, Vernice	49	50	52	50	51		
4. Rackley, Raymond	59	62	60	61	60		
5. Sinnett, Patrica	62	60	62	61	61		
6. Trinh, Tuyet	58	57	57	56	60		
7. Wagner, Cheryl	52	51	53	50	52		
8. Young, Merle	39	40	37	37	36		

C. Complete the monthly payroll sheet.

The Platt Realty Company pays its salespeople a straight $3\frac{1}{2}$% commission.

Name	Total Sales	Gross Pay
1. Flickinger, Jeffrey	$ 65,590	
2. Greer, Robyn	92,650	
3. Marshall, Pamela	135,900	
4. Perdue, Melissa	42,750	
5. Scaramucci, Todd	85,220	
6. Thompson, Cheryl	152,890	
7. Wallace, LaVonna	95,500	
8. Wehba, David	55,555	

D. Complete the following weekly payroll sheet.

The Phyle Office Products Company pays its salespeople on the following weekly schedule:

$1\frac{1}{2}$% on sales up to $18,000
2%   on sales from over $18,000 to $25,000
$2\frac{1}{2}$% on sales over $25,000

Name	Total Sales	Gross Pay
1. Cook, Michael	$18,952	
2. Forth, Christy	22,765	
3. Gaines, Phillip	15,691	
4. Ishmael, Jonathan	25,772	
5. Leonard, James	16,973	
6. Newell, Pamela	32,782	
7. Reed, Linda	21,105	
8. Sounders, Jeff	28,098	

E. Complete the weekly payroll sheet.

Radical Radials pays its salespeople a weekly salary plus a $5\frac{1}{4}\%$ commission on their sales.

Name	Salary	Sales	Commission	Gross Pay
1. Greg, Kristy	$180	$3527		
2. Miller, Margaret	145	2365		
3. Nevels, James	130	2450		
4. Quiroz, Marica	145	2935		
5. Robertson, Richard	180	3285		
6. Southern, Dawn	160	2962		
7. Taliaferro, Beatrice	145	2658		
8. Wolf, Mark	190	3502		

F. Solve these problems.

1. Claudia pays her operators in Claudia's Clip-Joint an hourly wage plus a 7% commission on product sales and a 3% commission on services. Joan, who receives $6.25 per hour, provided services totalling $735 and sold products worth $152 last week, while working 37 hours. What was Joan's gross pay?

2. Sally, another of Claudia's operators, receives $6.52 per hour, with time-and-a half for overtime (over 40 hours per week), along with the 7% commission on sales and the 3% on services. Last week Sally worked 43.5 hours, provided services worth $896, and sold products totalling $205. What was Sally's gross pay?

WORKSPACE

Key Terms

commission
differential piecework
gross pay
hourly wages
overtime

piecework
salary
salary plus commission
sliding scale commission
straight commission

Chapter 6 at a Glance

Page	Topic	Key Point	Example
212	Computing salary payments	To find salary payments for periods of less than 1 year, divide annual salary by the number of such periods per year.	Annual salary: $24,960 Weekly: $\frac{\$24,960}{52} = \$480$ Biweekly: $\frac{\$24,960}{26} = \$960$ Semimonthly: $\frac{\$24,960}{24} = \$1040$ Monthly: $\frac{\$24,960}{12} = \$2080$
213	Calculating hourly pay	Gross pay = no. hours × rate per hour	What is Jane's gross pay if she works 35 hours and earns $6.50 per hour? Gross pay = 35 × $6.50 = $227.50
214	Calculating pay with overtime	Multiply regular rate by 40 to find regular pay; subtract 40 from hours worked to find overtime hours; multiply regular rate by 1.5 to find overtime rate and multiply this rate by overtime hours to find overtime pay; add regular and overtime pay together.	What is Tim's gross pay if he works 45 hours and earns $6.50 per hour regularly? Regular pay = 40 × $6.50 = $260 Overtime hours = 45 − 40 = 5 Overtime rate = $6.50 × 1.5 = $9.75 Overtime pay = 5 × $9.75 = $48.75 Gross pay = $260 + $48.75 = $308.75
223	Calculating piecework	Gross pay = no. pieces × rate per piece	What is Ann's gross pay if she earns $2.30 per piece and makes 237 pieces? Gross pay = $2.30 × 237 = $545.10
224	Calculating differential piecework	Multiply each number of pieces by the appropriate rate; add the results.	Ray earns $2.90 per chair for the first 100 chairs he makes, $3.10 for the next 50 chairs, and $3.40 for each additional one. If he makes 162 chairs, what is his gross pay? 100 × $2.90 = $290.00 50 × $3.10 = 155.00 12 × $3.40 = 40.80 $485.80

Page	Topic	Key Point	Example
226	Calculating straight commission	Gross pay = rate × total sales	What is May's gross pay if she earns a 6% commission on sales and has sales of $4847.65? Gross pay = 6% × $4847.65 = 0.06 × $4847.65 = $290.859 = $290.86 rounded
227	Calculating sliding scale commissions	Multiply each level of sales reached by the appropriate rate; add the results.	Lee earns commissions as follows: 7% up to $8000; 7.5% from over $8000 to $10,000; 8.25% over $10,000. What is Lee's gross pay on sales of $10,846.29? 0.07 × $8000 = $560.00 0.075 × $2000 = 150.00 0.0825 × $846.29 = 69.82 $779.82
228	Calculating salary plus commission	Calculate salary (see page 212); calculate commission (see page 226); add commission and salary together.	Lou earns a salary of $26,000 per year plus a 2% commission on total sales. What is Lou's gross pay for a week with $6000 in sales? Salary = $26,000/52 = $500 Commission = 0.02 × $6000 = $120 Gross pay = $500 + $120 = $620

Name

Date

Course/Section

The following problems test your understanding of payroll.

1. Complete the following weekly payroll sheet.

Name	Hours					Total Hours	Regular Hours	Regular Rate	O.T. Hours	O.T. Rate	Regular Pay	O.T. Pay	Gross Pay
	M	T	W	T	F								
1. Burns, Tracy	8	9	9	9	9			$7.26					
2. Gower, Charles	9	8	9	8	8			7.65					
3. Jeter, Sandra	9	8	9	8	7			8.17					
4. Lee, Russell	9	8	9	7	4			7.92					
5. Mandrell, Carolyn	9	9	9	9	10			7.75					
6. Nguyen, Emmanuelle	9	8	9	7	9			8.09					
7. Palmer, Johnny	9	9	9	9	9			7.93					
8. Roberts, Linda	10	9	8	7	9			8.11					
Totals													

2. Complete the following weekly payroll sheet.

Name	Hours					Total Hours	Regular Hours	Regular Rate	O.T. Hours	O.T. Rate	Regular Pay	O.T. Pay	Gross Pay
	M	T	W	T	F								
1. Dunn, Lenora	8	9	8	9	8			$8.12					
2. Fisher, Ralph	8	9	9	9	8			9.27					
3. Hustad, Michael	9	8	7	6	5			7.95					
4. McChristian, Peggy	9	9	9	9	9			8.55					
5. Raffael, Larry	10	9	9	9	9			8.24					
6. Steward, Terry	8	9	10	9	8			9.12					
7. Suitor, Johnny	8	7	9	9	8			8.57					
8. Webb, Thomas	8	9	9	8	8			9.19					
Totals													

3. If your annual salary is $35,100, what is your (a) monthly pay, (b) semimonthly pay, (c) biweekly pay, (d) weekly pay?

4. If your annual salary is $26,208, what is your (a) monthly pay, (b) semimonthly pay, (c) biweekly pay, (d) weekly pay?

Complete the following payroll sheet.

	Name	Pieces Completed					Total Pieces	Rate	Gross Pay
		M	T	W	T	F			
5.	Eberhart, Martin	52	50	49	51	51		$1.95	
6.	Grumbein, Kenneth	54	55	56	55	53		1.82	
7.	Norman, Emma	49	48	48	49	48		1.85	
8.	Rounds, Rick	49	51	51	50	50		1.92	

Complete the following payroll sheet.

The Microwidget Company pays its employees on a piecework basis using the following schedule:

$$1–200 @ \$1.40 \text{ each}$$
$$201–225 @ \quad 1.48 \text{ each}$$
$$226 \text{ and up } @ \quad 1.55 \text{ each}$$

	Name	Microwidgets per Day					Total	Gross Wages
		M	T	W	T	F		
9.	Greg, Kristin	42	43	45	45	44		
10.	Milvo, Norm	38	39	40	42	39		
11.	Robertson, Richard	45	46	47	47	46		
12.	Taliaferro, Beatrice	46	46	45	46	46		

Complete the following monthly payroll sheet.

The Littletown Steel Company pays its salespeople a straight $4\frac{1}{2}\%$ commission.

	Name	Total Sales	Gross Pay
13.	Hannan, Scott	$ 75,650	
14.	Miller, Margaret	82,900	
15.	Owens, Mary	126,500	
16.	Shafer, Roy	98,800	

Complete the following weekly payroll sheet.

Wired for Sound pays its salespeople on the following weekly schedule:

2% on sales up to $12,000
$2\frac{1}{2}$% on sales from over $12,000 to $15,000
3% on sales over $15,000

	Name	Total Sales	Gross Pay
17.	Mobley, Sarah	$14,855	
18.	Quinton, Sonja	16,738	
19.	Tapp, Jeffrey	11,952	
20.	White, Mark	17,023	

Complete the following weekly payroll sheet.

Maria's Mannequins pays its salespeople a weekly salary plus a $2\frac{1}{2}$% commission on their sales.

	Name	Salary	Sales	Commission	Gross Pay
21.	Nevels, James	$205	$3752		
22.	Smith, Tammy	210	3985		
23.	Willis, Stephanie	195	3228		
24.	Young, Tammy	212	4176		

WORKSPACE

7

The Mathematics of Buying

CHAPTER OBJECTIVES

When you complete this chapter successfully, you will be able to:

1. Calculate discount rates, including single and chain discounts.

2. Determine the best times to pay bills with different cash discount dates.

■ Calculate the discount and net cost of a portable subcompact computer that has a list price of $835 and a 15% discount.

■ Kennywood Amusement Park ordered new clown costumes. The invoice was for $781, dated June 2, with terms 3/10, 30X. If the bill is paid on July 7, calculate the amount paid.

In the last chapter, we saw how companies pay for one of the most important items they buy: the services of their employees. But every business also buys things: raw materials to make products, finished products to resell—at least office supplies. And obviously, the less companies have to pay for the products they buy, the happier (and more profitable) they are. In this chapter, we will consider some of the ways in which companies lower their bills.

If you "buy into" this chapter, you'll be amply repaid.

Can You:

Calculate a discount?
Determine a trade discount?
Calculate a chain discount?
Find the single equivalent discount rate?
. . . if not, you need this section.

Calculating Discounts

Discount An amount deducted from the list price of a product or service.

You've probably heard of "discount" stores and seen advertisements promising you a "discount" if you buy now or buy so many or so much of something. A **discount** is an amount deducted from the list price. It is the product of the discount rate and the list price. The **list price** is the suggested retail price determined by the manufacturer or distributor and listed in its product list or catalog.

List Price The suggested retail price determined by the manufacturer or distributor.

> Discount = discount rate × list price

For example, if a textbook lists for $35.90 and is discounted 30%, the discount is

$$\text{Discount} = 30\% \times \$35.90$$
$$= 0.30 \times \$35.90 = \$10.77$$

The discount is $10.77. Using a business calculator with a percent key,

35.9 × 30 % ⟶ [10.77]

When you use a percent calculator, enter the list price number first.

If you do not use a percent calculator, remember to change the percent to a decimal number before multiplying. If you had trouble changing the percent to a decimal number, turn to Chapter 4, Section 4.1 for a review.

Net Price The difference between the list price and the discount.

If you are purchasing one of the textbooks in this example, you will want to calculate the actual cost. The actual or **net price** is the difference between the list price and the discount. The net is the cost after any deductions. It is sometimes called the "reduced price" or the "sale price."

> Net price = list price − discount

In the case of the textbook,

$$\text{Net price} = \text{list price} - \text{discount}$$
$$= \$35.90 - 10.77 = \$25.13$$

So you would pay only $25.13 for the book—a bargain!

YOUR TURN

WORK THIS PROBLEM

The Question:

Calculate the discount and net price in each of the following cases:

1. A pair of jeans that originally cost $35.95 is now on sale for 20% off. What is the reduced price?
2. A stereo system is reduced 32%. If the list price is $749.50, what is the sale price?

✔ YOUR WORK

The Solution:

1. Discount = 20% × $35.95 = 0.20 × $35.95 = $7.19
   Net price = $35.95 − 7.19 = $28.76
2. Discount = 32% × $749.50 = 0.32 × $749.50 = $239.84
   Net price = $749.50 − 239.84 = $509.66

Frequently the discount is only an intermediate result. Usually you are interested only in the amount you must pay, the net price. In this case, there is a shortcut method for finding the net price directly.

Complement The rate equal to 100% − discount rate; it allows you to calculate directly the net price. Also called *payment rate*.

If the discount rate is 30%, the amount you must pay or net price will be 100% − 30%, or 70% of the list price. That is, we can break the list price (100%) into two parts: the discount (30%) and the net price (70%). The 70% rate is called the **complement** of 30%.

> Complement = 100% − discount rate

The complement of the discount rate is often called the *payment rate*.

100%

Discount	Net Price
Discount rate = 30%	Complement or Payment rate = 70%

In the example of the textbook that lists for $35.90 but is on sale for 30% off, the net price is

100% − 30%

Net price = 70% × $35.90 = 0.70 × $35.90 = $25.13

The formula for the shortcut method is

> Net price = (100% − discount rate) × list price
> Payment rate

Be Careful!

The parentheses in the formula are used to indicate which arithmetic operation (− or ×) is done first. *Always* do the operation in the parentheses first.

YOUR TURN

WORK THIS PROBLEM

The Question:

Work the following problems using the shortcut method.

1. A pair of jeans that originally cost $35.95 is now on sale for 20% off. What is the reduced price?
2. A stereo system is reduced 32%. If the list price is $749.50, what is the sale price?

✔ YOUR WORK

The Solution:

1. Net price = (100% − 20%) × $35.95
   = 80% × $35.95 = 0.80 × $35.95 = $28.76
2. Net price = (100% − 32%) × $749.50
   = 68% × $749.50 = 0.68 × $749.50 = $509.66

Notice that these are the same answers we got earlier. Only our method has changed.

Determining Trade Discounts

Individuals are not the only ones who buy products and receive discounts. Actually, individuals are at the end of the so-called *distribution chain:*

Manufacturer ⟶ wholesaler ⟶ retailer ⟶ individual customer

That is, manufacturers make goods and sell them to wholesalers in large quantities. **Wholesalers** resell smaller quantities of these goods either to other businesses for their use or to retailers. **Retailers** resell individual items to individual customers.

Wholesaler A firm that resells smaller quantities of goods either to other businesses for their use or to retailers.

Retailer A firm that resells individual items to individual customers.

Trade Discount A discount given to a wholesaler by a manufacturer or to a retailer by a manufacturer or wholesaler.

Wholesalers and retailers often receive discounts—called **trade discounts**—from the manufacturers and wholesalers who supply them. Like a regular discount, a trade discount is an amount deducted from the list price.

Did You Know?

How did trade discounts get their name? What is traded?

Nothing is traded. Trade discounts were originally discounts given to members of the same or a similar business or trade.

Sometimes trade discounts are a flat rate one company charges another on all items, depending on the importance of the customer or the size of the sale. In other cases, the discount depends on the item ordered—is it new or about to be discontinued; the economic climate—are these products selling well or not; and even the season—are you trying to sell bathing suits in February?

Manufacturers and wholesalers frequently send their customers catalogs of available products with list prices. Accompanying the catalog is a sheet of discounts. Why not just put the discount prices in the catalog? It's relatively easy and cheap to change the discount sheet to reflect new prices and encourage sales of some items. Reprinting an entire catalog—often in full color—could be much more costly and time consuming.

Single Trade Discount
A trade discount equal to rate times list price.

Whether offered through a catalog or not, trade discounts may take either of two forms: single trade or chain. A **single trade discount** is like the individual discount we discussed earlier in this chapter:

> Single trade discount = rate × list price

Calculating Chain Discounts

Chain Discount A series of trade discounts, with each discount after the first based on an intermediate price.

Sometimes businesses "layer" the discounts they offer. For example, a manufacturer may offer a 15 percent discount on a specific item and also a 10 percent discount on the whole order. To find the price of an item given such a **chain discount,** follow these steps:

Steps: Calculating Two Trade Discounts

STEP 1. *Calculate* the first discount.

STEP 2. *Subtract* the discount from the list price. The result is called the **intermediate price.**

Intermediate Price In a chain discount, the result of deducting a discount from the initial price or from another intermediate price.

STEP 3. *Calculate* the second discount based on the intermediate price, *not* the list price.

STEP 4. *Subtract* the second discount from the intermediate price to obtain the net price.

For example, suppose an order of $500 offers trade discounts of 15% and 10%. What is the net price?

Step 1.

First discount = 15% × $500 = 0.15 × $500 = $75

Step 2.

Intermediate price = $500 − 75 = $425

Or using a business calculator,

 500 × 15 % ⟶ 75. +/− + 500 = ⟶ 425.

Remember, punch in the base (500) first.

Step 3.

Second discount = 10% × $425 = 0.10 × $425 = $42.50

Step 4. | Intermediate price | | Second discount |

Net cost = $425 − 42.50 = $382.50

Remember, always use the intermediate price to calculate the second discount. Do *not* use the original list price when you are doing step 3.

 Successive trade discounts of 15% and 10% are *not* equal to a discount of 25%. A discount rate of 25% on $500 yields a discount and net price of

$$\text{Discount} = 25\% \times \$500 = 0.25 \times \$500 = \$125$$
$$\text{Net price} = \$500 - 125 = \$375$$

Successive trade discounts of 15% and 10% on $500 yield a net price of $382.50.

Multiple trade discounts of 15% and 10% are specified as 15%/10% or simply 15/10.

 [WORK THIS PROBLEM] ─────────────────────────────

YOUR TURN

The Question:

An item lists for $726.50 but offers trade discounts of 20/15. What is the net price?

✔ YOUR WORK ─────────────────────────────

The Solution:

Step 1. | First rate | | List price |

First discount = 20% × $726.50 = 0.20 × $726.50 = $145.30

Step 2. | List price | | First discount |

Intermediate price = $726.50 − 145.30 = $581.20

Step 3. | Second rate | | Intermediate price |

Second discount = 15% × $581.20 = 0.15 × $581.20 = $87.18

Step 4. | Intermediate price | | Second discount |

Net price = $581.20 − 87.18 = $494.02

Using a business calculator,

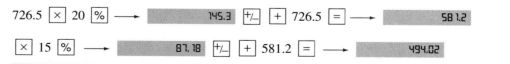

726.5 × 20 % ⟶ 145.3 +/− + 726.5 = ⟶ 581.2

× 15 % ⟶ 87.18 +/− + 581.2 = ⟶ 494.02

Now that you can calculate a chain of two trade discounts, try three. You guessed it; there are three prices beyond the list price: the first intermediate price, the second intermediate price, and the net price.

Steps: Calculating Three Trade Discounts

STEP 1. *Calculate* the first discount.

STEP 2. *Subtract* the first discount from the list price to find the first intermediate price.

STEP 3. *Calculate* the second discount using the first intermediate price, *not* the list price.

STEP 4. *Subtract* the second discount from the first intermediate price to find the second intermediate price.

STEP 5. *Calculate* the third discount using the second intermediate price, *not* the list or first intermediate price.

STEP 6. *Subtract* the third discount from the second intermediate price to find the net price.

In other words, when calculating a chain discount, *always use the previous intermediate price.* For example, if an item lists for $247.25 but carries multiple trade discounts of 15/10/5, what is the net price?

Step 1.

First rate		List price

First discount = 15% × $247.25 = $37.09 rounded

Step 2.

List price		First discount

First intermediate price = $247.25 − 37.09 = $210.16

Step 3.

Second rate		First intermediate price

Second discount = 10% × $210.16 = $21.02 rounded

Step 4.

First intermediate price		Second discount

Second intermediate price = $210.16 − 21.02 = $189.14

Step 5.

Third rate		Second intermediate price

Third discount = 5% × $189.14 = $9.46 rounded

Step 6.

Second intermediate price		Third discount

Net price = $189.14 − 9.46 = $179.68

**YOUR
TURN**

WORK THIS PROBLEM

The Question:

Find the net price for each of the following:

(a) An order lists for $357 and carries trade discounts of 30/20/10.

(b) An item lists for $756.75 and carries trade discounts of 15/10/7$\frac{1}{2}$.

✔ YOUR WORK

The Solution:

(a) **Step 1.**   First discount = 30% × $357 = 0.30 × $357 = $107.10

    Step 2.   First intermediate price = $357 − 107.10 = $249.90

    Step 3.   Second discount = 20% × $249.90 = 0.20 × $249.90 = $49.98

    Step 4.   Second intermediate price = $249.90 − 49.98 = $199.92

    Step 5.   Third discount = 10% × $199.92 = 0.10 × $199.92 = $19.99 rounded

    Step 6.   Net price = $199.92 − 19.99 = $179.93

(b) **Step 1.**   First discount = 15% × $756.75 = 0.15 × $756.75 = $113.51 rounded

    Step 2.   First intermediate price = $756.75 − 113.51 = $643.24

    Step 3.   Second discount = 10% × $643.24 = 0.10 × $643.24 = $64.32 rounded

    Step 4.   Second intermediate price = $643.24 − 64.32 = $578.92

    Step 5.   Third discount = $7\frac{1}{2}\%$ × $578.92 = 7.5% × $578.92 = 0.075 × $578.92
                = $43.42 rounded

    Step 6.   Net price = $578.92 − 43.42 = $535.50

Of course, in the real world, businesspeople use computers or calculators to find these discounts.

Finding Chain Discounts with a Calculator—The Long Way

A calculator reduces but does not eliminate the multiple steps involved in finding multiple trade discounts. To solve example problem (a) with a business calculator, press

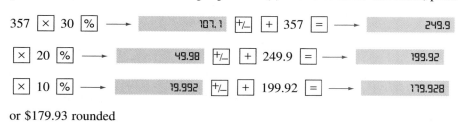

or $179.93 rounded

Whew! That's a lot of work, even using a calculator. Fortunately, there is an easier way using the shortcut method of calculating the net price by subtracting and multiplying, much as we did earlier.

Steps: Calculating Chain Discounts: The Shortcut

STEP 1.   *Calculate* the complements of each discount rate in the chain discount. Complement = 100% − discount rate.

STEP 2.   *Multiply* the list price times the complements from step 1. The product is the net price.

For example, let's find the net price if the chain discount is 15/10/5 and the list price is $500.

Step 1. The complement of the first trade discount is $100\% - 15\% = 85\%$.
The complement of the second discount is $100\% - 10\% = 90\%$.
The complement of the third discount is $100\% - 5\% = 95\%$.

Step 2. The net price $= 85\% \times 90\% \times 95\% \times \$500$
$= .85 \times .90 \times .95 \times 500 = \$363.38$ rounded

Calculating Multiple Chain Discounts—The Shortcut

You can easily punch in the equation for multiple trade discounts (with percents converted to decimals). Return to the example of a $357 item with multiple trade discounts of 30/20/10.

Step 1. Find the complements.
$$1.0 - 0.3 = 0.7$$
$$1.0 - 0.2 = 0.8$$
$$1.0 - 0.1 = 0.9$$
Jot down answers.

Step 2. .7 ☒ .8 ☒ .9 ☒ 357 ☐ ⟶ ▢ 179.928

Or, if your calculator has a percent key, you can solve this problem in one step. The complements of the chain discount 30/20/10 are 70%, 80%, and 90%. Then,

357 ☒ 70 ☒ ☒ 80 ☒ ☒ 90 ☒ ⟶ ▢ 179.928

YOUR TURN

WORK THIS PROBLEM

The Question:

Use the shorter method on the following problems.

(a) An order lists for $357 and carries trade discounts of 30/20/10. Calculate the net price.
(b) An item lists for $756.75 and carries trade discounts of $15/10/7\frac{1}{2}$. What is the net price?

✔ YOUR WORK

The Solution:

(a)

| 100% − 20% | 100% − 30% |
| 100% − 10% | List price |

Net price $= 90\% \times 80\% \times 70\% \times \$357$
$= 0.90 \times 0.80 \times 0.70 \times 357 = \$179.93$ rounded

Using the percent key on a business calculator, it's even quicker:

357 ☒ 90 ☒ ☒ 80 ☒ ☒ 70 ☒ ⟶ ▢ 179.928 or $179.93 rounded

(b)

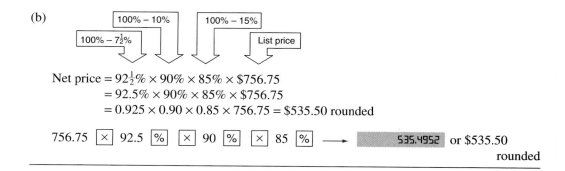

$$\text{Net price} = 92\tfrac{1}{2}\% \times 90\% \times 85\% \times \$756.75$$
$$= 92.5\% \times 90\% \times 85\% \times \$756.75$$
$$= 0.925 \times 0.90 \times 0.85 \times 756.75 = \$535.50 \text{ rounded}$$

$$756.75 \;\boxed{\times}\; 92.5 \;\boxed{\%}\; \boxed{\times}\; 90 \;\boxed{\%}\; \boxed{\times}\; 85 \;\boxed{\%}\; \longrightarrow \boxed{535.4952} \text{ or } \$535.50$$
$$\text{rounded}$$

Wasn't that much easier?

Single Equivalent Discount Rate

Single Equivalent Discount Rate A single rate equivalent to a series of trade discounts.

In problem (a), the discounts were 30/20/10, and the net price was $90\% \times 80\% \times 70\% \times$ list price or $50.4\% \times$ list price. The company must pay 50.4% of the list price. The total discount rate is $100\% - 50.4\% = 49.6\%$. This is called the **single equivalent discount rate** and is equal to the three trade discounts of 30%, 20%, and 10%. To calculate the single equivalent discount rate, first calculate the rate at which the company must pay. Then subtract that equivalent payment rate from 100%.

Steps: Calculating the Single Equivalent Discount Rate

STEP 1.   *Calculate* each complement by subtracting the discount rate from 100%.

STEP 2.   *Multiply* the complements and express the product as a percent. This number is called the *equivalent payment rate*.

STEP 3.   *Find* the complement of the equivalent payment rate from step 2 by subtracting this rate from 100%. This difference gives the single equivalent discount rate.

For example, consider an item that lists for $756.75 and carries trade discounts of $15/10/7\tfrac{1}{2}$. What is the single equivalent discount rate for this item? What is the net price?

First determine the complements and the equivalent payment rate.

$$(100\% - 15\%) \times (100\% - 10\%) \times (100\% - 7\tfrac{1}{2}\%) = 85\% \times 90\% \times 92\tfrac{1}{2}\%$$
$$= 85\% \times 90\% \times 92.5\%$$
$$= 0.85 \times 0.90 \times 0.925$$
$$= 0.707625 = 70.7625\%$$

$$\text{Single equivalent discount rate} = 100\% - 70.7625\% = 29.2375\%$$

$$\text{Net price} = 70.7625\% \times \$756.75 = \$535.50 \text{ rounded}$$

YOUR TURN

[WORK THIS PROBLEM]

The Question:

An item lists for $249 and carries trade discounts of 30/20/15. Calculate the net price. What is the single equivalent discount rate?

[✔ YOUR WORK]

The Solution:

The equivalent payment rate is

$$(100\% - 30\%) \times (100\% - 20\%) \times (100\% - 15\%) = 70\% \times 80\% \times 85\%$$
$$= 0.70 \times 0.80 \times 0.85$$
$$= 0.476 = 47.6\%$$

Net price = 47.6% × $249 = $118.52 rounded

Single equivalent discount rate = 100% − 47.6% = 52.4%

If you solved the problem correctly, you're ready for the questions in Section Test 7.1. If not, review this section again.

WORKSPACE

Name

Date

Course/Section

The following problems test your understanding of Section 7.1, Taking Advantage of Trade Discounts.

A. Calculate the discount and net price.

	List Price	Discount Rate	Discount	Net Price
1.	$450	12%		
2.	$560	5%		
3.	$1375	7%		
4.	$982	27%		
5.	$735.95	9%		
6.	$274.37	14%		
7.	$846.39	$5\frac{1}{4}\%$		
8.	$837.84	$7\frac{1}{2}\%$		

B. Calculate the net price.

	List Price	Trade Discounts	Net Price
1.	$472	10/5	
2.	$895	8/4	
3.	$2648	10/8/5	
4.	$4624	12/7/3	
5.	$475.05	8/6/3	
6.	$735.27	15/10/5	
7.	$527.38	20/15/12	
8.	$294.83	30/33/18	

C. Calculate the single equivalent discount rate and the net price.

List Price	Trade Discounts	Single Equivalent Discount Rate	Net Price
1. $635	12/8		
2. $792	15/9		
3. $729	20/18/14		
4. $3746	25/22/15		
5. $385.63	18/15/12		
6. $860.08	35/20/10		
7. $3900.55	25/20/12		
8. $3801.32	30/20/5		

D. Solve these problems.

1. Calculate (a) the discount and (b) the net cost of a pair of computerized, self-adjusting athletic shoes that list for $75.25 with a 25% discount.

2. Calculate (a) the discount and (b) net cost of a portable subcompact computer that has a list price of $835 and a 15% discount.

3. The portable computer that lists for $835 has trade discounts of 25/20/15. Calculate the net price.

4. A laser printer lists for $1275 with trade discounts of 12/10/5. Calculate the net price.

5. Calculate the net price for a designer watch listed for $568 with trade discounts of 10/10/5.

6. Calculate the net price for a power drill which has a list price of $95.50 with trade discounts 20/15/10.

7. A refrigerator lists for $761 with trade discounts 10/8/5. Calculate the net price.

8. An answering machine has a list price of $95 with trade discounts 30/25/20. Calculate the net price.

9. Kool, a refrigerator manufacturer, offers trade discounts of 30/25/10. On Ice, a second company, offers discounts of 35/25/5. Calculate the single equivalent discounts for (a) Kool and (b) On Ice, and (c) determine which company offers the larger discount.

10. Intense Tents offers trade discounts of 20/14/10. Warm and Dry, a second tent company, offers discounts of 25/15/4. Calculate the single equivalent discounts for (a) Intense Tents, and (b) Warm and Dry, and (c) determine which company offers the larger discount.

11. Wonderful Widgets offers trade discounts of 20/15/10. Wayne's Widgets offers discounts of 25/12/8. Calculate the single equivalent discounts for (a) Wonderful Widgets, and (b) Wayne's Widgets, and (c) determine which company offers the larger discount.

12. Bits 'n' Bytes, a local computer wholesaler, is pricing printers from various manufacturers. MBI Computers offers trade discounts of 30/20/15, while CED Computers offers discounts of 25/24/16. Calculate the single equivalent discounts for (a) MBI Computers and (b) CED Computers and (c) determine which manufacturer offers the larger discount.

13. Calculate the single equivalent discount rate for (a) 30/25/10 and (b) 25/10/30. (c) Are they the same?

14. Calculate the single equivalent discount rate for (a) 25/20/15 and (b) 15/25/20. (c) Are they the same?

15. The Broadbeam Department store advertised all clothing at 40% off. Preferred customers received an additional 30% discount on the sale price. What did Lyn, a preferred customer, pay for a dress originally priced at $149.95?

WORKSPACE

Can You:

Determine when a bill is due?

Calculate cash discounts?

Determine when to pay bills with EOM and Prox dating?

Decide when to pay bills with ROG dating?

Use extra dating to your company's advantage?

Find the remaining balance after a partial payment?

. . . If not, you need this section.

Determining the Due Date

Knuckle Method A method useful in calculating due dates in which each knuckle represents a 31-day month and each space between two knuckles represents a 30-day month, except February.

The first thing almost anyone, whether in business or not, wants to know about a bill is "how much?" The second is "when is it due?" In order to calculate the due date, you must know how many days are in each month. One easy method is called the **knuckle method.** Each knuckle and space between two knuckles represents a month. Each month on a knuckle has 31 days. Each month on a space has 30 days except February. February has 28 days except in leap years, when it has 29 days.

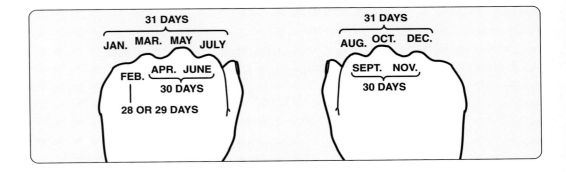

Steps: Calculating the Due Date

STEP 1. *Calculate* the number of days until the end of the month by subtracting the invoice date from the number of days in the month.

STEP 2. *Calculate* the days remaining in the discount period by subtracting the result of step 1 from the number of days in the discount period. If this answer is less than or equal to the number of days in the next month, it is the due date and you do not need to proceed to step 3.

STEP 3. If the result from step 2 is greater than the number of days in the month, *subtract* the number of days in the next month. Repeat step 3 until the result is less than or equal to the number of days in the month. This result is the due date.

For example, let's calculate the due date for an invoice dated July 18 that is due in 30 days.

Step 1. Since July has 31 days, there are $31 - 18 = 13$ days until the end of the month.

Step 2.

| Discount period | | from Step 1 |

30 − 13 = 17   The due date is August 17.

What if the discount period is more than one month? For example, what's the due date on an invoice dated August 7 that is due in 90 days?

Step 1.   August has 31 days. 31 − 7 = 24 days.

Step 2.   90 − 24 = 66 days.

Step 3.   September has 30 days. 66 − 30 = 36 days.
October has 31 days. 36 − 31 = 5 days.
The due date is November 5.

YOUR TURN

☐ **WORK THIS PROBLEM**

The Question:

(a) An invoice dated March 12 is due in 30 days. Calculate the due date.
(b) An invoice dated May 9 is due in 90 days. What is the due date?

☐ **✔ YOUR WORK**

The Solution:

(a) *Step 1.*   March has 31 days. 31 − 12 = 19 days.

Step 2.   30 − 19 = 11. The due date is April 11.

(b) *Step 1.*   May has 31 days. 31 − 9 = 22 days

Step 2.   90 − 22 = 68 days.

Step 3.   June has 30 days. 68 − 30 = 38 days.
July has 31 days. 38 − 31 = 7 days.
The due date is August 7.

Calculating Cash Discounts

If your business receives a bill that is due in 30 days, should you pay it early? Probably not. The money could be put to work for your business. Firms handling thousands of dollars in purchases each month find that leaving the money in an interest-bearing account is more profitable than paying the bills early, unless there is some incentive to pay early.

Cash Discount A discount extended to a customer in return for payment of a bill within a specified number of days from the invoice date.

To encourage early payment, many companies offer **cash discounts.** A discount written "2/10, n/30" means that a 2% discount is given if the bill is paid within 10 days, and the net (gross-non-discounted amount) is due in 30 days.

| 2% discount rate | | if paid within 10 days | | Net | | due within 30 days |

2 / 10   or   n / 30

The payment date is usually calculated from the date of the invoice—the bill sent by the supplier. For example, a cash discount of 2/10 with an invoice date of March 2 offers a discount of 2% for 10 days *after* the invoice date. The last day of the discount period is March 12. *Do not count the first day.*

As long as a company pays the amount on the invoice within the discount period, it can discount its payment but *only* for goods purchased—*not* for any handling or shipping charges.

For example, an invoice for $500, dated October 17 offers discount terms 2/10, n/30. If the bill is paid October 25, what amount should be paid?

Because the bill is paid within 10 days of the invoice date, a 2% discount is allowed.

$$\text{Discount} = 2\% \times \$500 = 0.02 \times \$500 = \$10$$
$$\text{Amount paid} = \$500 - 10 = \$490$$

If the bill is not paid within the specified period, a late penalty may be assessed. The penalty is usually a percent of the unpaid balance. For example, if the discount terms are 2/10, n/30, a late penalty may be assessed (provided the penalty is stipulated on the invoice) if the bill is not paid within 30 days.

YOUR TURN

| WORK THIS PROBLEM |

The Question:

An invoice for $278, dated November 5, offers discount terms 2/10, n/30. If the bill is paid November 10, what is the amount paid?

| ✔ YOUR WORK |

The Solution:

Because November 10 is within 10 days of the invoice date, the discount is 2%.

$$\text{Discount} = 2\% \times \$278 = 0.02 \times \$278 = \$5.56$$
$$\text{Amount paid} = \$278 - 5.56 = \$272.44$$

A business may offer several cash discount periods such as 4/10, 2/30, n/60. That means a 4% discount if paid within 10 days or a 2% discount if paid between 10 days and 30 days, and the net is due between 30 days and 60 days. If the bill is paid within 10 days, only a 4% discount is taken, *not* both a 4% and 2% discount.

For example, if an invoice for $155 dated February 5 offers discount terms 4/10, 2/30, n/60, what amount is due if the bill is paid on February 21?

February 21 is not within 10 days, but it is within 30 days, so the discount is 2%.

$$\text{Discount} = 2\% \times \$155 = 0.02 \times \$155 = \$3.10$$
$$\text{Amount paid} = \$155 - 3.10 = \$151.90$$

YOUR TURN

| WORK THIS PROBLEM |

The Question:

(a) An invoice for $1237.50 dated October 12 offers discount terms of 3/10, 2/30, n/60. Calculate the amount due if the bill is paid on November 7.

(b) A bill for $257.83, dated April 27, offers discount terms 3/10, 2/30, n/60. What amount is due if the bill is paid on June 5?

| ✔ YOUR WORK |

The Solution:

(a) Since November 7 is not within 10 days, but it is within 30 days, the discount is 2%.

$$\text{Discount} = 2\% \times \$1237.50 = 0.02 \times \$1237.50 = \$24.75$$
$$\text{Amount paid} = \$1237.50 - 24.75 = \$1212.75$$

(b) June 5 is more than 30 days past the billing date. No discount is given. The amount due is $257.83.

EOM and Prox Dating

End-of-Month (EOM) Dating (Prox Dating) A form of cash discount in which the discount applies if the bill is paid within a specified number of days after the end of the month in which the invoice is issued.

Some types of cash discount are monthly—for instance, **EOM (end-of-month) dating.** An invoice with a dating of 2/10EOM means a 2% discount applies if the bill is paid within 10 days after the end of the month or by the tenth of the next month.

For example, suppose an invoice for $275 dated May 17 offers discount terms 2/10EOM, n/60. What amount is due if the bill is paid on June 7?

A 2% discount is offered until June 10.

$$\text{Discount} = 2\% \times \$275 = 0.02 \times \$275 = \$5.50$$
$$\text{Amount paid} = \$275 - 5.50 = \$269.50$$

Another form of monthly dating is **Prox dating.** "Prox" is short for the Latin word *proximo*, which means the *next* month. A discount of 2/10prox means that a 2% discount applies if the bill is paid by the tenth of the next month. The discount 2/10prox is the same as 2/10EOM.

YOUR TURN

WORK THIS PROBLEM

The Question:

(a) A bill for $356, dated November 12, offers discount terms 3/10prox, n/60. What amount is due if the bill is paid on December 5?

(b) An invoice for $759, dated December 18, offers discount terms 2/10EOM, n/60. What amount is due if the bill is paid on February 7?

✔ YOUR WORK

The Solution:

(a) A 3% discount is offered until December 10.

$$\text{Discount} = 3\% \times \$356 = 0.03 \times \$356 = \$10.68$$
$$\text{Amount paid} = \$356 - \$10.68 = \$345.32$$

(b) A 2% discount is offered until January 10. February 7 is beyond the discount period. Since no discount is allowed, the amount due is $759.

With both EOM and prox dating, bills dated on or after the 26th of the month are granted a month's extension. For example, an invoice dated August 26 with terms 2/10EOM offers a discount of 2% until October 10.

YOUR TURN

WORK THIS PROBLEM

The Question:

(a) An invoice for $472, dated July 29, offers discount terms 3/10prox, n/90. When is the end of the discount period? What amount is due if the bill is paid on August 25?

(b) A bill for $56.75, dated October 25, offers discount terms 2/10EOM, n/60. What amount is due if the bill is paid on December 5?

✔ YOUR WORK

The Solution:

a) Since the billing date is after the 26th, an extra month's extension is added to the discount period. The discount period ends September 10. If the bill is paid August 25, a 3% discount is taken.

$$\text{Discount} = 3\% \times \$472 = 0.03 \times \$472 = \$14.16$$
$$\text{Amount paid} = \$472 - 14.16 = \$457.84$$

(b) The 2% discount period ends November 10. Remember, only bills dated on or after the 26th are granted a month's extension. If the bill is paid on December 5, no discount is allowed. The amount due is $56.75.

(Prox dating is an older business term. Today, most businesses use EOM dating instead.)

ROG Dating

Receipt-of-Goods (ROG) Dating A form of cash discount in which the discount applies if the bill is paid within a specified number of days from the delivery date.

Not all discounts start with the invoice date. With **ROG** or **receipt-of-goods dating,** the discount period starts with the date of receipt. This form of dating allows extra time when transportation is slow.

Suppose your company receives an invoice for $257.86, dated July 15, that offers discount terms 3/10ROG. If the goods were received August 7, when is the end of the discount period? What amount is due if the bill is paid August 15?

The discount period starts upon the receipt-of-goods on August 7. The discount period is 10 days, or until August 17. If the bill is paid on August 15, a 3% discount is allowed.

$$\text{Discount} = 3\% \times \$257.86 = 0.03 \times \$257.86 = \$7.74 \text{ rounded}$$
$$\text{Amount paid} = \$257.86 - 7.74 = \$250.12$$

YOUR TURN

| WORK THIS PROBLEM |

The Question:

(a) An invoice for $132.75, dated October 12, offers discount terms 4/15ROG. The goods were received on November 12. What amount is due if the bill is paid November 30?

(b) An invoice for $576.89, dated September 5, offers discount terms 2/10ROG. The goods were received on October 2. What amount is due if the bill is paid October 10?

| ✔ YOUR WORK |

The Solution:

(a) The discount period starts November 12 and ends 15 days later on November 27. Since November 30 is beyond the discount period, the amount due is $132.75.

(b) The discount period starts October 2 and ends October 12. Since October 10 is within the discount period, a 2% discount is allowed.

$$\text{Discount} = 2\% \times \$576.89 = 0.02 \times \$576.89 = \$11.54$$
$$\text{Amount due} = \$576.89 - 11.54 = \$565.35$$

Extra Dating

Extra Dating A form of cash discount in which the discount period may be extended by a specified number of days.

"Extra," "ex," or "X" are used to denote extra dating. **Extra dating** indicates that the discount period is extended for a certain number of days. Businesses sometimes use extra dating to encourage purchases. Seasonal businesses often use extra dating. For example, an air-conditioner supplier may use extra dating during the winter months to encourage off-seasonal purchases.

Suppose you receive an invoice for $846.23, dated October 3, that offers discount terms 3/10-60X. What amount is due if the bill is paid December 5?

The discount of 3% is offered for 10 days plus 60 extra days for a total of 70 days. December 5 is within the discount period.

$$\text{Discount} = 3\% \times \$846.23 = 0.03 \times \$846.23 = \$25.39 \text{ rounded}$$
$$\text{Amount paid} = \$846.23 - 25.39 = \$820.84$$

WORK THIS PROBLEM

The Question:

(a) An invoice for $152.98, dated January 15, offers discount terms 2/15-30X. What amount is due if the bill is paid February 20?

(b) An invoice for $576, dated March 12, offers discount terms 3/10-60X. What amount is due if the bill is paid May 15?

✔ **YOUR WORK**

The Solution:

(a) The discount period is for $15 + 30 = 45$ days. February 20 is within the discount period.

$$\text{Discount} = 2\% \times \$152.98 = 0.02 \times \$152.98 = \$3.06 \text{ rounded}$$
$$\text{Amount paid} = \$152.98 - 3.06 = \$149.92$$

(b) The discount period is for $10 + 60 = 70$ days, or until May 21. May 15 is within the discount period.

$$\text{Discount} = 3\% \times \$576 = 0.03 \times \$576 = \$17.28$$
$$\text{Amount paid} = \$576 - 17.28 = \$558.72$$

That wasn't too difficult, was it?

Crediting Partial Payments

Partial Payment Payment of less than the full amount due.

Often, a company cannot pay the entire invoice amount in time to obtain the cash discount. If a firm makes a **partial payment** during the discount period, most sellers allow the company to receive a cash discount. The cash discount is calculated *only* on the amount of the partial payment. You can calculate the amount still due after a partial payment using the following steps:

Steps: Determining the Balance after a Partial Payment

STEP 1. *Calculate* the amount credited.

$$\text{Amount credited} = \frac{\text{partial payment}}{100\% - \text{discount rate}}$$

STEP 2. *Calculate* the remaining balance.

$$\text{Remaining balance} = \text{invoice amount} - \text{amount credited}$$

To see how these steps work, consider an invoice dated March 12 for $450 and offering discount terms of 2/10, n/30. If a purchasing firm makes a partial payment of $200 on March 20, how much does it still have to pay?

Step 1. Amount credited $= \dfrac{\$200}{100\% - 2\%} = \dfrac{\$200}{98\%} = \dfrac{\$200}{0.98} = \$204.08$ rounded

Step 2. Remaining balance $= \$450 - 204.08 = \$245.92$

YOUR TURN

WORK THIS PROBLEM

The Question:

Kate's Delicacies just received an invoice for $1250 worth of snails. The invoice is dated June 23 and offers discount terms of 3/10, n/30. If Kate makes a partial payment of $600 on July 1, how much does she still owe?

✔ YOUR WORK

The Solution:

Since Kate made the partial payment within 10 days, the cash discount of 3% applies.

Step 1. Amount credited $= \dfrac{\$600}{100\% - 3\%} = \dfrac{\$600}{97\%} = \dfrac{\$600}{0.97} = \$618.56$

Step 2. Remaining balance $= \$1250 - 618.56 = \$631.44$

Calculators can help you determine all kinds of cash discounts.

Calculating Balance after a Partial Payment with a Calculator

Exactly how you find a cash discount using your calculator depends on your calculator's abilities. The simplest calculators require you basically to follow the two-step procedure just described. For example, the preceding problem can be solved by

Step 1. 1 $\boxed{-}$ .03 $\boxed{=}$ $\longrightarrow$ ▨ 0.97 (Jot down or put in calculator memory.)

Step 2. With a business calculator,

600 $\boxed{\div}$ 97 $\boxed{\%}$ $\longrightarrow$ ▨ 618.5567

$\boxed{+/-}$ $\boxed{+}$ 1250 $\boxed{=}$ $\longrightarrow$ ▨ 631.4433 or $631.44 rounded

---

For more practice on cash discounts, work the problems in Section Test 7.2.

SECTION TEST 7.2   The Mathematics of Buying

The following problems test your understanding of Section 7.2, Using Cash Discounts.

A. Calculate the cash discount and the net price or amount paid.

Invoice Date	Invoice Amount or List Price	Terms	Date Paid	Discount	Amount Paid or Net Price
1. 4/7	$750	3/10, n/30	4/15		
2. 9/23	825	2/10, n/30	10/15		
3. 5/12	95	4/10, 2/30, n/60	6/5		
4. 4/12	104	3/10, 1/30, n/60	4/20		
5. 9/16	632	2/10EOM	10/15		
6. 7/18	1950	3/10EOM	8/5		
7. 3/21	293	3/10prox	4/2		
8. 5/19	385	4/10prox	6/7		
9. 7/5	472	3/10-30X	8/10		
10. 10/2	86	2/10-30X	11/8		
11. 4/7	708	3/10, 2/30, n/60	5/2		
12. 5/20	975	4/10, 2/30, n/60	5/26		
13. 10/19	2815	2/10-60X	12/5		
14. 9/5	3209	3/10-60X	11/9		
15. 6/27	366	2/10EOM	8/5		
16. 8/28	529	3/10EOM	10/2		

B. Calculate the cash discount and the amount paid or net price.

Invoice Date	Invoice Amount or List Price	Terms	Goods Received	Date Paid	Discount	Amount Paid or Net Price
1. 9/10	$875	2/10ROG	9/15	9/23		
2. 12/2	1225	3/10ROG	12/15	12/22		
3. 3/7	130	2/10ROG	3/20	3/27		
4. 2/15	282	4/10	2/26	2/28		
5. 8/17	734	4/10ROG	8/28	9/1		
6. 1/5	98	2/15ROG	1/19	1/26		
7. 4/12	238	3/10	4/24	4/26		
8. 6/12	927	3/15ROG	6/28	7/2		

C. Calculate the amount credited and remaining balance for the following partial payments.

Invoice Date	Invoice Amount or List Price	Terms	Date Paid	Amount Paid	Amount Credited	Remaining Balance
1. 8/12	$850	2/10, n/30	8/20	$350		
2. 2/7	265	3/10, n/30	2/15	150		
3. 9/2	470	3/10-30X	10/5	200		
4. 4/6	775	2/10-30X	5/10	450		
5. 10/15	230	3/10, n/30	10/28	120		
6. 9/5	460	4/10, n/30	9/20	250		
7. 12/12	872	3/10EOM	1/7	500		
8. 8/17	927	2/10EOM	9/5	600		

D. Solve these problems.

1. Bob's Beeflike Burgers received a $75 invoice for mustard dated May 17, with terms 2/10, n/30. If the bill is paid on May 25, calculate the amount paid.

2. Bits 'n' Bytes received an $885 invoice for floppy disks dated March 7, with terms 3/10, n/30. Calculate the amount paid if the bill is paid on March 15.

3. Frame-Up received a $1273 invoice for picture matting dated November 27, with terms 3/10, 2/30, n/60. If the bill is paid on December 24, calculate the amount paid.

4. Big Ben's Bolts paid a bill for machine parts on January 25. The invoice was for $275.50, dated January 18, with terms 2/10, 1/30, n/60. Calculate the amount paid.

5. Happy Feet, a shoe store, ordered a quantity of cheerful shoelaces. Happy Feet received the invoice for $129.75, dated March 12, with terms 2/10-30X. If the bill is paid on April 19, calculate the amount paid.

6. Kennywood Amusement Park ordered new clown costumes. The invoice was for $781, dated June 2, with terms 3/10-30X. If the bill is paid on July 7, calculate the amount paid.

7. Scott Township replaced a damaged lifeguard chair at the township pool. The township received an invoice for $557, dated July 21, with terms 2/10EOM. If the lifeguard chair bill is paid on August 7, calculate the amount paid.

8. John Dewey Junior High School is undergoing summer repairs. One invoice was for $267.35, dated August 19, with terms 2/10prox. If the bill is paid on September 8, calculate the amount paid.

9. Petal Pushers received a $95.50 invoice for ribbon dated April 29, with terms 2/10prox. If the bill is paid on June 6, calculate the amount paid.

10. Lew recently ordered green-lipped mussels to serve at his trendy restaurant. With them came an invoice for $374, dated June 27, with terms 3/10EOM. Lew paid the bill on August 9. Calculate the amount paid.

11. Ava's Air Tours ordered new floatational seat cushions. She received an invoice for $689, dated March 4, with terms 3/10ROG. The seat cushions were received on March 20. If the bill is paid on March 27, calculate the amount paid.

12. Wired for Sound received an $842 invoice for speaker wire dated May 23, with terms 2/10ROG. The wire was received on June 5. If the bill is paid on June 12, calculate the amount paid.

13. Maria's Mannequins ordered two new preppy mannequins. The business received the invoice for $887, dated February 15, with terms 2/10, n/30. If $500 is paid on February 23, calculate (a) the amount credited and (b) the remaining balance.

14. The Garden of Earthly Delights paid $750 on April 9 toward a $1271 invoice for clay pots dated April 1, with terms 2/10, n/30. Calculate (a) the amount credited and (b) the remaining balance.

15. The Char Valley Grill received a $472 invoice for charcoal dated January 5, with terms 3/10EOM. If $250 is paid on February 7, calculate (a) the amount credited and (b) the remaining balance.

16. By the Byte, a local software store, received an invoice for $982, dated May 12, with terms 3/10-30X. If $500 is paid on June 15, calculate (a) the amount credited and (b) the remaining balance.

17. On January 23, 1996, Alex, the manager of the sporting goods department of Dynamite Deals, ordered 2 dozen bats with a list price of $25.95 each, and 6 dozen baseballs at $28.59 per dozen. He received trade discounts of 12%, 15%, and 23% on the order. Terms of trade were 3/5, 1/15, n/30EOM. The order was delivered on January 31, 1996, and payment was made on February 5, 1996. How much did Dynamite Deals pay for this order?

18. On February 26, 1996, Alex wanted to order Sam Snead golf club sets and had bids from two distributors. Global Supplier offered the clubs with list price $257, trade discounts of 15% and 24%, and terms of trade 2/10, n/45ROG. Sport Shorts had the same clubs listed at $262, with trade discounts of 14% and 27%, and terms 3/5, 2/15, n/30. If delivery could be arranged for March 5, 1996, and payment could be made on March 13, 1996, which company would offer the best actual price?

Key Terms

cash discount
chain discount
complement
discount
end-of-month (EOM) dating (prox dating)
extra dating
intermediate price
knuckle method
list price

net price
partial payment
receipt-of-goods (ROG) dating
retailer
single equivalent discount rate
single trade discount
trade discount
wholesaler

Chapter 7 at a Glance

Page	Topic	Key Point	Example
242	Discounts (single trade discount)	Discount = rate × price	List price = \$35.90 Discount rate = 30% Discount = $0.30 \times \$35.90$ $= \$10.77$
242	Net price	Net price = list price − discount	List price = \$35.90 Discount = \$10.77 Net price = $\$35.90 - 10.77$ $= \$25.13$
245	Chain discounts	Calculate the complement of each discount rate in the chain discount. Complement = 100% − discount rate Net cost = list price × each complement	List price = \$500 Discount rates = 15/10/5 $100\% - 15\% = 85\%$ $100\% - 10\% = 90\%$ $100\% - 5\% = 95\%$ Net = $\$500 \times 85\% \times 90\% \times 95\%$ $= \$363.375 = \$363.38$ rounded
250	Single equivalent discount rate	Calculate each complement by subtracting the discount rate from 100%. Multiply the complements and express as a percent. Subtract the result from 100%.	Discount rates = 15/10/5 $100\% - 15\% = 85\%$ $100\% - 10\% = 90\%$ $100\% - 5\% = 95\%$ $0.85 \times 0.90 \times 0.95 = 0.72675$ $= 72.675\%$ $100\% - 72.675\% = 27.325\%$
257	Due date	Number days until month end = number days in month − invoice date Number days left in discount period = number days in discount period − number days until month end If number days left in month is greater than number days in month, subtract number days in next month.	Invoice date: August 7, due in 90 days The days until month end = $31 - 7$ $= 24$ Number days left in discount period = $90 - 24 = 66$ 30 days in September: $66 - 30 = 36$ 31 days in October: $36 - 31 = 5$ Due date: November 5

Page	Topic	Key Point	Example
258	Cash discounts	Apply single discount rate if bill paid within indicated number of days after invoice date.	Invoice date: March 2 Terms: 2/10, n/30 Deduct 2% through March 12 and pay full amount thereafter.
260	EOM dating (prox dating)	Apply single discount rate if bill paid within indicated number days after end of month.	Invoice date: May 17 Terms: 2/10EOM, n/30 Deduct 2% through June 10 and pay full amount thereafter.
262	ROG dating	Apply single discount rate if bill paid within indicated number of days after delivery date.	Invoice date: June 15 Terms: 2/10ROG Delivered: July 7 Deduct 2% through July 17 and pay full amount thereafter.
262	Extra dating	Apply single discount rate if bill paid within indicated period after invoice date.	Invoice date: October 3 Terms 2/10-60X Deduct 2% through December 12 $(10 + 60 = 70$ days) and pay full amount thereafter.
263	Balance left after partial payments	$\text{Amount credited} = \dfrac{\text{partial payment}}{(100\% - \text{discount rate})}$ New balance = amount due − amount credited	Invoice date: March 12 Terms: 2/10, n/30 Amount due: $450 Payment made March 20: $200 $\text{Amount credited} = \dfrac{\$200}{(100\% - 2\%)}$ $= \dfrac{\$200}{98\%} = \dfrac{\$200}{0.98}$ $= \$204.08$ New balance $= \$450 - \$204.08$ $= \$245.92$

SELF-TEST   The Mathematics of Buying

The following problems test your understanding of the mathematics of buying.

1. A typist's chair has a list price of $275. If the discount is 23%, (a) calculate the discount and (b) the net price.

2. A monthly organizer lists for $98 with a discount of 32%. Calculate (a) the discount and (b) the net cost.

3. Calculate (a) the discount and (b) the net cost of a compact disc player with a list price of $235 and a 17% discount.

4. Calculate (a) the discount and (b) the net cost of an executive desk with a list price of $755 and a 21% discount.

5. A television set lists for $585 with trade discounts of 25/10. Calculate the net cost.

6. An adding machine lists for $72.50 with trade discounts of 15/10. Calculate the net cost.

7. A 15 drawer filing system has a list price of $908 with trade discounts of 25/15/10. Calculate the net cost.

8. A calculator has a list price of $22.50 with trade discounts of 15/8/5. Calculate the net cost.

9. The list price of a floating seat cushion is $172 with trade discounts of 20/15/5. Calculate (a) the net cost and (b) the single equivalent discount rate.

10. The list price of a receptionist desk is $752 with trade discounts of 15/15/10. Calculate (a) the net cost and (b) the single equivalent discount rate.

11. By the Byte, a local software store, needs to purchase new computers. Two companies have the same list price. However, Company A offers trade discounts of 25/10/5, and Company B offers trade discounts of 20/14/6. Calculate the single equivalent discount rate for (a) Company A and (b) Company B, and (c) determine which company offers the larger discount.

12. MBI Computers offers trade discounts of 20/10/10. CEB Computers offers trade discounts of 30/5/5. Calculate the single equivalent discount rate for (a) MBI Computers and (b) CEB Computers, and (c) determine which company offers the larger discount.

13. Bob's Beeflike Burgers ordered a new cash register and received an invoice for $470 dated March 12, with terms 3/10, n/30. If the bill is paid on March 20, calculate the amount paid.

14. Crystal's Diamonds received a $785 invoice for an engraving machine dated May 2, with terms 2/10, n/30. If the bill is paid on May 10, calculate the amount paid.

15. Kate's Delicacies received an $82.50 invoice for saffron dated November 7, with terms 3/10, 2/30, n/60. If the bill is paid on December 5, calculate the amount paid.

16. Scott Township bought new snow removal equipment and received the invoice for $1275 dated January 12, with terms 2/10, 1/30, n/60. If the bill is paid on January 20, calculate the amount paid.

17. The Char Valley Grill received an $89.25 invoice for placemats dated March 21, with terms 3/10-30X. If the bill is paid on April 25, calculate the amount paid.

18. Bits 'n' Bytes received a $973 invoice for keyboard vacuums dated July 12, with terms 2/10-30X. If the bill is paid on August 7, calculate the amount paid.

19. The Butter-em-Up Bakery ordered new pastry brushes and received the invoice for $35.75 dated January 23, with terms 3/10EOM. If the bill is paid on February 7, calculate the amount paid.

20. Intense Tents received an invoice for $178.75 dated August 17, with terms 2/10prox. If the bill is paid on September 5, calculate the amount paid.

21. The Fins and Feathers Pet Store purchased a shipment of saltwater fish. They received the invoice for $275 dated March 27, with terms 2/10prox. If the bill is paid on May 6, calculate the amount paid.

22. Tour De Countryside, a local bicycle shop, received an invoice for $482 dated May 29, with terms 3/10EOM, If the bill is paid on July 7, calculate the amount paid.

23. Extravagant Utensils bought a crate of cooking gadgets for $196. The invoice for $196 was dated March 22, with terms 3/10ROG. The merchandise was received on March 30. If the bill is paid on April 8, calculate the amount paid.

24. Trim and Green purchased six lawn mowers. It received an invoice for $1724 dated May 13, with terms 2/10ROG. The merchandise was received on May 27. If the bill is paid on June 5, calculate the amount paid.

25. Ava's Air Tours received a $595 invoice for flyers dated February 5, with terms 3/10, n/30. If $250 is paid on February 12, calculate (a) the amount credited and (b) the remaining balance.

26. Wreck-a-Mended Auto received an $889 invoice for oil dated April 15, with terms 3/10, n/30. If $450 is paid on April 23, calculate (a) the amount credited and (b) the remaining balance.

WORKSPACE

The Mathematics of Selling

CHAPTER OBJECTIVES

When you complete this chapter successfully, you will be able to:

1. Determine the markup on a product based on its cost.

2. Calculate the markup on a product based on its selling price.

3. Find the markdown and markdown rate on goods.

■ It costs Wreck-a-Mended Auto $362.12 to do a two-tone paint job on a four-door automobile. The firm uses a 28% rate of markup based on cost. What are the markup and the selling price?

■ The Image Sharpener sells a telephone with a voice synthesizer and answering machine. This combination costs $247.50 and has a markup of $202.50. What are the selling price and the rate of markup based on selling price?

■ The Appliance Warehouse is having trouble moving the fashion-smart neon color appliances that its former sales manager ordered. The bright pink dryer originally sold for $545. It was first marked down 30%, then further reduced 25%, and finally marked down an additional 20%. What is the final reduced price of the bright dryer?

In the last chapter, we looked at the way firms pay for the things they need in order to do business. But the ultimate aim of any company, of course, is to *sell* its goods and services. In this chapter, we will consider the ways in which retail businesses price their products. You'll be sold on how easy it is!

Can You:

Determine the markup on a product, given its cost and markup rate?
Calculate the markup rate for a product, given its costs and markup?
Find the cost of a product, given its markup and markup rate?
Determine the cost of a product, given its selling price and markup rate?
. . . If not, you need this section.

Comparing Cost, Markup, and Selling Price

Cost   Among retailers, the original price of merchandise paid by the retailer.

Markup   Among retailers, the amount a retailer adds to the cost to cover costs and provide a profit. It may be calculated based on cost or on desired selling price.

Selling Price   Among retailers, the cost of an item plus a markup on it—the price paid by the buyer.

An important part of any retail business is the proper pricing of its merchandise. The **cost** is the original price of the merchandise paid by the business. To the cost a retailer must add an additional amount called the **markup** to cover its expenses and give it a profit. The sum of cost plus the markup is the **selling price,** which is the retailer's price to individual consumers.

Selling price = cost + markup

For example, consider a radio whose cost is \$27.50 and markup is \$8.25.

Cost   Markup

Selling price = \$27.50 + \$8.25 = \$35.75

YOUR TURN

WORK THIS PROBLEM

The Question:

The cost of a suit is \$75.26 and is marked up \$41.39. Calculate the selling price.

✔ YOUR WORK

The Solution:

Selling price = cost + markup
= \$75.26 + 41.39 = \$116.65

In markup problems where you know the cost and selling price, just calculate the markup using the following form of the basic markup equation:

Markup = selling price − cost

For example, if the selling price is \$129.95 and the cost is \$95.50, the markup is

Selling price   Cost

Markup = \$129.95 − 95.50 = \$34.45

Finally, if you know the selling price and the markup, you can calculate the cost using the following markup equation.

> Cost = selling price − markup

For example, if the selling price is $32.95 and the markup is $9.88, the cost is

Selling price   Markup

Cost = $32.95 − 9.88 = $23.07

YOUR
TURN

WORK THIS PROBLEM

The Question:

(a) The cost of a calculator is $32.50. If the selling price is $41.95, calculate the markup.
(b) The cost of a power saw is $195.25 and the markup is $54.72. What is the selling price?
(c) The selling price of vinyl flooring is $13.75 per square yard. If the markup is $5.90, what is the cost?

✔ YOUR WORK

The Solution:

(a) Markup = selling price − cost
    = $41.95 − 32.50 = $9.45
(b) Selling price = cost + markup
    = $195.25 + 54.72 = $249.97
(c) Cost = selling price − markup
    = $13.75 − 5.90 = $7.85

When a business calculates markup, it cannot just subtract cost from the selling price. Instead, it must decide on a markup rate and apply it to either cost or selling price.

Markup *based on cost* is generally used by manufacturers, wholesalers, and some retailers who take inventory at cost. We will consider markup based on selling price in the next section.

There are four basic markup types of problems using markup based on cost: problems to find the markup, problems to find the markup rate, and two kinds of problems to find the cost. To solve each of these types of problems, however, you must have some information about the product's cost, selling price, markup, and/or markup rate.

Finding the Markup When Cost and Rate Are Known

To calculate the markup when you know the rate of markup and the cost, use the following equation.

> Markup = rate × cost   (based on cost)

This type of problem is the same type as calculating percents in Chapter 4. The cost is the base, the rate is the percent, and the markup is the part. If you need a review of the basic percent equation, return to Chapter 4. Section 4.2.

Suppose a new bicycle costs $75.30 and is marked up 32% based on cost. What are the markup and selling price?

First, use the formula to find the markup.

Rate   Cost

$$\text{Markup} = 32\% \times \$75.30 = 0.32 \times \$75.30 = \$24.096 = \$24.10 \text{ rounded}$$

Now you can apply the selling price formula.

$$\text{Selling price} = \text{cost} + \text{markup}$$
$$= \$75.30 + 24.10 = \$99.40$$

YOUR TURN

| WORK THIS PROBLEM |

The Question:

(a) A tool set costs $129.00 and is marked up 45% based on cost. Calculate the markup and selling price.

(b) A radio's cost is $59.26 and is marked up 37.5% based on cost. What are the markup and selling price?

| ✔ YOUR WORK |

The Solution:

(a) Markup = rate × cost
    = 45% × $129 = 0.45 × $129 = $58.05
    Selling price = cost + markup
    = $129 + 58.05 = $187.05

(b) Markup = rate × cost
    = 37.5% × $59.26 = 0.375 × $59.26 = $22.2225 = $22.22 rounded
    Selling price = cost + markup
    = $59.26 + 22.22 = $81.48

Finding the Rate When Markup and Cost Are Known

To find the rate when you know the cost and markup, use the formula:

$$\text{Rate} = \frac{\text{markup}}{\text{cost}} \qquad \text{(based on cost)}$$

Such problems are %-type percent problems from Chapter 4.

For example, if a stereo system costs $365 and is marked up $131.40, what is the rate of markup based on cost?

$$\text{Rate} = \frac{\text{markup}}{\text{cost}} = \frac{\$131.40}{\$365} = 0.36 = 36\%$$

YOUR TURN

$\boxed{\text{WORK THIS PROBLEM}}$

The Question:

(a) A lamp costs $28.30 and is marked up $8.49. What is the rate of markup based on cost?
(b) A freezer costs $320 and sells for $400. Calculate the rate of markup based on cost. (Hint: First calculate the markup.)

$\boxed{\checkmark \text{ YOUR WORK}}$

The Solution:

(a) $\text{Rate} = \dfrac{\text{markup}}{\text{cost}} = \dfrac{\$8.49}{\$28.30} = 0.30 = 30\%$

(b) Markup = selling price − cost = $400 − 320 = $80

$\text{Rate} = \dfrac{\text{markup}}{\text{cost}} = \dfrac{\$80}{\$320} = 0.25 = 25\%$

Note: You must first find markup.

Finding the Cost When Markup and Rate Are Known

To calculate the cost when you know the markup and the rate of markup, use the following formula:

$$\text{Cost} = \frac{\text{markup}}{\text{rate}} \qquad \text{(based on cost)}$$

This formula—like the formula for solving *B*-type problems in Section 4.2—is easy to find using a circle diagram.

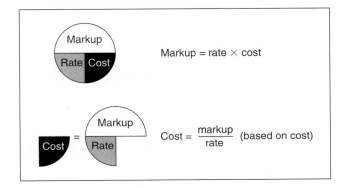

For example, if a place setting of china is marked up $35.40 and the rate of markup is 60% based on cost, what are the cost and selling price?

First, use the formula for cost.

$$\text{Cost} = \frac{\text{markup}}{\text{rate}} = \frac{\$35.40}{60\%} = \frac{\$35.40}{0.60} = \$59$$

Now you can apply the selling price formula.

$$\text{Selling price} = \text{cost} + \text{markup}$$
$$= \$59 + 35.40 = \$94.40$$

WORK THIS PROBLEM

YOUR TURN

The Question:

(a) A new chair is marked up $40.60. If the rate of markup is 28% based on cost, calculate the cost and selling price.

(b) A desk is marked up $36.40. If the rate of markup is 65% based on cost, what are the cost and selling price?

✔ YOUR WORK

The Solution:

(a) $\text{Cost} = \dfrac{\text{markup}}{\text{rate}} = \dfrac{\$40.60}{28\%} = \dfrac{\$40.60}{0.28} = \$145.00$

Or, using a calculator,   40.6 $\boxed{\div}$ 28 $\boxed{\%}$ ⟶ ▨▨▨ 145.

$$\text{Selling price} = \text{cost} + \text{markup}$$
$$= \$145.00 + \$40.60 = \$185.60$$

(b) $\text{Cost} = \dfrac{\text{markup}}{\text{rate}} = \dfrac{\$36.40}{65\%} = \dfrac{\$36.40}{0.65} = \$56.00$

Or, using a calculator,   36.4 $\boxed{\div}$ 65 $\boxed{\%}$ ⟶ ▨▨▨ 56.

$$\text{Selling price} = \text{cost} + \text{markup}$$
$$= \$56.00 + \$36.40 = \$92.40$$

Finding the Cost When the Selling Price and Rate Are Known

The last type of problem using markup based on cost, calculating the cost when you know the selling price and rate, calls for a new formula.

$$\text{Cost} = \frac{\text{selling price}}{100\% + \text{rate}} \qquad \text{(based on cost)}$$

For example, if a weight lifting set sells for $24.30 and is marked up 35% based on cost, what are the cost and markup?

$$\text{Cost} = \frac{\text{selling price}}{100\% + \text{rate}} = \frac{\$24.30}{100\% + 35\%} = \frac{\$24.30}{135\%} = \frac{\$24.30}{1.35} = \$18.00$$

Once you know the cost, it's easy to find the markup.

$$\text{Markup} = \text{selling price} - \text{cost}$$
$$= \$24.30 - 18.00 = \$6.30$$

YOUR TURN

[WORK THIS PROBLEM]

The Question:

(a) A chair sells for $278.46 and is marked up 56% based on cost. Calculate the cost and markup.

(b) A desk sells for $349.83 and is marked up 38% based on cost. Find the cost and markup.

[✔ YOUR WORK]

The Solution:

(a) $\text{Cost} = \dfrac{\text{selling price}}{100\% + \text{rate}} = \dfrac{\$278.46}{100\% + 56\%} = \dfrac{\$278.46}{156\%} = \dfrac{\$278.46}{1.56} = \$178.50$

   Or, using a calculator,   278.46 $\div$ 156 $\boxed{\%}$ ⟶ | 178.5 |

   $\text{Markup} = \text{selling price} - \text{cost}$
   $= \$278.46 - 178.50 = \$99.96$

(b) $\text{Cost} = \dfrac{\text{selling price}}{100\% + \text{rate}} = \dfrac{\$349.83}{100\% + 38\%} = \dfrac{\$349.83}{138\%} = \dfrac{\$349.83}{1.38} = \$253.50$

   $\text{Markup} = \text{selling price} - \text{cost}$
   $= \$349.83 - 253.50 = \$96.33$

It may seem like a lot to remember, but really there are only two basic formulas. The first formula

$$\text{Markup} = \text{rate} \times \text{cost}$$

or one of its equivalent forms is used in three of the four different problems. In the case where the first formula will not work, use the second formula based on cost.

$$\text{Cost} = \frac{\text{selling price}}{100\% + \text{rate}}$$

Using a calculator can speed the process of solving problems involving markups based on cost.

Calculating Markup Based on Cost with a Calculator

To see how you can use a calculator in markup problems, consider a tool set that costs $129 and is marked up 45% based on cost (a $58.05 markup) for a selling price of $187.05.

If you know the cost and the rate, you can find the markup and the selling price by pressing

129 ⌈×⌉ 45 ⌈%⌉ ⟶　　　　58.05　⌈+⌉ 129 ⌈=⌉ ⟶　　　　187.05

If you know the markup and cost, you can find the rate by pressing

58.05 ⌈÷⌉ 129 ⌈=⌉ ⟶　　　　0.45

If you know the markup and rate, you can find the cost by pressing

58.05 ⌈÷⌉ 45 ⌈%⌉ ⟶　　　　129.

If you know the selling price and markup rate, you can find the cost by pressing

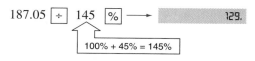

187.05 ⌈÷⌉ 145 ⌈%⌉ ⟶　　　　129.

100% + 45% = 145%

It's really easy, but if you don't think so yet, go back through this section again. When you feel comfortable with markups based on cost, try your hand at the problems in Section Test 8.1.

SECTION TEST 8.1 The Mathematics of Selling

Name _____

Date _____

Course/Section _____

The following problems test your understanding of Section 8.1, Calculating Markup Based on Cost.

A. Solve for the unknowns using markup based on cost.

	Cost	Markup	Selling Price	Rate
1.	$458.50			14%
2.	$29.85			20%
3.	$126.75	$40.56		
4.	$298.50	$47.76		
5.		$223.02		45%
6.		$309.30		25%
7.			$948.08	12%
8.			$1218.51	35%
9.	$2386.50	$1002.33		
10.	$192.40			85%
11.		$175.77		62%
12.		$492.47		55%
13.			$936.47	48%
14.	$872.15			40%
15.	$56.50	$13.56		
16.			$60.80	28%

B. Solve these problems.

1. The Appliance Warehouse can purchase ice crusher attachments for $78.50. The sales manager marked up each ice crusher attachment $20.41. What are (a) the selling price and (b) the rate of markup based on cost?

2. A case of artichokes costs $56.90. The produce manager at P&A Supermarkets marked up each case $11.38. What are (a) the selling price and (b) the rate of markup based on cost?

3. Crystal's Diamonds sells a brooch that has a markup of $117.39. Its rate of markup based on cost is 35%. What are (a) the cost and (b) the selling price?

4. Tredwell's Tire Shop sells a truck tire that has a markup of $53.55 and a 42% rate of markup based on cost. What are (a) the cost and (b) the selling price of a truck tire?

5. It costs Wreck-a-Mended Auto $362.12 to do a two-tone paint job on a four-door automobile. The firm uses a 28% rate of markup based on cost. What are (a) the markup and (b) the selling price?

6. Intense Tents buys a deluxe model tent for $735.28 and uses a 16% rate of markup based on cost to determine the selling price. What are (a) the markup and (b) the selling price of the deluxe model tent?

7. Trim and Green's riding lawn mower has a selling price of $2730 and a 12% rate of markup based on cost. What are (a) the cost and (b) the markup?

8. Bits 'n' Bytes buys a printer stand that costs $92.75 and uses a 19% rate of markup based on cost to determine the selling price. What are (a) the markup and (b) the selling price?

9. The Kitschy Kitchen sells an omelet pan for $45.26 that has a 24% rate of markup based on cost. What are (a) the cost and (b) the markup?

10. By the Byte sells a desktop publishing software package for $746.13. This includes a 32% rate of markup based on cost. What are (a) the cost and (b) the markup?

11. Phyle Office Products sells a briefcase that has a markup of $43.96 and the rate of markup based on cost is 20%. What are (a) the cost and (b) the selling price of the briefcase?

12. Clothes Make the Man sells a coat that costs $307.50 and has a markup of $86.10. What are (a) the selling price of the coat and (b) the rate of markup based on cost?

13. Usable Art bought a wooden, engraved cane for $235.75. The store will sell the cane with a 32% rate of markup based on cost. What are (a) the markup and (b) the selling price of the cane?

14. The Butter-em-Up Bakery has a markup of $4.81 on a dozen croissants. The rate of markup based on cost is 26%. What are (a) the cost and (b) the selling price?

15. Designer Discounts sells a name label handbag for $138.25. A 40% rate of markup based on cost is used to calculate the selling price. What are (a) the cost of the handbag and (b) the markup?

16. Sun Fun, Inc., purchases sun deck kits for $625 and sells the kit with a markup of $168.75. What are (a) the selling price of a sun deck kit and (b) the rate of markup based on cost?

WORKSPACE

Can You:

Determine the markup on a product, given its selling price and markup rate?
Find the markup rate on a product, given its markup and selling price?
Calculate the selling price of a product, given its markup and markup rate?
Determine the selling price of a product, given its cost and markup rate?
. . . If not, you need this section.

The second basic method to calculate markup is based on *selling price*. This method is used by many retailers.

There are four types of markup problems using markup based on selling price: problems to find the markup, problems to find the markup rate, and two types of problems to find the selling price.

All but the last of these problems use the fundamental markup formula based on selling price:

> Markup = rate × selling price   (based on selling price)

This formula is a percent equation where the base is the selling price, the rate is the percent, and the markup is the part.

Finding the Markup When the Selling Price and Rate Are Known

The easiest problem is calculating the markup when the selling price and rate are known. For example, if a synthetic sleeping bag sells for $125.90 and its markup rate is 30% based on selling price, what are the markup and cost?

Rate   Selling price

Markup = 30% × $125.90 = 0.30 × $125.90 = $37.77

Once we know the markup, it's easy to find the cost.

$$\text{Cost} = \text{selling price} - \text{markup}$$
$$= \$125.90 - 37.77 = \$88.13$$

WORK THIS PROBLEM

YOUR TURN

The Question:

(a) A wastepaper basket sells for $7.95 and is marked up 45% based on selling price. What are the markup and cost?

(b) A carabiner has a selling price of $4.00 and is marked up $34\frac{1}{2}\%$ based on selling price. Calculate the markup and cost.

✔ YOUR WORK

The Solution:

(a) Markup = rate × selling price
     = 45% × $7.95 = 0.45 × $7.95 = $3.5775 = $3.58 rounded
     Cost = selling price − markup
     = $7.95 − 3.58 = 4.37
(b) Markup = rate × selling price
     = $34\frac{1}{2}$% × $4.00 = 34.5% × $4.00 = 0.345 × $4.00 = $1.38
     Cost = selling price − markup
     = $4.00 − 1.38 = $2.62

Finding the Rate When the Markup and Selling Price Are Known

The second type of problem, calculating the rate of markup based on selling price when the markup and selling price are known, is a %-type percent problem.

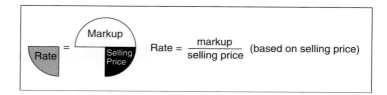

$$\text{Rate} = \frac{\text{markup}}{\text{selling price}} \quad \text{(based on selling price)}$$

For example, if a calculator is marked up $3.85 and sells for $17.50, what's the rate of markup based on selling price?

$$\text{Rate} = \frac{\text{markup}}{\text{selling price}} = \frac{\$3.85}{\$17.50} = 0.22 = 22\%$$

WORK THIS PROBLEM

YOUR TURN

The Question:

(a) A desk costs $53.97 and sells for $89.95. What is the rate of markup based on selling price? (Hint: First find the markup.)
(b) A dress is marked up $25.68 and sells for $42.80. Calculate the rate of markup based on selling price.

✔ YOUR WORK

The Solution:

(a) Markup = selling price − cost
     = $89.95 − 53.97 = $35.98

$$\text{Rate} = \frac{\text{markup}}{\text{selling price}} = \frac{\$35.98}{\$89.95} = 0.40 = 40\%$$

(b) $$\text{Rate} = \frac{\text{markup}}{\text{selling price}} = \frac{\$25.68}{\$42.80} = 0.60 = 60\%$$

Finding the Selling Price When the Markup and Rate Are Known

A third type of problem, calculating the selling price when you know the markup and rate of markup, is a *B*-type percent problem with selling price as the base.

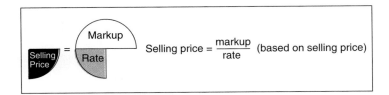

$$\text{Selling price} = \frac{\text{markup}}{\text{rate}} \quad \text{(based on selling price)}$$

For example, if a paperback book is marked up $2.20, which is 40% based on selling price, what are the selling price and cost?

$$\text{Selling price} = \frac{\text{markup}}{\text{rate}} = \frac{\$2.20}{40\%} = \frac{\$2.20}{0.40} = \$5.50$$

$$\text{Cost} = \text{selling price} - \text{markup}$$
$$= \$5.50 - 2.20 = \$3.30$$

YOUR TURN

─── WORK THIS PROBLEM ───────────────────────────────

The Question:

(a) A sewing machine is marked up $157.50, which is 35% based on selling price. Calculate the selling price and cost.

(b) A Tricam is marked up $8.28. If the rate of markup is 18% based on selling price, find the selling price and cost.

─── ✔ YOUR WORK ───────────────────────────────

The Solution:

(a) Selling price $= \dfrac{\text{markup}}{\text{rate}} = \dfrac{\$157.50}{35\%} = \dfrac{\$157.50}{0.35} = \$450$

 Cost = selling price − markup
  $= \$450 - 157.50 = \$292.50$

(b) Selling price $= \dfrac{\text{markup}}{\text{rate}} = \dfrac{\$8.28}{18\%} = \dfrac{\$8.28}{0.18} = \$46.00$

 Cost = selling price − markup
  $= \$46.00 - 8.28 = \$37.72$

Finding the Selling Price When the Cost and Rate Are Known

The last type of markup problem, calculating the selling price when you know the cost and rate of markup based on selling price, calls for a new formula:

$$\text{Selling price} = \frac{\text{cost}}{100\% - \text{rate}} \quad \text{(based on selling price)}$$

Suppose a new workbook costs the bookstore $9.24 and is marked up 23% based on selling price. What are the selling price and markup?

$$\text{Selling price} = \frac{\text{cost}}{100\% - \text{rate}} = \frac{\$9.24}{100\% - 23\%} = \frac{\$9.24}{77\%} = \frac{\$9.24}{0.77} = \$12.00$$

$$\text{Markup} = \text{selling price} - \text{cost} = \$12.00 - 9.24 = \$2.76$$

Using a calculator can speed the process of solving problems involving markups based on selling price.

Calculating Markup Based on Selling Price with a Calculator

To see how you can use a calculator in markup problems, let's consider a blouse that has a selling price of $42.80, a 60% rate of markup based on selling price, a cost of $17.12, and a markup of $25.68.

If you know the selling price and rate, you can find the markup and cost by pressing

42.8 ⨯ 60 % ⟶ [25.68] +/− + 42.8 = ⟶ [17.12]

If you know the markup and the selling price, you can find the rate by pressing

25.68 ÷ 42.8 = ⟶ [0.6] or 60%

If you know the markup and markup rate, you can find the selling price by pressing

25.68 ÷ 60 % ⟶ [42.8] or $42.80

If you know the cost and rate, you can find the selling price by pressing

17.12 ÷ 40 % ⟶ [42.8]

100% − 60% = 40%

WORK THIS PROBLEM

YOUR TURN

The Question:

(a) A microwave oven costs $260.64 and is marked up 28% based on selling price. Find the selling price and markup.

(b) A computer costs $2695.00 and is marked up 16% based on selling price. Calculate the selling price and markup.

✔ YOUR WORK

The Solution:

(a) $\text{Selling price} = \dfrac{\text{cost}}{100\% - \text{rate}} = \dfrac{\$260.64}{100\% - 28\%} = \dfrac{\$260.64}{72\%}$

$= \dfrac{\$260.64}{0.72} = \$362.00$

Markup = selling price − cost
= $362.00 − 260.64 = $101.36

(b) Selling price = $\dfrac{\text{cost}}{100\% - \text{rate}} = \dfrac{2695.00}{100\% - 16\%} = \dfrac{\$2695.00}{84\%}$

$= \dfrac{\$2695.00}{0.84} = \$3208.33$ rounded

Markup = selling price − cost
= $3208.33 − 2695.00 = $513.33

Like markup based on cost, then, markup based on selling price always involves one of two equations (though these equations are different from those for cost-based markup).

Markup Based on Selling Price	Markup Based on Cost
Markup = rate × selling price	Markup = rate × cost
Selling price = $\dfrac{\text{cost}}{100\% - \text{rate}}$	Cost = $\dfrac{\text{selling price}}{100\% + \text{rate}}$

When working markup problems first, always check whether the markup is based on cost or based on selling price. Otherwise your calculations will be useless.

The problems in Section Test 8.2 will test your understanding of markups based on selling price.

WORKSPACE

SECTION TEST 8.2   The Mathematics of Selling

The following problems test your understanding of Section 8.2, Calculating Markup Based on Selling Price.

A. Solve for the unknowns using markup based on selling price.

	Cost	Markup	Selling Price	Rate
1.			$238.50	26%
2.			$229.60	45%
3.		$789.92	$2468.50	
4.		$136.14	$907.60	
5.		$91.74		12%
6.		$586.85		25%
7.	$217.69			45%
8.	$418.74			16%
9.			$256.75	36%
10.		$3.97		20%
11.		$48.33		18%
12.			$75.50	24%
13.		$352.94	$882.35	
14.	$124.15			48%
15.	$103.46			65%
16.		$114.66	$409.50	

B. Solve these problems

1. Bits 'n' Bytes buys computer monitors for $263.68 and sells them with a markup of $65.92. What are (a) the selling price and (b) the rate of markup based on selling price?

2. The Appliance Warehouse purchases refrigerators for $513.91 and sells them with a markup of $241.84. What are (a) the selling price and (b) the rate of markup based on selling price?

3. Kate's Delicacies sells Black Forest Cakes by the dozen. Kate has a markup of $28.13 per dozen to cover spoilage costs. The rate of markup based on selling price is 58%. What are (a) the cost and (b) the selling price of a dozen of Black Forest Cakes?

4. Cycling helmets have a markup of $15.95 and a 20% rate of markup based on selling price. What are (a) the cost and (b) the selling price of one cycling helmet?

5. The Ancient Mariner sells a bass boat that costs $3042.60 and has a 12% rate of markup based on selling price. What are (a) the markup and (b) the selling price?

6. Ava's Air Tours sells a two-hour tour package that costs $200.09 per seat. She uses a 15% markup based on selling price. What are (a) the markup and (b) the selling price?

7. Crystal's Diamonds has a pair of earrings that sell for $162.20 and a 15% rate of markup based on selling price. What are (a) the cost and (b) the markup?

8. A Splotch watch costs $160.95 and has a 42% rate of markup based on selling price. What are (a) the markup and (b) the selling price?

9. A "punkette" dummy from Maria's Mannequins has a selling price of $245.40 and a 35% rate of markup based on selling price. What are (a) the cost and (b) the markup?

10. Managers at the By the Byte software store have decided that the most a customer will pay for a tax preparation program is a selling price of $76.90. With a 20% rate of markup based on selling price, what are (a) the cost and (b) the markup?

11. Trim and Green sells a lawn mower that has a markup of $69.81 and a 26% rate of markup based on selling price. What are (a) the cost and (b) the selling price of the lawn mower?

12. The Kitschy Kitchen sells a set of gourmet pans that costs the business $143 to purchase. The pans are priced with a markup of $132. What are (a) the selling price of the pans and (b) the rate of markup based on selling price?

13. A 6-ounce package of smoked Alaskan salmon costs $14.43 and has a 26% rate of markup based on selling price. What are (a) the markup and (b) the selling price?

14. The Swell Telephone Company sells a cordless telephone with a markup of $36.70 and a 40% rate of markup based on selling price. What are (a) the cost and (b) the selling price?

15. The Swell Telephone Company also sells a cordless telephone that has its own answering machine for $186. A 37% rate of markup based on selling price was used to calculate the price. What are (a) the cost and (b) the markup?

16. The Image Sharpener sells a telephone with a voice synthesizer and answering machine. This combination costs $247.50 and has markup of $202.50. What are (a) the selling price and (b) the rate of markup based on selling price?

17. Dynamite Deals had purchased wine glasses at a cost of $15.00 each. These were marked up the normal 85% based on cost. In June, the wine glasses were marked down 25% for a wedding special. Then July 1, they were marked back up 15%. (a) What was the selling price of the wine glasses in July? (b) What was the actual markup rate based on the original cost in July?

18. In the hardware department of Dynamite Deals, manager Sue Morgan offered The Handy-Dandy fertilizer spreader for sale with a markup of 67% based on the selling price. Sue's cost for these fertilizer spreaders was $19.14 each. What did Sue charge for each fertilizer spreader?

WORKSPACE

SECTION 8.3: Setting Markdowns

Can You:

Find a markdown?
Calculate a series of markdowns?
Determine a markdown rate?
. . . If not, you need this section.

Markdown A discount on the selling price of an item designed to spur sales of that item.

Oftentimes, it is necessary for a business to reduce the price or *mark down* its merchandise. Such a **markdown**—a discount on the selling price—may be necessary to keep prices competitive, to move old merchandise, or to close out a line of merchandise due to changes in style or models.

Finding Markdown

If the markdown rate is given, the markdown is calculated the same as a discount.

> Markdown = rate × selling price
>
> Reduced selling price = selling price − markdown

Suppose a watch originally sold for $129.90 but has been marked down 30%. Calculate the markdown and the reduced selling price.

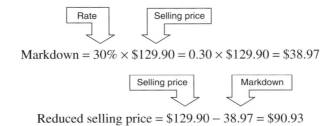

Markdown = 30% × $129.90 = 0.30 × $129.90 = $38.97

Reduced selling price = $129.90 − 38.97 = $90.93

WORK THIS PROBLEM

The Question:

(a) A new television lists for $575.68 but is marked down 18%. What is the markdown and reduced selling price?

(b) A coffee maker originally sold for $29.95 but has been marked down 28%. Calculate the markdown and sale price.

✔ YOUR WORK

The Solution:

(a) Markdown = 18% × $575.68 = 0.18 × $575.68
        = $108.6224 = $103.62 rounded
  Reduced selling price = $575.68 − 103.62 = $472.06

(b) Markdown = 28% × \$29.95 = 0.28 × \$29.95
                     = \$8.386 = \$8.39 rounded
      Reduced selling price = \$29.95 − 8.39 = \$21.56

Finding a markdown with a calculator is a snap.

Finding Markdowns with a Calculator

It's easy to find the markdown if you know the selling price and markdown rate. In part
(a) of the previous problem,

575.68 ☒ × ☒ 18 ☒ % ☒ ⟶   103.6224

575.68 ☒ − ☒ 103.6224 ☒ = ☒ ⟶   472.0576

Calculating a Series of Markdowns

Intermediate Reduced Price In calculating markdowns, the result of deducting a markdown from the initial price or from another intermediate reduced price.

When a business is trying to sell slow-moving or close-out merchandise, it is often necessary to mark down prices several times. In calculating a series of markdowns, you base each markdown on the previous selling price and calculate an **intermediate reduced price.** (This is the same process we used in calculating multiple trade discounts in Section 7.1.)

Steps: Calculating a Series of Markdowns

STEP 1. *Calculate* the first markdown.

STEP 2. *Subtract* the markdown from the original selling price. The result is an intermediate reduced price.

STEP 3. *Calculate* the next markdown based on the intermediate reduced price, *not* the original selling price.

STEP 4. *Subtract* the new markdown from the intermediate reduced price.

STEP 5. If you have more than two markdowns, *repeat* steps 3 and 4 for each additional markdown.

For example, suppose an executive desk chair originally sold for \$650. This price was first marked down 25%, then reduced 20% more, and finally marked down an additional 30%. What is the final reduced price?

Step 1.   First markdown = 25% × \$650 = 0.25 × \$650 = \$162.50

Step 2.   Intermediate reduced price = \$650 − \$162.50 = \$487.50

Step 3.   Second markdown = 20% × \$487.50 = 0.20 × \$487.50 = \$97.50

Step 4.   Intermediate reduced price = \$487.50 − 97.50 = \$390

Step 5.   Repeat steps 3 and 4.

　　　　Step 3.   Third markdown = 30% × \$390 = 0.30 × \$390 = \$117

　　　　Step 4.   Final reduced price = \$390 − 117 = \$273

WORK THIS PROBLEM

The Question:

A Splotch watch originally sold for $150. The price was first marked down 20%, then reduced 15%, and finally marked down an additional 30%. What is the final reduced price?

✔ YOUR WORK

The Solution:

Step 1. 20% × $150 = 0.20 × $150 = $30

Step 2. $150 − 30 = $120

Step 3. 15% × $120 = 0.15 × $120 = $18

Step 4. $120 − 18 = $102

Step 5. Repeat steps 3 and 4.

   Step 3. 30% × $102 = 0.30 × $102 = $30.60

   Step 4. $102 − 30.60 = $71.40

Yes, that is a lot of work. But fortunately, as with chain discounts, there is an easier method.

Steps: Calculating a Series of Markdowns: The Shortcut

STEP 1. *Calculate* the complement of each markdown rate by subtracting the markdown rate from 100%.

STEP 2. *Multiply* the original selling price times each complement. The product is the final reduced price.

For example, a stereo system from Wired for Sound originally sold for $450. The price was first marked down 20%, then an additional 25%, and finally another 30%. What is the final reduced price?

Step 1. The complement of the first markdown = 100% − 20% = 80%.
The complement of the second markdown = 100% − 25% = 75%.
The complement of the third markdown = 100% − 30% = 70%.

Step 2. Final reduced price = $450 × 80% × 75% × 70%
= $450 × 0.80 × 0.75 × 0.70 = $189

With a calculator,   450 ⊠ 80 % ⊠ 75 % ⊠ 70 % ⟶ 189.

WORK THIS PROBLEM

The Question:

(a) A table originally sold for $750. The price was first marked down 30%, then an additional 25%, and finally another 20%. What is the final reduced price?

(b) A microcomputer system originally sold for $2500. The price was first marked down 15%, then an additional 10%, and finally another 10%. What is the final reduced price?

The Solution:

(a) Find the complements of the markdown rates.

$$100\% - 30\% = 70\%; \quad 100\% - 25\% = 75\%; \quad 100\% - 20\% = 80\%$$

$$\text{Final reduced price} = \$750 \times 70\% \times 75\% \times 80\%$$
$$= \$750 \times 0.70 \times 0.75 \times 0.80 = \$315$$

(b) $100\% - 15\% = 85\%; \quad 100\% - 10\% = 90\%; \quad 100\% - 10\% = 90\%$

$$\text{Final reduced price} = \$2500 \times 85\% \times 90\% \times 90\%$$
$$= \$2500 \times 0.85 \times 0.90 \times 0.90 = \$1721.25$$

Determining Markdown Rate

A second type of markdown problem, calculating the markdown rate when you know the markdown, is simply a %-type percent problem.

For example, if a chair originally sold for $65 but has been marked down $13, what is the markdown rate based on selling price?

Markdown

$$\text{Rate} = \frac{\$13}{\$65} = 0.20 = 20\%$$

Selling price

YOUR TURN

The Question:

(a) A lamp originally sold for $25 but has been marked down $3. Find the rate of markdown.
(b) A used automobile's first price was $6545 but has been marked down to $6021.40. Calculate the markdown rate. (Hint: First find the markdown.)

The Solution:

(a) $\text{Rate} = \dfrac{\text{markdown}}{\text{selling price}} = \dfrac{\$3}{\$25} = 0.12 = 12\%$

(b) Markdown = selling price − reduced price

$$= \$6545 - 6021.40 = \$523.60$$

$$\text{Rate} = \frac{\text{markdown}}{\text{selling price}} = \frac{\$523.60}{\$6545} = 0.08 = 8\%$$

Note: You must first calculate markdown.

Problems to test your grasp of markdowns are in Section Test 8.3.

WORKSPACE

SECTION TEST 8.3 The Mathematics of Selling

The following problems test your understanding of Section 8.3, Setting Markdowns.

A. Solve for the unknowns.

	Original Price	Rate	Markdown	Reduced Price
1.	$29.50	26%		
2.	$276.95	40%		
3.	$82.75		$19.86	
4.	$3278.00		$229.46	
5.	$863.25	28%		
6.	$235.50	36%		
7.	$525.60		$26.28	
8.	$1576.00		$118.20	
9.	$98.50			$84.71
10.	$16.75			$15.41
11.			$85.86	$486.54
12.			$2.73	$42.77

B. Solve these problems.

1. Clothes Make the Man has taken a cashmere scarf and cap set that originally sold for $76 and marked it down by $16\frac{1}{2}$%. What are (a) the markdown and (b) the reduced price?

2. Maria's Mannequins has a unisex dummy on sale for 20% off. The original selling price was $395.25. What are (a) the markdown and (b) the reduced price?

3. The Happy Feet Shoe Store has self-propelled athletic shoes that originally sold for $168.40, marked down by $42.10. What are (a) the markdown rate and (b) the reduced price?

4. Bits 'n' Bytes has a computer stand that has an original price of $59.95 and is marked down $23.98. What are (a) the markdown rate and (b) the reduced price?

5. A robotic vacuum cleaner's original selling price of $372.50 is marked down 12%. What are (a) the markdown and (b) the reduced price?

6. Pat's Pastries normally sells Chocolate Wonders for $12.95 each. There is a special this week and the price is marked down 20%. What are (a) the markdown and (b) the reduced price on a Chocolate Wonder?

7. The speedboat that Eddie wants to buy is on sale. The original price of $4250 is marked down $552.50. What are (a) the markdown rate and (b) the reduced price?

8. Al Lumpkins has his heart set on a snowblower. The original price of $849.50 is marked down $67.96. What are (a) the markdown rate and (b) the reduced price?

9. Angela has been waiting to buy a new set of speakers for her stereo. Wired for Sound has the set she wants. The original selling price of $475 is marked down 7%. What are (a) the markdown and (b) the reduced price?

10. Tour De Countryside has a bicycle on sale. The bicycle that originally sold for $208.25 is marked down 16%. What are (a) the markdown and (b) the reduced price?

11. The Sporting Life has a warm-up suit that originally sold for $98. It was first marked down 20%, then further reduced 15%, and finally reduced an additional 25%. What is the final reduced price of the warm-up suit?

12. Trim and Green has been trying to sell the ultimate lawn mower, but sales have been slow. The mower originally sold for $2250. It was first marked down 10%, then further reduced 15%, and finally reduced an additional 10%. What is the final reduced price of the ultimate lawn mower?

13. The Appliance Warehouse is having trouble moving the fashion-smart neon color appliances that its former sales manager ordered. The bright pink dryer originally sold for $545. It was first marked down 30%, then further reduced 25%, and finally reduced an additional 20%. What is the final reduced price of the bright dryer?

14. The green neon freezers are also not selling. The freezer originally sold for $782. It was first marked down 20%, then further reduced 25%, and finally reduced an additional 20%. What is the final reduced price?

15. Foxy's Furs has experienced problems moving its coats. One mink jacket that originally sold for $650 was first marked down 15%, then further reduced 10%, and finally reduced an additional 20%. What is the final reduced price?

16. A circular bed originally sold for $1370. It was first marked down 20%, then further reduced 25%, and finally reduced an additional 20%. What is the final reduced price?

Key Terms

cost   markup
intermediate reduced price   selling price
markdown

Chapter 8 at a Glance

Page	Topic	Key Point	Example
276	Selling price, cost, and markup	Selling price = cost + markup Cost = selling price − markup Markup = selling price − cost	Cost = $27.50 Markup = $8.25 Selling price = $35.75 $35.75 = $27.50 + $8.25 $27.50 = $35.75 − $8.25 $8.25 = $35.75 − $27.50
277	Calculating markup based on cost when cost and rate are known	Markup = rate × cost	What is the markup on a bicycle with a cost of $200 and a markup rate of 35% based on cost? Markup = 0.35 × $200 = $70
278	Calculating rate when markup based on cost and cost are known	$\text{Rate} = \dfrac{\text{markup}}{\text{cost}}$	What is the markup rate based on cost for a bicycle with a cost of $200 and a $70 markup? $\text{Rate} = \dfrac{\$70}{\$200} = .35 = 35\%$
279	Calculating cost when markup based on cost and rate are known	$\text{Cost} = \dfrac{\text{markup}}{\text{rate}}$	What is the cost on a bicycle with a 35% markup based on cost of $70? $\text{Cost} = \dfrac{\$70}{0.35} = \$200$
280	Finding cost when selling price and rate based on cost are known	$\text{Cost} = \dfrac{\text{selling price}}{100\% + \text{rate}}$	What is the cost on a bicycle that sells for $270, including a 35% markup rate based on cost? $\text{Cost} = \dfrac{\$270}{100\% + 35\%} = \dfrac{\$270}{135\%}$ $= \dfrac{\$270}{1.35} = \$200$
287	Finding markup based on selling price when selling price and rate are known	Markup = rate × selling price	What is the markup based on selling price of a sleeping bag that sells for $110 including a 30% markup? Markup = 0.30 × $110 = $33
288	Finding rate when markup based on selling price and selling price are known	$\text{Rate} = \dfrac{\text{markup}}{\text{selling price}}$	What is the markup rate on a sleeping bag that sells for $110, including a $33 markup based on selling price? $\text{Rate} = \dfrac{\$33}{\$110} = 0.30 = 30\%$

Page	Topic	Key Point	Example
289	Finding selling price when markup based on selling price and rate are known	Selling price $= \dfrac{\text{markup}}{\text{rate}}$	What is the selling price of a sleeping bag with a 30% markup of $33 based on selling price? Selling price $= \dfrac{\$33}{0.30} = \$110$
289	Finding selling price when cost and rate based on selling price are known	Selling price $= \dfrac{\text{cost}}{100\% - \text{rate}}$	What is the selling price of a sleeping bag that costs $77 and is marked up 30% based on selling price? Selling price $= \dfrac{\$77}{100\% - 30\%}$ $= \dfrac{\$77}{70\%} = \dfrac{\$77}{0.70} = \$110$
297	Markdowns	Markdown = rate × selling price Reduced selling price = selling price − markdown	What are the markdown and reduced selling price for a $300 watch marked down 30%? Markdown $= 0.30 \times \$300 = \$90$ Reduced selling price $= \$300 - \$90 = \$210$
298	Multiple markdowns	Calculate first markdown. First intermediate reduced price (IRP) = original price − 1st markdown Second markdown = rate × first IRP Second IRP = first IRP − second markdown Third markdown = rate × second IRP Final price = second IRP − third markdown	What is the final price of a $2000 used car first marked down 25%, then 20% more, then 25% more? First markdown $= 0.25 \times \$2000 = \$500$ First IRP $= \$2000 - \$500$ $= \$1500$ Second markdown $= 0.20 \times \$1500 = \$300$ Second IRP $= \$1500 - \$300$ $= \$1200$ Third markdown $= 0.25 \times \$1200 = \$300$ Final price $= \$1200 - \$300 = \$900$
299	Shortcut to multiple markdowns	Calculate the complements by subtracting each markdown rate from 100%. Complement = 100% − markdown rate Multiply the initial price by the complements.	What is the final price of a $2000 used car first marked down 25%, then 20% more, then 25% more? $100\% - 25\% = 75\%$ $100\% - 20\% = 80\%$ $100\% - 25\% = 75\%$ Final price $= \$2000 \times 0.75 \times 0.80 \times .75$ $= \$900$
300	Finding markdown rate when markdown and selling price are known	Rate $= \dfrac{\text{markdown}}{\text{selling price}}$	What is the markdown rate on a $65 chair marked down $13? Rate $= \dfrac{\$13}{\$65} = 0.20 = 20\%$

SELF-TEST The Mathematics of Selling

The following problems test your understanding of the mathematics of selling.

1. The Appliance Warehouse was able to purchase a large washer for $423.75 and sells the washer with a markup of $101.70. What are (a) the selling price and (b) the rate of markup based on cost?

2. A case of kiwi fruits at P&A Supermarkets costs $78.50 and has a markup of $32.97. What are (a) the selling price and (b) the rate of markup based on cost?

3. The Ancient Mariner sells a fishing boat that has a markup of $697.50 and a 30% rate of markup based on cost. What are (a) the cost and (b) the selling price of the fishing boat?

4. Kate's Delicacies sells a cheesecake that has a markup of $7.74 and a 30% rate of markup based on cost. What are (a) the cost and (b) the selling price of the cheesecake?

5. Trim and Green has a luxury lawn mower that costs $527.50 and has a 56% rate of markup based on cost. What are (a) the markup and (b) the selling price?

6. Trim and Green also has a lawn edger that costs $92.25 and has a 40% rate of markup based on cost. What are (a) the markup and (b) the selling price?

7. By the Byte has a desktop publishing package that sells for $944.28 at a 29% rate of markup based on cost. What are (a) the cost and (b) the markup?

8. Phyle's Office Products sells a calculator for $16.50. The rate of markup based on cost is 32%. What are (a) the cost and (b) the markup?

9. Designer Discounts has a key chain that costs $12.64 and has markup of $7.11. What are (a) the selling price and (b) the rate of markup based on selling price?

10. Astute Accents has a purse that costs $149.35 and has a markup of $108.15. What are the (a) selling price and (b) the rate of markup based on selling price?

11. The Fins and Feathers Pet Shop sells a saltwater aquarium that has a markup of $263.83 and a 28% rate of markup based on selling price. What are (a) the cost and (b) the selling price?

12. The Happy Feet Shoe Store uses a markup of $54.23 on alpine-quality hiking boots. The rate of markup based on selling price is 55%. What are (a) the cost and (b) the selling price?

13. High Flyers purchases hang gliders for $1952.50 and sells the hang gliders with a 45% rate of markup based on selling price. What are (a) the markup and (b) the selling price of a hang glider?

14. Wreck-a-Mended Auto sells a bumper that costs $85.15. If the rate of markup is based on selling price of 48%, what are (a) the markup and (b) the selling price?

15. Models Unlimited sells a functional, scale model F-15 for $481. The markup rate is based on selling price and is 27%. What are (a) the cost and (b) the markup?

16. A Splotch watch has a selling price of $313.05 and a 40% rate of markup based on selling price. What are (a) the cost and (b) the markup?

17. One of the canaries at the Fly-By-Night Aviary won't make a sound. To sell the bird, the store has taken its original selling price of $27.50 and marked it down by 26%. What are (a) the markdown and (b) the reduced price of the canary?

18. For Your Eyes Only is selling a signature pair of glasses that originally sold for $245.95 and is now marked down 40%. What are (a) the markdown and (b) the reduced price?

19. Tour De Countryside is selling a recumbent bicycle that was originally priced at $475 and is marked down $57. What are (a) the markdown rate and (b) the reduced price?

20. The Image Sharpener has a talking alarm clock on sale. The original price of $97.60 is marked down $24.40. What are (a) the markdown rate and (b) the reduced price?

21. To decorate his restaurant foyer, Lew bought a statue that originally sold for $450. It was first marked down 20%, then further reduced 25%, and finally marked down an additional 30%. What is the final reduced price that Lew paid?

22. Big Boys Toys has a remote-controlled toy sports car that originally sold for $58. It was first marked down 20%, then further reduced 10%, and finally marked down an additional 20%. What is the final reduced price?

Simple Interest

CHAPTER OBJECTIVES

When you complete this chapter successfully, you will be able to:

1. Calculate the number of days between two dates using exact time and ordinary or 30-day-month time.

2. Calculate simple interest using the following methods:
   a. Accurate interest
   b. Ordinary interest at ordinary or 30-day-month time
   c. Ordinary interest at exact time

3. Solve a simple interest problem for principal, interest rate, or time.

■ J.R.'s Construction Company has taken out a loan to buy a new crane. If the loan period runs from May 5, 1998 and finishes October 18, 1998, how many days are involved at (a) exact time and (b) ordinary time?

■ Friendly Savings and Loan lent J.R.'s Construction Company $2950 on July 15, 1998 at 21.3% interest. The due date on the loan is October 13, 1998. What is the interest on the loan using (a) accurate simple interest, (b) ordinary simple interest at ordinary time, and (c) ordinary simple interest at exact time?

■ Wired for Sound needs to take out a loan at 18% interest to expand its listening room. If the loan is for 72 days with ordinary simple interest at ordinary time of $270, what is the principal of the loan?

Many times an individual or business needs additional capital for improvements. In return for the loan of the money, lending institutions receive a payment called "interest" for the amount borrowed.

This chapter is designed to aid you in computing the amount of interest charged for the use of a bank loan according to the amount of the loan, time, and interest rate.

Is your pencil handy? Good! This chapter is simple but interesting, so let's get started.

SECTION 9.1: Calculating Time

Can You:

Calculate the number of days between two given dates using exact time?
Find the number of days between two given dates using ordinary or 30-day-month time?
. . . If not, you need this section.

Ordinary or 30-Day-Month Time A method of counting days in which each month is considered to have 30 days and the year is thus considered to have 360 days.

Exact Time A method of counting days in which the precise number of days is counted in each month, resulting in a 365-day or 366-day year.

"Time? I can count days, this sounds easy." It is easy, but there are a few details to cover. In business transactions such as loans there are two methods of counting days: exact time and ordinary or 30-day-month time. In **ordinary** or **30-day-month time,** we count each month as having 30 days. With **exact time,** we count the actual number of days in the month. We discussed exact time briefly in Section 7.2. Here you will learn how to calculate the exact time between two given dates.

With both methods we do not count the starting day. For example, the number of days from October 12 to October 15 will be three days. One way to determine the number of days is to count them individually.

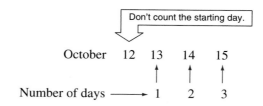

YOUR TURN

WORK THIS PROBLEM

The Question:

Find the number of days from September 5 to September 19.

✔ YOUR WORK

The Solution:

14 days. Was that your answer? Count them. September 6, 7, 8, 9, 10, 11, 12, 13, 14, 15, 16, 17, 18, 19. A total of 14 days. Remember, don't count the starting day (in this case, September 5).

Did you notice an easier way to do this problem? The best method is to subtract the starting day from the last day. This gives the difference between the two dates or the number of days between them. As required, this method doesn't count the starting day. In the problem, the number of days will be 19 − 5 = 14 days.

WORK THIS PROBLEM

The Question:

Use the shortcut to find the number of days from June 9 to June 27.

✔ YOUR WORK

The Solution:

$$
\begin{array}{r}
27 \\
-\,9 \\
\hline
18 \text{ days}
\end{array}
$$

27 ◄——— June 27, the ending date
−9 ◄——— June 9, the starting date

Using Exact Time

To calculate the number of days between *two different months* using exact time, you must first remember the number of days in each month. As an aid, you can use the knuckle method described in Chapter 7, Section 7.2, or the chart that follows.

January	31		July	31
February	28	(29 in leap year)	August	31
March	31		September	30
April	30		October	31
May	31		November	30
June	30		December	31

A leap year is one in which the year number is evenly divisible by 4. (Years ending in "00" are leap years only if the year number is divisible by 400.) The year 1996 was a leap year. But 2010 will not be a leap year. The year 2000 is a leap year, but 2100 will not be.

To find exact time, follow these steps:

Steps: Calculating Exact Time

STEP 1. *Calculate* the number of days for each individual month between the beginning and ending months.

STEP 2. *Add* the days for each month to get the exact number of days.

For example, to calculate the number of days from March 16 to July 4, find the number of days in March, April, May, June, and July.

Step 1.   March 16 to March 31   15 days ◄——— 31 − 16 = 15 days
                                All April   30
                                 All May   31
                                 All June   30
                                 to July 4   <u>4</u>
Step 2.   Total ———► 110 days

In July, we count all 4 days, but in March we do not count March 16. Remember, never count the starting day.

Some business calculators can determine ordinary or exact time automatically.

YOUR TURN

WORK THIS PROBLEM

The Question:

Find the number of days between the two dates using exact time. Remember to count February as 29 days in leap year.

(a) April 14 to November 5, 1997   (b) January 27 to August 17, 1996

✔ YOUR WORK

The Solution:

(a) ***Step 1.***   April 14 to April 30   16 days ◄——— 30 − 14 = 16 days
                                   May   31
                                   June   30
                                   July   31
                                  August   31
                            September 30
                               October   31
                     to November 5   <u>5</u>
           Step 2.   Total ———► 205 days

(b) ***Step 1.***   January 27 to January 31   4 days ◄——— 31 − 27 = 4 days
                                  February   29 ◄——— 1996, a leap year
                                  March   31
                                  April   30
                                  May   31
                                  June   30
                                  July   31
                                  August   <u>17</u>
           Step 2.   Total ———► 203 days

Finding Exact Time by the Calendar Method

An alternate method of calculating exact time uses the days-in-the-year calendar. Each entry in the table gives the day of the year for any date.

Days-in-the-Year Calendar (Nonleap Year*)

Day of month	Jan.	Feb.	Mar.	Apr.	May	June	July	Aug.	Sept.	Oct.	Nov.	Dec.
1	1	32	60	91	121	152	182	213	244	274	305	335
2	2	33	61	92	122	153	183	214	245	275	306	336
3	3	34	62	93	123	154	184	215	246	276	307	337
4	4	35	63	94	124	155	185	216	247	277	308	338
5	5	36	64	95	125	156	186	217	248	278	309	339
6	6	37	65	96	126	157	187	218	249	279	310	340
7	7	38	66	97	127	158	188	219	250	280	311	341
8	8	39	67	98	128	159	189	220	251	281	312	342
9	9	40	68	99	129	160	190	221	252	282	313	343
10	10	41	69	100	130	161	191	222	253	283	314	344
11	11	42	70	101	131	162	192	223	254	284	315	345
12	12	43	71	102	132	163	193	224	255	285	316	346
13	13	44	72	103	133	164	194	225	256	286	317	347
14	14	45	73	104	134	165	195	226	257	287	318	348
15	15	46	74	105	135	166	196	227	258	288	319	349
16	16	47	75	106	136	167	197	228	259	289	320	350
17	17	48	76	107	137	168	198	229	260	290	321	351
18	18	49	77	108	138	169	199	230	261	291	322	352
19	19	50	78	109	139	170	200	231	262	292	323	353
20	20	51	79	110	140	171	201	232	263	293	324	354
21	21	52	80	111	141	172	202	233	264	294	325	355
22	22	53	81	112	142	173	203	234	265	295	326	356
23	23	54	82	113	143	174	204	235	266	296	327	357
24	24	55	83	114	144	175	205	236	267	297	328	358
25	25	56	84	115	145	176	206	237	268	298	329	359
26	26	57	85	116	146	177	207	238	269	299	330	360
27	27	58	86	117	147	178	208	239	270	300	331	361
28	28	59	87	118	148	179	209	240	271	301	332	362
29	29		88	119	149	180	210	241	272	302	333	363
30	30		89	120	150	181	211	242	273	303	334	364
31	31		90		151		212	243		304		365

*In a leap year, such as 1996 or 2000, add 1 if February 29 is in the time period.

Steps: Exact Time: The Calendar Method

STEP 1. *Find* the day in the year for the beginning and ending dates.

STEP 2. *Subtract* the day in the year of the beginning date from the day in the year of the ending date. The result is the exact time for a nonleap year.

STEP 3. *For leap years only:* If February 29 is included in the period, *add* 1 to the number of days.

STEP 4. If the beginning and ending dates are in different years, *calculate* the number of days from the beginning date to year's end and add the number of days from January 1 to the ending date.

For example, let's use the day-in-the-year calendar to calculate the exact time from May 12, 1995 to October 15, 1995.

Step 1.   October 15 = 288; May 12 = 132.

Step 2.   Exact time (nonleap year) = ending − beginning = 288 − 132 = 156 days.

February is not included in the period, so we do not need to worry about leap years.

| WORK THIS PROBLEM |

The Question:

Use the day-in-the-year calendar to calculate the exact time from (a) January 17, 1996 to April 15, 1996 and (b) October 10, 1997 to February 1, 1998.

| ✔ YOUR WORK |

The Solution:

(a) ***Step 1.***   April 15 = 105; January 17 = 17.

    Step 2.   Exact time (nonleap year) = 105 − 17 = 88 days.

    Step 3.   Because 1996 is a leap year, leap-year exact time = 88 + 1 = 89 days.

(b) ***Step 1.***   October 10 = 283; February 1 = 32

    Step 2.   365 − 283 = 81 days to year end
                  32 days to February 1
                  Exact time = 82 + 32 = 114 days

Note that during a leap year, if February 29 is included in the time period, you must use step 3 in order to calculate the time.

Calculating Ordinary or 30-Day-Month Time

The second method of counting days is ordinary or 30-day-month time. This method assumes each month has 30 days and the year has 360 days. Ordinary time is commonly used by businesses extending loans (credit) to customers.

The following procedure will help you find ordinary time:

Steps: Calculating Days Using Ordinary or 30-Day-Month Time

STEP 1.   *Find* the number of months between the starting date and the same day in the ending month. Multiply the number of months by 30.

STEP 2.   *Identify* the ending date and the last day in step 1. Generally, these are different dates. Now, find the difference between these dates. With any additional days, *add* them to the total. If you go past the due date, *subtract* the extra days.

For example, let's find the number of days from March 13 to August 27 using ordinary time.

Step 1.   March 13 to August 13 = 5 months

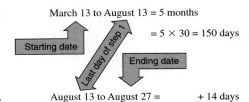

= 5 × 30 = 150 days

Step 2.   August 13 to August 27 =   + 14 days
                                    Total ⟶ 164 days

WORK THIS PROBLEM

The Question:

Find the number of days between the following two dates using ordinary time.

(a) May 9 to July 23
(b) January 6 to December 18

✔ YOUR WORK

The Solution:

(a) **Step 1.** May 9 to July 9 = 2 months = 2 × 30 days =   60 days
    Step 2. July 9 to July 23 =   + 14 days
                                       Total ⟶   74 days

(b) **Step 1.** January 6 to December 6 = 11 months = 11 × 30 days =   330 days
    Step 2. December 6 to December 18 =   + 12 days
                                               Total ⟶   342 days

When using ordinary time, we assume each year has 360 days. It doesn't matter whether it's a leap year.

In the preceding problems, the day in the ending month was always "later" than the day in the first month. In this case, we always added the additional days (step 2). Let's look at an example where there are "extra" or "left-over" days.

Find the number of days from February 9 to July 2 using ordinary time.

Step 1. February 9 to July 9 = 5 months = 5 × 30 days =   150 days
Step 2. July 2 to July 9 =   − 7 days
                                   Total ⟶   143 days

Note that in step 1 we have gone past the ending date. Hence, we subtract the *extra* days in step 2.

WORK THIS PROBLEM

The Question:

Find the number of days between the two dates using ordinary time.

(a) May 15 to October 4   (b) July 27 to September 27
(c) October 25 to December 9   (d) April 24 to October 25

✔ YOUR WORK

The Solution:

(a) **Step 1.** May 15 to October 15 = 5 months = 5 × 30 days =   150 days
    Step 2. October 4 to October 15 =   − 11 days
                                                  Total ⟶ 139 days

(b) **Step 1.** July 27 to September 27 = 2 months = 2 × 30 days =   60 days
                                                  Total ⟶ 60 days

(c) **Step 1.** October 25 to December 25 = 2 months = 2 × 30 days =   60 days
    Step 2. December 9 to December 25 =   −16 days
                                                  Total ⟶ 44 days

(d) **Step 1.** April 24 to October 24 = 6 months = 6 × 30 days = 180 days
    Step 2. October 24 to October 25 =   +1 day
                                                  Total ⟶ 181 days

In (d), the ending date, 25th, is later in the month than the beginning date, 24th. Since we have an additional day, we add. In (c), the ending date, 9th, is earlier in the month than the beginning date, 25th. Here we subtract the extra days. Remember, add for additional days and subtract for extra or left-over days.

Now continue to Section Test 9.1 for some practice problems on calculating the number of days.

SECTION TEST 9.1 Simple Interest ━━━━━

The following problems test your understanding of Section 9.1, Calculating Time.

A. Calculate the exact number of days between the two dates in the same year. Assume it is not a leap year.

1. March 15 to July 27 _____

2. June 21 to August 5 _____

3. September 29 to December 14 _____

4. January 3 to April 15 _____

5. February 26 to June 30 _____

6. July 13 to December 2 _____

7. February 3 to November 5 _____

8. June 6 to December 10 _____

9. March 7 to July 16 _____

10. January 28 to November 17 _____

11. March 2 to June 14 _____

12. April 16 to September 23 _____

B. Calculate the number of days between the two dates in the same year using ordinary or 30-day-month time. Assume it is not a leap year.

1. August 3 to October 11 _____

2. September 17 to December 23 _____

3. March 23 to June 5 _____

4. April 30 to July 5_____

5. February 2 to April 11 _____

6. January 13 to April 2 _____

7. May 29 to September 11 _____

8. August 17 to October 15 _____

9. January 26 to March 29 _____

10. January 11 to April 20 _____

11. February 5 to December 18 _____

12. February 18 to October 30 _____

C. Solve these problems.

1. Wired for Sound needs to take out a loan to expand its listening room. If the period of the loan is from March 12, 1998 to August 27, 1998, how many days are involved at (a) exact time and (b) ordinary (30-day-month) time?

2. J.R.'s Construction Company has taken out a loan to buy a new crane. If the loan period runs from May 5, 1998 and finishes October 18, 1998, how many days are involved at (a) exact time and (b) ordinary (30-day-month) time?

3. Petal Pushers borrowed money to acquire a new flower-refrigeration unit. The loan period runs from April 15, 1998 to December 18, 1998. How many days will Petal Pushers borrow the money calculated as (a) exact time and (b) ordinary time?

4. Woody's Lumber Mill needs to take out a loan to buy a new log splitter. If the period of the loan is from January 15, 1997 to April 3, 1997, how many days are involved at (a) exact time and (b) ordinary time?

5. Maria's Mannequins took out a loan to add to its stock of "unisex" models. If the loan period is from January 3, 1996 to March 15, 1996, how many days are involved at (a) exact time and (b) ordinary time?

6. The Fly-by-Night Aviary has taken a loan to build a special room for tropic birds. If the loan period is from February 21, 1996 to September 11, 1996, how many days are involved at (a) exact time and (b) ordinary time?

7. Ava's Air Tours took out a loan to buy a new plane on December 9, 1996 and repaid it March 13, 1997. How many days are involved at (a) exact time and (b) ordinary time?

8. When he started his restaurant, Lew needed a loan to furnish it. He took out the loan on November 25, 1996 and repaid it April 15, 1997. How many days are involved at (a) exact time and (b) ordinary time?

9. Kate's Delicacies borrowed money to send three of its cooks for cake-decorating lessons. The loan period ran from January 24, 1996 to May 12, 1996. How many days are involved at (a) exact time and (b) ordinary time?

10. Crystal's Diamonds took out a loan to buy new display cases. The loan ran from December 26, 1995 to March 15, 1996. How many days are involved at (a) exact time and (b) ordinary time?

11. Bits 'n' Bytes borrowed money to expand its showroom. If the loan ran from October 2, 1997 to March 20, 1998, how many days are involved at (a) exact time and (b) ordinary time?

12. The Garden of Earthly Delights took out a loan to buy a large consignment of Dutch tulip bulbs. The loan period was from December 19, 1996 to May 2, 1997. How many days are involved at (a) exact time and (b) ordinary time?

WORKSPACE

Can You:

Calculate simple interest and maturity value using accurate simple interest?
Find simple interest and maturity value using ordinary simple interest at ordinary or 30-day-month time?
Determine simple interest and maturity value using ordinary simple interest at exact time?
. . . If not, you need this section.

Principal   The amount borrowed.

Interest   The amount charged by a bank or other lender in return for lending a sum of money.

Simple Interest Loan   A loan in which the principal and interest are repaid in one payment on the due date; interest on such loans is always equal to the *product* of principal, interest rate, and time period of the loan.

Time is not the only factor in calculating what a loan will cost you. Whether you borrow money as an individual or to start a new business or to expand a current business, the amount borrowed is called the **principal.** The amount charged by the bank or other lending institution is called the **interest.**

A **simple interest loan** is one in which the principal is borrowed for a specified time period and the principal and interest repaid in one payment on the due date. Simple interest is always the *product* of principal, interest rate, and time. That is,

$$I = P \times r \times t$$

where I = interest
   P = principal
   r = interest rate (converted to a decimal number)
   t = time (as a fractional portion of a year)

Businesspeople often abbreviate this formula as $I = Prt$.

All three of the commonly used methods of calculating simple interest use the formula $I = Prt$. The difference between these methods lies in the procedure for calculating the time. Two of the forms use exact time and the third uses ordinary time.

Determining Accurate Simple Interest

Accurate Simple Interest (exact interest)   A form of simple interest in which the time period is defined as exact time divided by the exact number of days in the year (365 or 366).

Use exact time to calculate **accurate simple interest** (also called **exact interest**). The time in the formula is given as a fraction of a year. Calculate it as the exact number of days between the starting and ending dates divided by the exact number of days in the year, 365 or 366 in a leap year.

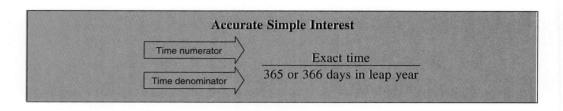

For example, to calculate a loan for 132 days in 1997, you would use time as $132 \div 365$ or $\frac{132}{365}$ year.

Steps: Calculating Simple Interest

STEP 1. *Calculate* the time. In the case of accurate simple interest, time is equal to the exact number of days divided by the number of days in that year.

STEP 2. *Insert* principal, rate, and time into the formula $I = Prt$.

STEP 3. *Convert* the interest rate to a decimal number and calculate the answer.

To see how these steps work, let's calculate the accurate simple interest on a loan of $2750 at 18% made on May 13 and due on October 6, 1997.

Step 1.
May 13 to May 31	18 days
June	30
July	31
August	31
September	30
to October 6	6
Total $\longrightarrow$	146 days

Step 2. $P = \$2750;$ $r = 18\%;$ $t = \dfrac{146}{365}$ ← Exact number of days / ← Exact number of days in 1997

so $I = Prt = \$2750 \times 18\% \times \dfrac{146}{365}$

Step 3. $18\% = 0.18;$ $I = \$2750 \times 0.18 \times \dfrac{146}{365} = \$198.00$

Using a calculator,

2750 ☒ 18 ☒ ☒ 146 ÷ 365 = ⟶ [198.]

If you need a reminder of how to change a percent to a decimal number, return to Section 4.1.

WORK THIS PROBLEM

The Question:

YOUR TURN

Calculate the accurate simple interest on a loan of $3500 at a 19% interest rate made on June 27, 1997 and due on September 8, 1997.

✔ YOUR WORK

The Solution:

Step 1.
$$\begin{aligned}
\text{June 27 to June 30} &= 3 \text{ days} \\
\text{July} &= 31 \\
\text{August} &= 31 \\
\text{to September} &= \underline{8} \\
\text{Total} \longrightarrow \quad & 73 \text{ days}
\end{aligned}$$

Step 2. $P = \$3500;\quad r = 19\%;\quad t = \dfrac{73}{365};\quad I = Prt = \$3500 \times 19\% \times \dfrac{73}{365}$

Step 3. $19\% = 0.19;\quad I = \$3500 \times 0.19 \times \dfrac{73}{365} = \$133.00$

Many problems are not as easy to calculate as the preceding one. Generally, using decimal numbers (and a calculator, if possible) is the best bet. If the answer is not a simple number, round to the nearest cents position. For a quick review of rounding, turn to Section 3.2.

Maturity Value On a simple interest loan, the total amount due when the loan "matures" (comes due); it is equal to the principal plus the interest.

When a loan is due, the borrower must pay back the principal plus the interest. This amount is called the **maturity value.** The maturity value is the total amount due when the loan is due (*matures*): maturity value = principal + interest. The abbreviation *MV* is used to represent maturity value, so

$$MV = P + I$$

where MV = maturity value
P = principal
I = interest

Steps: Calculating Maturity Value

STEP 1. *Calculate* the interest.

STEP 2. *Insert* principal and interest into the formula $MV = P + I$ and calculate the answer.

For example, let's find the maturity value of the problem following the steps for calculating simple interest, where $P = \$2750$, $r = 18\%$, and $t = \frac{146}{365}$. We already completed **step 1** in finding maturity value when we determined that $I = \$198.00$.

Step 2. $MV = P + I = \$2750 + 198 = \$2948$

YOUR TURN

WORK THIS PROBLEM

The Question:

Solve the following problems using accurate simple interest. Be careful when calculating the time.

(a) Calculate the interest and maturity value on a loan of $4500 at $17\frac{1}{2}\%$ made on March 12 and due on October 17, 1997.

(b) What is the interest and maturity value on a loan of $3750 at 17.3% made on April 5 and due on September 17, 1997?

(c) Calculate the interest and maturity value on a loan of $1200 at 19% made on January 16 and due on April 5, 1996.

✔ YOUR WORK

The Solution:

(a) *Step 1.*   March 12 to March 31   19 days
                              April   30
                               May   31
                              June   30
                              July   31
                           August   31
                       September   30
                   to October 17   17
                           Total ⟶ 219 days

$$I = Prt = \$4500 \times 17\tfrac{1}{2}\% \times \frac{219}{365} = \$4500 \times 0.175 \times \frac{219}{365} = \$472.50$$

Step 2. $MV = P + I = \$4500 + 472.50 = \$4972.50$

(b) *Step 1.*   April 5 to April 30   25 days
                               May   31
                              June   30
                              July   31
                           August   31
                   to September 17   17
                           Total ⟶ 165 days

$$I = Prt = \$3750 \times 17.3\% \times \frac{165}{365} = \$3750 \times 0.173 \times \frac{165}{365}$$

$$= \$293.270 \ldots = \$293.27 \text{ rounded}$$

Step 2. $MV = P + I = \$3750 + 293.27 = \$4043.27$

(c) *Step 1.*   January 16 to January 31   15 days
                           February   29 ⟵ 1996 is a leap year.
                              March   31
                         to April 5   5
                           Total ⟶ 80 days

$$I = Prt = \$1200 \times 19\% \times \frac{80}{366} \quad \longleftarrow \quad 1996 \text{ is a leap year.}$$

$$= \$1200 \times 0.19 \times \frac{80}{366} = \$49.836 \ldots = \$49.84 \text{ rounded}$$

Step 2. $MV = P + I = \$1200 + 49.84 = \$1249.84$

Ordinary Simple Interest at Ordinary or 30-Day-Month Time

Ordinary Simple Interest at Ordinary Time   A form of simple interest in which the time period is defined as ordinary time divided by 360.

Another way to calculate simple interest is **ordinary simple interest at ordinary time.** The difference between this method and accurate interest is the time fraction. In ordinary simple interest at ordinary time, the numerator is the number of days between the starting and ending dates using ordinary time. The denominator is 360.

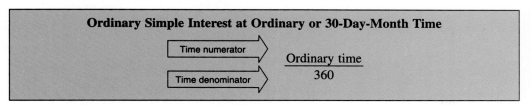

For example, let's calculate the simple interest and maturity value on a loan of $2750 at 18% made on May 13 and due on October 6, 1997.

Step 1.   May 13 to October 13 = 5 months = 5 × 30 days = 150 days
October 6 to October 13 =   − 7 days
Total ⟶ 143 days

$$I = Prt = \$2750 \times 18\% \times \frac{143}{360}$$

$$= \$2750 \times 0.18 \times \frac{143}{360} = \$196.625 = \$196.63 \text{ rounded}$$

Step 2. $MV = P + I = \$2750 + 196.63 = \$2946.63$

Remember, with ordinary simple interest at ordinary time, always use ordinary time in the numerator and 360 in the denominator of the time fraction.

Did you notice that this is the same problem that was previously used to calculate accurate simple interest? The accurate simple interest is $198.00. The ordinary simple interest at 30-day-month time is $196.63.

**YOUR
TURN**

The Question:

Work the following problems using ordinary simple interest at ordinary time.

(a) What is the interest and maturity value on a loan of $4500 at $17\frac{1}{2}\%$ made on March 12 and due on October 17, 1997?

(b) What is the interest and maturity value on a loan of $3750 at 17.3% made on April 5 and due on September 17, 1997?

(c) Calculate the interest and maturity value on a loan of $1200 at 19% made on January 16 and due on April 5, 1996.

✔ YOUR WORK

The Solution:

(a) *Step 1.*   March 12 to October 12 = 7 months = 7 × 30 days = 210 days
October 12 to October 17 =   + 5 days
Total ⟶ 215 days

$$I = Prt = \$4500 \times 17\tfrac{1}{2}\% \times \frac{215}{360} = \$4500 \times 0.175 \times \frac{215}{360}$$

$$= \$470.3125 = \$470.31 \text{ rounded}$$

  Step 2. $MV = P + I = \$4500 + 470.31 = \$4970.31$

(b) *Step 1.*   April 5 to September 5 = 5 months = 5 × 30 days = 150 days
September 5 to September 17 =   + 12 days
Total ⟶ 162 days

$$I = Prt = \$3750 \times 17.3\% \times \frac{162}{360} = \$3750 \times 0.173 \times \frac{162}{360}$$

$$= \$291.9375 = \$291.94 \text{ rounded}$$

  Step 2. $MV = P + I = \$3750 + 291.94 = \$4041.94$

(c) *Step 1.*   January 16 to April 16 = 3 months = 3 × 30 days =   90 days
April 5 to April 16 =   −11 days
Total ⟶ 79 days

$$I = Prt = \$1200 \times 19\% \times \frac{79}{360} = \$1200 \times 0.19 \times \frac{79}{360}$$

$$= \$50.033 \ldots = \$50.03 \text{ rounded}$$

  Step 2. $MV = P + I = \$1200 + 50.03 = \$1250.03$

Notice that the fact that 1996 is a leap year does not affect the calculation here. You can totally ignore the year when using ordinary simple interest at ordinary time because we assume each month has 30 days and each year has 360 days.

Ordinary Simple Interest at Exact Time (Banker's Interest)   A form of simple interest in which the time period is defined as exact time divided by 360.

The three problems just presented were the same ones used earlier to find accurate simple interest. The answers are different; in some cases, the interest is greater with accurate interest, in others, it is less.

In the third method, **ordinary simple interest at exact time (banker's interest),** the interest is always greater than or equal to the other two methods.

Calculating Ordinary Simple Interest at Exact Time

Ordinary simple interest at exact time is a combination of the other methods. The numerator of the time fraction uses exact time. The denominator is 360. Note that in both types of "ordinary" interest, the denominator is *always* 360.

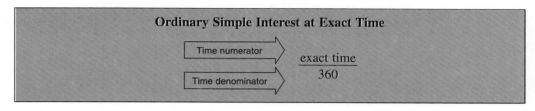

Again, we use the same steps as for accurate simple interest, changing only the fraction for exact time. For example, let's calculate the interest and maturity value on a loan of $2750 at 18% made on May 13 and due on October 6, 1997. Use ordinary simple interest at exact time.

Step 1.

May 13 to May 31	18 days
June	30
July	31
August	31
September	30
to October 6	6
Total ⟶	146 days

$$I = Prt = \$2750 \times 18\% \times \frac{146}{360} = \$2750 \times 0.18 \times \frac{146}{360} = \$200.75$$

Step 2. $MV = P + I = \$2750 + 200.75 = \$2950.75$

The calculated interest for this problem was $198.00 using accurate simple interest and $196.63 with ordinary simple interest at ordinary time. When using ordinary simple interest at exact time, the interest is *always* more.

WORK THIS PROBLEM

The Question:

Work the following problems. Use ordinary simple interest at exact time.

(a) What is the interest and maturity value on a loan of $1500 at 16% starting April 14 and due on September 5, 1997?

(b) Calculate the interest and maturity value on a loan of $1275 at 13.8% made on March 17 and due on May 5, 1997.

(c) What is the interest and maturity value on a loan of $3825 at 19% made on January 2 and due on April 15, 1996.

✔ YOUR WORK

The Solution:

(a) *Step 1.*   April 14 to April 30   16 days
                         May   31
                         June   30
                         July   31
                      August   31
            to September 5   5
            Total ⟶ 144 days

$$I = Prt = \$1500 \times 16\% \times \frac{144}{360} = \$1500 \times 0.16 \times \frac{144}{360} = \$96.00$$

Step 2. $MV = P + I = \$1500 + 96.00 = \$1596.00$

Using a calculator,

1500 ☒ 16 ☒% ☒ 144 ☒÷ 360 ☒+ 1500 ☒= ⟶ [1596.]

(b) *Step 1.*   March 17 to March 31   14 days
                            April   30
                        to May 5   5
                  Total ⟶ 49 days

$$I = Prt = \$1275 \times 13.8\% \times \frac{49}{360} = \$1275 \times 0.138 \times \frac{49}{360}$$

$$= \$23.94875 = \$23.95 \text{ rounded}$$

Step 2. $MV = P + I = \$1275 + 23.95 = \$1298.95$

Using a calculator,

1275 ☒ 13.8 ☒% ☒ 49 ☒÷ 360 ☒= ⟶ [23.94875]
                                                    └⟶ $23.95

☒+ 1275 ☒= ⟶ [1298.9487]
                        └⟶ $1298.95

(c) *Step 1.*   January 2 to January 31   29 days
                         February   29 ⟵ 1996 is a leap year
                            March   31
                      to April 15   15
                  Total ⟶ 104 days

$$I = Prt = \$3825 \times 19\% \times \frac{104}{360} = \$3825 \times 0.19 \times \frac{104}{360}$$

$$= \$209.95$$

Step 2. $MV = P + I = \$3825 + 209.95 = \$4034.95$

Note that 1996 is a leap year. When using exact time, you may find that leap year makes a difference.

Recapping Simple Interest

All three simple interest methods use the same basic formula, $I = Prt$, but each method uses a different time fraction. It is *very* important for you to remember the numerator and denominator for each method.

	Accurate Simple Interest	**Ordinary Simple Interest at Ordinary Time**	**Ordinary Simple Interest at Exact Time**
Time numerator	Exact time	Ordinary time	Exact time
Time denominator	365 or 366 in leap year	360	360
	Exact time Exact interest	Ordinary time Ordinary interest	Exact time Ordinary interest

To check your memory—and your skills—try the practice problems in Section Test 9.2.

Note that there is no one "best" method. Banker's interest yields the highest return for the lender but is the worst from the loan taker's perspective. If it comes down to making or not making a loan, banks are often willing to settle for the lesser interest of exact time.

WORKSPACE

SECTION TEST 9.2 Simple Interest ━━━

The following problems test your understanding of Section 9.2, Calculating Simple Interest.

A. In the following problems, calculate the time, accurate simple interest and the maturity value.

	Principal	Rate	Beginning Date	Due Date	Time	Interest	Maturity Value
1.	$5700	18%	March 4	September 15, 1997			
2.	1250	16%	August 2	November 15, 1997			
3.	875	19%	May 17	August 20, 1998			
4.	2570	17%	June 21	October 5, 1998			
5.	1600	20%	February 4	June 30, 1997			
6.	2400	21%	January 23	April 15, 1997			
7.	3700	16%	January 5	May 12, 1996			
8.	5750	17.5%	February 9	May 11, 1996			
9.	2500	19.5%	March 13	September 15, 1996			
10.	4050	21.3%	May 23	October 10, 1996			

B. In the following problems, calculate the time, ordinary simple interest at ordinary or 30-day-month time and maturity value.

	Principal	Rate	Beginning Date	Due Date	Time	Interest	Maturity Value
1.	$4500	16%	May 12	August 15			
2.	2750	18%	July 5	October 10			
3.	1800	21%	March 16	July 6			
4.	4550	19%	July 17	September 8			
5.	6700	21%	February 7	July 3			
6.	8050	17%	January 19	May 12			
7.	1750	22%	January 8	June 3			
8.	975	18.5%	February 3	May 1			
9.	3400	20.5%	March 29	September 5			
10.	4450	19.7%	May 9	October 6			

C. In the following problems, calculate the time, ordinary simple interest at exact time and the maturity value.

	Principal	Rate	Beginning Date	Due Date	Time	Interest	Maturity Value
1.	$1800	19%	March 24	September 30, 1997			
2.	4260	21%	August 9	November 12, 1997			
3.	2880	20%	May 7	August 17, 1998			
4.	9525	18%	June 9	October 5, 1998			
5.	4800	17%	February 7	June 19, 1997			
6.	6800	20%	January 9	April 16, 1997			
7.	6700	17%	January 3	May 19, 1996			
8.	5250	18.5%	February 5	May 28, 1996			
9.	7500	20.5%	May 22	October 7, 1996			
10.	3360	19.6%	May 17	October 23, 1996			

D. Solve these problems.

1. On March 12, 1997, Alice Wellington borrowed $3500 to open a health food store. The interest rate was 19%. The due date on the loan is June 15, 1997. What is the interest on the loan using (a) accurate simple interest, (b) ordinary simple interest at ordinary (30-day-month) time and (c) ordinary simple interest at exact time?

2. In order to buy a computer, the Fly-By-Night Aviary borrowed $2575 on May 28, 1997 at 18% interest. The due date on the loan is July 17, 1997. What is the interest on the loan using (a) accurate simple interest, (b) ordinary simple interest at ordinary (30-day-month) time, and (c) ordinary simple interest at exact time?

3. The Dorys borrowed $7500 on January 5, 1998, at 17% interest to purchase a pontoon boat. The due date on the loan is June 11, 1998. Find the interest on the loan using (a) accurate simple interest, (b) ordinary simple interest at ordinary (30-day-month) time, and (c) ordinary simple interest at exact time.

4. Lew decided to finance the new carpet for his restaurant. He borrowed $1700 on January 21, 1997 at 21% interest. The due date on the loan is April 15, 1997. What is the interest on the loan using (a) accurate simple interest, (b) ordinary simple interest at ordinary (30-day-month) time, and (c) ordinary simple interest at exact time?

5. To buy new display cases, Crystal's Diamonds borrowed $9600 on February 7, 1996, at 19.5% interest. The due date on the loan is May 12, 1996. Find the interest on the loan using (a) accurate simple interest, (b) ordinary simple interest at ordinary (30-day-month) time, and (c) ordinary simple interest at exact time.

6. To cover the cost of a new flower refrigerator, on February 19, 1996, Petal Pushers borrowed $4760 at 18.7% interest. The due date on the loan is June 11, 1996. Calculate the interest on the loan using (a) accurate simple interest, (b) ordinary simple interest at ordinary (30-day-month) time, and (c) ordinary simple interest at exact time.

7. Maria's Mannequins is expanding and borrowed $5300 on October 17, 1997, at 20% interest to buy child-sized mannequins. The due date on the loan is December 15, 1997. What is the interest on the loan using (a) accurate simple interest, (b) ordinary simple interest at ordinary (30-day-month) time, and (c) ordinary simple interest at exact time?

8. Friendly Savings and Loan lent J.R.'s Construction Company $2950 on July 15, 1998, at 21.3% interest. The due date on the loan is October 13, 1998. Find the interest on the loan using (a) accurate simple interest, (b) ordinary simple interest at ordinary (30-day-month) time, and (c) ordinary simple interest at exact time.

9. Klaus has decided to build a Bavarian beer hall. He borrowed $7250 on January 11, 1996 at 17.8% interest to build the new micro brewery. The due date on the loan is April 16, 1996. Calculate the interest on the loan using (a) accurate simple interest, (b) ordinary simple interest at ordinary (30-day-month) time, and (c) ordinary simple interest at exact time.

10. Woody's Lumber Mill borrowed $7950 on February 5, 1996, at 19.9% interest in order to pay for a new lathe. The due date on the loan is June 7, 1996. What is the interest on the loan using (a) accurate simple interest, (b) ordinary simple interest at ordinary (30-day-month) time, and (c) ordinary simple interest at exact time?

11. Marlena borrowed $4000 from her Aunt Sue to help pay for college education. Aunt Sue agreed that Marlena could pay back the $4000, together with 7% simple interest, in a lump sum in $5\frac{1}{2}$ years. How much will Marlena owe Aunt Sue at that time?

12. Joel borrowed $250 on August 10 from the local loan shark. He agreed to repay the money, along with simple interest of 23.9%, on December 23. How much will he have to repay if this is an ordinary (30-day-month) time, ordinary interest loan?

WORKSPACE

Solving for the Interest Variables

In the basic formula for simple interest, $I = Prt$, if you know three of the variables, you can easily find the fourth. In the previous section, you calculated I when given P, r, and t. In this section, you will learn to solve for the other three quantities. The procedure is similar to that used in solving percent problems in Chapter 4.

If you are good at doing algebra, solving the interest formula for each of the variables will be easy. If you are not good at algebra, you must either memorize the following formulas or learn how to use the *Interest Equation Finder* that follows:

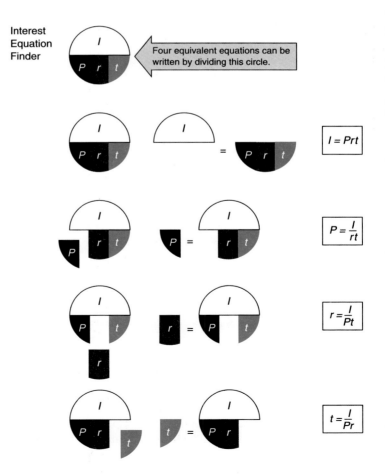

All four equations, $I = Prt$, $P = \dfrac{I}{rt}$, $r = \dfrac{I}{Pt}$, and $t = \dfrac{I}{Pr}$, are equivalent. Now let's put this interest equation finder to use. Use the following procedure to solve for any one of the four interest variables.

Steps: Solving for Other Interest Variables

STEP 1.   *Label* the parts you know—*I, P, r,* and/or *t*—being careful to write *t* as a fractional portion of a year.

STEP 2.   *Use* the interest equation finder to find the formula for the unknown quantity. Substitute the known parts of the formula.

STEP 3.   *Perform* the arithmetic (see the calculator box later in this section).

STEP 4.   *Only if solving for time (t): Multiply* your answer by the number of days in the year in order to express your answer in days.

Solving for Principal

If you know the interest, interest rate, and time, you can calculate the principal. As the interest equation finder shows, $P = I/rt$.

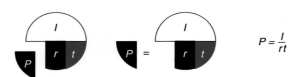

To calculate the principal, simply substitute the interest, interest rate, and time in the formula. For example, what is the principal on a loan with an interest rate of 18%, time of 146 days, and accurate simple interest of $198 (for a 365-day year).

Step 1. $I = \$198$; $r = 18\% = 0.18$; $t = \dfrac{146}{365}$ $\longleftarrow$ 365 for accurate interest

Step 2. $P = \dfrac{I}{rt} = \dfrac{\$198}{18\% \times \dfrac{146}{365}}$

Step 3. $P = \dfrac{I}{rt} = \dfrac{\$198}{\underbrace{0.18 \times \dfrac{146}{365}}} = \dfrac{\$198}{0.072} = \$2750$

⬆ Simplify this first.

Always change the percent to a decimal number before multiplying. The easiest way to complete the calculation is to simplify the denominator first. Then, divide the numerator by the denominator. Remember, the time fraction will always depend on the interest method used.

WORK THIS PROBLEM

The Question:

Work the following problems. Be careful with the time fraction.

(a) What principal is necessary to yield ordinary simple interest of $50 at 16% in 45 days?

(b) Find the principal on a loan with interest rate 18%, time 73 days, and accurate simple interest $130.50? (Assume a 365-day year.)

✔ YOUR WORK

The Solution:

(a) **Step 1.** $I = \$50$; $r = 16\%$; $t = \dfrac{45}{360}$

Ordinary interest—use 360 days

Step 2. $P = \dfrac{I}{rt} = \dfrac{\$50}{16\% \times \dfrac{45}{360}}$

Step 3. $P = \dfrac{\$50}{0.16 \times \dfrac{45}{360}} = \dfrac{\$50}{0.02} = \$2500$

(b) **Step 1.** $I = \$130.50$; $r = 18\%$; $t = \dfrac{73}{365}$

Accurate interest—use 365 days

Step 2. $P = \dfrac{I}{rt} = \dfrac{\$130.50}{18\% \times \dfrac{73}{365}}$

Step 3. $P = \dfrac{\$130.50}{0.18 \times \dfrac{73}{365}} = \dfrac{\$130.50}{0.036} = \$3625$

Solving for Interest Rate

If you know the interest, principal, and time, you can compute the interest rate. As the interest equation finder shows, $r = I/Pt$.

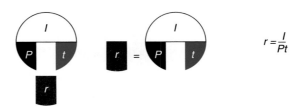

To calculate the interest rate, just substitute the interest, principal, and time into the formula and simplify. (As before, the time fraction will depend on the interest method.) When you divide the numerator by the denominator, the answer will be a decimal number which must be changed to a percent. The basic steps remain the same as for finding principal.

For example, what is the interest rate on a loan of $2700 for 65 days with ordinary simple interest at ordinary time of $87.75?

Step 1. $I = \$87.75$; $P = \$2700$; $t = \dfrac{65}{360}$

> Ordinary interest—use 360 days

Step 2. $r = \dfrac{I}{Pt} = \dfrac{87.75}{2700 \times \dfrac{65}{360}}$

Step 3. $r = \dfrac{87.75}{487.50} = 0.18 = 18\%$

When solving for the interest rate, *always* convert the answer to a percent. Otherwise your answer will be meaningless.

YOUR TURN

WORK THIS PROBLEM

The Question:

Solve the following interest rate problems.

(a) What interest rate will earn accurate simple interest of $525 on $5000 in 219 days (not a leap year)?

(b) Find the interest rate on a loan of $2500 made on March 6 and due on July 21, with ordinary simple interest at ordinary time of $150.

✔ YOUR WORK

The Solution:

(a) **Step 1.** $I = \$525$; $P = \$5000$; $t = \dfrac{219}{365}$

> Accurate interest—use 365 days

Step 2. $r = \dfrac{I}{Pt} = \dfrac{525}{5000 \times \dfrac{219}{365}}$

Step 3. $r = \dfrac{525}{3000} = 0.175 = 17.5\%$ or $17\frac{1}{2}\%$

(b) **Step 1.**   March 6 to July 6 = 4 months   =   120 days
                   July 6 to July 21   = + 15 days
                                            Total ⟶ =   135 days

$I = \$150$; $P = \$2500$; $t = \dfrac{135}{360}$

> Ordinary interest—use 360 days

Step 2. $r = \dfrac{I}{Pt} = \dfrac{150}{2500 \times \dfrac{135}{360}}$

Step 3. $r = \dfrac{150}{937.50} = 0.16 = 16\%$

Solving for Time

The final type of interest problem involves solving for the time. If the interest, principal, and interest rate are known, you can calculate the time. As the interest equation finder shows, $t = I/Pr$.

To calculate the time, first substitute the interest, principal, and interest rate in the formula and simplify. The result will be a (common or decimal) fraction that represents years. To arrive at the number of days, multiply the fraction by the number of days in the year. Multiply by 365 or 366 for exact time, and by 360 for ordinary (30-day-month) time.

Once again, we refer to the four-step procedure. Now, we need step 4. For example, what is the time (in days) on a loan of $4000 at 17.7% if the interest is $88.50 (ordinary simple interest at ordinary time)?

Step 1. $P = \$4000; \quad r = 17.7\%; \quad I = \$88.50$

Step 2. $t = \dfrac{I}{Pr} = \dfrac{88.50}{4000 \times 17.7\%}$

Step 3. $t = \dfrac{88.50}{4000 \times 0.177} = \dfrac{88.50}{708} = 0.125$

Step 4. $0.125 \times 360 = 45$ days

Ordinary time

With a calculator,

88.5 ÷ 4000 ÷ 0.177 =   0.125 × 360 = ⟶   45.

YOUR TURN

WORK THIS PROBLEM

The Question:

Work the following problems:

(a) How much time (in days) will it take a principal of $2500 to earn $50 interest at 18% (ordinary simple interest at ordinary time)?

(b) What is the time (in days) on a loan of $5250 at 18% with $189 interest (accurate simple interest—not a leap year)?

✔ YOUR WORK

The Solution:

(a) *Step 1.* $P = \$2500;$ $I = \$50;$ $r = 18\%$

   Step 2. $t = \dfrac{I}{Pr} = \dfrac{50}{2500 \times 18\%}$

   Step 3. $t = \dfrac{50}{2500 \times 0.18} = \dfrac{50}{450} = 0.111 \ldots$
   $= 0.111 \text{ rounded}$

   Step 4. $0.111 \times 360 = 39.96 = 40 \text{ days rounded}$

   Ordinary time

Getting a repeated decimal number like 0.111 . . . may seem a little strange. But in this case, just round. Using three decimal places is sufficient, even for bankers. If you're using a calculator, use all the decimal places your calculator displays. To obtain the number of days, multiply $0.1111111 \times 360 = 39.999996 = 40$ rounded.

(b) *Step 1.* $P = \$5250;$ $I = \$189;$ $r = 18\%$

   Step 2. $t = \dfrac{I}{Pr} = \dfrac{189}{5250 \times 18\%}$

   Step 3. $t = \dfrac{189}{945} = 0.2$

   Step 4. $0.2 \times 365 = 73 \text{ days}$

   Exact time

Solving for Other Interest Variables with a Calculator

Calculators are very useful in solving for other interest variables. Let's complete the following problem:

$$P = \frac{I}{rt} = \frac{\$198}{18\% \times \dfrac{146}{365}} = \frac{\$198}{0.18 \times \dfrac{146}{365}}$$

Step 1.   Calculate the denominator. The following keystroke sequence is for a business type calculator:

   146 ÷ 365 × 18 % ⟶   0.072

Step 2.   Divide $198 by 0.072. If your calculator does not have an inverse key $\boxed{\tfrac{1}{x}}$, you must remember, write down, or store 0.072 in your calculator memory. Then divide.

   198 ÷ 0.072 = ⟶   2750.

The answer: $P = 2750$

Note: If you have an inverse key $\boxed{\frac{1}{x}}$, you can complete the calculation without storing or remembering the denominator. Use the following sequence to solve for *P:*

146 $\boxed{\div}$ 365 $\boxed{\times}$ 18 $\boxed{\%}$ $\boxed{=}$ $\boxed{\frac{1}{x}}$ $\boxed{\times}$ 198 $\boxed{=}$ ⟶ | 2750. |

Another example: From the interest equation finder, calculate the rate in this problem:

$$r = \frac{I}{Pt} = \frac{122.72}{5000 \times \dfrac{127}{365}}$$

Step 1.   Calculate the denominator.

5000 $\boxed{\times}$ 127 $\boxed{\div}$ 365 $\boxed{=}$ ⟶ | 1739.726 |

Step 2.   Divide the numerator by the denominator and round.

122.72 $\boxed{\div}$ 1739.726 $\boxed{=}$ ⟶ | 0.0705398 |

The answer: $r = 7.1\%$ rounded

Note: If your calculator has an inverse $\boxed{\frac{1}{x}}$ key or storage $\boxed{STO}$ key, this calculation can be completed in one step. For example, to find *r,*

5000 $\boxed{\times}$ 127 $\boxed{\div}$ 365 $\boxed{=}$ $\boxed{\frac{1}{x}}$ $\boxed{\times}$ 122.72 $\boxed{=}$ ⟶ | 0.0705398 |

Recapping Simple Interest Problems

Be sure you remember the three formulas used to solve for principal, interest rate, and time.

$$P = \frac{I}{rt}$$

$$r = \frac{I}{Pt}$$

$$t = \frac{I}{Pr}$$

When you've got them down pat, continue to Section Test 9.3.

WORKSPACE

SECTION TEST 9.3   Simple Interest

The following problems test your understanding of Section 9.3, Solving for Principal, Interest Rate, and Time.

A. In the following problems, calculate the unknown using ordinary simple interest at ordinary or 30-day-month time.

	Principal	Rate	Time (in days)	Interest
1.		20%	60	$240
2.		18%	90	292.50
3.	6800		72	231.20
4.	6400		60	128
5.	4000	15%		50
6.	5000	16%		80
7.	2800		80	112
8.	3500		60	105
9.	4800	18%		144
10.	3200	17%		68
11.		18%	45	75
12.		16%	72	224
13.	4500		60	135
14.	7500		72	330
15.		21%	40	105
16.		18%	180	607.50
17.	5000	22%		440
18.	4800	21%		84
19.	6000		90	300
20.	6000		40	80
21.	4500	22%		198
22.	5000	16%		100
23.		18%	60	180
24.		22%	90	132

B. Solve these problems.

1. Wired for Sound needs to take out a loan at 18% interest to expand its listening room. If the loan is for 72 days with ordinary simple interest of $270, what is the principal on the loan?

2. Maria's Mannequins took out a loan at 19% interest to add to its stock of "unisex" models. If the loan is for 30 days with ordinary interest of $95, what is the principal on the loan?

3. Petal Pushers borrows $3000 to acquire a new flower-refrigeration unit. If the loan is for 60 days with ordinary interest of $85, what is the interest rate on the loan?

4. Woody's Lumber Mill needs to take out a loan of $4000 to buy a new log splitter. If the loan is for 60 days with ordinary interest of $100, what is the interest rate on the loan?

5. J.R.'s Construction Company has taken out a $6300 loan at 17% interest to buy a new crane. If the ordinary interest at ordinary time is $238, what is the time period of the loan?

6. The Fly-By-Night Aviary has taken on an $8000 loan at 19% interest to build a special room for tropical birds. If the ordinary interest at ordinary time is $152, what is the time period of the loan?

7. Ava's Air Tours took out a loan at 18% interest to redecorate its waiting room. If the loan is for 30 days with ordinary interest of $67.50, what is the principal on the loan?

8. When he started his restaurant, Lew needed a $2500 loan to buy an expresso maker. If the loan is for 90 days with ordinary interest of $100, what is the interest rate on the loan?

9. The Garden of Earthly Delights took out a loan at 22% to buy a large consignment of Dutch tulip bulbs. If the loan is for 73 days with accurate simple interest of $220, what is the principal on the loan?

10. Bits 'n' Bytes borrowed money to expand its showroom. If the loan is at 21% interest for 65 days with accurate simple interest of $218.40, what is the principal on the loan?

11. Crystal's Diamonds took out a $5475 loan to buy new display cases. If the loan is for 47 days with accurate simple interest of $141, what is the interest rate on the loan?

12. Kate's Delicacies borrowed $2750 to send three of its cooks for cake decorating lessons. If the loan is for 73 days with accurate simple interest of $104.50, what is the interest rate on the loan?

13. Musselmann's Gym took out a $4700 loan at 17% to buy more weights. If the accurate simple interest is $319.60, what is the time period of the loan?

14. The Butter-em-Up Bakery took out a $3285 loan at 19% for a new oven. If the accurate simple interest is $145.35, what is the time period of the loan?

15. Wreck-a-Mended Auto got a loan at 21% to buy a new dent straightener. If the loan is for 95 days with accurate simple interest of $383.04, what is the principal on the loan?

Key Terms

accurate simple interest (exact interest)
exact time
interest
maturity value
ordinary or 30-day-month time

ordinary simple interest at ordinary time
ordinary simple interest at exact time
   (banker's interest)
principal
simple interest loan

Chapter 9 at a Glance

Page	Topic	Key Point	Example
310	Finding no. of days within a month	No. days = end date − start date	How many days are there from June 9 to June 27? No. days = 27 − 9 = 18
311	Exact time by hand	Find no. of days in each month.	How many days are there from April 14 to June 5? April: 30 − 14 = 16 May $\qquad$ = 31 to June 5: $\qquad$ = $\underline{5}$ $\qquad\qquad\qquad$ 52
312	Exact time by table	No. days = end date no. in table − beginning date number in table Add 1 if the period includes February 29 in a leap year.	How many days are there from January 14 to March 5, 1996? No. days = 64 − 14 = 50 $\qquad\qquad$ 50 + 1 = 51
314	Ordinary or 30-day-month time	Multiply the no. of full months from the starting date by 30. Add in remaining no. days in last month.	How many days are there using ordinary time from March 13 to July 27? 4 months × 30 = 120 days No. days left = 27 − 13 = 14 120 + 14 = 134
321	Accurate simple interest	Interest = principal × rate × time $(I = P \times r \times t = Prt)$ $t = \dfrac{\text{exact time}}{365 \text{ or } 366}$	What is the accurate simple interest on a $2750 loan at 18% for 146 days (not leap year)? $I = \$2750 \times 0.18 \times \dfrac{146}{365} = \$198$
323	Maturity value	Maturity value = principal + interest $(MV = P + I)$	What is the maturity value of an 18% loan of $2750 for 146 days at simple interest? $I = \$198$ (from above) $MV = \$2750 + 198 = \$2948$
325	Ordinary simple interest at ordinary or 30-day-month time	$I = P \times r \times t$, where $t = \dfrac{\text{ordinary time}}{360}$	Find ordinary simple interest on an 18% loan of $2750 for 143 days. $I = \$2750 \times 0.18 \times \dfrac{143}{360}$ $= \$196.63$ rounded

Page	Topic	Key Point	Example
327	Ordinary simple interest at exact time	$I = P \times r \times t$ where $t = \dfrac{\text{exact time}}{360}$	Find ordinary simple interest on an 18% loan of $2750 from May 13 to October 6. Exact time = 146 days $I = \$2750 \times 0.18 \times \dfrac{146}{360}$ $= \$200.75$
336	Solving for principal	$P = \dfrac{I}{rt}$	Find the principal of an 18% loan for 146 days with accurate interest of $198 for a 365-day year. $P = \dfrac{\$198}{0.18 \times \dfrac{146}{365}} = \$2750$
337	Solving for interest rate	$r = \dfrac{I}{Pt}$	Find the interest rate on a $2700 loan for 65 days with ordinary interest of $87.75. $r = \dfrac{\$87.75}{\$2700 \times \dfrac{65}{360}} = 18\%$
339	Solving for time	$t = \dfrac{I}{Pr} \times \text{days per year}$	Find the time (in days) on a $4000 loan at 17.7% for $88.50 in ordinary simple interest at ordinary time. $t = \dfrac{\$88.50}{4000 \times 0.177} \times 360$ $= 0.125 \times 360 = 45$

SELF-TEST   Simple Interest ━━━━━━━━

The following problems test your understanding of simple interest.

1. Calculate the number of days from April 9, 1996 to October 15, 1996 using (a) exact time and (b) ordinary or 30-day-month time.

2. Calculate the number of days from June 17, 1996 to November 12, 1996 using (a) exact time and (b) ordinary time.

3. Calculate the number of days from January 12, 1996 to April 3, 1996 using (a) exact time and (b) ordinary time.

4. Calculate the number of days from February 5, 1996 to July 6, 1996 using (a) exact time and (b) ordinary time.

5. On March 17, 1997, Lew borrowed $4700 at 18% interest to redo his restaurant's foyer. The due date on the loan is July 14, 1997. What is the interest on the loan using (a) accurate simple interest, (b) ordinary simple interest at ordinary or 30-day-month time, and (c) ordinary simple interest at exact time?

6. Ava's Air Tours borrowed $7200 to expand its waiting room on June 29, 1997 at 19% interest. The due date on the loan is August 12, 1997. What is the interest on the loan using (a) accurate simple interest, (b) ordinary simple interest at ordinary time and (c) ordinary simple interest at exact time?

7. Wreck-a-Mended Auto borrowed $5600 on January 24, 1996 at 17% interest to buy a spray painter. The due date on the loan is June 19, 1996. What is the interest on the loan using (a) accurate simple interest, (b) ordinary simple interest at ordinary time, and (c) ordinary simple interest at exact time?

8. On February 3, 1996, Klaus's Bavarian beer hall borrowed $1900 at 21% interest, The due date on the loan is April 15, 1996. What is the interest on the loan using (a) accurate simple interest (b) ordinary simple interest at ordinary time, and (c) ordinary simple interest at exact time?

9. To buy more weights, Musselmann's Gym took out a 20% loan for 72 days with ordinary interest of $116. What is the principal on the loan?

10. To buy new display cases, Crystal's Diamonds took out an 18% loan for 108 days with ordinary interest of $243. What is the principal on the loan?

11. The Fly-By-Night Aviary got a $3200 loan to buy a special mating cage. The loan carried ordinary interest of $64 for the 45-day loan period. What was the interest rate?

12. Bits 'n' Bytes got a $2400 loan to install piped-in music. The loan carried ordinary interest of $114 for the 90-day loan period. What was the interest rate?

13. In order to buy a new flower-refrigeration unit, Petal Pushers took out a $3700 loan at 18% with ordinary interest at ordinary time of $74. What was the time period (in days) of the loan?

14. In order to repave its parking lot, the Butter-em-Up Bakery took out a $6300 loan at 19% with ordinary interest at ordinary time of $266. What was the time period (in days) of the loan?

15. To reglaze part of one of its greenhouses. The Garden of Earthly Delights took out a 20% loan for 73 days with accurate simple interest of $316. What is the principal on the loan?

16. To buy a new lathe, Woody's Lumber Mill took out an 18% loan for 40 days with accurate simple interest of $136.80. What is the principal on the loan?

17. Wired for Sound got a $3285 loan to redo its sign. The loan carried accurate simple interest of $59.85 for the 35-day loan period. What was the interest rate?

18. Zippy Delivery Service got a $5500 loan to buy new uniforms for its drivers. The loan carried accurate simple interest of $231 for the 73-day loan period. What was the interest rate?

19. In order to send some of its staff to cake decorating courses. Kate's Delicacies took out a $3600 loan at 20% with accurate simple interest of $288. What was the time period (in days) of the loan?

20. In order to buy some high-pressure cleaning equipment, Sparkle Cleaning took out a $7300 loan at 17% with accurate simple interest of $255. What was the time period (in days) of the loan?

Bank Discount Loans

CHAPTER OBJECTIVES

When you complete this chapter successfully, you will be able to:

1. Calculate the net proceeds and bank discount of a noninterest-bearing promissory note using ordinary interest at ordinary or 30-day-month time and banker's interest.

2. Determine the bank discount and net proceeds of an interest-bearing note using ordinary interest at ordinary or 30-day-month time and banker's interest.

■ Fern's Organic Farms accepted a $2500 60-day note from Lew's restaurant on March 7. The note was discounted on April 5 at 17%. What are the bank discount and net proceeds using ordinary simple interest at ordinary (30-day-month) time?

■ Boxed In accepted a $2000, 16%, 90-day note from Crystal's Diamonds on October 22. The note was discounted on November 5 at 15%. What are the bank discount and net proceeds using ordinary simple interest at exact time—banker's interest?

■ The Wild Child Recording Company accepted a $4500, 16%, 90-day note from Compound Discs on March 10. The note was discounted on April 4 at 15%. What are the maturity value, bank discount, and net proceeds, using ordinary simple interest at ordinary (30-day-month) time?

Part of doing business today is extending credit to your customers—whether they are individuals or other companies. Another part of doing business is borrowing money from other companies and from banks.

In this chapter, we'll look at how companies use the credit they extend to other firms to get loans from their own banks.

Pencil in hand? Get ready for some information you can "bank" on.

Can You:

Calculate the net proceeds on a bank discount loan?
Determine the bank discount using ordinary interest at 30-day-month time and banker's interest?
. . . If not, you need this section.

Promissory Note A form of simple interest loan agreement that specifies the principal, interest, and due date.

From time to time, most businesses borrow money from banks or other businesses. When the loan is for a short time period (less than a year) and is paid in one payment, a **promissory note** is often used to formalize the loan. The promissory note states the conditions of the loan—the principal, interest, and repayment schedule.

For example, Bits 'n' Bytes ordered disk drives from Disks 'R' Us and agreed to pay $2500 in 60 days with interest at 12%. The following is the promissory note signed by Bits 'n' Bytes' treasurer, Ms. Karen M. Paisley.

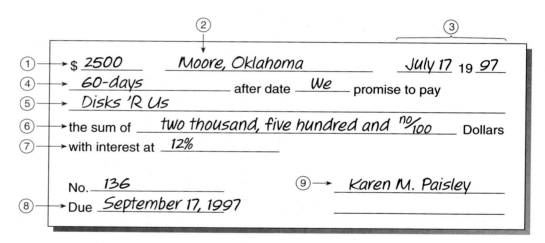

Information required on a promissory note includes

Face Value The principal (amount borrowed) on a promissory note.

① Principal or **face value** of the note—the amount borrowed written in numerals.

② City and state.

③ Date of the note.

④ Length of time of the note.

Interest-Bearing Note A promissory note in which the lender charges interest to the borrower.

⑤ Payee—the company or institution giving the credit.

⑥ Principal written in words.

⑦ The interest rate for an interest-bearing note. On an **interest-bearing note**, the maturity value (principal + interest) must be paid. A note where no interest is charged is called a **noninterest-bearing note.** This promissory note is an interest-bearing note.

Noninterest-Bearing Note A promissory note in which the lender does *not* charge interest to the borrower.

⑧ Due date of the note—when it must be repaid.

⑨ Signature(s) of the payer—the person borrowing the money.

350

Collateral An asset pledged to the lender to assure payment.

Bank Discount Method A form of loan in which a bank lends a firm a sum of money based on a promissory note held by the firm, first deducting the interest on the loan.

Net Proceeds In a bank discount loan, the loan amount minus the interest—the amount the borrower actually receives.

Bank Discount In a bank discount loan, the amount of interest retained by the bank before making a loan.

Discount Period The time period remaining on a promissory note during which a bank is willing to make a loan to the lender on that note.

Once Disks 'R' Us has accepted the promissory note, it may use the note as collateral to borrow money from its bank. **Collateral** is an asset pledged to assure repayment—if Disks 'R' Us doesn't repay the bank, then the bank—not Disks 'R' Us—will get the money from Bits 'n' Bytes.

In making a loan to Disks 'R' Us, the company's bank will probably use the **bank discount method.** With bank discount, the interest is subtracted from the principal before it is given to the firm taking out the loan. The actual amount borrowed, called the **net proceeds,** will be the original principal "discounted" or decreased by the amount of the interest (the **bank discount**). The period from the day the note is discounted until it is due—the time for which the bank loans the money—is called the **discount period.**

An example will make it clear. A loan of $5000 at 12% for 60 days will cost $100 interest (ordinary simple interest at ordinary (30-day-month time). Under the bank discount procedure, the lender would subtract the interest from the amount loaned and the borrower would actually receive the net proceeds of $5000 − $100 = $4900.

Although a business may accept a note on a particular date, there may be a significant time lapse before that business borrows the cash from a bank. In order to calculate the bank discount, you must first figure the amount of time for which the money will be borrowed. Generally, the length of the note and the amount of time the money is borrowed from the bank are not the same.

Calculating Net Proceeds Under Bank Discount

Finding the bank discount when you know the value of the note, the discount interest rate, and the discount period is quite easy. Just follow these steps:

Steps: Bank Discount on Noninterest-Bearing Notes

STEP 1. *Calculate* the discount period:

$$\text{Discount period} = \frac{\text{total length}}{\text{of loan}} - \frac{\text{no. of days on loan}}{\text{already elapsed (if any)}}$$

STEP 2. *Calculate* the bank discount (interest) using the following formula:

$$\text{Discount} = I = Prt$$

where P = value of the note
r = discount interest rate
t = discount period

STEP 3. *Calculate* net proceeds:

$$\text{Net proceeds} = \text{principal} - \text{bank discount}$$

For example, Figaro, the barber, bought some equipment for his shop using a promissory note. Wolfgang's Barber Supply Company holds Figaro's 90-day note for $1800 made on August 13. Wolfgang's needs cash to meet current expenses, so, on September 13, Wolfgang's bank discounted the note at 12% (ordinary simple interest at ordinary (30-day-month) time). What are the discount period, bank discount, and net proceeds?

Step 1. The note was made on August 13 but wasn't taken to the bank until September 13. The bank will lend Wolfgang's the money only for the time remaining on the note—the discount period.

August 13 to September 13 = 1 month = 30 days
Discount period = 90 − 30 = 60 days

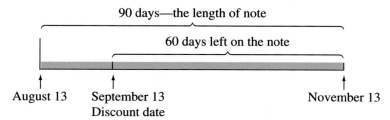

The total time of the note was 90 days. The note was made on August 13 but not discounted until September 13, 30 days later. This left 60 days on the note. Remember, the money was borrowed from the bank only for the time remaining on the note.

Step 2. Next, you must calculate the bank discount. *Bank discount is always equal to the interest on the loan.*

$$P = \$1800; \quad r = 12\%; \quad t = \frac{60}{360} \quad \text{Ordinary interest}$$

$$\text{Discount} = I = Prt = \$1800 \times 12\% \times \frac{60}{360} = \$36$$

Step 3. Net proceeds = principal − bank discount
= $1800 − $36 = $1764

In this example, Wolfgang's bank buys Figaro's note from Wolfgang's company for $1764, and 60 days later, Figaro must pay the bank $1800. Everyone wins: Figaro has his equipment. Wolfgang's has the cash it needed, and the bank made a profit by lending some of its cash.

The promissory note in the preceding problem is a *noninterest-bearing* note. Wolfgang's did not charge Figaro interest. (The bank did, of course, charge Wolfgang's interest.) Interest-bearing promissory notes are discussed in Section 10.2.

**YOUR
TURN**

| WORK THIS PROBLEM |

The Question:

Find the bank discount and net proceeds for the following problems. Use ordinary simple interest at 30-day-month time.

(a) What are the bank discount and net proceeds on a $5000, 60-day note dated May 18 and discounted June 3 at 12%?
(b) Wreck-a-Mended Auto holds a 60-day note for $1217.86 made on March 23. On April 5 the note was discounted at the bank at 11.7%. What are the bank discount and net proceeds?

| ✔ YOUR WORK |

The Solution:

(a) ***Step 1.*** May 18 to June 3 = 15 days
Discount period = 60 − 15 = 45 days

Step 2. $P = \$5000;\quad r = 12\%;\quad t = \dfrac{45}{360}$

Discount $= I = Prt = \$5000 \times 12\% \times \dfrac{45}{360} = \$75$

Step 3. Net proceeds $= \$5000 - \$75 = \$4925$

(b) *Step 1.* March 23 to April 5 = 12 days
Discount period = 60 − 12 = 48 days

Step 2. $P = \$1217.86;\quad r = 11.7\%;\quad t = \dfrac{48}{360}$

Discount $= I = Prt = \$1217.86 \times 11.7\% \times \dfrac{48}{360}$

$= 18.998616 = \$19.00$ rounded

Step 3. Net proceeds $= \$1217.86 - \$19.00 = \$1198.86$

As you might expect, a calculator can come in handy when calculating bank discount on a non-interest-bearing note.

> ## *Calculating Bank Discount on Noninterest-Bearing Notes*
>
> To see how a calculator can help, let's use one to solve part (a) in the preceding "Your Turn", which involves a $5000, 60-day note dated May 18 and discounted June 3 at 12%.
>
> *Step 1.* You may find it easier to calculate the discount period by hand than to do it on the calculator, since the periods are generally short. As you saw, the discount period in this case is 45 days.
>
> *Step 2.* You'll need to mentally convert the rate *r* from a percent to a decimal, 0.12, and then enter the following:
>
> 5000 × .12 × 45 ÷ 360 = +/− + 5000 = ⟶ 4925.
>
> If you are using a business calculator,
>
>
>
> 5000 × 12 % × 45 ÷ 360 = ⟶ 75. +/− + 5000
>
> = ⟶ 4925.   to subtract 75

Determining Bank Discount Using Banker's Interest

Banker's Interest
Another name for ordinary simple interest at exact time.

Although we used ordinary interest at 30-day-month time to calculate bank discounts, any of the three simple interest methods can be used. Another commonly used method is ordinary interest at *exact time,* which is usually called **banker's interest.** Banker's interest uses the time fraction

$$\frac{\text{Exact time}}{360}$$

For example, a $2500 noninterest-bearing note dated May 18 and due on August 18 is discounted June 2 at 15%. Let's calculate the bank discount and net proceeds using banker's interest.

Step 1.   The time left on the note is from June 2 to August 18.

$$
\begin{aligned}
\text{June 2 to June 30} &= 30 - 2 = 28 \text{ days} \\
\text{July 31} &= 31 \text{ days} \\
\text{to August 18} &= \underline{18 \text{ days}} \\
\text{Discount period} = \textit{exact time left} &= 77 \text{ days}
\end{aligned}
$$

Step 2. $P = \$2500; \quad r = 15\%; \quad t = \dfrac{77}{360}$

$$
\text{Discount} = I = Prt = \$2500 \times 15\% \times \frac{77}{360} = \$80.21 \text{ rounded}
$$

Step 3.   Net proceeds = principal − discount
$$
= \$2500 - 80.21 = \$2419.79
$$

YOUR TURN

| WORK THIS PROBLEM |

The Question:

A $3800 noninterest-bearing note dated August 13 and due on November 13 is discounted September 2 at 14%. What are the discount and net proceeds using banker's interest?

| ✔ YOUR WORK |

The Solution:

Step 1.
$$
\begin{aligned}
\text{September 2 to September 30} &= 28 \text{ days} \\
\text{October} &= 31 \text{ days} \\
\text{to November 13} &= \underline{13 \text{ days}} \\
\text{Discount Period} &= 72 \text{ days}
\end{aligned}
$$

Step 2. $P = \$3800; \quad r = 14\%; \quad t = \dfrac{72}{360}$

$$
\text{Discount} = \$3800 \times 14\% \times \frac{72}{360} = \$106.40
$$
$$
\text{Net proceeds} = \$3800 - \$106.40 = \$3693.60
$$

If you feel comfortable with these kinds of calculations, turn to Section Test 10.1 and try the exercises there. Otherwise, go back and review this section.

SECTION TEST 10.1 Bank Discount Loans

The following problems test your understanding of Section 10.1, Promissory Notes and Bank Discount on Noninterest-Bearing Notes.

A. Calculate the bank discount and net proceeds for the following noninterest-bearing notes. Use ordinary simple interest at ordinary or 30-day-month time.

	Principal	Time	Date of Note	Discount Date	Discount Rate	Bank Discount	Net Proceeds
1.	$3000	90 days	3/6	4/6	12%		
2.	5000	90 days	5/12	7/12	15%		
3.	2500	60 days	9/3	9/18	18%		
4.	4800	60 days	10/28	11/13	16%		
5.	1250	75 days	9/5	10/20	13%		
6.	2800	75 days	3/17	5/2	17%		
7.	1550	45 days	4/27	5/6	15%		
8.	3200	45 days	6/15	6/20	16%		
9.	4000	90 days	8/2	8/15	19%		
10.	2400	90 days	3/6	3/20	21%		

B. Calculate the bank discount and net proceeds for the following noninterest-bearing notes. Use ordinary simple interest at exact time (banker's interest).

	Principal	Time	Date of Note	Discount Date	Discount Rate	Bank Discount	Net Proceeds
1.	$4500	90 days	4/2	5/4	15%		
2.	2400	90 days	6/15	8/17	16%		
3.	3200	60 days	8/12	9/8	17%		
4.	3500	60 days	10/12	11/3	15%		
5.	1850	75 days	8/9	9/20	13%		
6.	2700	75 days	4/12	5/20	18%		
7.	1600	45 days	5/19	6/7	16%		
8.	3700	45 days	6/5	6/18	17%		
9.	4200	90 days	9/8	10/4	19%		
10.	2700	90 days	3/7	3/23	20%		

C. Solve these problems.

1. The Woody's Lumber Mill accepted a $3000 90-day note from Build-It-Yourself on March 12. The note was discounted on April 15 at 14%. What are (a) the bank discount and (b) the net proceeds? (Use ordinary simple interest at ordinary (30-day-month) time.)

2. Big Ben's Bolts accepted a $4500 90-day note from Oliver's Hardware Store on May 17. The note was discounted on June 15 at 18%. What are (a) the bank discount and (b) the net proceeds? (Use ordinary simple interest at ordinary (30-day-month) time.)

3. Maria's Mannequins accepted a $1500 60-day note from Needless-Markup Department Store on June 2. The note was discounted on June 15 at 16%. What are (a) the bank discount and (b) the net proceeds? (Use ordinary simple interest at ordinary (30-day-month) time.)

4. Fern's Organic Farms accepted a $2500 60-day note from Lew's restaurant on March 7. The note was discounted on April 5 at 17%. What are (a) the bank discount and (b) the net proceeds? (Use ordinary simple interest at ordinary (30-day-month) time.)

5. Fit-to-Print accepted a $5000 60-day note from Zippy Delivery Service on July 7. The note was discounted on July 21 at 19%. What are (a) the bank discount and (b) the net proceeds? (Use ordinary simple interest at ordinary (30-day-month) time.)

6. Wired for Sound accepted an $1800 60-day note from the Easy Street Diner on August 23. The note was discounted on September 5 at 12%. What are (a) the bank discount and (b) the net proceeds? (Use ordinary simple interest at exact time—banker's interest.)

7. Sparkle Cleaning accepted a $1200 90-day note from Petal Pushers on March 21. The note was discounted on April 8 at 17%. What are (a) the bank discount and (b) the net proceeds? (Use ordinary simple interest at exact time—banker's interest.)

8. The Beauty Supply Warehouse accepted a $2800 90-day note from Helen's Beauty Salon on May 15. The note was discounted on May 23 at 16%. What are (a) the bank discount and (b) the net proceeds? (Use ordinary simple interest at exact time—banker's interest.)

9. Bits 'n' Bytes accepted a $6500 60-day note from Bob's Beeflike Burgers on April 9. The note was discounted on April 24 at 17%. What are (a) the bank discount and (b) the net proceeds? (Use ordinary simple interest at exact time—banker's interest.)

10. J.R.'s Construction Company accepted a $2300 60-day note from Crystal's Diamonds on March 28. The note was discounted on April 11 at 19%. What are (a) the bank discount and (b) the net proceeds? (Use ordinary simple interest at exact time—banker's interest.)

Can You:

Calculate the bank discount and net proceeds of an interest-bearing note using ordinary interest?

Determine the bank discount and net proceeds of an interest-bearing note using banker's interest?

. . . If not, you need this section.

The examples in Section 10.1 all used noninterest-bearing notes—no interest on the original note.

Discounting noninterest-bearing notes is simple, but not widely practiced. Most businesses can't afford to lend money to their customers and not charge interest. Generally, they will accept only *interest-bearing* notes.

Using Ordinary Interest to Calculate Discount

The steps used to calculate the bank discount of an interest-bearing note are similar to the method used for noninterest-bearing notes. The difference is that the bank discount is calculated on the *maturity value* of the note.

Steps: Bank Discount on Interest-Bearing Notes

STEP 1.   *Calculate* the interest and maturity value of the promissory note using the following formulas:

$$I = Prt \quad \text{and} \quad MV = P + I$$

where P = original value of the note
r = interest rate on the note
t = entire time of the note
MV = maturity value

STEP 2.   *Calculate* the discount period.

STEP 3.   *Calculate* the bank discount on the loan from the bank:

$$\text{Bank discount} = I = Prt$$

where $P = MV$ from step 1
r = discount interest rate
t = discount period from step 2

STEP 4.   *Calculate* net proceeds:

$$\text{Net proceeds} = \text{maturity value} - \text{bank discount}$$

For example, the Hillary Climbing Company holds a 60-day, 18% note for $2000 made on June 5. The note is discounted at the bank on June 20 at 12%. Calculate the bank discount and net proceeds using ordinary simple interest at ordinary (30-day-month) time.

Step 1. $P = \$2000$; $r = 18\%$; $t = \dfrac{60}{360}$

$I = Prt = \$2000 \times 18\% \times \dfrac{60}{360} = \$60$

$MV = P + I = \$2000 + \$60 = \$2060$

Step 2.   June 5 to June 20 = 15 days
Discount period = 60 − 15 = 45 days

Step 3. $MV = \$2060$; $r = 12\%$; $t = \dfrac{45}{360}$

Bank discount $= MV \cdot rt = \$2060 \times 12\% \times \dfrac{45}{360} = \$30.90$

Step 4.   Net proceeds = maturity value − bank discount
$= \$2060 − \$30.90 = \$2029.10$

YOUR TURN

| WORK THIS PROBLEM |

The Question:

(a) Big Ben's Bolts holds a 90-day, 15% note for $1200 made by a construction company on April 17 and discounted May 2 at 13%. What are the maturity value, bank discount, and net proceeds? Use ordinary interest at ordinary (30-day-month) time.

(b) The Garden of Earthly Delights holds a 60-day, 18% note for $5000 made by a local landscaper on July 5 and discounted August 13 at $12\frac{1}{2}\%$. What are the maturity value, bank discount, and net proceeds? Use ordinary interest at ordinary (30-day-month) time.

| ✔ YOUR WORK |

The Solution:

(a) **Step 1.** $P = \$1200$; $r = 15\%$; $t = \dfrac{90}{360}$

$I = Prt = \$1200 \times 15\% \times \dfrac{90}{360} = \$45$

$MV = P + I = \$1200 + 45 = \$1245$

Step 2.   April 17 to May 2 = 15 days
Discount period = 90 − 15 = 75 days

Step 3. $MV = \$1245$; $r = 13\%$; $t = \dfrac{75}{360}$

Bank discount $= \$1245 \times 13\% \times \dfrac{75}{360} = \$33.72$ rounded

Step 4.   Net proceeds $= \$1245 − \$33.72 = \$1211.28$

(b) **Step 1.** $P = \$5000$; $r = 18\%$; $t = \dfrac{60}{360}$

$I = Prt = \$5000 \times 18\% \times \dfrac{60}{360} = \$150$

$MV = P + I = \$5000 + 150 = \$5150$

Step 2.   July 5 to August 13 = 38 days
Discount period = 60 − 38 = 22 days

Step 3. $\quad MV = \$5150; \quad r = 12\frac{1}{2}\%; \quad t = \dfrac{22}{360}$

$$\text{Bank discount} = \$5150 \times 12\frac{1}{2}\% \times \dfrac{22}{360} = \$39.34 \text{ rounded}$$

Step 4. $\quad$ Net proceeds = $\$5150 - \$39.34 = \$5110.66$

Calculators can simplify the task of finding the maturity value, bank discount, and net proceeds using ordinary interest.

Calculating Bank Discount on Interest-Bearing Notes

To see how your calculator can help, let's use it to solve (b) in the preceding "Your Turn" in which we have a \$5000, 60-day, 18% note made on July 5 and discounted August 13 at $12\frac{1}{2}\%$.

Step 1. $\quad$ You can find *MV* in one smooth calculation as follows:

$$5000 \;\boxed{\times}\; 18 \;\boxed{\%}\; \boxed{\times}\; 60 \;\boxed{\div}\; 360 \;\boxed{=}\; \boxed{+}\; 5000 \;\boxed{=}\; \longrightarrow \qquad \boxed{5\,150.}$$

Step 2. $\quad$ Calculating the discount period by hand is generally as fast as doing it by calculator, though you can use the calculator to first add up the time on the note already past and then subtract it from the total note period of 60 days. In either case, the answer is 22.

Step 3. $\quad$ Punch in the following:

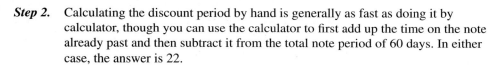

$$5150 \;\boxed{\times}\; 12.5 \;\boxed{\%}\; \boxed{\times}\; 22 \;\boxed{\div}\; 360 \;\boxed{=}\; \longrightarrow \qquad \boxed{39.340277}$$

Jot down or store in your calculator's memory the rounded form 39.34 or use the $\boxed{+/-}$ key to subtract directly.

$$\boxed{+/-}\; \boxed{+}\; 5150 \;\boxed{=}\; \longrightarrow \qquad \boxed{5\,110.66}$$

Using Banker's Interest to Calculate Discount

In the previous problems, we used ordinary interest at ordinary (30-day-month) time. But sometimes bank discount loans involve banker's interest instead.

For example, a \$2500, 18% note dated May 18 and due on August 18 is discounted June 2 at 12%. Calculate the maturity value, bank discount, and net proceeds. Use banker's interest (ordinary interest at exact time).

Step 1. $\quad$ The full time on the note is from May 18 to August 18.

$$
\begin{aligned}
\text{May 18 to May 31} &= 31 - 18 = 13 \text{ days} \\
\text{June} &= 30 \text{ days} \\
\text{July} &= 31 \text{ days} \\
\text{to August 18} &= \underline{18 \text{ days}} \\
\text{Total exact time} &= 92 \text{ days}
\end{aligned}
$$

$$P = \$2500; \quad r = 18\%; \quad t = \frac{92}{360}$$

$$I = Prt = \$2500 \times 18\% \times \frac{92}{360} = \$115$$

$$MV = P + I = \$2500 + \$115 = \$2615$$

Step 2.   The discount period is from June 2 to August 18.

June 2 to June 30 = 30 − 2	= 28 days
July	= 31 days
to August 18	= 18 days
Discount period	= 77 days

Step 3. $MV = \$2615; \quad r = 12\%; \quad t = \dfrac{77}{360}$

$$\text{Bank discount} = MV \cdot rt = \$2615 \times 12\% \times \frac{77}{360}$$
$$= \$67.12 \text{ rounded}$$

Step 4.   Net proceeds = maturity value − bank discount
$$= \$2615 − \$67.12 = \$2547.88$$

YOUR TURN

| WORK THIS PROBLEM |

The Question:

A \$1500, 16% note dated April 14 and due on July 14 is discounted on May 5 at 15%. Calculate the maturity value, bank discount, and net proceeds using banker's interest.

| ✔ YOUR WORK |

The Solution:

Step 1.   The full time on the note is from April 14 to July 14.

April 14 to April 30	= 16 days
May	= 31 days
June	= 30 days
to July 14	= 14 days
Total exact time	= 91 days

$$P = \$1500; \quad r = 16\%; \quad t = \frac{91}{360}$$

$$I = \$1500 \times 16\% \times \frac{91}{360} = \$60.67 \text{ rounded}$$

$$MV = \$1500 + \$60.67 = \$1560.67$$

Step 2.   The discount period is from May 5 to July 14.

May 5 to May 31	= 26 days
June	= 30 days
to July 14	= 14 days
Discount period	= 70 days

Step 3. $MV = \$1560.67; \quad r = 15\%; \quad t = \dfrac{70}{360}$

$$\text{Bank discount} = \$1560.67 \times 15\% \times \frac{70}{360} = \$45.52 \text{ rounded}$$

Step 4.   Net proceeds = \$1560.67 − \$45.52 = \$1515.15

Now try the questions in Section Test 10.2.

SECTION TEST 10.2   Bank Discount Loans

The following problems test your understanding of Section 10.2, Bank Discount of an Interest-Bearing Note.

A. Calculate the maturity value, bank discount, and net proceeds for the following interest-bearing notes. Use ordinary simple interest at ordinary or 30-day-month time.

	Principal	Interest Rate	Time	Date of Note	Discount Date	Discount Rate	Maturity Value	Bank Discount	Net Proceeds
1.	$4000	15%	90 days	4/5	5/5	12%			
2.	4800	16%	90 days	6/3	7/3	18%			
3.	3200	18%	60 days	3/12	3/17	15%			
4.	4200	21%	60 days	9/7	10/22	17%			
5.	3750	12%	75 days	8/5	9/20	15%			
6.	2200	16%	75 days	4/12	4/27	15%			
7.	1850	20%	45 days	5/12	6/7	17%			
8.	2700	19%	45 days	7/18	8/2	15%			
9.	4300	17%	90 days	8/9	8/18	19%			
10.	2500	16%	90 days	4/8	4/21	18%			

B. Calculate the maturity value, bank discount, and net proceeds for the following interest-bearing notes. Use ordinary simple interest at exact time (banker's interest).

	Principal	Interest Rate	Time	Date of Note	Discount Date	Discount Rate	Maturity Value	Bank Discount	Net Proceeds
1.	$2500	16%	90 days	5/6	6/7	14%			
2.	2400	17%	90 days	7/7	8/2	19%			
3.	5000	15%	60 days	4/13	5/16	17%			
4.	4600	20%	60 days	8/11	9/3	16%			
5.	3500	15%	75 days	9/15	10/2	12%			
6.	5200	18%	75 days	5/13	6/22	16%			
7.	3450	21%	45 days	5/2	6/1	18%			
8.	2300	18%	45 days	6/14	7/5	17%			
9.	3800	20%	90 days	7/12	8/15	18%			
10.	2800	18%	90 days	5/11	5/25	16%			

C. Solve these problems.

1. The Liki Valve Company accepted a $1200, 16%, 90-day note from Norton's Plumbing Supplies on August 5. The note was discounted on August 15 at 14%. What are (a) the maturity value, (b) the bank discount, and (c) the net proceeds? (Use ordinary simple interest at ordinary (30-day-month) time.)

2. Fins and Feathers Wholesalers accepted a $3000, 18%, 90-day note from the Fly-By-Night Aviary on March 7. The note was discounted on April 2 at 14%. What are (a) the maturity value, (b) the bank discount, and (c) the net proceeds? (Use ordinary simple interest at ordinary (30-day-month) time.)

3. L. A. Mountaineering Tools accepted a $2700, 17%, 60-day note from Intense Tents on May 21. The note was discounted on June 3 at 15%. What are (a) the maturity value, (b) the bank discount, and (c) the net proceeds? (Use ordinary simple interest at ordinary (30-day-month) time.)

4. All Cycles Big and Small accepted a $900, 18%, 60-day note from Going In Circles on March 7. The note was discounted on March 19 at 17%. What are (a) the maturity value, (b) the bank discount, and (c) the net proceeds? (Use ordinary simple interest at ordinary (30-day-month) time.)

5. Maria's Mannequins accepted a $2400, 16%, 90-day note from Needless-Makeup Department Stores on March 5. The note was discounted on March 18 at 18%. What are (a) the maturity value, (b) the bank discount, and (c) the net proceeds? (Use ordinary simple interest at ordinary (30-day-month) time.)

6. Payne's Glass Works accepted an $1800, 15%, 60-day note from Oliver's Hardware Store on October 7. The note was discounted on November 1 at 18%. What are (a) the maturity value, (b) the bank discount, and (c) the net proceeds? (Use ordinary simple interest at exact time—banker's interest.)

7. Boxed In accepted a $2000, 17%, 90-day note from Crystal's Diamonds on October 22. The note was discounted on November 5 at 15%. What are (a) the maturity value, (b) the bank discount, and (c) the net proceeds? (Use ordinary simple interest at exact time—banker's interest.)

8. Sparkle Cleaning accepted an $8000, 14%, 90-day note from Petal Pushers on June 2. The note was discounted on June 15 at 13%. What are (a) the maturity value, (b) the bank discount, and (c) the net proceeds? (Use ordinary simple interest at exact time—banker's interest.)

9. Fit-to-Print accepted a $1900, 15%, 60-day note from Zippy Delivery Service on March 3. The note was discounted on March 18 at 16%. What are (a) the maturity value, (b) the bank discount, and (c) the net proceeds? (Use ordinary simple interest at exact time—banker's interest.)

10. J.R.'s Construction Company accepted a $2300, 20%, 90-day note from Hole-in-the-Wall Bagels on May 21. The note was discounted on June 11 at 16%. What are (a) the maturity value, (b) the bank discount, and (c) the net proceeds? (Use ordinary simple interest at exact time—banker's interest.)

11. The Beauty Supply Warehouse accepted a $7500 note from Claudia's Clip-Joint on August 3. This noninterest-bearing note was due on November 12. The note was discounted on September 23 at 16%, using exact time, ordinary interest. (a) How much was the bank discount? (b) What were the net proceeds paid to the Beauty Supply Warehouse?

12. On February 15, 1996, General Wholesale accepted a $21,000 120-day note at 15% annual interest rate from Dynamite Deals. This exact time, exact interest note was discounted on March 8, 1996, at 16% (exact time, exact interest). (a) When is this note due? (b) What is the total that Dynamite Deals will pay? (c) What was the bank discount amount? (d) How much did General Wholesale receive (net proceeds) on March 8?

WORKSPACE

Key Terms

bank discount	discount period	net proceeds
bank discount method	face value	noninterest-bearing note
banker's interest	interest-bearing note	promissory note
collateral		

Chapter 10 at a Glance

Page	Topic	Key point	Example
351	Net proceeds under bank discount	Discount period = loan length − days elapsed Discount = $I = Prt$, where P = note value, r = discount rate, t = discount period Net proceeds = principal − discount	What are the bank discount and net proceeds on a $5000, 60-day note dated May 18 and discounted June 3 at 12%? Discount period = 60 − 15 = 45 days Discount = $\$5000 \times 0.12 \times \dfrac{45}{360}$ = $75 Net proceeds = $5000 − 75 = $4925
353	Bank discount using banker's interest	Same as preceding except that $t = \dfrac{\text{exact time}}{360}$	Find the bank discount and net proceeds on a $2500 note dated May 18 due August 18 and discounted on June 2 at 15%. Discount period = 77 days left on note Discount = $\$2500 \times 0.15 \times \dfrac{77}{360}$ = $80.21 rounded Net proceeds = $2500 − 80.21 = $2419.79
357	Bank discount on interest-bearing notes at ordinary interest	I of promissory note = Prt where P = note's original value; r = note's interest rate; t = whole time of note MV of promissory note = $P + I$ of note Calculate the bank discount and net proceeds as in the above note using MV of promissory note as P.	Find the bank discount and net proceeds on a 60-day, 18% note for $2000 made on June 5 and discounted June 20 at 12%. $I = \$2000 \times 0.18 \times \dfrac{60}{360} = \$60$ $MV = \$2000 + 60 = \$2060$ Discount period = 45 Discount = $\$2060 \times 0.12 \times \dfrac{45}{360}$ = $30.90 Net proceeds = $2060 − 30.90 = $2029.10

Page	Topic	Key point	Example
359	Bank discount on interest-bearing notes at banker's interest	Follow steps in the previous note using banker's interest time of exact time/360.	Find the bank discount and net proceeds on a $2500, 18% note dated May 18, due on August 18, and discounted June 2 at 12%. Exact time = 92 days $$I = \$2500 \times 0.18 \times \frac{92}{360} = \$115$$ $MV = \$2500 + 115 = \$2615$ Discount period = 77 days $$\text{Discount} = \$2615 \times 0.12 \times \frac{77}{360}$$ $$= \$67.12 \text{ rounded}$$ Net proceeds = $2615 − 67.12 = $2547.88

SELF-TEST Bank Discount Loans

The following problems test your understanding of bank discount loans.

1. Fern's Organic Farms accepted an $1100 90-day note from Lew's restaurant on May 11. The note was discounted on June 6 at 15%. What are (a) the bank discount and (b) the net proceeds? (Use ordinary simple interest at ordinary (30-day-month) time.)

2. Bits 'n' Bytes accepted a $7000 90-day note from Micro Systems Support on March 8. The note was discounted on April 3 at 19%. What are (a) the bank discount and (b) the net proceeds? (Use ordinary simple interest at ordinary (30-day-month) time.)

3. Pizza Ovens by Mini accepted a $3500 60-day note from the Pizza Palace on April 19. The note was discounted on April 29 at 13%. What are (a) the bank discount and (b) the net proceeds? (Use ordinary simple interest at ordinary (30-day-month) time.)

4. The Timekeeper Company accepted a $1200 60-day note from J.R.'s Construction on March 5. The note was discounted on April 2 at 17%. What are (a) the bank discount and (b) the net proceeds? (Use ordinary simple interest at ordinary (30-day-month) time.)

5. The Crystal Pure Water Company accepted a $1600 90-day note from the Easy Street Diner on November 8. The note was discounted on December 14 at 19%. What are (a) the bank discount and (b) the net proceeds? (Use ordinary simple interest at exact time—banker's interest.)

6. Intense Tents Company accepted a $2600 60-day note from the local Boy Scouts Council on March 3. The note was discounted on March 17 at 17%. What are (a) the bank discount and (b) the net proceeds? (Use ordinary simple interest at exact time—banker's interest.)

7. Boxed In accepted a $7500 60-day note from Crystal's Diamonds on May 6. The note was discounted on May 17 at 18%. What are (a) the bank discount and (b) the net proceeds? (Use ordinary simple interest at exact time—banker's interest.)

8. The Wild Child Recording Company accepted a $4500, 16%, 90-day note from Compound Discs on March 10. The note was discounted on April 4 at 15%. What are (a) the maturity value, (b) the bank discount, and (c) the net proceeds? (Use ordinary simple interest at ordinary (30-day-month) time.)

9. Fins and Feathers Wholesalers accepted a $3200, 19%, 90-day note from the Fly-By-Night Aviary on May 17. The note was discounted on June 5 at 16%. What are (a) the maturity value, (b) the bank discount, and (c) the net proceeds? (Use ordinary simple interest at ordinary (30-day-month) time.)

10. Fit to Print accepted a $1600, 17%, 90-day note from Kate's Delicacies on November 2. The note was discounted on December 5 at 14%. What are (a) the maturity value, (b) the bank discount, and (c) the net proceeds? (Use ordinary simple interest at ordinary (30-day-month) time.)

11. HAL Computers accepted a $1600, 16%, 60-day note from Bits 'n' Bytes on March 7. The note was discounted on March 11 at 15%. What are (a) the maturity value, (b) the bank discount, and (c) the net proceeds? (Use ordinary simple interest at ordinary (30-day-month) time.)

12. The Write Stuff accepted a $5000, 18%, 60-day note from Emma Eyre on April 2. The note was discounted on April 19 at 19%. What are (a) the maturity value, (b) the bank discount, and (c) the net proceeds? (Use ordinary simple interest at exact time—banker's interest.)

13. The Attic Cleaners accepted a $3700, 16%, 90-day note from Nora's Collectibles on March 28. The note was discounted on April 5 at 18%. What are (a) the maturity value, (b) the bank discount, and (c) the net proceeds? (Use ordinary simple interest at exact time—banker's interest.)

14. Dairy King Wholesalers accepted a $1500, 17%, 90-day note from the Butter-em-Up Bakery on August 13. The note was discounted on September 11 at 15%. What are (a) the maturity value, (b) the bank discount, and (c) the net proceeds? (Use ordinary simple interest at exact time—banker's interest.)

More Complex Loans

CHAPTER OBJECTIVES

When you complete this chapter successfully, you will be able to:

1. Determine the finance charges and annual percentage rate on an installment loan.

2. Find the sum of the digits of a number of payments and apply the Rule of 78 for early loan payoffs.

3. Determine interest and balances on credit card accounts with different interest payment terms.

■ Upward Bound bought a hang glider for $1200, making a down payment of $150 and financing the remainder for 18 months at $65.65 per month. What are the (a) total installment cost, (b) finance charge, (c) loan amount, (d) annual percentage rate (APR) using the formula, and (e) APR using the table?

■ Tick Tock Clocks sold a grandfather clock to Annie Thyme, who financed it for 18 months. The finance charge was $374.50. If she pays off the loan 7 months early, what is the finance charge rebate?

■ J.R.'s Construction received its Woody Lumber bill with a balance of $562 as of April 1. Woody's Lumber calculates the finance charge as 15% of the average daily balance. During the month, J.R.'s had the following Woody's Lumber transactions:

April 4	New Purchase:	$45
April 12	Payment:	$75
April 17	New Purchase:	$50
April 26	New Purchase	$82

What are (a) the average daily balance, (b) the finance charge, and (c) new balance for the next month? Use a 30-day billing cycle.

s you can see, not all loans are as simply structured—or repaid—as the loans we discussed earlier. Fortunately, a few simple rules make it easy to solve problems involving complex loans.

So "lend" us your attention for this "interesting" discussion.

SECTION 11.1: Using Installment Loans

Can You:

Determine the finance charges on an installment loan?
Calculate the annual percentage rate on an installment loan?
. . . If not, you need this section.

So far, all the loans we have discussed have been simple interest loans, where the loan is repaid in one payment. But both businesses and individuals often prefer the terms of an installment loan.

Installment Loan A loan repaid in fixed regular payments ("installments") for a specified length of time.

An **installment loan** is a loan to be repaid in fixed regular payments ("installments") for a specified length of time. This type of loan is frequently used for purchases ranging from household appliances to trucks.

For example, Bits 'n' Bytes recently purchased a delivery truck for $18,000. The company made a down payment of $4,500 and financed the remaining amount as an installment loan. The terms of the loan require Bits 'n' Bytes to make 36 monthly payments of $456.50.

Determining Finance Charges

Finance Charges The interest on an installment loan, calculated as the difference between the total installment cost and the cash cost.

How much extra will Bits 'n' Bytes pay for the convenience of an installment loan? The answer depends on the total **finance charges**—the interest, which is the cost of borrowing the money. You can determine the finance charges by following two simple steps.

Steps: Calculating Finance Charges on an Installment Loan

STEP 1. *Calculate* the total installment cost:

$$\text{Total installment cost} = \left(\begin{array}{c}\text{number of}\\\text{payments}\end{array} \times \begin{array}{c}\text{amount of}\\\text{each payment}\end{array}\right) + \text{down payment}$$

STEP 2. *Calculate* the finance charge:

$$\text{Finance charge} = \text{total installment cost} - \text{cash cost or cash price}$$

To see how these steps work, let's apply them to the Bits 'n' Bytes example:

Step 1. Total installment cost = $\left(\begin{array}{c}\text{number of}\\\text{payments}\end{array} \times \begin{array}{c}\text{amount per}\\\text{payment}\end{array}\right)$ + down payment

$$= (36 \times \$456.50) + \$4500 = \$16{,}434 + \$4500 = \$20{,}934$$

Step 2. Finance charge = total installment cost − cash cost

$$= \$20{,}934 - 18{,}000 = \$2934$$

The finance charge is $2934.

YOUR TURN

WORK THIS PROBLEM

The Question:

Calculate the total installment cost and finance charges for the following loans:

(a) Intense Tents sold a five-season tent to Don Reed, Yosemite Mountain guide extraordinaire, for $450. Don made a down payment of $50 and financed the balance for 12 months with payments of $35.65.
(b) Bob's Beeflike Burgers bought a new cholesterol-free frying system for $17,000. Bob made a down payment of $3500 and financed the remainder for 24 months at $665 per month.

✔ YOUR WORK

The Solution:

(a) ***Step 1.*** Total installment cost = $(12 \times \$35.65) + 50$
$$= \$427.80 + 50 = \$477.80$$

  Step 2. Finance charge = $\$477.80 - 450 = \$27.80$

(b) ***Step 1.*** Total installment cost = $(24 \times \$665) + 3500$
$$= \$15{,}960 + 3500 = \$19{,}460$$

  Step 2. Finance charge = $\$19{,}460 - 17{,}000 = \$2460$

Using a calculator can make the task of finding installment costs and finance charges go faster.

Finding Finance Charges with a Calculator

Using a calculator lets you solve these problems in one step. For example, consider problem (a) in the preceding "Your Turn," which involves a $450 purchase with a down payment of $50 and 12 monthly payments of $35.65 each. Just punch in the following:

12 ⨯ 35.65 = ⟶ 427.8 + 50 − 450 = ⟶ 27.8

Calculating the Annual Percentage Rate

Annual Percentage Rate (APR) The true annual rate charged to borrowers, it is equal to the monthly interest rate × 12.

The finance charges on an installment loan depend on the interest rate on the loan. The Federal Reserve Board, in *Regulation Z,* also known as the *Truth-in-Lending Act,* requires all lenders to state the *effective interest rate.* This rate, the **annual percentage rate (APR)** is the true annual rate charged to borrowers.

Interest rates are set by individual states and may vary widely depending on the type of lending institution (for example, bank, credit card company, finance company, or pawn shop) and the state.

There are several ways to calculate the APR. The following method gives a quick approximation:

Steps: Approximating the APR

STEP 1.   *Calculate* the total installment cost.

$$\text{Total installment cost} = \binom{\text{number of}}{\text{payments}} \times \binom{\text{amount of}}{\text{payment}} + \text{down payment}$$

STEP 2.   *Calculate* the finance charges.

$$\text{Finance charge} = \text{total installment cost} - \text{cash cost}$$

STEP 3.   *Calculate* the loan amount (the amount borrowed).

$$\text{Loan amount} = \text{cash cost} - \text{down payment}$$

STEP 4.   *Estimate* the APR as follows:

$$\text{APR} = \frac{2 \times \text{number of payments per year} \times \text{finance charge}}{\text{loan amount} \times (\text{total number of payments} + 1)}$$

Loan Amount   The principal (the amount borrowed) in an installment loan.

The **loan amount** is the principal, the amount borrowed. The number of payments per year is 12 if paid monthly and 52 if paid weekly.

Let's use these steps to calculate the APR for our Bits 'n' Bytes truck. We already found the answers to steps 1 and 2 earlier in this chapter.

Step 1.   Total installment cost = $20,934

Step 2.   Finance charge = $2934

Step 3.   Loan amount = cash cost − down payment
              = $18,000 − 4500 = $13,500

Step 4. $\text{APR} = \dfrac{2 \times \text{number of payments per year} \times \text{finance charge}}{\text{loan amount} \times (\text{total number of payments} + 1)}$

$$= \frac{2 \times 12 \times \$2934}{\$13,500 \times (36 + 1)} = \frac{\$70,416}{\$13,500 \times 37} = \frac{\$70,416}{\$499,500}$$

$$= 0.1409 \ldots = 14.1\% \text{ rounded}$$

Does 14.1% sound high? Shop around and see if you can find a better interest rate.

WORK THIS PROBLEM

The Question:

Use the formula in step 4 of the preceding steps to estimate the APR for the two problems in the previous "Your Turn."

(a) Don Reed's tent
(b) Bob's Beeflike Burgers

✔ YOUR WORK

The Solution:

(a) *Step 1.* Total installment cost = $477.80

    Step 2. Finance charges = $27.80

    Step 3. Loan amount = $450 − $50 = $400

    Step 4. $\text{APR} = \dfrac{2 \times 12 \times \$27.80}{\$400 \times (12 + 1)} = \dfrac{\$667.20}{\$400 \times 13} = \dfrac{\$667.20}{\$5200}$

           $= 0.1283 \ldots = 12.8\%$ rounded

(b) *Step 1.* Total installment cost = $19,460

    Step 2. Finance charges = $2460

    Step 3. Loan amount = $17,000 − 3500 = $13,500

    Step 4. $\text{APR} = \dfrac{2 \times 12 \times \$2460}{\$13,500 \times (24 + 1)} = \dfrac{\$59,040}{\$13,500 \times 25} = \dfrac{\$59,040}{\$337,500}$

           $= 0.1749 \ldots = 17.5\%$ rounded

A more accurate method of calculating the APR uses the table on pages 374–375 and the following steps:

Steps: Calculating the APR by the Table

STEP 1. *Calculate* the table value.

$$\text{Table value} = \frac{\text{finance charges}}{\text{loan amount}} \times \$100$$

STEP 2. *Find* the table value in the APR table. Use the number of payments column (left column) to locate the correct row. Read across the row to find the value closest to the table value. Read the APR at the top of the column.

APR Table Per $100

Number of Payments	12.00%	12.25%	12.50%	12.75%	13.00%	13.25%	13.50%	13.75%	14.00%	14.25%	14.50%	14.75%	15.00%
1	1.00	1.02	1.04	1.06	1.08	1.10	1.21	1.15	1.17	1.19	1.21	1.23	1.25
2	1.50	1.53	1.57	1.60	1.63	1.66	1.69	1.72	1.75	1.78	1.82	1.85	1.88
3	2.01	2.05	2.09	2.13	2.17	2.22	2.26	2.30	2.34	2.38	2.43	2.47	2.51
4	2.51	2.57	2.62	2.67	2.72	2.78	2.83	2.88	2.93	2.99	3.04	3.09	3.14
5	3.02	3.08	3.15	3.21	3.27	3.34	3.40	3.46	3.53	3.59	3.65	3.72	3.78
6	3.53	3.60	3.68	3.75	3.83	3.90	3.97	4.05	4.12	4.20	4.27	4.35	4.42
7	4.04	4.12	4.21	4.29	4.38	4.47	4.55	4.64	4.72	4.81	4.89	4.98	5.06
8	4.55	4.65	4.74	4.84	4.94	5.03	5.13	5.23	5.32	5.42	5.51	5.61	5.71
9	5.07	5.17	5.28	5.39	5.49	5.60	5.71	5.82	5.92	6.03	6.14	6.25	6.35
10	5.58	5.70	5.82	5.94	6.05	6.17	6.29	6.41	6.53	6.65	6.77	6.88	7.00
11	6.10	6.23	6.36	6.49	6.62	6.75	6.88	7.01	7.14	7.26	7.40	7.53	7.65
12	6.62	6.76	6.90	7.04	7.18	7.32	7.46	7.60	7.74	7.89	8.03	8.17	8.31
13	7.14	7.29	7.44	7.59	7.75	7.90	8.05	8.20	8.36	8.51	8.66	8.81	8.97
14	7.66	7.82	7.99	8.15	8.31	8.48	8.64	8.81	8.97	9.13	9.30	9.46	9.63
15	8.19	8.36	8.54	8.71	8.88	9.06	9.23	9.41	9.59	9.76	9.94	10.11	10.29
16	8.71	8.90	9.08	9.27	9.46	9.64	9.83	10.02	10.20	10.39	10.58	10.77	10.95
17	9.24	9.44	9.63	9.83	10.03	10.23	10.43	10.63	10.82	11.02	11.22	11.42	11.62
18	9.77	9.98	10.19	10.40	10.61	10.82	11.03	11.24	11.45	11.66	11.87	12.08	12.29
19	10.30	10.52	10.74	10.96	11.18	11.41	11.63	11.85	12.07	12.29	12.52	12.74	12.97
20	10.83	11.06	11.30	11.53	11.76	12.00	12.23	12.47	12.70	12.93	13.17	13.41	13.64
21	11.36	11.61	11.85	12.10	12.34	12.59	12.84	13.08	13.33	13.58	13.82	14.07	14.32
22	11.90	12.16	12.41	12.67	12.93	13.19	13.44	13.70	13.96	14.22	14.48	14.74	15.00
23	12.44	12.71	12.97	13.24	13.51	13.78	14.05	14.32	14.59	14.87	15.14	15.41	15.68
24	12.98	13.26	13.54	13.82	14.10	14.38	14.66	14.95	15.23	15.51	15.80	16.08	16.37
25	13.52	13.81	14.10	14.40	14.69	14.98	15.28	15.57	15.87	16.17	16.46	16.76	17.06
26	14.06	14.36	14.67	14.97	15.28	15.59	15.89	16.20	16.51	16.82	17.13	17.44	17.75
27	14.60	14.92	15.24	15.56	15.87	16.19	16.51	16.83	17.15	17.47	17.80	18.12	18.44
28	15.15	15.48	15.81	16.14	16.47	16.80	17.13	17.47	17.80	18.13	18.47	18.80	19.14
29	15.70	16.04	16.38	16.72	17.07	17.41	17.75	18.10	18.45	18.79	19.14	19.49	19.83
30	16.24	16.60	16.95	17.31	17.66	18.02	18.38	18.74	19.10	19.45	19.81	20.17	20.54
31	16.79	17.16	17.53	17.90	18.27	18.63	19.00	19.38	19.75	20.12	20.49	20.87	21.24
32	17.35	17.73	18.11	18.49	18.87	19.25	19.63	20.02	20.40	20.79	21.17	21.56	21.94
33	17.90	18.29	18.69	19.08	19.47	19.87	20.26	20.66	21.06	21.46	21.85	22.25	22.65
34	18.46	18.86	19.27	19.67	20.08	20.49	20.90	21.31	21.72	22.13	22.54	22.95	23.36
35	19.01	19.43	19.85	20.27	20.69	21.11	21.53	21.95	22.38	22.80	23.23	23.65	24.08
36	19.57	20.00	20.43	20.87	21.30	21.73	22.17	22.60	23.04	23.48	23.92	24.35	24.79
37	20.13	20.57	21.02	21.46	21.91	22.36	22.81	23.26	23.70	24.16	24.61	25.06	25.51
38	20.69	21.15	21.61	22.07	22.52	22.99	23.45	23.91	24.37	24.84	25.30	25.77	26.24
39	21.26	21.73	22.20	22.67	23.14	23.61	24.09	24.57	25.04	25.52	26.00	26.48	26.96
40	21.82	22.30	22.79	23.27	23.76	24.25	24.73	25.22	25.71	26.20	26.70	27.19	27.69
41	22.39	22.88	23.38	23.88	24.38	24.88	25.38	25.88	26.39	26.89	27.40	27.91	28.41
42	22.96	23.47	23.98	24.49	25.00	25.51	26.03	26.55	27.06	27.58	28.10	28.62	29.15
43	23.53	24.05	24.57	25.10	25.62	26.15	26.68	27.21	27.74	28.27	28.81	29.34	29.88
44	24.10	24.63	25.17	25.71	26.25	26.79	27.33	27.88	28.42	28.97	29.52	30.07	30.62
45	24.67	25.22	25.77	26.32	26.88	27.43	27.99	28.55	29.11	29.67	30.23	30.79	31.36
46	25.25	25.81	26.37	26.94	27.51	28.08	28.65	29.22	29.79	30.36	30.94	31.52	32.10
47	25.82	26.40	26.98	27.56	28.14	28.72	29.30	29.89	30.48	31.07	31.66	32.25	32.84
48	26.40	26.99	27.58	28.18	28.77	29.37	29.97	30.57	31.17	31.77	32.37	32.98	33.59
49	26.98	27.59	28.19	28.80	29.41	30.02	30.63	31.24	31.86	32.48	33.09	33.71	34.34
50	27.56	28.18	28.80	29.42	30.04	30.67	31.29	31.92	32.55	33.18	33.82	34.45	35.09
51	28.15	28.78	29.41	30.05	30.68	31.32	31.96	32.60	33.25	33.89	34.54	35.19	35.84
52	28.73	29.38	30.02	30.67	31.32	31.98	32.63	33.29	33.95	34.61	35.27	35.93	36.60
53	29.32	29.98	30.64	31.30	31.97	32.63	33.30	33.97	34.65	35.32	36.00	36.68	37.36
54	29.91	30.58	31.25	31.93	32.61	33.29	33.98	34.66	35.35	36.04	36.73	37.42	38.12
55	30.50	31.18	31.87	32.56	33.26	33.95	34.65	35.35	36.05	36.76	37.46	38.17	38.88
56	31.09	31.79	32.49	33.20	33.91	34.62	35.33	36.04	36.76	37.48	38.20	38.92	39.65
57	31.68	32.39	33.11	33.83	34.56	35.28	36.01	36.74	37.47	38.20	38.94	39.68	40.42
58	32.27	33.00	33.74	34.47	35.21	35.95	36.69	37.43	38.18	38.93	39.68	40.43	41.19
59	32.87	33.61	34.36	35.11	35.86	36.62	37.37	38.13	38.89	39.66	40.42	41.19	41.96
60	33.47	34.23	34.99	35.75	36.52	37.29	38.06	38.83	39.61	40.39	41.17	41.95	42.74

Number of Payments	15.25%	15.50%	15.75%	16.00%	16.25%	16.50%	16.75%	17.00%	17.25%	17.50%	17.75%	18.00%	18.25%
1	1.27	1.29	1.31	1.33	1.35	1.38	1.40	1.42	1.44	1.46	1.48	1.50	1.52
2	1.91	1.94	1.97	2.00	2.04	2.07	2.10	2.13	2.16	2.19	2.22	2.26	2.29
3	2.55	2.59	2.64	2.68	2.72	2.76	2.80	2.85	2.89	2.93	2.97	3.02	3.06
4	3.20	3.25	3.30	3.36	3.41	3.46	3.51	3.57	3.62	3.67	3.73	3.78	3.83
5	3.84	3.91	3.97	4.04	4.10	4.16	4.23	4.29	4.35	4.42	4.48	4.54	4.61
6	4.49	4.57	4.64	4.72	4.79	4.87	4.94	5.02	5.09	5.17	5.24	5.32	5.39
7	5.15	5.23	5.32	5.40	5.49	5.58	5.66	5.75	5.83	5.92	6.00	6.09	6.17
8	5.80	5.90	6.00	6.09	6.19	6.29	6.38	6.48	6.58	6.67	6.77	6.87	6.96
9	6.46	6.57	6.68	6.78	6.89	7.00	7.11	7.22	7.32	7.43	7.54	7.65	7.76
10	7.12	7.24	7.36	7.48	7.60	7.72	7.84	7.96	8.08	8.19	8.31	8.43	8.55
11	7.79	7.92	8.05	8.18	8.31	8.44	8.57	8.70	8.83	8.96	9.09	9.22	9.35
12	8.45	8.59	8.74	8.88	9.02	9.16	9.30	9.45	9.59	9.73	9.87	10.02	10.16
13	9.12	9.27	9.43	9.58	9.73	9.89	10.04	10.20	10.35	10.50	10.66	10.81	10.97
14	9.79	9.96	10.12	10.29	10.45	10.62	10.78	10.95	11.11	11.28	11.45	11.61	11.78
15	10.47	10.64	10.82	11.00	11.17	11.35	11.53	11.70	11.88	12.06	12.24	12.42	12.59
16	11.14	11.33	11.52	11.71	11.90	12.09	12.28	12.46	12.65	12.84	13.03	13.22	13.41
17	11.82	12.02	12.22	12.42	12.62	12.83	13.03	13.23	13.43	13.63	13.83	14.04	14.24
18	12.50	12.72	12.93	13.14	13.35	13.57	13.78	13.99	14.21	14.42	14.64	14.85	15.06
19	13.19	13.41	13.64	13.86	14.09	14.31	14.54	14.76	14.99	15.22	15.44	15.67	15.90
20	13.88	14.11	14.35	14.59	14.82	15.06	15.30	15.54	15.78	16.01	16.25	16.49	16.73
21	14.57	14.82	15.06	15.31	15.56	15.81	16.06	16.31	16.56	16.81	17.07	17.32	17.57
22	15.26	15.52	15.78	16.04	16.30	16.57	16.83	17.09	17.36	17.62	17.88	18.15	18.41
23	15.96	16.23	16.50	16.78	17.05	17.33	17.60	17.87	18.15	18.43	18.70	18.98	19.26
24	16.65	16.94	17.23	17.51	17.80	18.09	18.37	18.66	18.95	19.24	19.53	19.82	20.11
25	17.35	17.65	17.95	18.25	18.55	18.85	19.15	19.45	19.75	20.05	20.36	20.66	20.96
26	18.06	18.37	18.68	18.99	19.30	19.62	19.93	20.24	20.56	20.87	21.19	21.50	21.82
27	18.76	19.09	19.41	19.74	20.06	20.39	20.71	21.04	21.37	21.69	22.02	22.35	22.68
28	19.47	19.81	20.15	20.48	20.82	21.16	21.50	21.84	22.18	22.52	22.86	23.20	23.55
29	20.18	20.53	20.88	21.23	21.58	21.94	22.29	22.64	22.99	23.35	23.70	24.06	24.41
30	20.90	21.26	21.62	21.99	22.35	22.72	23.08	23.45	23.81	24.18	24.55	24.92	25.29
31	21.61	21.99	22.37	22.74	23.12	23.50	23.88	24.26	24.64	25.02	25.40	25.78	26.16
32	22.33	22.72	23.11	23.50	23.89	24.28	24.68	25.07	25.46	25.86	26.25	26.65	27.04
33	23.06	23.46	23.86	24.26	24.67	25.07	25.48	25.88	26.29	26.70	27.11	27.52	27.93
34	23.78	24.19	24.61	25.03	25.44	25.86	26.28	26.70	27.12	27.54	27.97	28.39	28.81
35	24.51	24.94	25.36	25.80	26.23	26.66	27.09	27.52	27.96	28.39	28.83	29.27	29.71
36	25.24	25.68	26.12	26.57	27.01	27.46	27.90	28.35	28.80	29.25	29.70	30.15	30.60
37	25.97	26.42	26.88	27.34	27.80	28.26	28.72	29.18	29.64	30.10	30.57	31.03	31.50
38	26.70	27.17	27.64	28.11	28.59	29.06	29.53	30.01	30.49	30.96	31.44	31.92	32.40
39	27.44	27.92	28.41	28.89	29.38	29.87	30.36	30.84	31.34	31.83	32.32	32.81	33.31
40	28.18	28.68	29.18	29.68	30.18	30.68	31.18	31.68	32.19	32.69	33.20	33.71	34.22
41	28.92	29.44	29.95	30.46	30.97	31.49	32.01	32.52	33.04	33.56	34.09	34.61	35.13
42	29.67	30.19	30.72	31.25	31.78	32.31	32.84	33.37	33.90	34.44	34.97	35.51	36.05
43	30.42	30.96	31.50	32.04	32.58	33.13	33.67	34.22	34.77	35.31	35.86	36.42	36.97
44	31.17	31.72	32.28	32.83	33.39	33.95	34.51	35.07	35.63	36.19	36.76	37.33	37.89
45	31.92	32.49	33.06	33.63	34.20	34.77	35.35	35.92	36.50	37.08	37.66	38.24	38.82
46	32.68	33.26	33.84	34.43	35.01	35.60	36.19	36.78	37.37	37.96	38.56	39.16	39.75
47	33.44	34.03	34.63	35.23	35.83	36.43	37.04	37.64	38.25	38.86	39.47	40.08	40.69
48	34.20	34.81	35.42	36.03	36.65	37.27	37.88	38.50	39.13	39.75	40.37	41.00	41.63
49	34.96	35.59	36.21	36.84	37.47	38.10	38.74	39.37	40.01	40.65	41.29	41.93	42.57
50	35.73	36.37	37.01	37.65	38.30	38.94	39.59	40.24	40.89	41.55	42.20	42.86	43.52
51	36.50	37.15	37.81	38.46	39.12	39.79	40.45	41.11	41.78	42.45	43.12	43.79	44.47
52	37.27	37.94	38.61	39.28	39.95	40.63	41.31	41.99	42.67	43.36	44.04	44.73	45.42
53	38.04	38.72	39.41	40.10	40.79	41.48	42.17	42.87	43.57	44.27	44.97	45.67	46.38
54	38.82	39.52	40.22	40.92	41.63	42.33	43.04	43.75	44.47	45.18	45.90	46.62	47.34
55	39.60	40.31	41.03	41.74	42.47	43.19	43.91	44.64	45.37	46.10	46.83	47.57	48.30
56	40.38	41.11	41.84	42.57	43.31	44.05	44.79	45.53	46.27	47.02	47.77	48.52	49.27
57	41.16	41.91	42.65	43.40	44.15	44.91	45.66	46.42	47.18	47.94	48.71	49.47	50.24
58	41.95	42.71	43.47	44.23	45.00	45.77	46.54	47.32	48.09	48.87	49.65	50.43	51.22
59	42.74	43.51	44.29	45.07	45.85	46.64	47.42	48.21	49.01	49.80	50.60	51.39	52.20
60	43.53	44.32	45.11	45.91	46.71	47.51	48.31	49.12	49.92	50.73	51.55	52.36	53.18

Let's use the table to calculate the APR for the Bits 'n' Bytes truck example. We already know that the finance charge was $2934, the loan amount was $13,500, and the number of payments was 36.

Step 1.   Table value $= \dfrac{\text{finance charge}}{\text{loan amount}} \times \$100$

$$= \frac{\$2934}{\$13,500} \times \$100 = \$21.7333 \ldots = \$21.73 \text{ rounded}$$

Step 2.   In the table, locate the row with 36 payments. Read right to find the value closest to $21.73. In this case, $21.73 is in the table. The APR at the top of this column—13.25%—is the answer. It is accurate to $\frac{1}{4}\%$.

Notice that the table's answer—13.25%—differs from our approximated answer—14.1% The table's answer is more accurate.

YOUR TURN

| WORK THIS PROBLEM |

The Question:

Use the APR table to calculate the APR for the problems in the previous "Your Turn:"

(a) Don Reed's tent—finance charges of $27.80, a loan amount of $400, and 12 payments.
(b) Bob's Beeflike Burgers—finance charges of $2460, a loan amount of $13,500, and 24 payments.

| ✔ YOUR WORK |

The Solution:

(a) **Step 1.**   Table value $= \dfrac{\$27.80}{\$400} \times \$100 = \$6.95$

   Step 2.   Read across row 12; $6.95 is between the two values in the table—6.90 and 7.04. The closest value is 6.90. The APR, as given at the top of this column, is thus 12.50%.

(b) **Step 1.**   Table value $= \dfrac{\$2460}{\$13,500} \times \$100 = \$18.222 \ldots = \$18.22 \text{ rounded}$

   Step 2.   Read across row 24; $18.22 is between the two values in the table—18.09 and 18.37. The closest value is 18.09. The APR, as given at the top of this column, is thus 16.50%.

Now turn to Section Test 11.1 and check to make sure you understand how to calculate finance charges and APRs.

SECTION TEST 11.1 More Complex Loans

The following problems test your understanding of Section 11.1, Using Installment Loans.

1. Woody's Lumber Yard bought a forklift for $6500. Woody made a down payment of $500 and financed the remainder for 24 months at $284.50 per month. What are the (a) total installment cost, (b) finance charge, (c) loan amount, (d) APR using the formula, and (e) APR using the table?

2. Tick Tock Clocks sold a grandfather clock to Annie Thyme for $3200. She made a down payment of $600 and financed the remainder for 18 months at $165.25 per month. What are the (a) total installment cost, (b) finance charge, (c) loan amount, (d) APR using the formula, and (e) APR using the table?

3. Lew's Restaurant bought a new sound system for $2500. Lew made a down payment of $700 and financed the remainder for 12 months at $161 per month. What are the (a) total installment cost, (b) finance charge, (c) loan amount, (d) APR using the formula, and (e) APR using the table?

4. The Pump You Up Weights Store sold a home weight center to Carlos Atlas for $4000. He made a down payment of $800 and financed the remainder for 24 months at $153.60 per month. What are the (a) total installment cost, (b) finance charge, (c) loan amount, (d) APR using the formula, and (e) APR using the table?

5. Upward Bound bought a hang glider for $1200, making a down payment of $150 and financing the remainder for 18 months at $65.65 per month. What are the (a) total installment cost, (b) finance charge, (c) loan amount, (d) APR using the formula, and (e) APR using the table?

6. The Garden of Earthly Delights bought a new greenhouse for $7300. The company made a down payment of $700 and financed the remainder for 36 months at $233.65 per month. What are the (a) total installment cost, (b) finance charge, (c) loan amount, (d) APR using the formula, and (e) APR using the table?

7. In Hot Water sold a hot tub to Mr. and Mrs. Soakes for $2450. The Soakes made a down payment of $550 and financed the remainder for 18 months at $119.50 per month. What are the (a) total installment cost, (b) finance charge, (c) loan amount, (d) APR using the formula, and (e) APR using the table?

377

8. J.R.'s Construction Company bought a new computer system for $5700. The company made a down payment of $1200 and financed the remainder for 24 months at $215.25 per month. What are the (a) total installment cost, (b) finance charge, (c) loan amount, (d) APR using the formula, and (e) APR using the table?

9. Kate's Delicacies bought a new oven for $4000. Kate's Delicacies made a down payment of $1400 and financed the remainder for 24 months at $124.50 per month. What are the (a) total installment cost, (b) finance charge, (c) loan amount, (d) APR using the formula, and (e) APR using the table?

10. On the Rocks, a store for climbing enthusiasts, had an indoor climbing wall built for $8000. On the Rocks made a down payment of $1500 and financed the remainder for 36 months at $216.15 per month. What are the (a) total installment cost, (b) finance charge, (c) loan amount, (d) APR using the formula, and (e) APR using the table?

11. Zippy Delivery Service bought a delivery truck for $17,500. Zippy made a down payment of $2800 and financed the remainder for 48 months at $414.75 per month. What are the (a) total installment cost, (b) finance charge, (c) loan amount, (d) APR using the formula, and (e) APR using the table?

12. Ava's Air Tours bought a glider for $19,200, making a down payment of $2500 and financing the remainder for 42 months at $524 per month. What are the (a) total installment cost, (b) finance charge, (c) loan amount, (d) APR using the formula, and (e) APR using the table?

> ### *Can You:*
>
> Find the sum of the digits of a number of payments?
> Calculate the finance charges refunded when making an early payoff of a loan using the Rule of 78?
> . . . If not, you need this section.

Rule of 78 A formula for calculating refunds to borrowers who pay off installment loans early; it is based on the sum of the digits.

Sum of the Digits A method for calculating the number of payments remaining and total in funding refunds to borrowers who pay off installment loans early; it is equal to $[n \times (n + 1)]/2$, where n is the number of payments.

If an installment loan is paid off early, a portion of the finance charge is often refunded to the borrower. The method used to calculate the refund is the **Rule of 78,** which uses the **sum of the digits (SOTD)** of the number of payments.

Finding the Sum of the Digits

To find the sum of digits, first number the loan periods; then add this segment of numbers. For example, the SOTD for six payments is

$$1 + 2 + 3 + 4 + 5 + 6 = 21$$

The SOTD for 12 monthly payments is

$$1 + 2 + 3 + 4 + 5 + 6 + 7 + 8 + 9 + 10 + 11 + 12 = 78$$

which is where the Rule of 78 got its name. You can also calculate the sum of the digits using the following formula:

$$\text{SOTD} = \frac{n \times (n + 1)}{2} \qquad \text{where } n \text{ is the number of payments}$$

For example, let's calculate the sum of the digits for 12 payments:

$$\text{SOTD} = \frac{12 \times (12 + 1)}{2} = \frac{12 \times 13}{2} = \frac{156}{2} = 78$$

You can find this answer with a calculator, too.

Finding the Sum of the Digits with a Calculator

How you calculate the sum of the digits depends on the type of calculator you have.

If yours is a simple calculator, punch in the following to find the sum of the digits for 12 payments.

12 $\boxed{+}$ 1 $\boxed{=}$ $\boxed{\times}$ 12 $\boxed{\div}$ 2 $\boxed{=}$ ⟶ 78.

Notice that we reversed the order of the multiplicand and multiplier in the numerator of the fraction.

The same method will work on a sophisticated calculator, but if yours has parentheses keys, you don't have to remember to reverse the elements in the numerator. Instead, punch in

12 $\boxed{\times}$ $\boxed{(}$ 12 $\boxed{+}$ 1 $\boxed{)}$ $\boxed{\div}$ 2 $\boxed{=}$ ⟶ 78.

Both methods give the same answer, but use of the formula is usually quicker than adding the sequence of numbers.

YOUR TURN

WORK THIS PROBLEM

The Question:

Calculate the sum of the digits for 18 payments by

(a) Adding the digits   (b) Using the formula

✔ YOUR WORK

The Solution:

(a) SOTD = 1 + 2 + 3 + 4 + 5 + 6 + 7 + 8 + 9 + 10 + 11 + 12 + 13 + 14 + 15 + 16 + 17 + 18
           = 171

(b) SOTD = $\dfrac{18 \times (18 + 1)}{2} = \dfrac{18 \times 19}{2} = \dfrac{342}{2} = 171$

Calculating Refunds with the Rule of 78

Now that you can calculate the sum of the digits, you can use the following steps to calculate the finance charges refunded under the Rule of 78.

Steps: Calculating Refunds with the Rule of 78

STEP 1. *Calculate* the sum of the digits for the number of payments left when the loan was paid off.

STEP 2. *Calculate* the sum of the digits for the total number of payments.

STEP 3. *Divide* the answer to step 1 by the answer to step 2 to find the refund fraction:

$$\text{Refund fraction} = \frac{\text{SOTD for no. of payments left}}{\text{SOTD for total no. of payments}}$$

STEP 4. *Calculate* the finance charge refund:

$$\text{Finance charge refund} = \text{total finance charge} \times \text{refund fraction}$$

To see how these steps work, let's return to our example of Bits 'n' Bytes' purchase of an $18,000 delivery truck with a down payment of $4,500 and a 36-month installment loan with payments of $456.50. If Bits 'n' Bytes pays off its loan 6 months early, what rebate will it get under the Rule of 78?

In Section 11.1, we calculated the total finance charges of $2934.

Step 1. SOTD for no. of payments left $= \dfrac{6 \times (6+1)}{2} = \dfrac{6 \times 7}{2} = \dfrac{42}{2} = 21$

Step 2. SOTD for total no. of payments $= \dfrac{36 \times (36+1)}{2} = \dfrac{36 \times 37}{2} = \dfrac{1332}{2} = 666$

Step 3. Refund fraction $= \dfrac{21}{666}$

Step 4. Finance charge refund $= \$2934 \times \dfrac{21}{666} = \$92.51$ rounded

YOUR TURN

| WORK THIS PROBLEM |

The Question:

(a) Don Reed, who bought a tent with total finance charges of $27.80 and 12 monthly payments, pays off his loan 4 months early. What rebate is due him under the Rule of 78?

(b) Bob's Beeflike Burgers, which bought a new cholesterol-free frying system with total finance charges of $2460 and 24 monthly payments, pays off its loan 8 months early. What rebate is due BBB under the Rule of 78?

| ✔ YOUR WORK |

The Solution:

(a) **Step 1.** SOTD for no. payments left $= \dfrac{4 \times (4+1)}{2} = \dfrac{4 \times 5}{2} = \dfrac{4 \times 5}{2} = \dfrac{20}{2} = 10$

Step 2. SOTD for total no. payments $= \dfrac{12 \times (12+1)}{2} = \dfrac{12 \times 13}{2} = \dfrac{156}{2} = 78$

Step 3. Refund fraction $= \dfrac{10}{78}$

Step 4. Finance charge refund $= \$27.80 \times \dfrac{10}{78} = \$3.56$ rounded

(b) **Step 1.** SOTD for no. payments left $= \dfrac{8 \times (8+1)}{2} = \dfrac{8 \times 9}{2} = \dfrac{72}{2} = 36$

Step 2. SOTD for total no. payments $= \dfrac{24 \times (24+1)}{2} = \dfrac{24 \times 25}{2} = \dfrac{600}{2} = 300$

Step 3.   Refund fraction $= \dfrac{36}{300}$

Step 4.   Finance charge refund $= \$2460 \times \dfrac{36}{300} = \$295.20$

It's really easy, isn't it? If you don't think so, go over this section until you're comfortable with these calculations. Otherwise go directly to Section Test 11.2.

SECTION TEST 11.2   More Complex Loans ▬▬▬

The following problems test your understanding of Section 11.2, Working with the Rule of 78.

1. Woody's Lumber Yard bought a forklift and financed it for 24 months. The finance charge was $828. If Woody pays off the loan 4 months early, what is the finance charge rebate?

2. Tick Tock Clocks sold a grandfather clock to Annie Thyme, who financed it for 18 months. The finance charge was $374.50. If she pays off the loan 7 months early, what is the finance charge rebate?

3. Lew's Restaurant bought a new sound system and financed it for 12 months. The finance charge was $132. If Lew pays off the loan 3 months early, what is the finance charge rebate?

4. The Pump You Up Weight Store sold a home weight center to Carlos Atlas, who financed it for 24 months. The finance charge was $486.40. If he pays off the loan 9 months early, what is the finance charge rebate?

5. Upward Bound bought a hang glider and financed it for 18 months. The finance charge was $131.70. If Upward Bound pays off the loan 6 months early, what is the finance charge rebate?

6. The Garden of Earthy Delights bought a greenhouse and financed it for 36 months. The finance charge was $1811.40. If the company pays off the loan 11 months early, what is the finance charge rebate?

7. The Soakes bought a hot tub and financed it for 18 months. The finance charge was $251. If they pay off the loan 5 months early, what is the finance charge rebate?

8. J.R.'s Construction Company bought a new computer system and financed it for 24 months. The finance charge was $666. If the company pays off the loan 8 months early, what is the finance charge rebate?

9. Kate's Delicacies bought a new oven and financed it for 24 months. The finance charge was $388. If Kate's pays off the loan 9 months early, what is the finance charge rebate?

10. On the Rocks had an indoor climbing wall built and financed it for 36 months. The finance charge was $1281.40. If On the Rocks pays off the loan 10 months early, what is the finance charge rebate?

11. Zippy Delivery Service bought a delivery truck and financed it for 48 months. The finance charge was $5208. If Zippy pays off the loan 13 months early, what is the finance charge rebate?

12. Ava's Air Tours bought a glider and financed it for 42 months. The finance charge was $5308. If Ava pays off the loan 15 months early, what is the finance charge rebate?

WORKSPACE

> ## Can You:
>
> Interchange the APR and the monthly interest rate on a credit card account?
> Calculate finance charges on a credit card account?
> Calculate the interest and new balance on a credit card account that assesses interest monthly?
> Calculate the interest and new balance on a credit card account that uses average daily balance?
> . . . If not, you need this section.

Open-End Credit A form of loan in which individuals and businesses borrow and repay money on an ongoing basis through the use of credit cards.

In addition to simple interest loans and installment loans, both individuals and businesses often use credit cards—also called *revolving charge plans* or **open-end credit**—to borrow money.

Credit cards are available from a variety of businesses, including VISA, MasterCard, and many national and local department stores. Cardholders may "charge" purchases to their accounts or get cash advances. Both activities result in an increase in the total amount due the credit card issuer.

Unpaid Balance The remaining amount due after a partial payment is made on an open-end credit account.

Cardholders are billed for any charges made during a month when they get the next month's statement. If a cardholder pays the entire balance by the due date, he or she will not be liable for any finance charges. Otherwise, interest is charged on the remainder—the **unpaid balance**—or the average daily balance. Many major credit card companies also charge cardholders an annual fee.

Since maximum interest rates are set by the states, the APR may vary from a low of 12% to a high of 24%. Major credit card companies often choose their headquarters based on the policies of the various states, preferring states like North Dakota, which permit high interest rates.

Converting Between APR and Monthly Rates

The interest rate for a credit card may be stated as an APR or as a monthly rate. Most credit card issuers state the interest rate using both methods. Thus you need to know how to convert an APR to a monthly rate and a monthly rate to an APR.

To convert APR to a monthly rate, use the following formula:

$$\text{Monthly rate} = \frac{\text{APR}}{12}$$

For example, let's convert an APR of 21% to a monthly rate:

$$\text{Monthly rate} = \frac{\text{APR}}{12} = \frac{21\%}{12} = \frac{7}{4}\% = 1\frac{3}{4}\%$$

Most credit card companies express the monthly rate as a fractional percent rather than as a decimal percent—that is, as $1\frac{3}{4}\%$ rather than 1.75%.

WORK THIS PROBLEM

The Question:

Convert 15% APR to a monthly rate and express your answer as a fractional percent.

✔ YOUR WORK

The Solution:

$$\text{Monthly rate} = \frac{15\%}{12} = \frac{5}{4}\% = 1\frac{1}{4}\%$$

Converting a monthly rate to an APR is equally simple. Just use the following formula:

> **APR = monthly rate × 12**

So, to convert a monthly rate of $1\frac{3}{4}\%$ to an APR.

$$\text{APR} = \text{monthly rate} \times 12$$

$$= 1\frac{3}{4}\% \times 12 = \frac{7}{\cancel{4}_1}\% \times \frac{\cancel{12}^{\,3}}{1} = 21\%$$

WORK THIS PROBLEM

The Question:

Convert a monthly rate of $1\frac{1}{4}\%$ to an APR.

✔ YOUR WORK

The Solution:

$$\text{APR} = 1\frac{1}{4}\% \times 12 = \frac{5}{\cancel{4}_1}\% \times \frac{\cancel{12}^{\,3}}{1} = 15\%$$

Finding Finance Charges

Being able to convert APRs to monthly rates allows you to find the finance charges on an account if you know the APR and the balance.

Steps: Finding Finance Charges on the Balance

STEP 1. *Convert* the APR to a monthly rate.

STEP 2. *Find* the finance charges using the following formula:

$$\text{Finance charges} = \text{balance} \times \text{monthly rate}$$

For example, if the balance on a credit card account is \$182.35 and the APR is 15%, what is the finance charge?

Step 1.   Monthly rate $= \dfrac{15\%}{12} = \dfrac{5}{4}\% = 1\dfrac{1}{4}\%$

Step 2.   Finance charges = balance × monthly rate

$$= \$182.35 \times 1\frac{1}{4}\%$$
$$= \$182.35 \times 1.25\%$$
$$= \$182.35 \times 0.0125 = \$2.28 \text{ rounded}$$

YOUR TURN

WORK THIS PROBLEM

The Question:

The balance on a revolving charge account is \$156.73. The APR is 18%. Calculate the finance charges.

✔ YOUR WORK

The Solution:

Step 1.   Monthly rate $= \dfrac{18\%}{12} = \dfrac{3}{2}\% = 1\dfrac{1}{2}\%$

Step 2.   Finance charges $= \$156.73 \times 1\dfrac{1}{2}\%$

$$= \$156.73 \times 1.5\% = \$156.73 \times 0.015 = \$2.35 \text{ rounded}$$

Using a calculator can save you a lot of time and energy in finding finance charges on balances.

> ## *Finding Finance Charges on a Balance with a Calculator*
>
> Let's return to the account with a balance of \$156.73 and an APR of 18%. Punch in the following:
>
> $18 \boxed{\div} 12 \boxed{\div} 100 \boxed{=} \longrightarrow \boxed{0.015} \boxed{\times} 156.73 \boxed{=} \longrightarrow \boxed{2.35095}$
>
> $\underbrace{}_{1.5\%}$
>
> (Note that dividing by 100 converts 1.5% to the decimal 0.015.)
>
> Using a business calculator,
>
> $156.73 \boxed{\times} 18 \boxed{\%} \boxed{\div} 12 \boxed{=} \longrightarrow \boxed{2.35095}$

Choosing a Payment Option

When you receive a monthly statement on a credit card account, you have three choices for payment:

1. Pay the entire balance.
2. Pay the minimum required amount—usually 5 to 10% of the balance, with a $10 minimum.
3. Pay an amount larger than the minimum but smaller than the total balance.

Option 1 allows you to avoid all finance charges. Options 2 and 3 mean that you will have to pay finance charges next month on the unpaid balance from this month.

Calculating Average Daily Balance and New Balance

Average Daily Balance
In an open-end credit account, the average of the account's balance over the course of a month.

Most major credit cards do not calculate the interest on the unpaid balance at the end of the month. The balance on the last day of the month may not reflect the actual amount of money on loan during the rest of the month. Thus many companies calculate the interest as a percent of the **average daily balance**—the average of the account's balance throughout the month.

You can easily calculate average daily balance by following these steps:

Steps: Calculating Average Daily Balance

STEP 1.   *Calculate* the daily balance for each day using the following formula:

$$\text{Daily balance} = \begin{bmatrix} \text{previous} \\ \text{balance} \end{bmatrix} + \begin{bmatrix} \text{new} \\ \text{purchases} \end{bmatrix} + \begin{bmatrix} \text{cash} \\ \text{advances} \end{bmatrix} - \begin{bmatrix} \text{payments} \end{bmatrix}$$

STEP 2.   *Multiply* each daily balance by the number of days the balance occurs.

STEP 3.   *Add* the results of step 2 to get the total daily balance.

STEP 4.   *Calculate* the average daily balance as follows:

$$\text{Average daily balance} = \frac{\text{total daily balance}}{\text{no. days in billing cycle}}$$

Once you know the average daily balance, you can easily calculate the new balance using the following steps:

Steps: Calculating New Balance Based on Average Daily Balance

STEP 1.   *Calculate* the monthly rate.

STEP 2.   *Calculate* the finance charges on the average daily balance as follows:

$$\text{Finance charge} = \text{average daily balance} \times \text{monthly rate}$$

STEP 3. *Calculate* the new balance as follows:

$$\begin{bmatrix} \text{New} \\ \text{balance} \end{bmatrix} = \begin{bmatrix} \text{previous} \\ \text{balance} \end{bmatrix} + \begin{bmatrix} \text{finance} \\ \text{charges} \end{bmatrix} + \begin{bmatrix} \text{new} \\ \text{purchases} \end{bmatrix} + \begin{bmatrix} \text{cash} \\ \text{advances} \end{bmatrix} - \begin{bmatrix} \text{payments} \end{bmatrix}$$

For example, Kent Clark's SuperCharge bill for October shows a balance of $278 as of September 1. SuperCharge calculates finance charges of 18% on the average daily balance. Kent also used his card during the preceding month to make the following transactions:

September 5	New purchase: $37
September 10	Payment: $50
September 16	Cash advance: $75
September 25	New purchase: $42

What are his average daily balance and his new balance? Use a 30-day billing cycle.

First, calculate average daily balance:

Step 1.

		Balance
September 1		$278
September 5	$278 + 37 =	$315
September 10	$315 − 50 =	$265
September 16	$265 + 75 =	$340
September 25	$340 + 42 =	$382

Step 2.

September 1–5	= 4 days	4 × $278 = $1112
September 5–10	= 5 days	5 × $315 = $1575
September 10–16	= 6 days	6 × $265 = $1590
September 16–25	= 9 days	9 × $340 = $3060
September 25– October 1	= 6 days	6 × $382 = $2292

Step 3. Total daily balance = $9629

Step 4. Average daily balance = $\dfrac{\$9629}{30} = \$320.97$ rounded

Now that we have the average daily balance, we can calculate the finance charges and the new balance:

Step 1. Monthly rate = $\dfrac{18\%}{12} = \dfrac{3}{2}\% = 1\dfrac{1}{2}\%$

Step 2. Finance charges = average daily balance × monthly rate

$$= \$320.97 \times 1\frac{1}{2}\%$$

$$= \$320.97 \times 1.5\% = \$320.97 \times 0.015 = \$4.81 \text{ rounded}$$

Step 3. $\begin{bmatrix} \text{New} \\ \text{balance} \end{bmatrix} = \begin{bmatrix} \text{previous} \\ \text{balance} \end{bmatrix} + \begin{bmatrix} \text{finance} \\ \text{charges} \end{bmatrix} + \begin{bmatrix} \text{new} \\ \text{purchases} \end{bmatrix} + \begin{bmatrix} \text{cash} \\ \text{advances} \end{bmatrix} - \begin{bmatrix} \text{payments} \end{bmatrix}$

$$= \$278 + 4.81 + (37 + 42) + 75 - 50 = \$386.81$$

YOUR TURN

$\boxed{\text{WORK THIS PROBLEM}}$

The Question:

Ella Fant's MassiveCard bill in July showed a balance of $532 on June 1, which is subject to a 15% finance charge on the average daily balance. She also used her card during the preceding month to make the following transactions:

June 10	Cash advance:	$120
June 16	Payment:	$ 75
June 20	New purchase:	$ 95
June 23	New purchase:	$ 27

What are her average daily balance and her new balance? Use a 30-day billing cycle.

$\boxed{\checkmark \text{ YOUR WORK}}$

The Solution:

First, calculate the average daily balance:

Step 1.   **Balance**

June 1		$532
June 10	$532 + 120 =	$652
June 16	$652 − 75 =	$577
June 20	$577 + 95 =	$672
June 23	$672 + 27 =	$699

Step 2.
June 1–10	= 9 days	9 × $532 =	$4788
June 10–16	= 6 days	6 × $652 =	$3912
June 16–20	= 4 days	4 × $577 =	$2308
June 20–23	= 3 days	3 × $672 =	$2016
June 23–			
July 1	= 8 days	8 × $699 =	$5592

Step 3.   Total daily balance = $18,616

Step 4.   Average daily balance = $\dfrac{\$18,616}{30}$ = $620.53 rounded

Next, use the average daily balance to calculate the finance charges and new balance:

Step 1.   Monthly rate = $\dfrac{15\%}{12} = \dfrac{5}{4}\% = 1\dfrac{1}{4}\%$

Step 2.   Finance charge = $620.53 × $1\dfrac{1}{4}\%$

                          = $620.53 × 1.25%

                          = $620.53 × 0.0125 = $7.76 rounded

Step 3.   New balance = $532 + 7.76 + 95 + 27 + 120 − 75 = $706.76

If you had no problems with the previous "Your Turn," you should have no problems answering the questions in Section Test 11.3. Otherwise go back and work through this section again.

SECTION TEST 11.3 More Complex Loans

The following problems test your understanding of Section 11.3, Using Open-End Credit.

1. Jill received her Outward Bound bill showing a balance of $653 as of November 1. Outward Bound calculates the finance charge as 18% of the average daily balance. During the month Jill had the following transactions:

November 7	New Purchase:	$126
November 15	Payment:	$100
November 19	New Purchase:	$73
November 23	New Purchase:	$273

What are (a) the average daily balance, (b) the finance charge, and (c) the new balance for the next month? Use a 30-day billing cycle.

2. J.R.'s Construction received its Woody's Lumber bill with a balance of $562 as of April 1. Woody's Lumber calculates the finance charge as 15% of the average daily balance. During the month, J.R.'s had the following Woody's Lumber transactions:

April 4	New Purchase:	$45
April 12	Payment:	$75
April 17	New Purchase:	$50
April 26	New Purchase:	$82

What are (a) the average daily balance, (b) the finance charge, and (c) the new balance for the next month? Use a 30-day billing cycle.

3. Dora O'Ken received her Gold Charge bill with a balance of $725 as of November 1. Gold Charge calculates the finance charge as 18% of the average daily balance. During the month, Dora had the following Gold Charge transactions:

November 3	New Purchase:	$75
November 18	Payment:	$125
November 20	Cash Advance:	$60
November 23	New Purchase:	$104

What are (a) the average daily balance, (b) the finance charge, and (c) the new balance for the next month? Use a 30-day billing cycle.

4. Gene Harlow received his Platinum Charge bill with a balance of $295 as of September 1. Platinum Charge calculates the finance charge as 21% of the average daily balance. During the month, Gene had the following Platinum Charge transactions:

September 7	Cash Advance:	$95
September 11	New Purchase:	$39
September 15	Payment:	$75
September 23	New Purchase:	$128

What are (a) the average daily balance, (b) the finance charge, and (c) the new balance for the next month? Use a 30-day billing cycle.

5. Petal Pushers received a charge account bill from the Garden of Earthly Delights with a balance of $937 as of September 1. The Garden of Earthly Delights calculates the finance charge as 15% of the average daily balance. During the month, Petal Pushers had the following "Garden" transactions:

September 4	New Purchase:	$50
September 9	New Purchase:	$105
September 17	Payment:	$95
September 22	New Purchase:	$47

What are (a) the average daily balance, (b) the finance charge, and (c) the new balance for the next month? Use a 30-day billing cycle.

6. Marco Rugby received his Ava's Air Tours charge bill with a balance of $129 as of April 1. Ava's Air Tours calculates the finance charge as 18% of the average daily balance. During the month, Marco had the following transactions with Ava's Air Tours:

April 5	New Purchase:	$79
April 14	Payment:	$90
April 19	New Purchase:	$100
April 22	New Purchase:	$38

What are (a) the average daily balance, (b) the finance charge, and (c) the new balance for the next month? Use a 30-day billing cycle.

7. Claudia's Clip-Joint borrowed $25,000 from the bank to remodel the salon in order to install a tanning bed. The loan was to be repaid in 60 monthly payments of $556.11 each. With just 14 payments remaining, Claudia paid off the balance on the loan. Find each of the following: (a) total installment cost, (b) total finance charge, (c) APR using the table, and (d) finance charge rebate, using the *Rule of 78*.

8. Claudia received a credit card bill with a balance of $1375.82 as of June 1. The APR on this card was 15%. During June, Claudia had made the following transactions:

June 6	New purchase:	$47.37
June 12	Payment:	$85.00
June 20	New purchase:	$127.54

Using a 30-day billing cycle, find each of the following: (a) average daily balance, (b) finance charge, and (c) new balance as of July 1.

Key Terms

Annual Percentage Rate (APR)   loan amount
average daily balance   open-end credit
balance   Rule of 78
finance charges   sum of the digits (SODT)
installment loan

Chapter 11 at a Glance

Page	Topic	Key Point	Example
370	Finance charges on installment loans	Total installment cost = (no. payments × amount per payment) + down payment	To buy a new $18,000 truck, a firm makes a $4500 down payment and takes a loan with 36 monthly payments of $456.50 each. What is the finance charge? Total installment cost = (36 × $456.50) + 4500 $$= \$20{,}934$$
		Finance charge = total installment cost − cash cost or cash price	Finance charge = $20,934 − 18,000 $$= \$2934$$
372	Approximate the annual percentage rate	Find total installment cost and finance charge as above. Loan amount = cash cost − down payment $$APR = \frac{\left(\begin{array}{c}2 \times \text{no. payments per year} \\ \times \text{ finance charge}\end{array}\right)}{\left(\begin{array}{c}\text{loan amount} \\ \times \text{ (total no. payments + 1)}\end{array}\right)}$$	Using the above example, find the APR. Total installment cost = $20,934; Finance charge = $2934 Loan amount = $18,000 − 4500 = $13,500 $$APR = \frac{2 \times 12 \times \$2934}{\$13{,}500 \times (36 + 1)}$$ $$= 14.1\% \text{ rounded}$$
373	Annual percentage rate by table	$$\text{Table value} = \frac{\text{finance charges}}{\text{loan amount}} \times \$100$$ Find the APR on the table.	Find the APR for the above example. $$\text{Table value} = \frac{\$2934}{\$13{,}500} \times \$100$$ $$= \$21.73 \text{ rounded}$$ $$= 13.25\% \text{ APR}$$
379	Sum of the digits	$$\text{Sum of the digits} = \frac{n \times (n+1)}{2}$$ where n = no. payments	Find the sum of the digits for 12 monthly payments: $$SOTD = \frac{12 \times (12 + 1)}{2}$$ $$= 78$$
380	Finance charge refunds for early repayment	$$\text{Refund fraction} = \frac{\text{SOTD no. payments left}}{\text{SOTD total no. payments}}$$ Finance charge refund = total finance charge × refund fraction	Finance charges = $2934 No. of payments = 36 Loan paid off 6 months early. $$\text{Refund fraction} = \frac{[6 \times (6+1)]/2}{[36 \times (36+1)]/2} = \frac{21}{666}$$ $$\text{Finance charge refund} = \$2934 \times \frac{21}{666}$$ $$= \$92.51 \text{ rounded}$$

Page	Topic	Key Point	Example
385	Converting between APR and monthly rates	Monthly rate $= \dfrac{APR}{12}$ APR = monthly rate $\times$ 12	What is the monthly rate if the APR is 24%? Monthly rate $= \dfrac{24\%}{12} = 2\%$ What is the APR if the monthly rate is 1.5%? APR $= 1.5\% \times 12 = 18\%$
386	Finding finance charges on the balance	Finance charges = balance $\times$ monthly rate	What are the finance charges on a balance of $200 at a 2% monthly rate? Finance charges $= \$200 \times 0.02 = \$4$
388	Finding the average daily balance and new balance	Daily balance = previous balance + new purchases + cash advances − payments	If you hold a credit card that charges a 1.5% monthly interest rate on the average daily balance, what is your average daily balance and new balance in a month in which you have a balance of $278 as of September 1 and make these transactions: 9/5, new purchase: $37 9/10, payment: $50 9/16, cash advance: $75 9/25, new purchase: $42 Daily balances: 9/1 = $278 9/5: $278 + 37 = $315 9/10: $315 − 50 = $265 9/16: $265 + 75 = $340 9/25: $340 + 42 = $382
		Multiply each daily balance by the no. of days it occurs.	9/1–9/5 = 4 × $278 = $1112 9/5–9/10 = 5 × $315 = $1575 9/10–9/16 = 6 × $265 = $1590 9/16–9/25 = 9 × $340 = $3060 9/25–10/1 = 6 × $382 = $2292
		To find average daily balance, add results from previous step and divide by no. days.	Total daily balance = $1112 + 1575 + 1590 + 3060 + 2292 = $9629 Average daily balance = $9629/30 = $320.97 rounded
		Finance charges = average daily balance $\times$ monthly rate	Finance charges = $320.97 × 0.015 = $4.81 rounded
		Find new balance using formula for new balance.	New balance = $278 + 4.81 + 37 + 42 + 75 − 50 = $386.81

Name

Date

Course/Section

The following problems test your understanding of more complex loans.

1. Shiring Investments bought a personal computer for $1200. The company made a down payment of $300 and financed the remainder for 12 months at $80.50 per month. What are the (a) total installment cost, (b) finance charge, (c) loan amount, (d) APR using the formula, and (e) APR using the table?

2. Rolling Hills Ranch bought a horse trailer for $9650. The ranch made a down payment of $750 and financed the remainder for 36 months at $315 per month. What are the (a) total installment cost, (b) finance charge, (c) loan amount, (d) APR using the formula, and (e) APR using the table?

3. The Garden of Earthly Delights had a hydroponic system built for $14,500, making a down payment of $2250 and financing the remainder for 48 months at $345.60 per month. What are the (a) total installment cost, (b) finance charge, (c) loan amount, (d) APR using the formula, and (e) APR using the table?

4. Annoying Telemarketing bought six desks for $1925. Annoying Telemarketing made a down payment of $425 and financed the remainder for 18 months at $94.25 per month. What are the (a) total installment cost, (b) finance charge, (c) loan amount, (d) APR using the formula, and (e) APR using the table?

5. Crystal's Diamonds bought a new display case and financed it for 12 months. The finance charge was $66. If Crystal's pays off the loan 5 months early, what is the finance charge rebate?

6. Big Ben's Bolts bought a new bolt-sizing machine and financed it for 36 months. The finance charge was $2440. If the company pays off the loan 9 months early, what is the finance charge rebate?

7. Wired for Sound had a new listening room built and financed it for 48 months. The finance charge was $4338.80. If Wired for Sound pays off the loan 14 months early, what is the finance charge rebate?

8. Annoying Telemarketing bought six desks and financed the purchase for 18 months. The finance charge was $196.50. If the company pays off the loan 8 months early, what is the finance charge rebate?

9. Loni Ranger received her Silver Charge bill with a balance of $736 as of September 1. Silver Charge calculates the finance charge as 18% of the average daily balance. During the month, Loni had the following Silver Charge transactions:

September 6	Cash Advance:	$85
September 8	New Purchase:	$123
September 13	Payment:	$75
September 21	New Purchase:	$81

What are (a) the average daily balance (b) the finance charge, and (c) the new balance for the next month? Use a 30-day billing cycle.

10. Randy Sayles received his UncoverCard bill with a balance of $815 as of April 1. UncoverCard calculates the finance charge as 15% of the average daily balance. During the month, Randy had the following UncoverCard transactions:

April 4	New Purchase:	$79
April 11	Payment:	$90
April 15	Cash Advance:	$100
April 23	New Purchase:	$38
April 26	Cash Advance:	$75

What are (a) the average daily balance (b) the finance charge, and (c) the new balance for the next month? Use a 30-day billing cycle.

Compound Interest and Present Value

> **What happens if I leave my money in the bank for more than a year?**

> **You get interest on your money, plus interest on your interest.**

CHAPTER OBJECTIVES

When you complete this chapter successfully, you will be able to:

1. Determine compound interest, maturity values under compound interest, and effective rates under compound interest.

2. Find the present value of a current investment.

■ The Friendly Savings Bank pays 6.5% compounded daily. Crystal's Diamonds invested $5200 on November 13, 1991. What is the value of the investment if it is withdrawn on November 13, 2001?

■ To prepare for his retirement when he closes his restaurant, Lew has an investment at 10% that is compounded semiannually. What is the effective interest rate on this investment?

■ The owners of Wired for Sound are saving to recarpet their listening room. They estimate that it will cost $6500 in two years. How much must Wired for Sound invest today at 6% compounded monthly in order to have the amount in two years?

The flip side to borrowing; the topic of the last chapter, is investing. Before you—or anyone else, including a business—invests money, you'll want to know "what's in it for you."

What's in it for you and for other investors is interest on your investment. In the next chapter, we'll look at some of the ways you can earn interest. First, though, you need to understand the effects of more complex interest rates than we've discussed so far.

Pencil ready? Get set for an "interesting" discussion you'll come to "value" presently.

SECTION 12.1: Compound Interest

Suppose Jeb "Trolly" Carr, a private in General George Washington's army, had invested $10 at 6% a year in 1776. If he allowed the interest to accumulate over the years, how much would his descendants have had in 1976 to celebrate the bicentennial? The answer is that the original $10 would have grown to $1,151,259.04 and would now be increasing at almost $8 per hour! Unfortunately Jeb spent the money.

Compound Interest A method of accumulating interest in which the interest from the previous period is added to the principal for the next period.

Compound interest is merely "interest on interest." An example will help clarify this concept. Let's calculate the interest on $500 at 6% for two years both as simple interest and as interest compounded annually.

As you learned in Chapter 9, the *simple* interest on $500 at 6% for two years would be $60.

$$I = Prt = \$500 \times 6\% \times 2 = \$60$$

Calculating Compound Interest

In contrast, calculating interest compounded annually is more complex.

Steps: Calculating Interest Compounded Annually

STEP 1. *Calculate* the interest and maturity value for the first year.

STEP 2. *Calculate* the interest and maturity values for subsequent years, always using the maturity value from the end of the previous year.

STEP 3. *Add* the interest amounts for all years together.

Thus we calculate the interest on $500 at 6% for two years, compounded annually, as follows:

Step 1. First Year: $I = Prt = \$500 \times 6\% \times 1 = \$30$
Maturity value $= MV = P + I = \$500 + 30 = \$530$

Step 2.

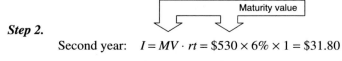

Maturity value

Second year: $I = MV \cdot rt = \$530 \times 6\% \times 1 = \$31.80$

Step 3. Total interest $= \$30 + 31.80 = \$61.80$

Notice that the compound interest ($61.80) and the simple interest ($60) differ by $1.80. This difference is the "interest on the interest." The $1.80 is the second year's interest on the first year interest of $30.

YOUR TURN

| WORK THIS PROBLEM |

The Question:

Calculate the interest on $200 at 5% compounded annually for three years.

| ✔ YOUR WORK |

The Solution:

Step 1.   First Year: $I = Prt = \$200 \times 5\% \times 1 = \$10$
Maturity value ⟶ $MV = P + I = \$200 + 10 = \$210$

Step 2a.   Second year: $I = MV \cdot rt = \$210 \times 5\% \times 1 = \$10.50$
New maturity value ⟶ $MV = MV + I = \$210 + 10.50 = \$220.50$

Previous maturity value

Step 2b.   Third year: $I = MV \cdot rt = \$220.50 \times 5\% \times 1 = \$11.025 = \$11.03$ rounded
New maturity value ⟶ $MV = MV + I = \$220.50 + 11.03 = \$231.53$

Previous maturity value

Step 3.   Total interest $= \$10 + 10.50 + 11.03 = \$31.53$

Finding Maturity Values Using Compound Interest

One way to find the maturity value of money under compound interest is to multiply each preceding year's principal by $1 + r$, where r is the annual interest rate. For example, $200 at 5% compounded annually for three years is

$200	original principal
×1.05	$1 + r$ ($r = 5\% = 0.05$)
$210.00	total amount after 1 year
×1.05	
$220.50	total amount after 2 years
×1.05	
$231.525	or $231.53 total amount after 3 years

This method can be quite long for many compound interest problems. To compute the interest compounded *monthly* for two years would require an interest calculation for each month—a total of 2×12 or 24 monthly calculations! This method is too time-consuming for such problems. A much easier and quicker method involves using the Compound Interest table or using a calculator.

COMPOUND INTEREST TABLE
Maturity Value of $1 at Compound Interest

n*	½%	1%	1½%	2%	3%	4%	5%	6%
1	1.0050000	1.0100000	1.0150000	1.0200000	1.0300000	1.0400000	1.0500000	1.0600000
2	1.0100250	1.0201000	1.0302250	1.0404000	1.0609000	1.0816000	1.1025000	1.1236000
3	1.0150751	1.0303010	1.0456784	1.0612080	1.0927270	1.1248640	1.1576250	1.1910160
4	1.0201505	1.0406040	1.0613636	1.0824322	1.1255088	1.1698586	1.2155062	1.2624770
5	1.0252513	1.0510100	1.0772840	1.1040808	1.1592741	1.2166529	1.2762816	1.3382256
6	1.0303775	1.0615202	1.0934433	1.1261624	1.1940523	1.2653190	1.3400956	1.4185191
7	1.0355294	1.0721354	1.1098449	1.1486857	1.2298739	1.3159318	1.4071004	1.5036303
8	1.0407070	1.0828567	1.1264926	1.1716594	1.2667701	1.3685690	1.4774554	1.5938481
9	1.0459106	1.0936853	1.1433900	1.1950926	1.3047732	1.4233118	1.5513282	1.6894790
10	1.0511401	1.1046221	1.1605408	1.2189944	1.3439164	1.4802443	1.6288946	1.7908477
11	1.0563958	1.1156684	1.1779489	1.2433743	1.3842339	1.5394541	1.7103394	1.8982986
12	1.0616778	1.1268250	1.1956182	1.2682418	1.4257609	1.6010322	1.7958563	2.0121965
13	1.0669862	1.1380933	1.2135524	1.2936066	1.4685337	1.6650735	1.8856491	2.1329283
14	1.0723211	1.1494742	1.2317557	1.3194788	1.5125897	1.7316764	1.9799316	2.2609040
15	1.0776827	1.1609690	1.2502321	1.3458683	1.5579674	1.8009435	2.0789282	2.3965582
16	1.0830712	1.1725786	1.2689856	1.3727857	1.6047064	1.8729812	2.1828746	2.5403517
17	1.0834865	1.1843044	1.2880203	1.4002414	1.6528476	1.9479005	2.2920183	2.6927728
18	1.0939289	1.1961475	1.3073406	1.4282462	1.7024331	2.0258165	2.4066192	2.8543392
19	1.0993986	1.2081090	1.3269508	1.4568112	1.7535061	2.1068492	2.5269502	3.0255995
20	1.1048956	1.2201900	1.3468550	1.4859474	1.8061112	2.1911231	2.6532977	3.2071355
21	1.1104201	1.2323919	1.3670578	1.5156663	1.8602946	2.2787681	2.7859626	3.3995636
22	1.1159722	1.2447159	1.3875637	1.5459797	1.9161034	2.3699188	2.9252607	3.6035374
23	1.1215520	1.2571630	1.4083772	1.5768993	1.9735865	2.4647155	3.0715238	3.8197497
24	1.1271598	1.2697346	1.4295028	1.6084372	2.0327941	2.5633042	3.2250999	4.0489346
25	1.1327956	1.2824320	1.4509454	1.6406060	2.0937779	2.6658363	3.3863549	4.2918707
26	1.1384596	1.2952563	1.4727095	1.6734181	2.1565913	2.7724698	3.5556727	4.5493830
27	1.1441519	1.3082089	1.4948002	1.7068865	2.2212890	2.8833686	3.7334563	4.8223459
28	1.1498726	1.3212910	1.5172222	1.7410242	2.2879277	2.9987033	3.9201291	5.1116867
29	1.1556220	1.3345039	1.5399805	1.7758447	2.3565655	3.1186514	4.1161356	5.4183879
30	1.1614001	1.3478489	1.5630802	1.8113616	2.4272625	3.2433975	4.3219424	5.7434912
40	1.2207942	1.4888637	1.8140184	2.2080397	3.2620378	4.8010206	7.0399887	10.2857179
50	1.2832258	1.6446318	2.1052424	2.6915880	4.3839060	7.1066834	11.4673998	18.4201543
60	1.3488502	1.8166967	2.4432198	3.2810308	5.8916031	10.5196274	18.6791859	32.9876908
70	1.4178305	2.0067634	2.8354563	3.9995582	7.9178219	15.5716184	30.4264255	59.0759302
80	1.4903386	2.2167152	3.2906628	4.8754392	10.6408906	23.0497991	49.5614411	105.7959935
90	1.5665547	2.4486327	3.8189485	5.9431331	14.3004671	34.1193333	80.7303650	189.4645112
100	1.6466685	2.7048138	4.4320457	7.2446461	19.2186320	50.5049482	131.5012578	339.3020835

*n represents the number of compounding periods.

Finding compound interest is usually done with a scientific calculator, but it is important that you understand the process. In the next few pages we will show how it is done "by hand." The calculator procedure is shown at the end of this section.

Steps: Calculating Compound Interest and Maturity Value— A Shortcut

STEP 1. *Find* the number of compounding periods. The number of compounding periods, *n*, is the number of periods per year times the number of years.

The following table lists the most common compounding periods:

Compounding Period	Periods per Year
Annual	1
Semiannual	2
Quarterly	4
Monthly	12
Daily	365 or 366 in leap year

STEP 2. *Find* the interest rate per compounding period. The interest rate per compounding period is the annual rate divided by the number of compounding periods per year.

STEP 3. *Use* the Compound Interest table. Find the number of time periods (step 1) and the interest rate per time period (step 2) to locate the appropriate value in the table. The number of time periods determines the row; the interest rate per time period determines the correct column.

STEP 4. *Multiply* the table value times the principal and round to get the maturity value. Subtract the principal from the maturity value to find the interest.

These steps sound harder than they are. For example, let's find the maturity value of $600 at 6% compound monthly for two years.

Step 1.

| Number of monthly periods per year | | Number of years |

$$n = 12 \times 2 = 24$$

Step 2. Rate $= \dfrac{6\%}{12}$ ⟵ Annual rate

⟵ Number of monthly periods per year

$= \dfrac{1}{2}\%$

Step 3. Using $n = 24$ and $\frac{1}{2}\%$, the table value is 1.1271598.

Step 4. $1.1271598 \times \$600 = \$676.29588 = \$676.30$ rounded

The maturity value of $600 at 6% compounded monthly for two years is $676.30. The interest charged on the loan is

$$\$676.30 - \$600 = \$76.30$$

YOUR TURN

[WORK THIS PROBLEM]

The Question:

(a) What is the maturity value of $2500 at 8% compounded quarterly for 5 years?
(b) Find the maturity value on $625 at 6% compounded quarterly for 5 years.
(c) What is the maturity value on $625 at 6% compounded monthly for 5 years?
(d) What is the maturity value on $2000 at 8% compounded semiannually for 25 years?

✔ YOUR WORK

The Solution:

(a) **Step 1.** $n = 4 \times 5 = 20$ (four quarters per year × 5 years)

   Step 2. Rate $= \dfrac{8\%}{4}$ ⟵ Annual rate

         ⟵ Number of quarter periods per year

      $= 2\%$

   Step 3. The table value is 1.4859474. ⟵ From row 20 in the 2% column

   Step 4. $1.4859474 \times \$2500 = \$3714.8685 = \$3714.87$ rounded

(b) **Step 1.** $n = 4 \times 5 = 20$ (4 quarters per year × 5 years)

   Step 2. Rate $= \dfrac{6\%}{4} = \dfrac{3}{2}\% = 1\dfrac{1}{2}\%$

   Step 3. Table value $= 1.3468550$ ⟵ From row 20 in the $1\frac{1}{2}\%$ column

   Step 4. $1.3468550 \times \$625 = \$841.784375 = \$841.78$ rounded

(c) **Step 1.** $n = 12 \times 5 = 60$ (12 months/year × 5 years)

   Step 2. Rate $= \dfrac{6\%}{12} = \dfrac{1}{2}\%$

   Step 3. Table value $= 1.3488502$ ⟵ From row 60 in the $\frac{1}{2}\%$ column

   Step 4. $1.3488502 \times \$625 = \$843.031375 = \$843.03$ rounded

(d) **Step 1.** $n = 2 \times 25 = 50$ (2 periods/year × 25 years)

   Step 2. Rate $= \dfrac{8\%}{2} = 4\%$

   Step 3. Table value $= 7.1066834$

   Step 4. $7.1066834 \times \$2000 = \$14{,}213.366 = \$14{,}213.37$ rounded

---

It's easy, if you follow the four steps.

The compound interest table included in this text isn't large enough to work all compound interest problems. For example, to compute the maturity value of any principal at $5\frac{3}{4}\%$, compounded monthly for 9 years, the table is too small. The number of compounding periods is 12 × 9 = 108 and the interest rate per time period is $5\frac{3}{4}\%/12 = 23/48\%$. Neither of these values appears in the table. To work these problems, you will have to obtain a book of interest tables or use another method involving a scientific calculator.

Determining the Effective Rate

Effective Rate   The simple annual interest rate equivalent to the compounded rate.

If your bank uses 6% interest compounded quarterly, the resulting interest is more than the simple interest at 6% for one year. The **effective rate** is a simple annual interest rate equivalent to the compound rate. The effective rate is useful in comparing interest rates.

You can find the effective rate by following these steps:

Steps: Calculating Effective Rates

STEP 1. *Calculate* the maturity value of $1000 using the compound rate. (You can use any principal, but $1000 will yield a more accurate calculation.)

STEP 2. *Calculate* the interest:

$$\text{Interest} = \text{maturity value} - \text{principal}$$

STEP 3. *Calculate* the effective rate:

$$\text{Effective} = \frac{\text{annual interest}}{\text{principal}}$$

For example, let's calculate the effective rate for 6% interest compounded quarterly.

Step 1. $P = \$1000$

$n = 4$ (quarters per year) $\times 1$ (year) $= 4$

$r = \dfrac{\text{annual rate}}{n} = \dfrac{6\%}{4} = 1\dfrac{1}{2}\%$

Table value is 1.0613636

$MV = 1.0613636 \times \$1000 = \$1061.36$ rounded

Step 2. $I = MV - P = \$1061.36 - 1000 = \$61.36$

Step 3. Effective rate $= \dfrac{\$61.36}{\$1000.00} = 0.06136 = 6.136\%$

WORK THIS PROBLEM

YOUR TURN

The Question:

Calculate the effective rate for 6% interest compounded monthly.

✔ YOUR WORK

The Solution:

Step 1. $P = \$1000$

$n = 12 \times 1 = 12$

$r = \dfrac{6\%}{12} = \dfrac{1}{2}\%$

Table value $= 1.0616778$

$MV = 1.0616778 \times \$1000 = \$1061.68$ rounded

Step 2. $I = \$1061.68 - 1000 = \$61.68$

Step 3. Effective rate $= \dfrac{\$61.68}{\$1000.00} = 0.06168 = 6.168\%$

Note that the effective rate of 6% compounded monthly—6.168%—is higher than the effective rate of 6% compounded quarterly—6.136%. The more frequently the rate is compounded, the higher the effective rate.

Calculating Interest Compounded Daily

Many banks use interest compounded daily. Unfortunately, our compound interest table is too limited for daily interest calculations. For example, to calculate 6% interest compounded daily for one year (using 360 days), the number of periods and rate are

$$n = 360 \times 1 = 360; \qquad r = \frac{6\%}{360} = 0.016666\ldots\%$$

Neither value is in our table. We need either a bigger table or a new table. We will use the Daily Compound Interest table.

To calculate daily interest, use the following steps:

Steps: Calculating Daily Interest

STEP 1. *Calculate* the time using ordinary or 30-day-month time (see Section 9.1, for a review).

STEP 2. Using the Daily Compound Interest table, *find* the row indicated by time from step 1 and the column using the annual rate.

STEP 3. *Multiply* the table value times the principal and round the answer.

To see how these steps work, let's calculate (1) the maturity value of $1000 at 6% compounded daily for one year and (2) the effective rate.

1. *Step 1.*   Time is one year.

   Step 2.   Table value is 1.06183124.

   Step 3. $MV = 1.06183124 \times \$1000 = \$1061.83$ rounded

2. Apply steps for finding effective rate.

   Step 1. $MV = \$1061.83$ (from step 3)

   Step 2. $I = MV - P = \$1061.83 - 1000 = \$61.83$

   Step 3.   Effective rate $= \dfrac{\$16.83}{\$1000.00} = 0.06183 = 6.183\%$

DAILY COMPOUND INTEREST TABLE
Maturity Value of $1

Time	5.0%	5.5%	6.0%	6.5%	7.0%	7.5%	8.0%
1 day	1.00013889	1.00015278	1.00016667	1.00018056	1.00019444	1.00020833	1.00022222
2 day	1.00027780	1.00030558	1.00033336	1.00036114	1.00038893	1.00041671	1.00044449
3 day	1.00041672	1.00045840	1.00050008	1.00054176	1.00058345	1.00062513	1.00066681
4 day	1.00055567	1.00061125	1.00066683	1.00072242	1.00077800	1.00083359	1.00088919
5 day	1.00069464	1.00076412	1.00083361	1.00090310	1.00097260	1.00104210	1.00111161
6 day	1.00083362	1.00091702	1.00100042	1.00108382	1.00116723	1.00125065	1.00133407
7 day	1.00097263	1.00106993	1.00116725	1.00126457	1.00136191	1.00145925	1.00155659
8 day	1.00111165	1.00122288	1.00133411	1.00144536	1.00155661	1.00166788	1.00177916
9 day	1.00125069	1.00137584	1.00150100	1.00162617	1.00175136	1.00187656	1.00200178
10 day	1.00138976	1.00152883	1.00166792	1.00180702	1.00194615	1.00208529	1.00222445
11 day	1.00152884	1.00168184	1.00183486	1.00198791	1.00214097	1.00229406	1.00244716
12 day	1.00166794	1.00183487	1.00200183	1.00216882	1.00233583	1.00250287	1.00266993
13 day	1.00180706	1.00198793	1.00216883	1.00234977	1.00253073	1.00271172	1.00289274
14 day	1.00194620	1.00214101	1.00233586	1.00253075	1.00272567	1.00292062	1.00311561
15 day	1.00208536	1.00229412	1.00250292	1.00271176	1.00292064	1.00312956	1.00333852
16 day	1.00222454	1.00244725	1.00267000	1.00289280	1.00311565	1.00333855	1.00356149
17 day	1.00236374	1.00260040	1.00283711	1.00307388	1.00331070	1.00354758	1.00378450
18 day	1.00250295	1.00275357	1.00300425	1.00325499	1.00350579	1.00375665	1.00400756
19 day	1.00264219	1.00290677	1.00317142	1.00343614	1.00370092	1.00396576	1.00423068
20 day	1.00278145	1.00305999	1.00333862	1.00361731	1.00389608	1.00417492	1.00445384
21 day	1.00292072	1.00321324	1.00350584	1.00379852	1.00409128	1.00438413	1.00467705
22 day	1.00306002	1.00336651	1.00367309	1.00397976	1.00428652	1.00459337	1.00490031
23 day	1.00319933	1.00351980	1.00384037	1.00416104	1.00448180	1.00480266	1.00512362
24 day	1.00333866	1.00367312	1.00400768	1.00434234	1.00467712	1.00501200	1.00534699
25 day	1.00347802	1.00382645	1.00417501	1.00452368	1.00487247	1.00522137	1.00557040
26 day	1.00361739	1.00397982	1.00434237	1.00470505	1.00506786	1.00543080	1.00579386
27 day	1.00375678	1.00413320	1.00450976	1.00488646	1.00526329	1.00564026	1.00601737
28 day	1.00389619	1.00428661	1.00467718	1.00506790	1.00545876	1.00584977	1.00624092
29 day	1.00403562	1.00444005	1.00484463	1.00524937	1.00565427	1.00605932	1.00646453
1 mo	1.00417507	1.00459350	1.00501210	1.00543087	1.00584981	1.00626892	1.00668819
2 mo	1.00836757	1.00920810	1.01004933	1.01089124	1.01173384	1.01257713	1.01342112
3 mo	1.01257757	1.01384390	1.01511180	1.01638126	1.01765229	1.01892490	1.02019907
4 mo	1.01680515	1.01850099	1.02019964	1.02190109	1.02360536	1.02531245	1.02702236
5 mo	1.02105039	1.02317948	1.02531298	1.02745091	1.02959326	1.03174005	1.03389128
6 mo	1.02531334	1.02787946	1.03045196	1.03303086	1.03561619	1.03820794	1.04080615
9 mo	1.03820929	1.04210932	1.04602394	1.04995321	1.05389718	1.05785592	1.06182947
1 yr	1.05126745	1.05653618	1.06183124	1.06715276	1.07250088	1.07787573	1.08327744
5 yr	1.28400313	1.31650302	1.34982507	1.38399004	1.41901927	1.45493459	1.49175841
10 yr	1.64866403	1.73318021	1.82202771	1.91542844	2.01361568	2.11683466	2.22534314
15 yr	2.11688977	2.28173698	2.45941867	2.65093389	2.85735945	3.07985596	3.31967434
20 yr	2.71809308	3.00391363	3.31978496	3.66886611	4.05464812	4.48098896	4.95215209

YOUR TURN

| WORK THIS PROBLEM |

The Question:

Calculate the maturity value of $1575 at $5\frac{1}{2}$% interest compounded daily for five years.

✔ YOUR WORK

The Solution:

Step 1.   Time is five years.

Step 2.   Table value is 1.31650302.

Step 3. $MV = 1.31650302 \times \$1575 = \$2073.49$ rounded

Calculators can make finding compound interest simple, regardless of the period.

Finding Compound Interest Using a Calculator

You can create tables of compound interest by using the following formula:

$$\text{Maturity value} = P(1 + r)^n$$

where P = principal
$\quad\quad r$ = rate per compounding period
$\quad\quad n$ = number of compounding periods

Note the position of the n in the equation. It indicates that you are to "raise" the expression in the parentheses to "the power of n." That is, you multiply the expression times itself n times. For example, $3^2 = 3 \times 3 = 9$

$$3^3 = 3 \times 3 \times 3 = 27$$
$$3^4 = 3 \times 3 \times 3 \times 3 = 81$$

Let's use the formula to calculate the maturity value of $1000 at 6% compounded quarterly for five years.

$$P = \$1000; \quad n = 4 \text{ (quarters)} \times 5 \text{ (years)} = 20; \quad r = \frac{6\%}{4} = 0.015$$
$$\text{Maturity value} = \$1000(1 + 0.015)^{20} = \$1000(1.015)^{20}$$

You could punch in 1.015×1.015 . . . until you have twenty 1.015s and then multiply by $1000. But if you have a scientific calculator with a $\boxed{y^x}$ key, just punch in

1.015 $\boxed{y^x}$ 20 $\boxed{=}$ $\boxed{\times}$ 1000 $\boxed{=}$ ⟶   1346.855

Clearly a calculator, even without a $\boxed{y^x}$ key, is faster than doing a lot of calculations by hand.

Speaking of hands, try yours at the problems in Section Test 12.1.

SECTION TEST 12.1   Compound Interest and Present Value

The following problems test your understanding of Section 12.1, Compound Interest.

A. Calculate the maturity value for the following compound interest problems using the Compound Interest table.

	Principal	Time	Annual Rate	Compounding Period	Maturity Value
1.	$2500	5 years	8%	Quarterly	
2.	3200	6 years	6%	Quarterly	
3.	2725	5 years	6%	Monthly	
4.	1850	2 years	12%	Monthly	
5.	650	12 years	6%	Annual	
6.	725	7 years	5%	Annual	
7.	2290	10 years	8%	Semiannual	
8.	6400	7 years	6%	Semiannual	
9.	2479	2 years	6%	Monthly	
10.	3828	5 years	8%	Quarterly	

B. Calculate the maturity value for the following daily compound interest problems using the Compound Interest table. Use ordinary or 30-day-month time.

	Principal	Beginning Date	Ending Date	Annual Rate	Maturity Value
1.	$2800	June 4	June 28	5.5%	
2.	3400	July 7	August 7	6%	
3.	3850	March 25	August 25	5%	
4.	1725	May 17	November 17	8%	
5.	3820	April 6	May 2	7.5%	
6.	5920	June 12	July 5	6.5%	
7.	580	May 5, 1996	May 5, 2001	7%	
8.	775	June 8, 1998	June 8, 2008	5.5%	
9.	2947	April 3, 1997	April 3, 2002	7%	
10.	3714	July 7, 1997	July 7, 2012	7.5%	

C. Calculate the effective rate for the following compound interest problems.

	Annual Rate	Compounding Period	Effective Rate
1.	8%	Semiannual	
2.	8%	Quarterly	
3.	8%	Daily	
4.	10%	Semiannual	
5.	12%	Semiannual	
6.	12%	Quarterly	
7.	12%	Monthly	
8.	7.5%	Daily	
9.	5.5%	Daily	
10.	7%	Daily	

D. Solve these problems.

1. Kate Shrewsbury wants to invest $2800 so that she can expand Kate's Delicacies in the future. RuraBank pays 8% compounded quarterly. What is the maturity value after five years?

2. If Kate (problem 1) invests $2800 at 12% compounded quarterly for five years, what will be the maturity value?

3. If Kate (problem 1) invests $2800 at 12% compounded monthly for five years, what will be the maturity value?

4. If RuraBank pays 12% compounded semiannually, what is the maturity value of $2800 after five years?

5. The Pursuit Bank will pay 8% compounded daily. If Kate invests her $2800 at Pursuit, what will be the maturity value in five years?

6. If Pursuit's rate drops to 6.5% compounded daily, what will be the maturity value of $2800 in five years?

7. The Friendly Savings Bank pays 6.5% interest compounded daily. Crystal's Diamonds invested $5200 on November 13, 1991. What is the maturity value of the investment if it is withdrawn on November 13, 2001?

8. Edith Wharton loaned Bob Frost $8000 on March 17, 1995 to open a New Age Bookstore. If Bob is to repay Edith at 7.5% compounded daily on March 17, 2000, how much will Bob pay Edith?

9. Crystal's Diamonds is considering investing at 6.5% compounded daily in order to install a new security system in three years. What is the effective interest rate of this investment?

10. Fit-to-Print, which hopes to buy a larger press in about 5 years, has a chance to invest at 5% compounded daily. What is the effective interest rate of this investment?

11. To prepare for his retirement when he closes his restaurant, Lew has an investment at 10% that is compounded semiannually. What is the effective interest rate on this investment?

12. Petal Pushers wants to install a new floral refrigeration unit in three years and so has invested at 6% compounded monthly. What is the effective interest rate of this investment?

WORKSPACE

Can You:

Calculate present value?
. . . If not, you need this section.

Finding Present Value

How much money must I invest today in order to have $500 ten years from now?

Present Value The amount that must be invested now (in the present) in order to obtain a desired maturity value.

You can solve this problem by using present value. **Present value** is the amount of money you must invest now (in the present) in order to obtain a given maturity value. Actually, present value problems are just compound interest problems worked backward, where the maturity value is known and the amount to be invested (principal) is unknown.

Solving present value problems is easy if you follow basically the same steps you used in compound interest problems.

Steps: Finding Present Value

STEP 1. *Find* the number of compounding periods. The number of compounding periods, *n,* is the number of periods per year times the number of years.

STEP 2. *Find* the interest rate per compounding period. The interest rate per compounding period is the annual rate divided by the number of compounding periods per year.

STEP 3. *Use* the Present Value table. *Find* the number of time periods (step 1) and the interest rate per time period (step 2) to locate the appropriate value in the Present Value table. The number of time periods determines the row; the interest rate per time period determines the correct column.

STEP 4. *Multiply* the table value times the maturity value and round.

Let's use these steps to find the present value of $5000 at 6% compounded monthly for five years. That is, how much must be invested today (present value) at 6% compounded monthly to yield $5000 in five years?

Step 1.

| Number of monthly periods per year | Number of years |

$$n = 12 \times 5 = 60$$

Step 2.   Rate $= \dfrac{6\%}{12}$ ← Annual rate
 ← Number of monthly periods per year

$ = \dfrac{1}{2}\%$

Step 3.   Using $n = 60$, and $\frac{1}{2}\%$, the table value is 0.7413722.

Step 4. $0.7413722 \times \$5000 = \$3706.861 = \$3706.86$ rounded

That is, if you invest $3706.86 at 6% compounded monthly for five years, you will have a maturity value of $5000.

411

PRESENT VALUE TABLE
Present Value of $1

n*	$\frac{1}{2}$%	1%	$1\frac{1}{2}$%	2%	3%	4%	5%	6%
1	0.9950249	0.9900990	0.9852216	0.9803922	0.9708738	0.9615385	0.9523810	0.9433962
2	0.9900745	0.9802960	0.9706617	0.9611688	0.9425959	0.9245562	0.9070295	0.8899964
3	0.9851488	0.9705902	0.9563170	0.9423223	0.9151417	0.8889964	0.8638376	0.8396193
4	0.9802475	0.9609803	0.9421842	0.9238454	0.8884870	0.8548042	0.8227025	0.7920937
5	0.9753707	0.9514657	0.9282603	0.9057308	0.8626088	0.8219271	0.7835262	0.7472582
6	0.9705181	0.9420452	0.9145422	0.8879714	0.8374843	0.7903145	0.7462154	0.7049605
7	0.9656896	0.9327180	0.9010268	0.8705602	0.8130915	0.7599178	0.7106813	0.6650571
8	0.9608852	0.9234832	0.8877111	0.8534904	0.7894092	0.7306902	0.6768394	0.6274124
9	0.9561047	0.9143398	0.8745922	0.8367553	0.7664167	0.7025867	0.6446089	0.5918985
10	0.9513479	0.9052870	0.8616672	0.8203483	0.7440939	0.6755642	0.6139132	0.5583948
11	0.9466149	0.8963237	0.8489332	0.8042630	0.7224213	0.6495809	0.5846793	0.5267875
12	0.9419053	0.8874492	0.8363874	0.7884932	0.7013799	0.6245970	0.5568374	0.4969694
13	0.9372192	0.8786626	0.8240270	0.7730325	0.6809513	0.6005741	0.5303214	0.4688390
14	0.9325565	0.8699630	0.8118493	0.7578750	0.6611178	0.5774751	0.5050680	0.4423010
15	0.9279169	0.8613495	0.7998515	0.7430147	0.6418619	0.5552645	0.4810171	0.4172651
16	0.9233004	0.8528213	0.7880310	0.7284458	0.6231669	0.5339082	0.4581115	0.3936463
17	0.9187068	0.8443775	0.7763853	0.7141626	0.6050164	0.5133732	0.4362967	0.3713644
18	0.9141362	0.8360173	0.7649116	0.7001594	0.5873946	0.4936281	0.4155206	0.3503438
19	0.9095882	0.8277399	0.7536075	0.6864308	0.5702860	0.4746424	0.3957340	0.3305130
20	0.9050629	0.8195445	0.7424704	0.6729713	0.5536758	0.4563870	0.3768895	0.3118047
21	0.9005601	0.8114302	0.7314980	0.6597758	0.5375493	0.4388336	0.3589424	0.2941554
22	0.8960797	0.8033962	0.7206876	0.6468390	0.5218925	0.4219554	0.3418499	0.2775051
23	0.8916216	0.7954418	0.7100371	0.6341559	0.5066917	0.4057263	0.3255713	0.2617973
24	0.8871857	0.7875661	0.6995439	0.6217215	0.4919337	0.3901215	0.3100679	0.2469786
25	0.8827718	0.7797684	0.6892058	0.6095309	0.4776056	0.3751168	0.2953028	0.2329985
26	0.8783799	0.7720480	0.6790205	0.5975793	0.4636947	0.3606892	0.2812407	0.2198100
27	0.8740099	0.7644039	0.6689857	0.5858620	0.4501891	0.3468166	0.2678483	0.2073680
28	0.8696616	0.7568356	0.6590993	0.5743746	0.4370768	0.3334775	0.2550936	0.1956301
29	0.8653349	0.7493421	0.6493589	0.5631123	0.4243464	0.3206514	0.2429463	0.1845567
30	0.8610297	0.7419229	0.6397624	0.5520709	0.4119868	0.3083187	0.2313774	0.1741101
40	0.8191389	0.6716531	0.5512623	0.4528904	0.3065568	0.2082890	0.1420457	0.0972222
50	0.7792861	0.6080388	0.4750047	0.3715279	0.2281071	0.1407126	0.0872037	0.0542884
60	0.7413722	0.5504496	0.4092960	0.3047823	0.1697331	0.0950604	0.0535355	0.0303143
70	0.7053029	0.4983149	0.3526769	0.2500276	0.1262974	0.0642194	0.0328662	0.0169274
80	0.6709885	0.4511179	0.3038902	0.2051097	0.0939771	0.0433843	0.0201770	0.0094522
90	0.6383435	0.4083912	0.2618522	0.1682614	0.0699278	0.0293089	0.0123869	0.0052786
100	0.6072868	0.3697112	0.2256294	0.1380330	0.0520328	0.0198000	0.0076045	0.0029472

*n represents the number of compounding periods.

YOUR TURN

WORK THIS PROBLEM

The Question:

(a) What is the present value of $2500 at 8% compounded quarterly for 5 years?
(b) How much must be invested today at 8% compounded semiannually to yield $2750 in 25 years?
(c) What is the present value of $11,467.40 at 5% compounded annually for 50 years?

✔ YOUR WORK

The Solution:

(a) **Step 1.** $n = 4 \times 5 = 20$

    Step 2.   Rate $= \dfrac{8\%}{4} = 2\%$

    Step 3.   Table value $= 0.6729713$

    Step 4. $0.6729713 \times \$2500 = \$1682.42825 = \$1682.43$ rounded

(b) **Step 1.** $n = 2 \times 25 = 50$

    Step 2.   Rate $= \dfrac{8\%}{2} = 4\%$

    Step 3.   Table value $= 0.1407126$

    Step 4. $0.1407126 \times \$2,750 = \$386.95965 = \$386.96$ rounded

(c) **Step 1.** $n = 1 \times 50 = 50$

    Step 2.   Rate $= \dfrac{5\%}{1} = 5\%$

    Step 3.   Table value $= 0.0872037$

    Step 4. $0.0872037 \times \$11,467.40 = \$999.9997094 = \$1000.00$ rounded

The Present Value table included in the text isn't large enough to work all present value problems. For other interest rates and compounding periods, you will have to obtain a book on interest and present value tables or use a scientific calculation.

Finding Present Value Using a Calculator

You can calculate present value using a modification of the formula we used for compound interest:

$$\text{Present value} = \frac{\text{maturity value}}{(1 + r)^n}$$

where r = rate per compounding period
     n = number of compounding periods

For example, the present value of $1000 at 5% compounded semiannually for six years is

$$n = 2 \text{ (periods per year)} \times 6 \text{ (years)} = 12$$

$$\text{Present value} = \frac{\$1000}{\left[\left(1 + \dfrac{0.05}{2}\right)\right]^{12}} = \frac{\$1000}{(1.025)^{12}} = \$743.56 \text{ rounded}$$

If your calculator has y^x and $\frac{1}{x}$ keys, punch in

1.025 y^x 12 $=$ $\frac{1}{x}$ $\times$ 1000 $=$ $\longrightarrow$ 743.5559

This answer is equal to $743.56 rounded.

The problems in Section Test 12.2 will give you (and possibly your calculator) a workout and make sure you understand present value.

SECTION TEST 12.2   Compound Interest and Present Value

The following problems test your understanding of Section 12.2, Present Value.

A. Calculate the present value for the following problems using the Present Value table.

	Maturity Value	Time	Annual Rate	Compounding Period	Present Value
1.	$2500	6 years	6%	Quarterly	
2.	3200	5 years	8%	Quarterly	
3.	2725	2 years	12%	Monthly	
4.	1850	5 years	6%	Monthly	
5.	650	7 years	5%	Annual	
6.	725	12 years	6%	Annual	
7.	2290	10 years	6%	Semiannual	
8.	6400	7 years	8%	Semiannual	
9.	2479	2 years	6%	Monthly	
10.	3828	5 years	8%	Quarterly	
11.	2500	5 years	6%	Monthly	
12.	3200	4 years	6%	Quarterly	
13.	2725	3 years	8%	Quarterly	
14.	1850	4 years	6%	Semiannual	
15.	650	6 years	5%	Annual	
16.	725	5 years	6%	Quarterly	

B. Solve these problems

1. Lew needs to have $4000 in five years to reroof his restaurant. How much must he invest today at 6% compounded monthly in order to have the amount in five years?

2. The owners of Wired for Sound are saving to recarpet their listening room. They estimate that it will cost $6500 in two years. How much must Wired for Sound invest today at 6% compounded monthly in order to have the amount in two years?

3. The By-A-Nose Horse Farm needs to have $2500 in six years to expand its stables. How much must Mr. Nose invest today at 8% compounded quarterly in order to have $2500 in six years?

4. Kuo Min Tang wants to expand his martial arts school. He estimates that he will need $3400 in three years for the expansion. How much must he invest today at 6% compounded quarterly in order to have the amount in three years?

5. Jim Dean is saving to buy a motorcycle. He needs to have $2000 in one year. How much must Jim invest today at 12% compounded monthly in order to have the amount in one year?

6. The More-Again Warranty Bank pays 6% compounded semiannually. How much does Maria, owner of Maria's Mannequins need to invest to have $6200 in seven years for a second mannequin store?

7. Ralph Green has started a lawn care business to save for his first car. He will be 16 in five years and estimates that he will need to have $3000 to purchase a used car. How much must Ralph need to invest today at 6% compounded monthly in order to have the amount in five years?

8. Ken and Barbie's Barbeque is planning to remodel in two years and wants to set aside the money now. If the firm wants to have $4400 in two years for remodeling, how much must it invest today at 12% compounded monthly in order to have the amount in two years?

9. The owners of the Concino De Mino Mexican Restaurant need to have $2800 in two years to repay their parents for money that they borrowed to start the restaurant. How much must the owners invest today at 8% compounded quarterly in order to have the amount in two years?

10. Canine Capers wants to purchase a male Russian wolfhound for breeding in two years. The price at that time will probably be $1500. How much must the firm invest today at 12% compounded quarterly in order to have $1500 in two years?

11. Jimmy Sexton wants to save $7000 in five years to buy a new bass boat. How much must he invest today at 6% compounded monthly in order to have $7000 in five years?

12. Dorian Gray needs to have $8200 in 12 years to open a second portrait studio. How much must he invest today at 5% compounded annually in order to have $8200 in 12 years?

13. Will Wright wants to purchase a glider. The Kitty Hawk Bank pays 12% compounded quarterly. He wants to have $2750 in three years. How much must Will invest today in order to have $2750 in three years?

14. Gib Ewell will need $4300 to expand his health food store in five years. How much must he invest today at 8% compounded quarterly in order to have $4300 in five years?

15. The Inside Story, an interior design firm, is growing and will need to have $9500 in four years for a new office. How much must the firm invest today at 6% compounded quarterly in order to have $9500 in four years?

16. Abby deposited $3500 in a savings account paying 5.5%, compounded quarterly. After one year, she deposited an additional $1800 in the account. Six months later, she withdrew $2000, leaving the rest for another three months before closing the account. How much was in the account when she closed it?

17. Shane inherited some money from a distant uncle, and he decided to save it until he had $15,000 in the bank. The account he opened paid an annual interest rate of 4.5%, compounded monthly. If he reached his goal in 5 years, how much did he originally invest?

WORKSPACE

Key Terms

compound interest
effective rate
present value

Chapter 12 at a Glance

Page	Topic	Key Point	Example
398	Annual compound interest	Find interest and maturity value for the first year.	To find interest on $500 at 6% for two years, compounded annually: First year $I = Prt$ $\qquad = \$500 \times 0.06 \times 1 = \$30$ $MV = P + I = \$500 + 30 = \$530$
		Find interest and maturity value for subsequent years, using the maturity value from previous year. Add interest amounts for all years together.	Second year $I = MV \cdot rt$ $\qquad = \$530 \times 0.06 \times 1$ $\qquad = \$31.80$ Total interest = $\$30 + 31.80 = \$61.80$
399	Maturity values with compound interest	Multiply the compounding periods per year by the number of years to find n. $\text{Rate per compounding period} = \dfrac{\text{annual rate}}{\text{compounding periods in one year}}$	To find the maturity value of $600 at 6%, compounded monthly for two years: $n = 12 \times 2 = 24$ $\text{Rate per compounding period} = \dfrac{6\%}{12} = \dfrac{1}{2}\%$
		Locate value in Compound Interest table. Maturity value = table value × principal	Table value = 1.1271598 Maturity value = $1.1271598 \times \$600$ $\qquad = \$676.30$ rounded
402	Effective rates	Calculate the maturity value of $1000 at the compound rate.	To find the effective rate for 6% interest compounded quarterly: $P = \$1000;\ n = 4 \times 1 = 4;$ $r = \dfrac{6\%}{4} = 1\dfrac{1}{2}\%$ Table value = 1.0613636 $MV = 1.0613636 \times \$1000$ $\qquad = \$1061.36$ rounded
		Interest = maturity value − principal $\text{Effective rate} = \dfrac{\text{annual interest}}{\text{principal}}$	$I = \$1061.36 - 1000 = \$61.36$ $\text{Effective rate} = \dfrac{\$61.36}{\$1000.00} = 0.06136$ $\qquad = 6.136\%$

Page	Topic	Key Point	Example
404	Interest compounded daily	Find ordinary or 30-day-month time. Locate value in Daily Compound Interest table. MV = table value × principal	To find the maturity value of $1000 at 6%, compounded daily: Time = 1 year Table value = 1.0618314 $MV = 1.0618314 \times \$1000$ $= \$1061.83$ rounded
411	Present value	n = no. periods per year × no. years Interest rate per period $= \dfrac{\text{annual rate}}{\text{compounding periods in one year}}$ Find the value in the Present Value table. Present value = table value × maturity value	To find the present value of $5000 at 6%, compounded monthly for five years: $n = 12 \times 5 = 60$ $\text{Rate} = \dfrac{6\%}{12} = \dfrac{1}{2}\%$ Table value = 0.7413722 Present value $= 0.7413722 \times \$5000$ $= \$3706.86$ rounded

SELF-TEST   Compound Interest and Present Value

Name

Date

Course/Section

The following problems test your understanding of compound interest and present value.

1. RuraBank pays 6% compounded quarterly. Kate Shrewsbury wants to invest $3500 so that she can expand Kate's Delicacies in 5 years. What is the maturity value after five years?

2. The Write Stuff wants to convert a storeroom to a showroom in a few years. How much will the firm have, if it invests $3500 for five years at 6% compounded monthly? (That is, what is the maturity value?)

3. To fund a new weight room, Musselmann's Gym has invested $4200 at 8% compounded semiannually for three years. What is the maturity value of this investment?

4. The Pursuit Bank has decided to loan Bebe $4200 to start a child care center. If Bebe repays the bank at 8% compounded quarterly for three years, how much will Bebe pay the bank?

5. Len Liki invested $2100 at 10% compounded semiannually for four years to help his daughter through college so she could take over the family firm, the Liki Valve Co. What is the maturity value of Len's investment?

6. Maria's Mannequins will need more storage space in five years. If the firm invests $3500 at 6% compounded daily for five years, what will the maturity value be?

7. Business at Bob's Beeflike Burgers has been booming. Bob wants to set aside money for expansion. What is the maturity value of $5100 invested at 7.5% compounded daily for five years?

8. Ms. Spade knew she would need a new kiln for her pottery shop in 2006. If she invested $1700 on December 3, 1995 at 5.5% compounded daily and withdraws her money December 3, 2005, what is the maturity value?

9. Juanita Verde received $6000 for a recycling process that she developed. She invested $6000 on May 7, 1996 at 6.5% compounded daily. What is the maturity value of her investment if she withdraws it on May 7, 2001?

10. On August 11, 1992, when Olivia turned 8, her parents invested $1600 inheritance that she received at 7% compounded daily. If she withdraws it on her eighteenth birthday, August 11, 2002, to enroll in a chef's course, how much will she have? (That is, what is the maturity value?)

11. The Butter-em-Up Bakery plans to add an eat-in area in four years. If it invests at 7% compounded daily, what is the effective interest rate?

12. Business is so good that Lew expects to need double the capacity in his trendy restaurant in five years. If he invests at 5.5% compounded daily, what is the effective interest rate?

13. New noise ordinances mean that Atenal Music will have to soundproof its store in three years. If it invests at 8% compounded semiannually, what is the effective interest rate?

14. Abby Lincoln wants to buy out her law partner, whose ethics worry her. If she invests at 8% compounded quarterly, what is the effective interest rate?

15. Bits 'n' Bytes expects to open three new stores two years from now and has invested at 12% compounded monthly. What is the effective interest rate of this investment?

16. Ham Andea, an overworked manager at Woody's Lumber, wants to build a summer cabin ten years from now. He needs to have $4500 to start construction. How much must he invest today at 10% compounded semiannually in order to have the amount in ten years?

17. Cherry needs to have $7000 in five years to open her own woodshop. How much must she invest today at 8% compounded quarterly in order to have that amount in five years?

18. Sew What? needs to have $2500 in two years to replace a fabric cutting machine. How much must the firm invest today at 6% compounded monthly in order to have that amount in two years?

19. The Garden of Earthly Delights needs to have $3100 in six years to install new glass in a small "tropical" greenhouse. How much must the firm invest today at 6% compounded quarterly in order to have that amount in six years?

20. Ace Skolar needs to have $12,000 in 15 years to offset education expenses. How much must his parents invest today at 8% compounded semiannually in order to have that amount in 15 years?

Investments

CHAPTER OBJECTIVES

When you complete this chapter successfully, you will be able to:

1. Determine the value of ordinary and due annuities.

2. Calculate the payments necessary to establish a sinking fund.

3. Find the yield of stocks and bonds.

■ Al Dumas invested $375 at the beginning of each semiannual period in an annuity with an interest rate of 6%. What is the amount of the annuity after 12 years?

■ Bob and Ted's California Experience wants to remodel a beach house it rents and is considering borrowing $12,000 and setting up a sinking fund to repay the loan in two years. The account will yield 12% interest compounded monthly. What are the required monthly payments?

■ Ms. Reed owns 75 shares of Wild Child Recording Company stock. If the company pays a $5\frac{1}{4}$% dividend on its preferred stock, based on an $80 par value, what is Ms. Reed's total dividend?

■ What is the annual interest and semiannual payment on a Bell Telephone of Pennsylvania $1000 bond with $9\frac{5}{8}$% interest?

People and businesses invest money in many ways and for many reasons. Individuals want to save money to buy a home or start a business, to educate their children, or to ensure a comfortable retirement, for example. Companies often must fund pension programs or save to cover future investments in facilities and equipment. Toward this end, many individuals and businesses invest in annuities and sinking funds.

In addition, individuals and businesses "meet" in the purchase of stocks and bonds. Individuals invest in these financial instruments. Businesses issue them as a way to acquire capital and expand their operations. Still other businesses specialize in helping individuals purchase the stocks and bonds of other firms.

Don't get a "sinking" feeling. Your "stock" in yourself will rise when you see how easy these investments can be to understand.

SECTION 13.1: Annuities: Due and Ordinary

Can You:

Calculate the amount of an annuity due?
Determine the amount of an ordinary annuity?
. . . If not, you need this section.

Annuity A series of investments or payments, usually of equal amounts and at regular intervals, into a compound interest account.

In Chapter 12, you learned how to calculate the maturity value of an investment made at *one* time. An **annuity** is a series of investments or payments, usually of equal amounts and at regular intervals, into a compound interest account. Although the term "annuity" sounds like "annual" and may imply that the payments are made annually, annuity payments are often made on a more frequent basis.

Annuities are used by individuals to save for some future need, such as retirement, a college fund, or the down payment on a house. Annuities called *sinking funds* (see Section 13.2) are used by businesses to pay for loans and to pay off company bonds (discussed in Section 13.3).

Annuity Due An annuity on which payments are made at the *beginning* of each time period.

Ordinary Annuity An annuity on which payments are made at the *end* of each time period.

There are two basic types of annuities. In an **annuity due,** payments are made at the *beginning* of each time period. In an **ordinary annuity,** payments are made at the *end* of each time period.

Calculating the Amount of an Annuity Due

We'll begin our explorations with a look at calculating the amount of an annuity due. One way is to use two simple but repetitive steps:

Steps: Calculating the Value of an Annuity Due

STEP 1. *Calculate* the interest for the first year and add it to the initial investment to find the value at the end of the first year.

STEP 2. *Perform* the same calculations for subsequent years, using the previous year-end value as the base.

For example, let's consider an *annuity due* where $500 is invested at the beginning of each year in an account that earns 6% interest, compounded annually.

Step 1. First year interest: $I = Prt = \$500 \times 6\% \times 1 = \$30$
Value at the end of the first year $= \$500 + 30 = \$530$

Step 2. At the beginning of the second year, a \$500 payment is added to the account. The new account balance, for the beginning of the second year, is $\$530 + 500 = \$1030$.
Second year interest: $I = Prt = \$1030 \times 6\% \times 1 = \$61.80$
Value at the end of the second year $= \$1030 + 61.80 = \$1091.80$

YOUR TURN

| WORK THIS PROBLEM |

The Question:

At the beginning of the third year, another \$500 payment is made into the account just discussed. Calculate the new account balance, the interest for the third year, and finally, the value at the end of the third year.

| ✔ YOUR WORK |

The Solution:

Step 2a. Value at the beginning of the third year $= \$1091.80 + 500 = \$1591.80$

Step 2b. Third year interest: $I = Prt = \$1591.80 \times 6\% \times 1 = \$95.508 = \$95.51$ rounded
Value at the end of the third year $= \$1591.80 + 95.51 = \$1687.31$

This method of calculating the value of an annuity can be long and tedious for most problems. Imagine calculating the value at the end of 20 years of monthly payments—you'd have to do $10 \times 12 = 120$ calculations! Fortunately, there is a shortcut method. Just follow these four steps and use the Annuity Due table.

Steps: Calculating the Value of an Annuity Due—A Shortcut

STEP 1. *Find* the number of compounding periods. The number of compounding periods, n, is the number of periods per year times the number of years.

STEP 2. *Find* the interest rate per compounding period.

$$\text{Rate per period} = \frac{\text{annual rate}}{\text{Number of periods per year}}$$

STEP 3. *Look* at the Annuity Due table. Using the number of time periods (from step 1) and the interest rate per time period (from step 2), locate the appropriate value in the table. The number of time periods determines the row; the interest rate per time period determines the correct column.

STEP 4. *Multiply* the table value times the amount of each payment and round.

To see how these steps work, let's calculate the value of an annuity due after five years with monthly payments of \$50 and 6% interest compounded monthly.

Step 1.

| Number of monthly periods per year | Number of years |

$$n = 12 \times 5 = 60$$

Step 2. Rate $= \dfrac{6\%}{12}$ ←——— Annual rate

  ←——— Number of monthly payments per year

$= \frac{1}{2}\%$

Step 3. Using $n = 60$ and $\frac{1}{2}\%$, the table value is 70.1188807.

Step 4. $70.1188807 \times \$50 = \$3505.944035 = \$3505.94$ rounded

If you invest \$50 per month in an annuity due with 6% interest compounded monthly, at the end of five years the annuity will be worth \$3505.94.

Annuity Due Table of \$1

n*	$\frac{1}{2}\%$	1%	$1\frac{1}{2}\%$	2%	3%	4%	5%	6%
1	1.0050000	1.0100000	1.0150000	1.0200000	1.0300000	1.0400000	1.0500000	1.0600000
2	2.0150250	2.0301000	2.0452250	2.0604000	2.0909000	2.1216000	2.1525000	2.1836000
3	3.0301001	3.0604010	3.0909034	3.1216080	3.1836270	3.2464640	3.3101250	3.3746160
4	4.0502506	4.1010050	4.1522669	4.2040402	4.3091358	4.4163226	4.5256312	4.6370930
5	5.0755019	5.1520151	5.2295509	5.3081210	5.4684099	5.6329755	5.8019128	5.9753185
6	6.1058794	6.2135352	6.3229942	6.4342834	6.6624622	6.8982945	7.1420085	7.3938376
7	7.1414088	7.2856706	7.4328391	7.5829691	7.8923360	8.2142263	8.5491089	8.8974679
8	8.1821158	8.3685273	8.5593317	8.7546284	9.1591061	9.5827953	10.0265643	10.4913160
9	9.2280264	9.4622125	9.7027217	9.9497210	10.4638793	11.0061071	11.5778925	12.1807949
10	10.2791665	10.5668347	10.8632625	11.1687154	11.8077957	12.4863514	13.2067872	13.9716426
11	11.3355624	11.6825030	12.0412114	12.4120897	13.1920296	14.0258055	14.9171265	15.8699412
12	12.3972402	12.8093280	13.2368296	13.6803315	14.6177904	15.6268377	16.7129828	17.8821377
13	13.4642264	13.9474213	14.4503820	14.9739382	16.0863242	17.2919112	18.5986320	20.0150659
14	14.5365475	15.0968955	15.6821378	16.2934169	17.5989139	19.0235876	20.5785636	22.2759699
15	15.6142303	16.2578645	16.9323698	17.6392853	19.1568813	20.8245311	22.6574918	24.6725281
16	16.6973014	17.4304431	18.2013554	19.0120710	20.7615877	22.6975124	24.8403664	27.2128798
17	17.7857879	18.6147476	19.4893757	20.4123124	22.4144354	24.6454129	27.1323847	29.9056525
18	18.8797169	19.8108950	20.7967164	21.8405586	24.1168684	26.6712294	29.5390039	32.7599917
19	19.9791154	21.0190040	22.1236671	23.2973698	25.8703745	28.7780786	32.0659541	35.7855912
20	21.0840110	22.2391940	23.4705221	24.7833172	27.6764857	30.9692017	34.7192518	38.9927267
21	22.1944311	23.4715860	24.8375799	26.2989835	29.5367803	33.2479698	37.5052144	42.3922903
22	23.3104032	24.7163018	26.2251436	27.8449632	31.4528837	35.6178886	40.4304751	45.9958277
23	24.4319552	25.9734649	27.6335208	29.4218625	33.4264702	38.0826041	43.5019989	49.8155774
24	25.5591150	27.2431995	29.0630236	31.0302997	35.4592643	40.6459083	46.7270988	53.8645120
25	26.6919106	28.5256315	30.5139690	32.6709057	37.5530423	43.3117446	50.1134538	58.1563827
26	27.8303701	29.8208878	31.9866785	34.3443238	39.7096335	46.0842144	53.6691264	62.7057657
27	28.9745220	31.1290967	33.4814787	36.0512103	41.9309225	48.9675830	57.4025868	67.5281116
28	30.1243946	32.4503877	34.9987009	37.7922345	44.2188502	51.9662863	61.3227119	72.6397983
29	31.2800166	33.7848915	36.5386814	39.5680792	46.5754157	55.0849378	65.4388475	78.0581862
30	32.4414167	35.1327404	38.1017616	41.3794408	49.0026782	58.3283353	63.7607899	83.8016774
40	44.3796415	49.3752371	55.0819123	61.6100228	77.6632975	98.8265363	126.8397630	164.0476836
50	56.9283888	65.1078140	74.7880705	86.2709895	116.1807733	158.7737670	219.8153955	307.7560589
60	70.1188807	82.4863666	97.6578715	116.3325702	167.9450399	247.5103126	371.2629038	565.1158717
70	83.9839360	101.6831002	124.1992092	152.9774694	237.5118856	378.8620771	617.9549362	1026.0080998
80	98.5580521	122.8882369	155.0015153	197.6473970	331.0039091	573.2947758	1019.7902624	1851.3958849
90	113.8774905	146.3119001	190.7488489	252.0997894	456.6493708	861.1026669	1674.3376660	3329.5396984
100	129.9803669	172.1861968	232.2350890	318.4769520	625.5063647	1287.1286528	2740.5264148	5976.6701421

*n represents the number of compounding periods

WORK THIS PROBLEM

The Question:

(a) Calculate the value of an annuity due after 5 years with quarterly payments of $200 and 8% interest compounded quarterly.

(b) Calculate the value of an annuity due after 25 years with semiannual payments of $625 and 8% interest compounded semiannually.

✔ YOUR WORK

The Solution:

(a) *Step 1.* $n = 4 \times 5 = 20$

    Step 2.   Rate $= \dfrac{8\%}{4} = 2\%$

    Step 3.   Annuity Due table value = 24.7833172

    Step 4. $24.7833172 \times \$200 = \$4956.66344 = \$4956.66$ rounded

(b) *Step 1.* $n = 2 \times 25 = 50$

    Step 2.   Rate $= \dfrac{8\%}{2} = 4\%$

    Step 3.   Annuity Due table value = 158.7737670

    Step 4. $158.7737670 \times \$625 = \$99{,}233.604375 = \$99{,}233.60$ rounded

Determining the Value of an Ordinary Annuity

For ordinary annuities, we use the same Annuity Due table but make two changes to our four-step procedure.

Steps: Calculating the Value of an Ordinary Annuity

STEP 1. *Find* the number of compounding periods. Since the payment is made at the end of the time period, there is one less compounding period than with an annuity due problem. The number of compounding periods, *n,* is the number of periods per year times the number of years minus 1.

STEP 2. *Find* the interest rate per compounding period. The interest rate per compounding period is the annual rate divided by the number of compounding periods per year.

STEP 3. *Use* the Annuity Due table. *Find* the number of time periods (step 1) and the interest rate per time period (step 2) to locate the appropriate value in the Annuity Due table. The number of time periods determines the row; the interest rate per time period determines the correct column. Since the annuity payment is made at the end of the time period, you must add 1 to the value from the Annuity Due table value.

STEP 4. *Multiply* the table value plus 1 times the amount of each payment and round.

Let's use these steps to calculate the value of an ordinary annuity after two years with monthly payments of $50 and 6% interest compounded monthly.

Step 1.

| Number of monthly periods per year | | Number of years |

$$n = (12 \times 2) - 1 = 24 - 1 = 23$$

Step 2.   Rate $= \dfrac{6\%}{12}$ ⟵ Annual rate
⟵ Number of monthly payments per year

$= \tfrac{1}{2}\%$

Step 3.   Using $n = 23$ and $\tfrac{1}{2}\%$, the Annuity Due table value is 24.4319552. Add 1 to this value.

$$24.4319552 + 1 = 25.4319552$$

Step 4. $25.4319552 \times \$50 = \$1271.59776 = \$1271.60$ rounded

If you invest \$50 per month in an ordinary annuity with 6% interest compounded monthly, at the end of two years the annuity will be worth \$1271.60.

YOUR TURN

| WORK THIS PROBLEM |

The Question:

Calculate the value of an ordinary annuity after five years with quarterly payments of \$200 and 8% interest compounded quarterly.

| ✔ YOUR WORK |

The Solution:

Step 1. $n = (4 \times 5) - 1 = 20 - 1 = 19$

Step 2.   Rate $= \dfrac{8\%}{4} = 2\%$

Step 3.   Annuity Due table value is 23.2973698. $23.2973698 + 1 = 24.2973698$

Step 4. $24.2973698 \times \$200 = \$4859.47396 = \$4859.47$ rounded

Businesspeople virtually always use calculators to determine annuity values.

Finding Annuity Values Using a Calculator

If you don't have a table, but you do have a scientific calculator with a $\boxed{y^x}$ key, you can find the value of an annuity using one of two formulas:

$$\text{Value of an ordinary annuity} = m \times \left[\frac{(1 + r)^n - 1}{r} \right]$$

$$\text{Value of an annuity due} = m \times \left[\frac{(1 + r)^{n + 1} - (1 + r)}{r} \right]$$

where, in both formulas, m = amount of payment
n = number of payments (or compounding periods)
r = rate per compounding period

For example, let's find the values of an ordinary annuity and an annuity due after five years with quarterly payments of $200 and 8% interest compounded quarterly.

$$m = \$200; \quad n = 4 \times 5 = 20; \quad r = \frac{8\%}{4} = 2\% = 0.02$$

So $1 + r = 1.02$. Thus,

$$\text{Value of an ordinary annuity} = \$200 \times \frac{(1 + 0.02)^{20} - 1}{0.02}$$

$$= \$200 \times \frac{1.02^{20} - 1}{0.02}$$

To find the value of an ordinary annuity, punch in

1.02 $\boxed{y^x}$ 20 $\boxed{=}$ $\boxed{-}$ 1 $\boxed{=}$ $\boxed{\div}$ 0.02 $\boxed{\times}$ 200 $\boxed{=}$ $\longrightarrow$ ▨ 4859.47396

To calculate the value of an annuity due, use the other formula.

$$\text{Value of an annuity due} = \$200 \times \frac{(1 + 0.02)^{20 + 1} - (1 + 0.02)}{0.02}$$

$$= \$200 \times \frac{1.02^{21} - 1.02}{0.02}$$

To find the value of an annuity due, punch in

1.02 $\boxed{y^x}$ 21 $\boxed{=}$ $\boxed{-}$ 1.02 $\boxed{=}$ $\boxed{\div}$ 0.02 $\boxed{\times}$ 200 $\boxed{=}$ $\longrightarrow$ ▨ 4956.66344

Rounded, the value of the ordinary annuity is $4859.47, while that of the annuity due is $4956.66. These are the same answers you calculated earlier using the Annuity Due table.

Whether you use a calculator or a table, be sure to check whether you are being asked for the annuity due or the ordinary annuity.

If you're sure you've got them straight, go to Section Test 13.1 for some practice with annuities due and ordinary annuities.

WORKSPACE

SECTION TEST 13.1 Investments

The following problems test your understanding of Section 13.1, Annuities: Due and Ordinary.

A. Calculate the value of the annuity due using the Annuity Due table.

	Payment Amount	Payment Period	Annual Rate	Time	Amount of Annuity
1.	$ 85	Monthly	6%	2 years	
2.	30	Monthly	6%	5 years	
3.	450	Semiannual	8%	3 years	
4.	625	Semiannual	6%	5 years	
5.	325	Quarterly	6%	2 years	
6.	90	Quarterly	8%	3 years	
7.	800	Annual	5%	8 years	
8.	120	Annual	6%	14 years	
9.	600	Semiannual	6%	10 years	
10.	825	Semiannual	8%	15 years	

B. Calculate the value of the ordinary annuity using the Annuity Due table.

	Payment Amount	Payment Period	Annual Rate	Time	Amount of Annuity
1.	$ 75	Monthly	6%	2 years	
2.	20	Monthly	6%	1 year	
3.	275	Quarterly	6%	2 years	
4.	320	Quarterly	8%	3 years	
5.	750	Annual	5%	14 years	
6.	825	Annual	6%	17 years	
7.	480	Semiannual	6%	12 years	
8.	650	Semiannual	8%	8 years	
9.	320	Quarterly	8%	4 years	
10.	80	Quarterly	6%	5 years	

C. Solve these problems. Use the Annuity Due table or a calculator.

1. Louise Alcott invested $55 at the beginning of each month in an annuity with an interest rate of 6%. What is the amount of the annuity after five years?

2. Hans Andersen invested $90 at the beginning of each month in an annuity with an interest rate of 6%. What is the amount of the annuity after two years?

3. Simone Beauvoir invested $600 at the end of each semiannual period in an annuity with an interest rate of 8%. What is the amount of the annuity after five years?

4. Bert Brecht invested $830 at the end of each semiannual period in an annuity with an interest rate of 6%. What is the amount of the annuity after 14 years?

5. Edgar Burroughs invested $45 at the end of each month in an annuity with an interest rate of 6%. What is the amount of the annuity after one year?

6. Carol Capek wants to withdraw her two-year-old annuity. She has invested $95 at the end of each month in the annuity with a 6% interest rate. What is the annuity worth?

7. For five years, Will Cather invested $160 at the beginning of each quarter in an annuity with an interest rate of 6%. What is the final amount of the annuity?

8. Gaby D'Annunzio has been investing $240 at the beginning of each quarter in an annuity with an interest rate of 8%. What is the amount of the annuity after ten years?

9. Art Doyle invests $290 at the beginning of each semiannual period in an annuity with an · interest rate of 8%. What is the amount of the annuity after seven years?

10. Al Dumas invested $375 at the beginning of each semiannual period in an annuity with an interest rate of 6%. What is the amount of the annuity after 12 years?

11. Annie France plans to invest $150 at the end of each quarter in an annuity with an interest rate of 6%. What will be the amount of the annuity after six years?

12. Gene Genet has been investing $230 at the end of each quarter in an annuity with an interest rate of 8% for the past seven years. What is the amount of the annuity?

13. When Beth and Scott's son Alan was 6 months old, they started an annuity to help finance his college education. Each quarter they contributed $75 to the annuity, which paid an annual interest rate of 6% compounded quarterly. What will this annuity be worth on Alan's 18th birthday? Calculate (a) as an annuity due, and (b) as an ordinary annuity. (c) If they have a choice, which type of annuity should Beth and Scott purchase?

Can You:

Calculate the required payment for a sinking fund?
Check your sinking fund payment using an ordinary annuity calculation?
. . . If not, you need this section.

Sinking Fund A type of annuity that assures a set total after a certain number of payments.

Just as we were concerned with the present value of a simple investment at compound interest, so investors sometimes want to set up **sinking funds** that assure a set total after a certain number of payments. Businesses use sinking funds to pay loans or to issue bonds (see Section 13.3).

Individuals also use sinking funds. For example, if you want to have $10,000 for a down payment on a house in five years and can invest at 6% compounded monthly, you would want a sinking fund.

Calculating the Payment for a Sinking Fund

The question in sinking fund problems, then, is to find the payment necessary to reap the desired sum at the end of the period. You can do this easily by following these steps:

Steps: Calculating the Payment for a Sinking Fund

STEP 1. *Find* the number of payments, *n*.

$$n = \text{no. payments per year} \times \text{no. years}$$

STEP 2. *Find* the interest rate per period.

$$r = \frac{\text{annual rate}}{\text{no. payments per year}}$$

STEP 3. Using the Sinking Fund table, *find* the row using *n* from step 1 and the column using *r* from step 2.

STEP 4. *Multiply* the table value times the required amount and round.

For example, Bits 'n' Bytes needs to raise $20,000 in five years. If its sinking fund account yields 12% interest compounded quarterly, what is the required quarterly payment?

Step 1. $n = \text{no. payments per year} \times \text{no. years} = 4 \times 5 = 20$

Step 2. $r = \dfrac{\text{annual rate}}{\text{no. payments per year}} = \dfrac{12\%}{4} = 3\%$

Step 3. Sinking Fund table value = 0.0372157

Step 4. Payment = $0.0372157 \times \$20,000 = \$744.31$ rounded

433

SINKING FUND TABLE
Sinking Fund of $1

n*	0.50%	1.00%	1.50%	2.00%	3.00%	4.00%	5.00%	6.00%
1	1.0000000	1.0000000	1.0000000	1.0000000	1.0000000	1.0000000	1.0000000	1.0000000
2	0.4987531	0.4975124	0.4962779	0.4950495	0.4926108	0.4901961	0.4878049	0.4854369
3	0.3316722	0.3300221	0.3283830	0.3267547	0.3235304	0.3203485	0.3172086	0.3141098
4	0.2481328	0.2462811	0.2444448	0.2426238	0.2390270	0.2354900	0.2320118	0.2285915
5	0.1980100	0.1960398	0.1940893	0.1921584	0.1883546	0.1846271	0.1809748	0.1773964
6	0.1645955	0.1625484	0.1605252	0.1585258	0.1545975	0.1507619	0.1470175	0.1433626
7	0.1407285	0.1386283	0.1365562	0.1345120	0.1305064	0.1266096	0.1228198	0.1191350
8	0.1228289	0.1206903	0.1185840	0.1165098	0.1124564	0.1085278	0.1047218	0.1010359
9	0.1089074	0.1067404	0.1046098	0.1025154	0.0984339	0.0944930	0.0906901	0.0870222
10	0.0977706	0.0955821	0.0934342	0.0913265	0.0872305	0.0832909	0.0795046	0.0758680
11	0.0886590	0.0864541	0.0842938	0.0821779	0.0780774	0.0741490	0.0703889	0.0667929
12	0.0810664	0.0788488	0.0766800	0.0745596	0.0704621	0.0665522	0.0628254	0.0592770
13	0.0746422	0.0724148	0.0702404	0.0681184	0.0640295	0.0601437	0.0564558	0.0529601
14	0.0691361	0.0669012	0.0647233	0.0626020	0.0585263	0.0546690	0.0510240	0.0475849
15	0.0643644	0.0621238	0.0599444	0.0578255	0.0537666	0.0499411	0.0463423	0.0429628
16	0.0601894	0.0579446	0.0557651	0.0536501	0.0496108	0.0458200	0.0422699	0.0389521
17	0.0565058	0.0542581	0.0520797	0.0499698	0.0459525	0.0421985	0.0386991	0.0354448
18	0.0532317	0.0509820	0.0488058	0.0467021	0.0427087	0.0389933	0.0355462	0.0323565
19	0.0503025	0.0480518	0.0458785	0.0437818	0.0398139	0.0361386	0.0327450	0.0296209
20	0.0476665	0.0454153	0.0432457	0.0411567	0.0372157	0.0335818	0.0302426	0.0271846
21	0.0452816	0.0430308	0.0408655	0.0387848	0.0348718	0.0312801	0.0279961	0.0250045
22	0.0431138	0.0408637	0.0387033	0.0366314	0.0327474	0.0291988	0.0259705	0.0230456
23	0.0411347	0.0388858	0.0367308	0.0346681	0.0308139	0.0273091	0.0241368	0.0212785
24	0.0393206	0.0370735	0.0349241	0.0328711	0.0290474	0.0255868	0.0224709	0.0196790
25	0.0376519	0.0354068	0.0332635	0.0312204	0.0274279	0.0240120	0.0209525	0.0182267
26	0.0361116	0.0338689	0.0317320	0.0296992	0.0259383	0.0225674	0.0195643	0.0169043
27	0.0346856	0.0324455	0.0303153	0.0282931	0.0245642	0.0212385	0.0182919	0.0156972
28	0.0333617	0.0311244	0.0290011	0.0269897	0.0232932	0.0200130	0.0171225	0.0145926
29	0.0321291	0.0298950	0.0277788	0.0257784	0.0221147	0.0188799	0.0160455	0.0135796
30	0.0309789	0.0287481	0.0266392	0.0246499	0.0210193	0.0178301	0.0150514	0.0126489
40	0.0226455	0.0204556	0.0184271	0.0165557	0.0132624	0.0105235	0.0082782	0.0064615
50	0.0176538	0.0155127	0.0135717	0.0118232	0.0088655	0.0065502	0.0047767	0.0034443
60	0.0143328	0.0122444	0.0103934	0.0087680	0.0061330	0.0042018	0.0028282	0.0018757
70	0.0119666	0.0099328	0.0081724	0.0066676	0.0043366	0.0027451	0.0016992	0.0010331
80	0.0101970	0.0082189	0.0065483	0.0051607	0.0031117	0.0018141	0.0010296	0.0005725
90	0.0088253	0.0069031	0.0053211	0.0040460	0.0022556	0.0012078	0.0006271	0.0003184
100	0.0077319	0.0058657	0.0043706	0.0032027	0.0016467	0.0008080	0.0003831	0.0001774

*n represents the number of payments.

You should always check your answer by trying it in the formula for any ordinary annuity since sinking fund payments are usually made at the end of the time period.

Check: **Step 1.** $n = (4 \times 5) - 1 = 20 - 1 = 19$

Step 2. $r = \dfrac{12\%}{4} = 3\%$

Step 3. Annuity Due table value is 25.8703745. $25.8703745 + 1 = 26.8703745$

Step 4. Amount = $26.8703745 \times \$744.31 = \$19,999.89$ rounded

The $0.11 difference between the desired $20,000 and the answer in step 4 of the check is the result of rounding.

YOUR TURN

| WORK THIS PROBLEM |

The Question:

Your company needs to raise $15,000 in six years.

(a) Calculate the required semiannual payments into a sinking fund that yields 10% compounded semiannually.

(b) Check your answer using an ordinary annuity calculation.

| ✔ YOUR WORK |

The Solution:

(a) *Step 1.* $n = 2 \times 6 = 12$

    Step 2. $r = \dfrac{10\%}{2} = 5\%$

    Step 3. Sinking Fund table value = 0.0628254

    Step 4. Payment = 0.0628254 × $15,000 = $942.38 rounded

(b) *Step 1.* $n = (2 \times 6) - 1 = 12 - 1 = 11$

    Step 2. $r = \dfrac{10\%}{2} = 5\%$

    Step 3. Annuity Due table value is 14.9171265.   14.9171265 + 1 = 15.9171265

    Step 4. Amount = 15.9171265 × $942.38 = $14,999.98 rounded

You can use a calculator to solve sinking fund problems, just as banks do.

Finding Sinking Fund Payments Using a Calculator

You can find the payments for a sinking fund without using a table if your calculator has $\boxed{y^x}$ and $\boxed{\tfrac{1}{x}}$ keys. Just use this formula:

$$\text{Payment} = \frac{A \times r}{(1 + r)^n - 1}$$

where A = amount required
      n = number of payments
      r = interest rate per payment period

For example, let's calculate the semiannual payment needed to yield $15,000 in six years with 10% interest compounded semiannually.

10% ÷ 2 = 5%

$$\text{Payment} = \frac{\$15,000 \times 0.05}{(1 + 0.05)^{12} - 1} = \frac{\$15,000 \times 0.05}{1.05^{12} - 1}$$

Just punch in

1.05 $\boxed{y^x}$ 12 $\boxed{=}$ $\boxed{-}$ 1 $\boxed{=}$ $\boxed{\frac{1}{x}}$ $\boxed{\times}$ 15000 $\boxed{\times}$ 0.05 $\boxed{=}$ $\longrightarrow$ 942.38115

Rounded, this comes to $942.38, the same answer we got using the Sinking Fund table earlier.

Now check your understanding of sinking funds with the problems in Section Test 13.2.

SECTION TEST 13.2 Investments

The following problems test your understanding of Section 13.2, Sinking Funds.

A. Calculate the payment required for the sinking fund using the Sinking Fund table.

	Required Amount	Payment Period	Annual Rate	Time	Required Payment
1.	$2000	Monthly	6%	2 years	
2.	5300	Monthly	6%	5 years	
3.	4500	Semiannual	8%	3 years	
4.	6250	Semiannual	12%	5 years	
5.	3500	Quarterly	8%	2 years	
6.	2900	Quarterly	8%	3 years	
7.	8000	Annual	6%	8 years	
8.	1200	Annual	6%	14 years	
9.	6000	Semiannual	10%	10 years	
10.	8250	Semiannual	8%	15 years	
11.	5800	Monthly	12%	2 years	
12.	3000	Monthly	6%	5 years	
13.	7500	Semiannual	8%	3 years	
14.	6250	Semiannual	10%	5 years	
15.	3250	Quarterly	8%	6 years	
16.	4900	Quarterly	12%	3 years	

B. Solve these problems. Use the Sinking Fund table or a calculator.

1. The Garden of Earthly Delights borrowed $12,000 to build a new greenhouse. The company is setting up a sinking fund to repay the loan in five years. The account yields 12% interest compounded monthly. What are the required monthly payments?

2. By the Byte Software wants to borrow $21,000 to build a new branch. If the company sets up a sinking fund to repay the loan in six years, and the account yields 12% interest compounded quarterly, what will be the required quarterly payments?

3. The Big Valley School District needs to pay off a $45,000 bond in 13 years. The interest rate is 10% compounded semiannually. What are the required semiannual payments?

4. Big Ben's Bolts borrowed $31,500 to add metric machinery. The company is setting up a sinking fund to repay the loan in ten years. The account yields 8% interest compounded quarterly. What are the required quarterly payments?

5. Ava's Air Tours is considering borrowing $28,000 to add a new aircraft to the fleet. If the company sets up a sinking fund to repay the loan in five years, and the account yields 8% interest compounded semiannually, what are the required semiannual payments?

6. The Bundy Shoe Store borrowed $7,000 for remodeling. The company is setting up a sinking fund to repay the loan in two years. The account yields 12% interest compounded monthly. What are the required monthly payments?

7. Roy and Dale are considering borrowing $32,000 to start Silver Hills Ranch. They want to set up a sinking fund to repay the loan in five years. If the account yields 8% interest compounded semiannually, what will the required semiannual payments be?

8. Tread on Us Shoe Manufacturers borrowed $18,000 to build a company day care center. The company is setting up a sinking fund to repay the loan in four years. The account yields 8% interest compounded semiannually. What are the required semiannual payments?

9. Bob's Beeflike Burgers wants to borrow $31,000 to build two new burger shops. The company will set up a sinking fund to repay the loan in six years. The account yields 8% interest compounded quarterly. What will be the required quarterly payments?

10. The Wild Child Recording Company borrowed $54,000 to build a second sound studio. The company is setting up a sinking fund to repay the loan in five years. The account yields 8% interest compounded semiannually. What are the required semiannual payments?

11. Maria's Mannequins wants to open a second outlet on the North Side. If the company borrows $27,000, and sets up a sinking fund to repay the loan in five years and the account yields 10% interest compounded semiannually, what will the required semiannual payments be?

12. The Stunning Lawn Ornament Company borrowed $13,000 to purchase a giant flamingo mold. What will the required monthly payments be if the company is setting up a sinking fund to repay the loan in five years with an account that yields 12% interest compounded monthly?

13. Intense Tents borrowed $8,000 to build an indoor camping display. The company is setting up a sinking fund to repay the loan in three years. The account yields 12% interest compounded quarterly. What are the required quarterly payments?

14. Water Babies, Inc., wants to borrow $24,000 to build a hydroponic garden. The company is setting up a sinking fund to repay the loan in six years. The account yields 10% interest compounded semiannually. What are the required semiannual payments?

15. Awesome Mountaineering Expeditions borrowed $12,000 for a Himalayan expedition. The group is setting up a sinking fund to repay the loan in two years. The account yields 12% interest compounded monthly. What are the required monthly payments?

SECTION 13.3: Stocks and Bonds

Can You:

Find dividends for preferred stocks?
Calculate dividends for cumulative preferred stocks?
Determine dividends for common stock?
Buy stock?
Determine whether a bond is selling at a discount or a premium?
Calculate stock dividends?
Find the current yield of a bond?
. . . If not, you need this section.

Understanding Stocks

Stock   A share of ownership in a corporation.

Corporations issue **stock** in order to raise money. Stock is issued in shares, with each share representing a portion of ownership in the company. For example, if the Widget Company has issued a total of 1000 shares of stock and you own 20 shares, you then own $\frac{20}{1000}$ or $\frac{1}{50}$ of the company. You share proportionally in company growth or decline, and in making corporate decisions.

Dividend   A portion of
the profits of a corporation
paid out to stockholders in
proportion to the number
of shares they own.

Stock ownership offers other benefits. When the company makes a profit, it may declare a **dividend** to the stockholders, passing the profit on. Stockholders are then entitled to a dividend for each share owned.

Par Value   A value
assigned to a stock when it
is issued; it is used only for
accounting purposes.

A company may issue a few hundred shares of stock or millions of shares. When the stock is issued, the company assigns a value called the **par value.** The par value is used only for accounting purposes. The actual value of the stock depends not on the par value but on the supply of and demand for the stock on the "stock market."

Common Stock   Voting
stock in a corporation that
also entitles the holder to
dividends, when declared
by the company.

There are two basic types of stock, common and preferred. Most stock issued is **common stock.** Each share of common stock entitles the owner to one vote during company elections. If you own 20 shares of a company's stock, then you are entitled to 20 votes. Stock ownership also entitles you to dividends, when declared by the company.

Preferred Stock   Nonvot-
ing stock in a corporation
that entitles the holder to
first claim on any dividends
and in "cashing in" the
stock.

However, **preferred stock** has precedence over common stock in the payment of dividends. In case of company *liquidation* (going out of business), preferred stockholders get preference over common share stockholders when "cashing in" their stock. Preferred stockholders usually do not have voting privileges, though.

Finding Dividends for Preferred Stock

The dividend rate of preferred stock is often fixed and stated as a percent of the par value. Dividends generally do not exceed the stated rate.

> Dividend = rate × par value

For example, Big Ben's Bolts gives a 6% dividend on its preferred stock, based on a $50 par value. What is the total dividend on ten shares of stock?

439

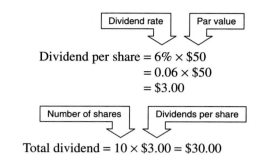

$$\text{Dividend per share} = 6\% \times \$50$$
$$= 0.06 \times \$50$$
$$= \$3.00$$

$$\text{Total dividend} = 10 \times \$3.00 = \$30.00$$

YOUR TURN

WORK THIS PROBLEM

The Question:

Bits 'n' Bytes gives a $6\frac{1}{2}\%$ dividend on its preferred stock, based on a $70 par value. Calculate the total dividend on 325 shares of stock.

✔ YOUR WORK

The Solution:

$$\text{Dividend per share} = 6\tfrac{1}{2}\% \times \$70 = 0.065 \times \$70 = \$4.55$$
$$\text{Total dividend} = 325 \times \$4.55 = \$1478.75$$

Calculating Dividends for Cumulative Preferred Stock

Cumulative Preferred Stock A form of preferred stock on which unpaid dividends must be made up before any other dividends may be dispersed to other shareholders.

When a company makes little or no profit, it may be unable to declare a dividend. But with **cumulative preferred stock,** unpaid dividends are made up in later years, if a profit is made then. For example, if a firm makes no profit one year but does make a profit the next, it must pay holders of cumulative preferred stock two years' worth of dividends then. Cumulative preferred stockholders are paid in full before common share stockholders receive any dividends.

Steps: Calculating Dividends for Cumulative Preferred Stock

STEP 1. *Calculate* the dividend per share.

$$\text{Dividend per share} = \text{rate} \times \text{par value}$$

STEP 2. *Calculate* the total dividend.

$$\text{Total dividend} = \text{no. of shares} \times \text{dividend per share} \times \text{no. of years}$$

For example, Bits 'n' Bytes didn't give a dividend last year to stockholders with cumulative preferred shares that normally pay $6\frac{1}{2}\%$ on a par value of \$70. This year, the company is giving two years' dividends. What is the total dividend for 140 shares of stock?

Step 1. Dividend per share = rate × par value
$$= 6\tfrac{1}{2}\% \times \$70 = 0.065 \times \$70 = \$4.55$$

Step 2. Total dividend = no. of shares × dividend per share × no. of years
$$= 140 \times \$4.55 \times 2 = \$1274$$

YOUR TURN

| WORK THIS PROBLEM |

The Question:

Woody's Lumber was unable to give a dividend last year to its stockholders with cumulative preferred shares that normally pay 5% on a par value of \$30. This year the company is giving two years' dividends. Calculate the total dividend for 250 shares of stock.

| ✔ YOUR WORK |

The Solution:

Step 1. Dividend per share = 5% × \$30 = 0.05 × \$30 = \$1.50

Step 2. Total dividend = 250 × \$1.50 × 2 = \$750

Determining Dividends for Common Stock

Stock dividends for common stock are not based on the par value or the market value. They are simply a certain amount per share.

$$\text{Total dividend} = \text{number of shares} \times \text{dividend per share}$$

For example, if you hold 100 shares of stock in the Widget Corp. and the company decides to pay a \$0.50 per share dividend, you get

$$\text{Total dividend} = 100 \times \$0.50 = \$50.00$$

YOUR TURN

> ┌─ WORK THIS PROBLEM ───
>
> *The Question:*
>
> Woody's Lumber has declared a common stock dividend of 12 cents per share. If you own 200 shares, what is your total dividend?
>
> ┌─ ✔ YOUR WORK ──
>
> *The Solution:*
>
> Total dividend = 200 × $0.12 = $24

Calculating Stock Dividends

Stock Dividend A dividend in the form of additional stock rather than cash.

Sometimes companies decide to retain their dividends to provide funds for company expansion. In such cases, they may declare a **stock dividend,** where each stockholder receives a dividend of additional stock rather than cash.

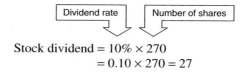

Stock dividend = dividend rate × number of shares

For example, the Rock Insurance Company has declared a 10% stock dividend. Ruby Reddy has 270 shares of stock. Calculate the number of additional shares of stock she is issued.

| Dividend rate | Number of shares |

Stock dividend = 10% × 270
= 0.10 × 270 = 27

YOUR TURN

> ┌─ WORK THIS PROBLEM ───
>
> *The Question:*
>
> Bob's Beeflike Burgers has declared a 5% stock dividend. If you have 380 shares of stock, how many additional shares will you receive?
>
> ┌─ ✔ YOUR WORK ──
>
> *The Solution:*
>
> Stock dividend = 5% × 380 = 0.05 × 380 = 19 shares

If the number of shares of stock to be issued isn't a whole number, the shares are rounded down. For example, if the number of shares calculated is 19.8, only 19 shares are issued. A cash amount is usually issued in lieu of the additional 0.8 shares.

Paying for Stock

Stock is usually purchased and sold by a *broker* at a *stock exchange,* such as the New York Stock Exchange. For these services, the broker charges a commission based on the number of shares of stock "traded" (bought or sold) and the total price.

Round Lot The purchase of stock in a lot that is a multiple of 100 shares.

Stocks are traded in either round lots or odd lots. A **round lot** is a multiple of a 100, such as 100, 200, 300, and so on. An **odd lot** is any number of shares not a multiple of 100.

Odd Lot The purchase of stock in a lot that is *not* a multiple of 100 shares.

Traditionally, stock prices are usually stated in eighths of a dollar, such as $42\frac{3}{8}$. You can find out the current value of a specific firm's stock from your broker or the current stock listing. The listing of all stocks on the New York Stock Exchange appears daily in most major newspapers. One of the most comprehensive listings of stock and bond transactions appears in *The Wall Street Journal.*

TYPICAL STOCK LISTING

① Yearly		②	③	④	⑤	⑥	⑦	⑧	⑨
High	**Low**	**Stock**	**Dividends**	**P/E**	**Sales in 100s**	**High**	**Low**	**Close**	**Change**
$56\frac{7}{8}$	$37\frac{1}{8}$	Coca-Cola	1.74	24	5493	$52\frac{3}{8}$	$50\frac{3}{4}$	$51\frac{1}{4}$	$+\frac{1}{4}$
$50\frac{3}{8}$	$30\frac{1}{8}$	General Motors	1.60	5	8782	$33\frac{7}{8}$	$33\frac{3}{8}$	$33\frac{1}{2}$	$+\frac{1}{8}$
$139\frac{5}{8}$	96	IBM	4.84	11	11712	$107\frac{3}{8}$	$103\frac{3}{4}$	$104\frac{1}{4}$	$-1\frac{7}{8}$
$38\frac{7}{8}$	$25\frac{3}{4}$	McDonald's	.34	15	7589	$34\frac{1}{2}$	$32\frac{3}{4}$	$33\frac{1}{8}$	$-\frac{1}{8}$

① The yearly high and low.

② This column contains the company's name, often abbreviated.

③ The annual dividend, if any. Coca-Cola's annual dividend for this year is $1.74.

④ P/E is the **price-earnings ratio.** It is the closing price of the stock divided by the earnings per share.

Price-Earnings Ratio The closing price of the stock on a given day divided by its earnings per share.

$$\text{P/E ratio} = \frac{\text{closing price}}{\text{earnings per share}}$$

The earnings per share is not the dividend per share. The P/E ratio for Coca-Cola is 24; that is, the current selling price is 24 times the annual earnings per share.

⑤ The day's sales in 100-lot shares. Today's sales in Coca-Cola stock totaled 5493×100 or 549,300 shares.

⑥ The high price of the day.

⑦ The low price of the day.

⑧ The closing price of the day.

⑨ The change from the previous day's closing price. Coca-Cola closed at $51\frac{1}{4}$, up $\frac{1}{4}$ from the previous day.

The total cost of a stock purchase (excluding commission and taxes) is the product of the number of shares and the cost per share.

$$\text{Total cost} = \text{number of shares} \times \text{cost per share}$$

For example, if you agreed to buy 25 shares of stock in Disney Corporation at $120\frac{1}{4}$ per share, your total cost for the stock (not counting any brokerage fees) would be

$$\text{Total cost} = \text{no. shares} \times \text{cost per share}$$
$$= 25 \times \$120\tfrac{1}{4}$$
$$= 25 \times \$120.25 = \$3,006.25$$

YOUR TURN

WORK THIS PROBLEM

The Question:

Calculate the cost of 300 shares of stock in McDonald's Corporation at $\$33\tfrac{1}{8}$ per share.

✔ YOUR WORK

The Solution:

$$\text{Total cost} = 300 \times \$33\tfrac{1}{8} = 300 \times \$33.125 = \$9,937.50$$

Understanding Bonds

Bond A long-term corporate or government debt that carries a stated interest rate and may be bought and sold.

In addition to selling stock, a company can raise money by selling bonds. A **bond** is a long-term debt with a stated interest rate.

Most bonds have a par value or face value of $1,000. As with stock, the value of a bond can fluctuate. The price of a bond is usually stated as a percent of the par value. Government bodies also issue bonds. Municipal (city) bonds are often issued at a lower interest rate than corporate bonds, but they are exempt from federal income tax and certain local taxes. These tax-exempt bonds can be a great savings, especially if you are in a high income tax bracket. And you may own U.S. (federal government) savings bonds.

The selling price of bonds is listed as a percent of their par value. When a bond is listed as selling for 82, it means the price is 82% of the par value, not necessarily $82.

Bond price = percent listing × par value

A percent figure over 100 indicates that the bond is selling at a *premium* because it is in high demand. A percent figure of less than 100 indicates that the bond is selling at a *discount* because investors think the part value is too high.

For example, the Rock Insurance Company has issued a corporate bond with a par value of $1000 that currently lists for 82. Calculate the current price of the bond.

Percent   Par value

$$Price = 82\% \times \$1000$$
$$= 0.82 \times \$1000 = \$820$$

YOUR TURN

WORK THIS PROBLEM

The Question:

Calculate the current price of an Evansville City municipal bond with a par value of $1000 and selling at $102\frac{1}{2}\%$. Is this bond selling at a discount or a premium?

✔ YOUR WORK

The Solution:

$$Price = 102\tfrac{1}{2}\% \times \$1000 = 1.025 \times \$1000 = \$1025$$

The bond is currently selling at a premium.

Bonds are a loan to the company and simple interest is paid based on the par value. The interest rate is fixed and specified on the bond. The interest on a bond is calculated using the simple interest formula $I = Prt$. (For a review of simple interest, see Chapter 9.)

Let's calculate the annual interest on a DuPont $1000 bond at 8%.

$$I = principal \times rate \times time$$
$$= \$1000 \times 8\% \times 1$$
$$= \$1000 \times 0.08 \times 1 = \$80$$

YOUR TURN

WORK THIS PROBLEM

The Question:

What is the annual interest on an Amoco $500 bond at $7\frac{1}{2}\%$?

✔ YOUR WORK

The Solution:

$$I = \$500 \times 7\tfrac{1}{2}\% \times 1$$
$$= \$500 \times 0.075 \times 1 = \$37.50$$

Registered Bond A bond whose ownership is registered with the issuing company, which pays the interest due directly to the owner.

Interest payments on bonds are usually made twice a year or semiannually. How you collect the interest depends on the bond type. With a **registered bond,** your name appears on the bond and

Coupon (Bearer) Bond
A bond from which the holder must periodically clip and mail a coupon to the issuer in order to receive the interest due.

is registered with the company issuing the bond. In this case, the interest is paid directly to you by the company. With a **coupon bond** (**bearer bond**), you "clip" a coupon attached to the bond and collect your interest due.

In the case of the DuPont bond, the semiannual payment is

Annual interest

$$\text{Semiannual interest} = \frac{\$80}{2} = \$40$$

YOUR TURN

| WORK THIS PROBLEM |

The Question:

What is the semiannual interest on the $500 Amoco bond at $7\frac{1}{2}\%$ for which we found an annual interest of $37.50 in the previous "Your Turn"?

| ✔ YOUR WORK |

The Solution:

$$\text{Semiannual interest} = \frac{\$37.50}{2} = \$18.75$$

Much like stocks, bonds are sold through a *bond market.* The listing of bonds appears in many newspapers including *The Wall Street Journal.* The following is a sample bond listing.

TYPICAL BOND LISTING

① Bond	② Sales in $1000	③ High	④ Low	⑤ Close	⑥ Change
DuPont 8s98	30	$101\frac{3}{8}$	$100\frac{3}{4}$	$100\frac{3}{4}$	$-\frac{5}{8}$
Goodyear 8.60s99	10	$98\frac{5}{8}$	$98\frac{3}{8}$	$98\frac{1}{2}$	$+\frac{1}{8}$
Texaco $8\frac{7}{8}$s05	19	102	101	102	-1

① The first column contains the company's name, bond interest rate, and the maturity date of the bond. DuPont's bond yields 8% interest, and the bond matures or is paid off by the company in 1998. The Texaco bonds pay $8\frac{7}{8}\%$ interest and mature in 2005.

② The daily sales in units of one thousand dollars. Goodyear sold $10,000 in bonds.

③ The high price of the day expressed as a percent.

④ The low price of the day expressed as a percent.

⑤ The closing price of the day expressed as a percent.

⑥ The change from the previous day's closing price expressed as a percent. Texaco bonds dropped 1% from yesterday's closing.

WORK THIS PROBLEM

The Question:

Using the closing price from our sample bond listing, calculate the price of a Texaco bond.

✔ YOUR WORK

The Solution:

$$\text{Price} = 102\% \times \$1000 = 1.02 \times \$1000 = \$1020$$

Determining Current Yield

Because the interest earned by the bond is based on the par value rather than the current price, we need a way to determine the actual earning capacity of the bond.

Current Yield The actual earning capacity of a bond.

One measure of earnings is a bond's **current yield.** You can find the current yield by following these steps:

Steps: Calculating Current Yield of a Bond

STEP 1. *Calculate* the annual interest on the bond.

STEP 2. *Calculate* the current price of the bond.

$$\text{Bond price} = \text{percent listing} \times \text{par value}$$

STEP 3. *Calculate* the current yield.

$$\text{Current yield} = \frac{\text{annual interest}}{\text{current price of bond}}$$

For example, let's calculate the current yield of a $1000 DuPont bond using the data in the Typical Bond Listing chart.

Step 1. $I = P \times r \times t$
$$= \$1000 \times 8\% \times 1 = \$1000 \times 0.08 \times 1 = \$80$$

Step 2. Bond price = closing percent listing × par value
$$= 100\tfrac{3}{4}\% \times \$1000 = 1.0075 \times \$1000 = \$1007.50$$

Step 3. Current yield = $\dfrac{\text{annual interest}}{\text{current price of bond}}$

$$= \dfrac{\$80.00}{\$1007.50} = 0.0794 \text{ rounded} = 7.94\%$$

The current yield of the bond is 7.9%, less than the annual interest rate of 8%, because the bond is selling for a premium.

YOUR TURN

WORK THIS PROBLEM

The Question:

Calculate the current yield of a $1000 Goodyear bond using the data in the Typical Bond Listing chart.

✔ **YOUR WORK**

The Solution:

Step 1. $I = \$1000 \times 8.6\% \times 1 = \$1000 \times 0.086 \times 1 = \$86$

Step 2. Price = $98\tfrac{1}{2}\% \times \$1000 = 0.985 \times \$1000 = \$985$

Step 3. Current yield = $\dfrac{\$86}{\$985} = 0.0873 \text{ rounded} = 8.73\%$

The current yield of 8.73% is higher than the bond's interest rate of 8.6% because the bond is selling at a discount.

As in so many areas of business, a calculator comes in handy when determining current yield on a bond.

Using a Calculator to Find Current Yield of a Bond

It's simpler to use the calculator to find the current yield of a bond.

For the DuPont bond example:

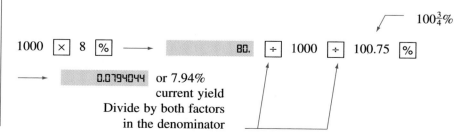

For the Goodyear bond example:

1000 ⊠ 8.6 % ⟶ 86. ÷ 1000 ÷ 98.5 %

⟶ 0.0873096 or 8.73% current yield

Divide by both factors in
the denominator

For a set of practice problems on stocks and bonds, see Section Test 13.3.

WORKSPACE

SECTION TEST 13.3   Investments

The following problems test your understanding of Section 13.3, Stocks and Bonds.

1. Experimental Book Technologies pays a $7\frac{1}{2}$% dividend on its preferred stock, based upon a $75 par value. If you own 120 shares of stock, what is your total dividend?

2. Ms. Reed owns 75 shares of Wild Child Recording Company stock. If the company pays a $5\frac{1}{4}$% dividend on its preferred stock, based on a $80 par value, what is Ms. Reed's total dividend?

3. Because of high wheat prices, Our Daily Food was unable to pay a dividend last year to its stockholders with cumulative preferred 6.2%, par value $50. This year wheat prices are down and the company is paying two years of dividends. What is the total dividend for 170 shares?

4. Because a new ultrasonic pasteurizing machine kept curdling the milk, the Moore Milk Company was unable to pay a dividend last year to its stockholders with cumulative preferred 7.3%, par value $120. This year the machine is working well and the company is paying two years of dividends. What is the total dividend for 175 shares?

5. American Express paid a $0.92 per share common stock dividend. What is the total dividend for 130 shares?

6. Pacific Gas and Electric paid a $1.64 per share common stock dividend. What is the total dividend for 250 shares?

7. AT&T paid a $1.32 per share common stock dividend. What is the total dividend for 175 shares?

8. Hilton paid a $1.20 per share common stock dividend. What is the total dividend for 80 shares?

9. The Patagonia Outfitters Company has declared a 10% stock dividend. If Nigel Olson owns 140 shares, how many additional shares of stock are issued with the dividend?

10. Bob Fulton owns 120 shares of Stanley Steamer stock. If the company declares a 25% stock dividend, how many additional shares of stock will be issued with the dividend?

11. The Oceangate Technology Company has declared a 15% stock dividend. If you own 80 shares, how many additional shares will be issued with the dividend?

12. Earl Sinclair owns 150 shares of North American Petroleum stock. The company has declared a 40% dividend. How many additional shares of stock will be issued with the dividend?

13. What are the annual interest and semiannual payment on an AT&T $1000 bond with $8\frac{3}{4}$% interest?

14. What are the annual interest and semiannual payment on a Pacific Gas and Electric $1000 bond with $9\frac{1}{8}$% interest?

15. What are the annual interest and semiannual payment on an International Business Machines $1000 bond with $9\frac{3}{8}$% interest?

16. What are the annual interest and semiannual payment on a Bell Telephone of Pennsylvania $1000 bond with $9\frac{5}{8}$% interest?

17. A Southwestern Bell $1000 bond with $8\frac{1}{4}$% interest has a current price of $91\frac{1}{2}$. What is the current yield? Round your answer to the nearest 0.01%.

18. An International Paper Company $1000 bond with 8.85% interest has a current price of $100\frac{1}{4}$. What is the current yield? Round your answer to the nearest 0.01%.

19. A Pacific Telephone and Telegraph $1000 bond with $8\frac{7}{8}$% interest has a current price of $97\frac{1}{2}$. What is the current yield? Round your answer to the nearest 0.01%.

20. A Service Merchandise $1000 bond with $11\frac{3}{4}$% interest has a current price of $95\frac{7}{8}$. What is the current yield? Round your answer to the nearest 0.01%.

21. Andi & Brian own 1250 shares of common stock in Glassride Enterprises. The company declared a 20% stock dividend. How many shares of stock will Andi & Brian own altogether now?

Key Terms

annuity
annuity due
bond
common stock
coupon (bearer) bond
cumulative preferred stock

current yield
dividend
odd lot
ordinary annuity
par value
preferred stock

price-earnings ratio
registered bond
round lot
sinking fund
stock
stock dividend

Chapter 13 at a Glance

Page	Topic	Key Point	Example
424	Annuities due	n = no. periods per year × no. years $$\text{Rate per period} = \frac{\text{annual rate}}{\text{no. of periods per year}}$$ Find the value in the Annuity Due table. Value = table value × amount per payment	To find the value of an annuity due after five years with monthly payments of $50 and 6% interest compounded monthly: $n = 12 \times 5 = 60$ Rate = 6%/12 = $\frac{1}{2}$% Table value = 70.1188807 Value = 70.1188807 × $50 = $3505.94 rounded
427	Ordinary annuities	n = no. periods per year × no. years $$\text{Rate per period} = \frac{\text{annual rate}}{\text{no. of periods per year}}$$ Find the value in the Annuity Due table and add 1. Value = (table value + 1) × amount per payment	To find the value of an ordinary annuity after two years with monthly payments of $50 and 6% interest compounded monthly: $n = (12 \times 2) - 1 = 24 - 1 = 23$ Rate = 6%/12 = $\frac{1}{2}$% Table value = 24.4319552 + 1 = 25.4319552 Value = 25.4319552 × $50 = $1271.60 rounded
433	Sinking funds	n = no. periods per year × no. years $$\text{Rate per period} = \frac{\text{annual rate}}{\text{no. payments per year}}$$ Find the value in the Sinking Fund table. Payment = table value × required amount	What is the required quarterly payment into a sinking fund that yields 12% interest compounded quarterly in order to wind up with $20,000 in five years? $n = 4 \times 5 = 20$ $r = 12\%/4 = 3\%$ Table value = 0.0372157 Payment = 0.0372157 × $20,000 = $744.31 rounded
439	Dividends on preferred stock	Dividend per share = rate × par value	What is the value of a 6% dividend on a stock with a par value of $50/share? Dividend per share = 0.06 × $50 = $3.00

Page	Topic	Key Point	Example
440	Dividends on cumulative preferred stock	Dividend per share = rate × par value Total dividend = no. shares × dividend per share × no. years	What is the total dividend due the holder of 140 shares of cumulative preferred stock with a par value of $70 on which two years' dividends of $6\frac{1}{2}\%$ are due? Dividend per share = 0.065 × $70 = $4.55 Total dividend = 140 × $4.55 × 2 = $1274.00
441	Dividends on common stock	Total dividend = no. shares × dividend per share	What is the total dividend due on 100 shares of common stock with a $0.50 per share dividend? Total dividend = 100 × $0.50 = $50
442	Stock dividends	Stock dividend = dividend rate × no. shares	How many shares of stock are due the holder of 270 shares if the firm declares a stock dividend of 10%? Stock dividend = 0.10 × 270 = 27
442	Stock purchases	Total cost = no. shares × cost per share	What is the total cost on a purchase of 25 shares at $120\frac{1}{4}$ per share? Total cost = 25 × $120.25 = $3006.25
444	Bond prices	Bond price = percent listing × par value	What is the current price of a bond with a par value of $1000 that currently lists for 82? Bond price = 0.82 × $1000 = $820
445	Interest on a bond	Bond interest = Prt	What is the interest on a $1000 bond at 8%? Bond interest = $1000 × 0.08 × 1 = $80
446	Semiannual interest on a bond	$\text{Semiannual interest} = \dfrac{\text{annual interest}}{2}$	What is the semiannual interest on a $1000 bond at 8%? Annual bond interest = $80 (see previous example) Semiannual interest = $\dfrac{\$80}{2}$ = $40

Page	Topic	Key Point	Example
447	Current yield of a bond	Annual interest = Prt Current bond price = percent list × par value Current yield = $\dfrac{\text{annual interest}}{\text{current price of bond}}$	What is the current yield on a $1000 bond at 8% that closed at $100\frac{3}{4}$? Annual interest = $80 (see previous example) Current bond price = 1.0075 × $1000 $= \$1007.50$ Current yield = $\dfrac{\$80.00}{\$1007.50}$ $= 0.0794$ rounded $= 7.94\%$

WORKSPACE

SELF-TEST Investments ━━━━━━━━

The following problems test your understanding of investments. Use the Annuity Due and Sinking Fund tables from the chapter.

1. Lily Hellman invested $250 at the beginning of each quarter in an annuity with an interest rate of 8%. What is the amount of her annuity after five years?

2. Irving Washington invested $345 at the beginning of each quarter in an annuity with an interest rate of 6%. What is the amount of the annuity after six years?

3. For six years, Frances Kafka has been investing $280 at the end of each semiannual period in an annuity with an interest rate of 6%. What is the amount of her annuity?

4. If Rachel Lindsay invests $325 at the end of each semiannual period in an annuity with an interest rate of 8% for nine years, what will be the amount of the annuity?

5. Edgar Masters invests $250 at the beginning of each quarter in an annuity with an interest rate of 6%. What is the amount of the annuity after three years?

6. Edna Millay invested $345 at the beginning of each quarter in an annuity with an interest rate of 8%. What is the amount of her annuity after ten years?

7. Marian Moore has invested $280 at the end of each semiannual period in an annuity with an interest rate of 6%. What is the amount of the annuity after ten years?

8. For nine years, Omar Kham has invested $325 at the end of each semiannual period in an annuity with an interest rate of 10%. What is the amount of the annuity after nine years?

9. The Garden of Earthly Delights borrowed $16,000 to put in a new watering system in one greenhouse. The company is setting up a sinking fund to repay the loan in four years. The account yields 12% interest compounded quarterly. What are the required quarterly payments?

10. The Titanic Water Company borrowed $12,000 to add a more thorough filtering system. The company is setting up a sinking fund to repay the loan in five years. The account yields 12% interest compounded monthly. What are the required monthly payments?

11. The Wild Child Record Company needs to soundproof the CEO's office. The company needs to borrow $86,000. If the company sets up a sinking fund to repay the loan in six years and the account yields 12% interest compounded quarterly, what will be the required quarterly payments?

12. The Fly-By-Night Aviary has decided to expand and so has borrowed $12,000. The company is setting up a sinking fund to repay the loan in five years. The account yields 10% interest compounded semiannually. What are the required semiannual payments?

13. Bob and Ted's California Experiences wants to remodel a beach house it rents and is considering borrowing $12,000 and setting up a sinking fund to repay the loan in two years. The account will yield 12% interest compounded monthly. What are the required monthly payments?

14. Woody's Lumber Mill is considering borrowing $9,000 to expand. If the company sets up a sinking fund to repay the loan in three years and the account yields 12% interest compounded quarterly, what will be the required quarterly payments?

15. When business fell off during a brief recession, J.R.'s Construction Company was unable to pay a dividend last year to its stockholders with cumulative preferred $6\frac{1}{2}$%, par value $60. This year the company is giving two years' dividend. What is the total dividend for 15 shares?

16. Owing to an invasion of moths, the Highlander Wool Company was unable to pay a dividend last year to its stockholders with cumulative preferred $4\frac{3}{4}$%, par value $90. This year the company is paying two years' dividend. What is the total dividend for 140 shares?

17. Disney Enterprises paid a $0.70 per share common stock dividend. What is the total dividend for 125 shares?

18. Southwestern Bell paid a $2.84 per share common stock dividend. What is the total dividend for 140 shares?

19. The Wild Child Recording Company has declared a 25% stock dividend. If you have 140 shares, how many additional shares of stock will you be issued?

20. Bill Derrick owns 150 shares of Tinker Toy stock. If the company declares a 20% stock dividend, how many additional shares of stock will be issued with the dividend?

21. What is the (a) annual interest and (b) semiannual payment on a Southwestern Bell $1000 bond with $7\frac{5}{8}$% interest?

22. What is the (a) annual interest and (b) semiannual payment on an Eckerd Drug $1000 bond with $11\frac{1}{8}$% interest?

23. A Wickes Lumber $1000 bond with $11\frac{7}{8}$% interest has a current price of $77\frac{3}{4}$. What is the current yield? Round your answer to the nearest 0.01%.

24. A Wang Computers $1000 bond with 9% interest has a current price of $55\frac{3}{4}$. What is the current yield? Round your answer to the nearest 0.01%.

14

Real Estate Mathematics

CHAPTER OBJECTIVES

When you complete this chapter successfully, you will be able to:

1. Determine mortgage payments and points.

2. Calculate the principal and interest portions of a mortgage payment and generate an amortization schedule.

■ If Shiring Associates finances a $230,000 office building for 15 years at 11%, what will be the monthly payments?

■ The McCoys are negotiating an FHA-insured mortgage for $85,000. The current market rate is $11\frac{3}{4}\%$, and the maximum federal rate is $11\frac{1}{4}\%$. What are the points and the discount amount?

■ Intense Tents is borrowing $175,000 for a new canoe wing. The mortgage will be at 11% for 25 years. Calculate the monthly payment and generate an amortization schedule for the first two payments.

In addition to needing to save money for the future, both businesses and individuals need a "place to live." To buy a home or store or factory, companies and families often need to take out a mortgage. And the happiest day in the life of many people is the day on which the mortgage is paid off.

In this chapter, we will look at the costs of taking out real estate loans over the many years they usually run. You should find this a fun chapter. Its "real" value will be apparent when you're ready to purchase your own "estate."

SECTION 14.1: Mortgage Payments and Points

Can You:

Calculate principal and interest payments on a mortgage?
Determine the number of points on a mortgage?
Find the amount of discount on a mortgage with points?
. . . If not, you need this section.

Mortgage A long-term loan on real estate.

Principal The amount borrowed. (see Section 9.2)

Interest The amount charged by the lender. (see Section 9.2)

With the high cost of real estate, many businesses—and even fewer individuals—can afford to pay the full price for new buildings or stores out of pocket. Fortunately, U.S. banks stand ready to make real estate loans—**mortgages**—as long as they retain title to the property until the mortgage is repaid.

Monthly payments for **principal** and **interest** (P&I) are easy to determine if you use the Mortgage Payment table. (Actually, you can use this table to calculate the payment on any type of installment loan.)

Monthly mortgage payments are often more complex, however. The bank wants to make sure the government doesn't take your property for unpaid taxes and that its investment is protected in the event of fire and other disasters. Thus, in many cases, a monthly "mortgage" payment also includes taxes and insurance fees that the bank then pays to the government or insurance company. These vary widely, however, so in this chapter we will concern ourselves only with mortgage P&I.

Steps: Calculating Mortgage P&I Payments

STEP 1. *Use* the Mortgage Payment table to find the monthly payment per $1000; the number of years is the row, and the interest rate is the column.

STEP 2. *Calculate* the monthly payment.

$$\text{Payment} = \text{table value} \times \frac{\text{principal}}{\$1000}$$

For example, Bits 'n' Bytes is buying a new store and is taking a $140,000 mortgage for 15 years at 11%. What is its monthly P&I payment?

Step 1.   Mortgage Payment table value = 11.3659693

Step 2. $\text{Payment} = \text{table value} \times \dfrac{\text{principal}}{\$1000}$

$= 11.3659693 \times \dfrac{\$140,000}{\$1000}$

$= 11.3659693 \times \$140 = \$1591.24 \text{ rounded}$

MORTGAGE PAYMENT TABLE
Mortgage (or Installment Loan) Monthly P&I Payment per $1000

Years	8%	9%	10%	11%	12%	13%	14%	15%	16%	18%
1	86.9884291	87.4514768	87.9158872	88.3816585	88.8487887	89.3172757	89.7871176	90.2583123	90.7308579	91.6799929
2	45.2272915	45.6847423	46.1449263	46.6078382	47.0734722	47.5418226	48.0128833	48.4866480	48.9631105	49.9241020
3	31.3363655	31.7997327	32.2671872	32.7387171	33.2143098	33.6939520	34.1776298	34.6653285	35.1570330	36.1523955
4	24.4129223	24.8850424	25.3625834	25.8455226	26.3338354	26.8274959	27.3264765	27.8307483	28.3402808	29.3749996
5	20.2763943	20.7583552	21.2470447	21.7424231	22.2444477	22.7530730	23.2682508	23.7899301	24.3180571	25.3934274
6	17.5332406	18.0255372	18.5258378	19.0340790	19.5501925	20.0741052	20.6057395	21.1450133	21.6918406	22.8077911
7	15.5862144	16.0890783	16.6011840	17.1224364	17.6527328	18.1919633	18.7400116	19.2967547	19.8620639	21.0178380
8	14.1366793	14.6502033	15.1741641	15.7084257	16.2528414	16.8072551	17.3715010	17.9454053	18.5287860	19.7232141
9	13.0187149	13.5429087	14.0786862	14.6258610	15.1842326	15.7535877	16.3337012	16.9243373	17.5252508	18.7568877
10	12.1327594	12.6675774	13.2150737	13.7750011	14.3470948	14.9310740	15.5266435	16.1334957	16.7513121	18.0185199
12	10.8245258	11.3803070	11.9507826	12.5355526	13.1341914	13.7462524	14.3712708	15.0087677	15.6582530	16.9911952
15	9.5565208	10.1426658	10.7460512	11.3659693	12.0016806	12.6524217	13.3174139	13.9958712	14.6870074	16.1042104
20	8.3644007	8.9972596	9.6502165	10.3218839	11.0108613	11.7157571	12.4352081	13.1678958	13.9125594	15.4331152
25	7.7181622	8.3919636	9.0870075	9.8011308	10.5322414	11.2783530	12.0376104	12.8083061	13.5888889	15.1742994
30	7.3376457	8.0462262	8.7757157	9.5232340	10.2861260	11.0619952	11.8487175	12.6444402	13.4475700	15.0708537

YOUR TURN

WORK THIS PROBLEM

The Question:

Calculate the monthly P&I payment for a mortgage of $175,000 at 12% for 30 years.

✔ YOUR WORK

The Solution:

Step 1. Mortgage Payment table value = 10.2861260

Step 2. $10.2861260 \times \dfrac{\$175,000}{\$1000}$

$10.2861260 \times \$175 = \$1800.07$ rounded

Fixed-Rate Mortgage
A mortgage on which the interest rate remains constant throughout the term of the loan.

Variable-Rate Mortgage
A mortgage on which the interest rate may fluctuate, depending on market interest rates, at least to some degree.

In the past, the most common type of mortgage has been the **fixed-rate mortgage,** in which the interest rate is fixed (cannot change) for the entire length of the loan. But the last few decades have seen increased popularity for **variable-rate mortgages,** in which interest rates may vary from year to year depending on overall interest rates.

Banks like variable rates when rates appear to be going up because they allow the bank to increase the interest it receives. Buyers like the lower initial rates offered on variable-rate mortgages and to some extent gamble that interest rates will not rise too dramatically.

If Bits 'n' Bytes had a variable-rate mortgage in the first example of this section, and the rate went up to 12%, its monthly mortgage payment would go up from $1591.24 to:

Step 1. Mortgage Payment table value is 12.0016806.

Step 2. Payment $= 12.0016806 \times \dfrac{\$140,000}{\$1000} = \$1680.24$ rounded

YOUR
TURN

WORK THIS PROBLEM

The Question:

Ms. Spode took out a variable-rate mortgage of $118,000 for 30 years to set up a pottery store.

(a) What are Ms. Spode's monthly payments the first year, when the rate is 11%?
(b) If the rate rises to 12%, what will be the new payments?

✔ YOUR WORK

The Solution:

(a) *Step 1.*   Mortgage Payment table value is 9.5232340.

    Step 2.   Payment $= 9.5232340 \times \dfrac{\$118,000}{\$1000}$

                    $= 9.5232340 \times \$118 = \$1123.74$ rounded

(b) *Step 1.*   Mortgage Payment table value is 10.2861260.

    Step 2.   Payment $= 10.2861260 \times \dfrac{\$118,000}{\$1000}$

                    $= 10.2861260 \times \$118 = \$1213.76$ rounded

The total amount of interest paid over the life of a mortgage loan may surprise you.

Total interest = (number of payments × P&I payment) − principal

For example, let's calculate the total interest for Bits 'n' Bytes' new store, assuming it has a $140,000 mortgage at 11% for 15 years and monthly P&I payments of $1591.24.

$$\text{No. payments} = 15 \times 12 = 180$$
$$\text{Total interest} = (180 \times \$1591.24) - \$140,000$$
$$= \$286,423.20 - \$140,000 = \$146,423.20$$

In this case, the interest is actually more than the amount initially borrowed.

YOUR
TURN

WORK THIS PROBLEM

The Question:

Calculate the total interest for the $118,000 mortgage for 30 years at 11% and monthly payments of $1123.74 taken by Ms. Spode to open her pottery shop.

✔ YOUR WORK

The Solution:

$$\text{No. payments} = 30 \times 12 = 360$$
$$\text{Total interest} = (360 \times \$1123.74) - \$118,000$$
$$= \$404,546.40 - \$118,000 = \$286,546.40$$

Paying Points

Market Rate The typical loan rate being charged by banks and other financial institutions at a given time.

Points On FHA or VA mortgages, an assessment by the lender, paid by the seller, that represents the difference between the market rate and maximum rate permitted by the FHA or VA; one point is charged for each $\frac{1}{8}$% the market rate exceeds the maximum federal rate.

A number of home mortgages are insured by the FHA (Federal Housing Administration) and the VA (Veteran's Administration). The FHA and VA do not lend money—they only guarantee the loans to the lender.

When the **market rate**—the loan rate charged by the banks or other financial institution—is higher than the maximum rate permitted by the FHA or VA, **points** are charged by the lending institution. Each point is 1% of the mortgage principal to be paid at settlement. Since federal regulations limit the amount of charges to the borrower on government-backed loans, the points are paid by the seller, who pays them to the buyer's mortgage company. (Banks may also charge points on nongovernment-backed loans, in which case the buyer may have to pay the points.)

> One point is charged for each $\frac{1}{8}$% the market rate exceeds the maximum federal rate on government-backed loans.

Calculating the number of points is easy.

Steps: Calculating the Number of Points

STEP 1. *Calculate* the difference in interest rates.

STEP 2. *Convert* the difference to eighths.

STEP 3. *Determine* the number of points. Each $\frac{1}{8}$% is equivalent to 1 point.

For example, if the current market rate is $12\frac{3}{4}$% and the maximum federal rate is $12\frac{1}{2}$%, how many points must you pay?

Step 1. $12\frac{3}{4}\% - 12\frac{1}{2}\% = \frac{1}{4}\%$

Step 2. $\frac{1}{4}\% = \frac{2}{8}\%$

Step 3. $\frac{2}{8}\% = 2$ points

YOUR TURN

| WORK THIS PROBLEM |

The Question:

With the market rate at 12% and the maximum federal rate at $11\frac{1}{4}$%, calculate the points.

| ✔ YOUR WORK |

The Solution:

Step 1. $12\% - 11\frac{1}{4}\% = \frac{3}{4}\%$

Step 2. $\frac{3}{4}\% = \frac{6}{8}\%$

Step 3. $\frac{6}{8}\% = 6$ points

Finding the Discount

After calculating the points, you need to find out what they mean in dollars and cents—the discount amount. To do so, just follow these steps.

Steps: Calculating the Discount

STEP 1.   *Calculate* the difference in interest rates.

STEP 2.   *Convert* the difference to eighths.

STEP 3.   *Determine* the number of points. Each $\frac{1}{8}$ equals 1 point.

STEP 4.   *Determine* the discount rate. Number of points = discount rate (expressed as a percent).

STEP 5.   *Calculate* the discount using the formula:

$$\text{Discount} = \text{discount rate} \times \text{mortgage principal}$$

Suppose the Duncan family is negotiating an FHA-insured mortgage for $59,500. The current market rate is $12\frac{3}{4}\%$ and the maximum federal rate is $12\frac{1}{4}\%$. Calculate the points and the discount amount.

Step 1. $12\frac{3}{4}\% - 12\frac{1}{4}\% = \frac{1}{2}\%$

Step 2. $\frac{1}{2}\% = \frac{4}{8}\%$

Step 3. $\frac{4}{8}\% = 4$ points

Step 4.   4 points $= 4\%$

Step 5.   Discount = discount rate $\times$ mortgage principal
$$= 4\% \times \$59,500 = 0.04 \times \$59,500 = \$2380$$

The discount of $2380 is paid by the seller to the Duncan's mortgage company.

**YOUR
TURN**

| WORK THIS PROBLEM |

The Question:

The Cawder family is negotiating a VA-insured loan for $87,250. The current market rate is $12\frac{3}{4}\%$ and the maximum federal rate is 12%. Calculate the points and discount amount.

| ✔ YOUR WORK |

The Solution:

Step 1. $12\frac{3}{4}\% - 12\% = \frac{3}{4}\%$

Step 2. $\frac{3}{4}\% = \frac{6}{8}\%$

Step 3. $\frac{6}{8}\% = 6$ points

Step 4.   6 points $= 6\%$

Step 5.   Discount = $6\% \times \$87,250 = 0.06 \times \$87,250 = \$5235$

To speed discount calculations, use your calculator.

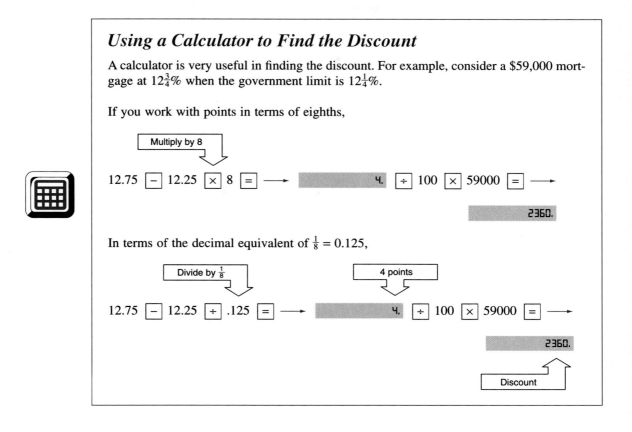

Using a Calculator to Find the Discount

A calculator is very useful in finding the discount. For example, consider a $59,000 mortgage at $12\frac{3}{4}\%$ when the government limit is $12\frac{1}{4}\%$.

If you work with points in terms of eighths,

Multiply by 8

$12.75 \; \boxed{-} \; 12.25 \; \boxed{\times} \; 8 \; \boxed{=} \longrightarrow \qquad 4. \quad \boxed{\div} \; 100 \; \boxed{\times} \; 59000 \; \boxed{=} \longrightarrow$

2360.

In terms of the decimal equivalent of $\frac{1}{8} = 0.125$,

Divide by $\frac{1}{8}$ 4 points

$12.75 \; \boxed{-} \; 12.25 \; \boxed{\div} \; .125 \; \boxed{=} \longrightarrow \qquad 4. \quad \boxed{\div} \; 100 \; \boxed{\times} \; 59000 \; \boxed{=} \longrightarrow$

2360.

Discount

For a set of practice problems on mortgages and points, go to Section Test 14.1.

WORKSPACE

SECTION TEST 14.1 Real Estate Mathematics

The following problems test your understanding of Section 14.1, Mortgage Payments and Points. Use the Mortgage Payment table in the chapter.

1. Ava's Air Tours is borrowing $150,000 for a customer waiting room. They will finance it for 20 years at 10%. What is the monthly payment?

2. If Ava's Air Tours financed the waiting room ($150,000) for 15 years at 10%, what would be the monthly payment?

3. If Shiring Associates finances a $230,000 office building for 25 years at 11%, what would be the monthly payment?

4. If Shiring Associates finances a $230,000 office building for 15 years at 11%, what will be the monthly payment?

5. Joanne Jett bought a new house for $80,000. The home has an adjustable-rate mortgage at 11% for 30 years. (a) What is the monthly payment? (b) If the interest rate increases to 12%, what will be the new monthly payment?

6. The Van Halens purchased a new home for $135,000. The home adjustable-rate mortgage is at 10% for 25 years. (a) What is the monthly payment? (b) If the rate increases to 11%, what will be the new monthly payment?

7. Lots of Plots, a real estate firm, is trying to sell a house for $105,000. (a) If the mortgage rate is 11% for 30 years, what will be the monthly payment? (b) What will be the total interest paid?

8. (a) If Lots of Plots wanted to sell the same $105,000 house, but calculated an 11% interest rate for 20 years, what would be the monthly payment? (b) What would be the total interest paid?

9. Musselmann's Gym financed a new spa area for $180,000 at 13% for 20 years. (a) What is the monthly payment? (b) What is the total interest paid?

10. (a) If Musselmann's financed the $180,000 at 13% for 25 years, what would be the total monthly payment? (b) What is the total interest paid?

11. The McCoys are negotiating an FHA-insured mortgage for $85,000. The current market rate is $11\frac{3}{4}\%$, and the maximum federal rate is $11\frac{1}{4}\%$. What are (a) the points and (b) the discount amount?

12. The current market rate is $10\frac{1}{4}\%$ and the maximum federal rate is $9\frac{3}{4}\%$. The Hatfields are negotiating an FHA-insured mortgage for $75,000. What are (a) the points and (b) the discount amount?

13. The Abbotts are trying to negotiate an FHA-insured house sale for $80,000. The current market rate is $11\frac{1}{2}\%$ and the maximum federal rate is $11\frac{1}{4}\%$. What are (a) the points and (b) the discount amount?

14. The Costellos are negotiating an FHA-insured mortgage for $70,000. The federal maximum is $11\frac{1}{2}\%$ and the current market rate is $11\frac{3}{4}\%$. What are (a) the points and (b) the discount amount?

15. The Woodwards are negotiating a VA mortgage for $78,000. The current market rate is $10\frac{3}{4}\%$. The maximum federal rate is $10\frac{1}{4}\%$. What are (a) the points and (b) the discount amount?

16. If the current market rate is $11\frac{1}{4}\%$ and the maximum federal rate is $10\frac{3}{4}\%$, what would be (a) the points and (b) the discount amount on a VA mortgage for $82,000?

17. The Bernsteins are negotiating a VA-insured loan for $92,000. The maximum federal rate is $12\frac{1}{4}\%$. The current market rate is $12\frac{1}{2}\%$. What are (a) the points and (b) the discount amount?

18. What are (a) the points and (b) the discount amount on a VA mortgage for $79,000 if the maximum federal rate is $12\frac{1}{2}\%$ and the current market rate is $12\frac{3}{4}\%$?

19. The current market rate is $12\frac{1}{2}\%$ and the maximum federal rate is 12%. What are (a) the points and (b) the discount amount on a $73,000 FHA-insured mortgage?

20. Charles Ray is negotiating an FHA-insured mortgage for $88,000. The maximum federal rate is $11\frac{1}{2}\%$. The current market rate is $12\frac{1}{4}\%$. What are (a) the points and (b) the discount amount?

21. Jimmy & Susan bought a new home to house their growing family. The purchase price was $93,500, and they made a 20% down payment, financing the rest with a VA-insured loan for 30 years. The current market rate was 10%, while the maximum federal rate was 9.5%.
    (a) How much did Jimmy & Susan have to borrow?
    (b) What was the monthly payment on the loan?
    (c) How much was the total finance charge over the life of the loan?
    (d) What was the total cost of their new home?
    (e) What were the points and the discount amount paid to the lender by the seller?

Can You:

Calculate the interest payment and principal payment of a mortgage?
Generate an amortization schedule?
. . . If not, you need this section.

Interest Payment The portion of a monthly loan payment that covers the month's interest charges.

Principal Payment The portion of a monthly loan payment that reduces the principal.

The total P&I of a mortgage payment will not change over the life of a fixed-rate mortgage and may change only every so many years on a variable-rate mortgage. However, the portion of a mortgage payment that represents principal and the portion that represents interest change from month to month. At the beginning of the mortgage, most of the payment is interest and a relatively small amount goes to reduce the principal.

The portion of the monthly payment that covers the month's interest charges is called the **interest payment**. The rest of the payment, which reduces the principal, is called the **principal payment**.

Determining Interest and Principal Payments

To calculate the interest and principal payments on a mortgage in any given month, follow these steps:

Steps: Determining Interest and Principal Payments

STEP 1. *Find* the interest payment by using the formula for simple interest at ordinary or 30-day-month time (see Section 9.1).

$$\text{Interest payment} = I = Prt$$

$$\text{where } t = \frac{30}{360} = \frac{1}{12}$$

STEP 2. *Subtract* the interest payment from the total payment to find the principal payment.

For example, on a mortgage of $84,000 at 14% for 30 years, the total monthly P&I payment is $995.29. What are the interest and principal payments the first month?

Step 1.

$$\text{Interest payment} = I = P \times r \times t = \$84{,}000 \times 14\% \times \frac{1}{12}$$

$$= \$84{,}000 \times 0.14 \times \frac{1}{12} = \$980.00$$

Step 2. Principal payment = P&I payment − interest payment
$$= \$995.29 - 980.00 = \$15.29$$

Paying off part of the principal means you have less borrowed on which to pay interest next month. That is, it explains why the interest portion of a P&I payment keeps declining.

To see why, let's find the new principal and the next month's interest and principal payments:

> New principal = previous principal − principal payment

In the case of the $84,000 mortgage just described,

$$\text{New principal} = \text{previous principal} - \text{principal payment}$$
$$= \$84,000 - 15.29 = \$83,984.71$$

Of the $995.29 payment, only $15.29 is applied to the principal!

Now we can apply our two-step method to the *new principal.*

Step 1.

New principal   Rate   One month

$$\text{Interest payment} = I = P \times r \times t = \$83,984.71 \times 14\% \times \frac{1}{12}$$
$$= \$83,984.71 \times 0.14 \times \frac{1}{12}$$
$$= \$979.821 \ldots = \$979.82 \text{ rounded}$$

Step 2.   Principal payment = P&I payment − interest payment
$$= \$995.29 - 979.82 = \$15.47$$

YOUR TURN

| WORK THIS PROBLEM |

The Question:

Find the new principal for the above example and calculate the interest and principal payments for the next month.

| ✔ YOUR WORK |

The Solution:

$$\text{New principal} = \text{previous principal} - \text{principal payment}$$
$$= \$83,984.71 - 15.47 = \$83,969.24$$

Step 1.   Interest payment $= I = P \times r \times t = \$83,969.24 \times 14\% \times \frac{1}{12}$
$$= \$83,969.24 \times 0.14 \times \frac{1}{12}$$
$$= \$979.641 \ldots = \$979.64 \text{ rounded}$$

Step 2.   Principal payment $= \$995.29 - 979.64 = \$15.65$

After the third payment, the new principal is

$$\text{New principal} = \text{previous principal} - \text{principal payment}$$
$$= \$83,969.24 - 15.65 = \$83,953.59$$

Note that after three payments, the principal has been reduced from $84,000 to $83,953.59. Only after many years will the principal be reduced to the point where most of the payment is applied to the principal.

Using a calculator speeds interest payment calculations.

Calculating Interest Payments with a Calculator

Once you set up the interest calculation, it is easy to punch it into your calculator. For example, to calculate the interest payment, the equation is

$$I = Prt = \$84,000 \times 14\% \times \frac{1}{12}$$
$$= \$84,000 \times 0.14 \times \frac{1}{12}$$

Using a calculator,

84000 $\boxed{\times}$ 14 $\boxed{\%}$ $\boxed{\div}$ 12 $\boxed{=}$ $\longrightarrow$ $\boxed{980.}$

Amortization Schedule
A complete listing of all interest payments and principal payments throughout the term of the loan.

A complete listing of all P&I payments is called an **amortization schedule.** To calculate a complete amortization schedule for a 30-year loan by our method would require 360 monthly calculations. Enough to test anyone's patience! At the time a real estate loan is made, the buyer is usually given a complete amortization schedule for the property.

The following is the first three months of the amortization schedule for our example loan.

Payment Number	Principal Balance	P&I Payment	Interest Payment	Principal Payment
1	$84,000.00	$995.29	$980.00	$15.29
2	$83,984.71	$995.29	$979.82	$15.47
3	$83,969.24	$995.29	$979.64	$15.65
	$83,953.59			

YOUR TURN

WORK THIS PROBLEM

The Question:

The Fly-By-Night Aviary purchased a new building and financed $150,000 at 11% for 20 years.

(a) Use the Mortgage Payment table from Section 14.1 to calculate the P&I payment.
(b) Calculate the interest and principal portions of the payment for the first two months.
(c) Summarize your calculations in an amortization schedule.

✔ YOUR WORK

The Solution:

(a) ***Step 1.*** Mortgage Payment table value is 10.3218839.

   Step 2. Payment $= 10.3218839 \times \dfrac{\$150,000}{\$1000}$

   $= 10.3218839 \times \$150 = \$1548.28$

(b) First Month

   Step 1. Interest payment $= \$150,000 \times 11\% \times \dfrac{1}{12} = \$1375$

   Step 2. Principal payment $= \$1548.28 - \$1375.00 = \$173.28$

   Second Month: New principal $= \$150,000 - \$173.28 = \$149,826.72$

   Step 1. Interest payment $= \$149,826.72 \times 11\% \times \dfrac{1}{12}$

   $= \$1373.41$ rounded

   Step 2. Principal payment $= \$1548.28 - \$1373.41 = \$174.87$

(c) Amortization schedule

Payment Number	Principal Balance	P&I Payment	Interest Payment	Principal Payment
1	$150,000.00	$1548.28	$1375.00	$173.28
2	$149,826.72	$1548.28	$1373.41	$174.87

For a set of practice problems, see Section Test 14.2.

SECTION TEST 14.2 Real Estate Mathematics

The following problems test your understanding of Section 14.2, Amortization. Use the Mortgage Payment table in this chapter.

1. Intense Tents is borrowing $175,000 for a canoe wing. The mortgage will be at 11% for 25 years. (a) Calculate the monthly payment. (b) Generate an amortization schedule for the first two payments.

2. The By the Byte Software House wants to borrow $210,000 to build a warehouse. It is considering a loan at 12% for 20 years. (a) Calculate the monthly payment. (b) Generate an amortization schedule for the first two payments.

3. Lew wants to build another gourmet restaurant. The company will be financing $195,000 at 13% for 15 years. (a) Calculate the monthly payment. (b) Generate an amortization schedule for the first two payments.

4. Crystal's Diamonds will be doing major construction to install a larger safe and a security system and to do other general remodeling. They want to borrow $270,000 at 11% for 20 years. (a) Calculate the monthly payment. (b) Generate an amortization schedule for the first two payments.

5. The Hopes are considering a $135,000 mortgage at 11% for 30 years. (a) Calculate the monthly payment. (b) Generate an amortization schedule for the first two payments.

6. The Hopes are also considering the $135,000 mortgage at 11% for 20 years. (a) Calculate the monthly payment. (b) Generate an amortization schedule for the first two payments.

7. The Wonders family is negotiating a 30-year, 12% mortgage for $73,000. (a) Calculate the monthly payment. (b) Generate an amortization schedule for the first three payments.

8. (a) Calculate the monthly payment and (b) generate an amortization schedule for the first three payments for the Wonders if they change their $73,000 mortgage at 12% from 30 to 20 years.

9. The Woody's Lumber Mill is considering borrowing $185,000 at 13% for 25 years to replace old equipment. (a) Calculate the monthly payment. (b) Generate an amortization schedule for the first three payments.

10. (a) If Woody's instead borrows the $185,000 at 13% for 15 years, what would be the monthly payment? (b) Generate an amortization schedule for the first three payments.

11. Fill out the portion of the amortization table shown below. The amount borrowed is $74,800 at 9.5% interest for 30 years. The monthly payment is $628.96.

Payment Number	Interest	Principal	Ending Balance
5	$590.99	$37.97	$74,613.11
6	(a)	(b)	(c)
7	(d)	(e)	(f)

Key Terms

amortization schedule	mortgage
fixed-rate mortgage	points
interest	principal
interest payment	principal payment
market rate	variable-rate mortgage

Chapter 14 at a Glance

Page	Topic	Key Point	Example
460	Mortgage P&I payments	Find payment per $1000 in Mortgage Payment table. $$\text{Payment} = \frac{\text{table}}{\text{value}} \times \frac{\text{principal}}{\$1000}$$	What is the P&I payment on a $140,000 mortgage for 15 years at 11%? Table value = 11.3659693 $$\text{Payment} = 11.3659693 \times \frac{\$140,000}{\$1000}$$ = $1591.24 rounded
462	Total interest on a mortgage	$$\frac{\text{Total}}{\text{interest}} = \left(\frac{\text{number}}{\text{of payments}} \times \frac{\text{P\&I}}{\text{payment}}\right) - \text{principle}$$	What is the total interest on a $140,000 mortgage at 11% for 15 years? $$\frac{\text{Total}}{\text{interest}} = (180 \times \$1591.24) - \$140,000 = \$146,423.20$$
463	Points	Find the difference in interest rates. Convert the difference to eighths. Each $\frac{1}{8}$% = 1 point.	How many points are due when the current market rate is $12\frac{3}{4}$% and the maximum federal rate is $12\frac{1}{2}$%? $12\frac{3}{4}$% − $12\frac{1}{2}$% = $\frac{1}{4}$% $\frac{1}{4}$% = $\frac{2}{8}$% $\frac{2}{8}$% = 2 points
464	Discount amount on mortgages	Find the number of points (see above). Use points as discount rate expressed as a percent. Discount = discount rate × principal	What is the discount on a $100,000 mortgage when the current market rate is $12\frac{3}{4}$% and the maximum federal rate is $12\frac{1}{2}$%? Number points = 2 (see above) Discount rate = 2% Discount = 0.02 × $100,000 = $2000
469	Interest and principal payments	Interest payment = Prt where $t = \frac{1}{12}$. $$\frac{\text{Principal}}{\text{payment}} = \frac{\text{total}}{\text{payment}} - \frac{\text{interest}}{\text{payment}}$$	What are the interest and principal payments in the first month of a $84,000 mortgage at 14% for 30 years on which monthly P&I payments are $995.29? Interest payment = $84,000 × 0.14 × $\frac{1}{12}$ = $980 Principal payment = $995.29 − 980 = $15.29

Page	Topic	Key Point	Example
470	Decreasing principal	$$\begin{array}{c}\text{New} \\ \text{principal}\end{array} = \begin{array}{c}\text{previous} \\ \text{principal}\end{array} - \begin{array}{c}\text{principal} \\ \text{payment}\end{array}$$	Find the new principal after the first month's P&I payment above. New principal = $84,000 − 15.29 = $83,984.71

Name

Date

Course/Section

The following problems test your understanding of real estate mathematics. Use the Mortgage Payment table in the chapter.

1. The New Wave Aquatica is borrowing $120,000 for a new sea lion basking area. It will finance the loan at 12% for 20 years. What is the monthly payment?

2. If the New Wave Aquatica finances the sea lion basking area ($120,000) for 15 years at 12%, what will be the monthly payment?

3. Rose Bonheur has a home adjustable-rate mortgage for $102,000 at 10% for 30 years. (a) What is the monthly payment? (b) If the interest rate increases to 11%, what will be the new monthly payment?

4. Michael Angelo bought a new house for $95,000. He financed it with an adjustable-rate mortgage at 12% for 20 years. (a) What is the monthly payment? (b) If the rate decreases to 11%, what will be the new monthly payment?

5. Donna Tello bought a new house for $125,000. Her mortgage rate is 11% and is for 30 years. What is (a) the monthly payment and (b) the total interest paid?

6. If Donna Tello's mortgage (for $125,000) is at 11% but only for 20 years, (a) what will be the monthly payment and, (b) what is the total interest paid?

7. The Garden of Earthly Delights is considering financing a new greenhouse for $225,000 at 12% for 20 years. What are (a) the monthly payments and (b) the total interest paid?

8. If the Garden of Earthly Delights instead finances the $225,000 at 12% for 25 years, what will be (a) the monthly payment and (b) the total interest paid?

9. The Palmer family is negotiating an FHA-insured mortgage for $78,000. The current market rate is $12\frac{1}{4}$% and the maximum federal rate is $11\frac{1}{2}$%. What are (a) the points and (b) the discount amount?

10. The Nicklaus family wants to purchase a new home for $95,000. The current market rate is $11\frac{3}{4}\%$ and the maximum federal rate is $11\frac{1}{4}\%$. What would be (a) the points and (b) the discount amount for an FHA-insured mortgage?

11. The federal maximum rate is $10\frac{1}{4}\%$ and the current market rate is $10\frac{1}{2}\%$. What would be (a) the points and (b) the discount amount for a VA mortgage for $89,000?

12. The Trevino family is negotiating a VA mortgage for $92,000. The federal maximum rate is $10\frac{3}{4}\%$. The current market rate is $11\frac{1}{2}\%$. What are (a) the points and (b) the discount amount?

13. Big Ben's Bolts is considering borrowing $280,000 at 12% for 25 years to expand its megabolt production area. (a) Calculate the monthly payments. (b) Generate an amortization schedule for the first two payments.

14. Big Ben's Bolts can also borrow the necessary $280,000 at 12% for only 15 years. (a) Calculate the monthly payments. (b) Generate an amortization schedule for the first two payments.

15. The Byers are considering buying a house for $155,000. The mortgage will be at 11% for 30 years. (a) Calculate the monthly payments. (b) Generate an amortization schedule for the first three payments.

16. The Byers can also get an 11% mortgage for 20 years on their $155,000 house. (a) Calculate the monthly payments. (b) Generate an amortization schedule for the first three payments.

Inventory and Overhead

CHAPTER OBJECTIVES

When you complete this chapter successfully, you will be able to:

1. Calculate inventory values using the specific identification, average cost, FIFO, and LIFO methods.

2. Determine inventory values using the retail method and allocate overhead by sales volume and floor space.

■ The Jolly Green Midget packages environmentally safe toys that are given away with children's meals at fast-food restaurants. During the last year, the firm purchased cases of toys in the following quantities:

January 14	580 @	$7.20
April 13	625 @	7.00
July 1	1060 @	7.10
September 23	750 @	7.12
November 27	985 @	7.20

At the end of the year, it had 1145 cases in stock total. What was its inventory value by the average cost, FIFO, and LIFO methods?

■ Two Wheels Are Enough, a bicycle shop, has a total monthly overhead of $11,500. The men's racer department had $11,205 in monthly sales, the tandem department $34,860, the touring department $8715, the children's department $12,450, and the women's racer department $15,770. What is the overhead for each department using the sales volume method?

$\mathbf{A}$ place in which to produce or from which to sell its goods and services is a first priority for any company. But once it begins to produce or retail goods, it has to keep track of them. The government requires that companies keep track of their investment in making and storing products. The goods sitting in the warehouses of manufacturers and wholesalers and on the shelves of retailers all represent costs to the business until they are sold.

In this chapter, we will look at various ways to calculate the size of such investments and other costs of doing business. Don't worry. You won't find yourself "over your head." Your "inventory" of business mathematics skills will stand you in good stead as you move through this chapter.

SECTION 15.1: Inventory Valuation

Can You:

Calculate an inventory value using the specific identification method?
Find the inventory value using the average cost method?
Determine the inventory value using the FIFO method?
Identify the inventory value using the LIFO method?
. . . If not, you need this section.

Inventory A procedure for listing individual items of merchandise in stock and their value.

An **inventory** is a procedure for listing the individual items of merchandise in stock and their value. This is an important process and is usually required at least once a year.

Keeping track of inventory enables a company to know at all times:

1. How much (or how little) of each item is in stock
2. How many items have been lost, damaged, or stolen
3. How fast merchandise is being sold

Inventory reports also enable companies to file financial reports (see Chapter 19) used by investors and tax returns for the government.

Perpetual Inventory System A computerized inventory system in which are entered listings of all goods received and sold, so that current inventory levels are always immediately available.

Today, most large businesses have computerized their inventories using the *Uniform Product Code (UPC)*. The UPC code, also called the *bar code,* is imprinted on most merchandise. Each item has a unique code. When the merchandise is sold, its code is "scanned" electronically and the information is recorded in the firm's computer system. When merchandise is received, the quantity and type of item is also entered into the system. This inventory method is called a **perpetual inventory system.**

Physical Inventories Hand counting items in inventory.

Inventory Valuation Methods Ways of assigning values to items in inventory; the most common are specific identification, average cost, FIFO, LIFO, and retail method.

Many smaller businesses lack computerized systems and instead use regular **physical inventories.** In a physical inventory, the quantity of each item on hand is counted. This total is then multiplied by the item's value to obtain the total value. Sounds simple, doesn't it? Well, there are a few details. A problem arises when determining the value for each item. If one item is purchased from several different suppliers at several times during the year, each time at a different price, which cost is used? The cost is determined by one of several **inventory valuation methods,** including specific identification, average cost, FIFO, and LIFO.

To see how these methods compare, we will apply each to the following situation. Count On Us made the following purchases of calculators:

February 17	150 @	$12.50
April 25	120 @	12.00
July 9	130 @	11.50
September 7	100 @	11.75

At its year-end inventory, Count On Us had 163 calculators in stock. What value does the store place on the calculators? The individual cost? The average cost? The first cost? The last cost? In the rest of this section, we will explore the effects of each of these approaches.

Calculating Inventory Value Using the Specific Identification Method

Specific Identification
A method of inventory valuation in which each item is valued at its actual purchase cost.

In the **specific identification** method, each item is valued at its *actual purchase cost*. Thus each item must be tagged with its original cost. Because of the time required with this method, it is generally used only by businesses with few items to inventory. It is not used by businesses with large physical inventories.

Steps: Inventory Valuation by Specific Identification

STEP 1. *Compile* data on quantities and costs.

STEP 2. *Multiply* the number of units by the cost of the unit and add the results to get the total inventory value.

Let's apply these steps to our Count On Us example.

Step 1. Calculators in stock break down as follows:

Calculators

	Number	Cost
February 17	5	$12.50
April 25	13	12.00
July 9	50	11.50
September 7	95	11.75
	163	

Step 2.

Calculators

Number		Cost		Total Cost
5	×	$12.50	=	$ 62.50
13	×	12.00	=	156.00
50	×	11.50	=	575.00
95	×	11.75	=	1116.25
In stock 163				$1909.75 Total value

The total inventory value of the calculators using the specific identification method is $1909.75.

YOUR TURN

| WORK THIS PROBLEM |

The Question:

The Couch Potato made the following purchases of Multivision televisions:

January 7	35 @ $350
April 11	50 @ 365
August 15	65 @ 360
October 5	50 @ 370

The year-end inventory revealed a total of 57 televisions in stock.

Number	Cost
10	$350
12	365
15	360
20	370
57	

Determine the inventory value of the televisions by the specific identification method.

✔ **YOUR WORK**

The Solution:

Step 1			Step 2
Number		**Cost**	**Total Cost**
10	×	$350 =	$ 3,500
12	×	365 =	4,380
15	×	360 =	5,400
20	×	370 =	7,400
57			$20,680

The inventory value by the specific identification method is $20,680.

The figures produced by the specific method are very accurate. But a lot of time is required to tag and check the purchase cost of each item. As you will see, the three other methods are quicker.

Using a Calculator in Inventory Valuation

Using a calculator with memory can simplify inventory calculations. For example, in the preceding inventory example using the specific identification method, you needed to calculate four products and add them. With your calculator, you can punch in the raw numbers and get a total very quickly.

To start, punch the CLR button twice to clear the memory of any previous data. Then punch in the following:

5 $\boxed{\times}$ 12.5 $\boxed{=}$ $\boxed{\text{M}^+}$ ⟶   62.5

13 $\boxed{\times}$ 12 $\boxed{=}$ $\boxed{\text{M}^+}$ ⟶   156.

50 $\boxed{\times}$ 11.5 $\boxed{=}$ $\boxed{\text{M}^+}$ ⟶   575.

95 $\boxed{\times}$ 11.75 $\boxed{=}$ $\boxed{\text{M}^+}$ ⟶   1116.25

The $\boxed{\text{M}^+}$ key adds each number to the memory as you go. To see the final total, just press the $\boxed{\text{MRC}}$ key and the number 1909.75 will pop up on the display.

Finding Inventory Value Using the Average Cost Method

Average Cost A method of inventory valuation in which costs of items in inventory are averaged.

A faster way to evaluate inventory uses the **average cost.** This method does not require finding the purchase cost of each item, but uses an *averaging* method.

Just follow these steps:

Steps: Inventory Valuation by Average Cost

STEP 1. *Multiply* the number of units of the item purchased by their cost to find the total cost.

STEP 2. *Divide* the total cost by the total number of units to find the average cost. Do not round.

$$\text{Average cost} = \frac{\text{total cost}}{\text{total no. of units}}$$

STEP 3. *Multiply* the number of units in stock times the average costs to find the inventory value and round the answer.

$$\text{Inventory value} = \text{no. of units in stock} \times \text{average cost}$$

Let's apply these steps to the Count On Us example. Again, the store made the following calculator purchases:

February 17	150 @ $12.50
April 25	120 @ 12.00
July 9	130 @ 11.50
September 7	100 @ 11.75

As before, at the year-end inventory, Count On Us had 163 calculators in stock.

Step 1.

	Number	Cost		Total Cost
February 17	150	× $12.50	=	$1875.00
April 25	120	× 12.00	=	1440.00
July 9	130	× 11.50	=	1495.00
September 7	100	× 11.75	=	1175.00
Total # of calculators	500			$5985.00 — Total cost

Step 2.

$$\text{Average cost} = \frac{\$5985.00}{500} = \$11.97$$

(Total cost / Total number of calculators)

Step 3.

In stock Average cost

$$\text{Inventory value} = 163 \times \$11.97 = \$1951.11$$

YOUR TURN

WORK THIS PROBLEM

The Question:

The Couch Potato made the following purchase of Multivision televisions:

January 7	35 @ $350
April 11	50 @ 365
August 15	65 @ 360
October 5	50 @ 370

The year-end inventory counted 57 televisions in stock. Determine the inventory value of the TVs by the average cost method.

✔ YOUR WORK

The Solution:

Step 1.

Number		Cost		Total Cost
35	×	$350	=	$12,250
50	×	365	=	18,250
65	×	360	=	23,400
50	×	370	=	18,500
200				$72,400

Step 2. Average cost $= \dfrac{\$72,400}{200} = \$362$

Step 3. Inventory value $= 57 \times \$362 = \$20,634$

Determining Inventory Value Using the FIFO Method

FIFO (first in, first out)
A method of inventory valuation in which items in stock at inventory time are considered to be (and are valued as if they were) the last ones received.

Another inventory valuation method is **FIFO,** which is short for **first in, first out.** This method assumes the *first items purchased are the first ones sold.* Thus the items in stock at inventory time are the last ones received. The individual value of the last received items is used in the inventory evaluation procedure.

Steps: Inventory Valuation by FIFO

STEP 1. *Determine* the number of units in stock and the price paid for each, assuming they were the last items purchased.

STEP 2. *Multiply* the units times their costs and add the results to find inventory value.

Again, let's apply the steps to the following calculator purchases made by Count On Us:

February 17	150 @ $12.50
April 25	120 @ 12.00
July 9	130 @ 11.50
September 7	100 @ 11.75

And again, the year-end inventory showed 163 calculators in stock.

Step 1. Since there are 163 calculators in stock, the *last* 100 were purchased September 7 at $11.75 apiece. The remaining 63 must be from the previous order on July 9 at $11.50 apiece.

September 7 order	100 @ $11.75
July 9 order	63 @ 11.50
Total # in stock	163

Step 2.

$$100 \times \$11.75 = \$1175.00$$
$$63 \times 11.50 = \underline{724.50}$$
$$\$1899.50 \quad \text{Total inventory value}$$

Note that the inventory value is based on the cost of the *last* 163 calculators purchased. As in the example preceding, the total inventory may exceed the last order. In this case, the purchase price of prior order(s) must be used. Notice that the inventory value found by this method is different from the value obtained in either the specific identification or the average cost method.

YOUR TURN

WORK THIS PROBLEM

The Question:

The Couch Potato had the following purchases of Multivision televisions.

January 7	35 @	$350
April 11	50 @	365
August 15	65 @	360
October 5	50 @	370

The year-end inventory counted 57 televisions in stock. Determine the inventory value by the FIFO method.

✔ YOUR WORK

The Solution:

	Step 1	**Step 2**
October 5 order	50 @ $370 =	50 × $370 = $18,500
August 15 order	7 @ $360 =	7 × $360 = 2,520
	57	$21,020

The inventory value using the FIFO method is $21,020.

Identifying Inventory Value Using the LIFO Method

LIFO (last in, first out) A method of inventory valuation in which items in stock at inventory time are considered to be (and valued as if they were) the first items received.

Yet another inventory evaluation method is **LIFO,** which is short for **last in, first out.** This is the opposite of the FIFO method. LIFO assumes the *last items purchased are the first ones sold.* Thus the items in stock at inventory time are the first items received. The inventory value is based on the cost of the first items received.

Steps: Inventory Valuation by LIFO

STEP 1.   *Determine* the number of units in stock and the price paid for each, assuming they were the first items purchased.

STEP 2.   *Multiply* the units times their costs and add the results to find inventory value.

Once again, we return to Count On Us and its purchases of calculators:

February 17	150 @ $12.50
April 25	120 @ 12.00
July 9	130 @ 11.50
September 7	100 @ 11.75

As before, the year-end inventory shows 163 calculators in stock.

Step 1.   Since there are 163 calculators in stock, the first 150 were purchased February 17 at $12.50 apiece. The remaining 13 were purchased in the second order on April 25 at $12.00 per item.

February 17 order	150 @ $12.50
April 25 order	13 @ 12.00
Total # in stock	163

Step 2. $150 \times \$12.50 = \$1875$
$ 13 \times 12.00 = 156$
$ \$2031$ ◁ Total inventory value

WORK THIS PROBLEM

The Question:

The Couch Potato made the following purchases of the Multivision televisions.

January 7	35 @ $350
April 11	50 @ 365
August 15	65 @ 360
October 5	50 @ 370

The year-end inventory counted 57 televisions in stock. Determine the inventory value by the LIFO method.

✔ YOUR WORK

The Solution:

	Step 1	**Step 2**
January 7 order	35 @ $350 = 35 × $350 =	$12,250
April 11 order	22 @ $365 = 22 × $365 =	8,030
	57	$20,280

The inventory value using the LIFO method is $20,280.

Comparing Inventory Methods

Although we worked the same problems, the total inventory value was different with each method.

	Calculators	Televisions
Specific identification method	$1909.75	$20,680
Average cost method	1951.11	20,634
FIFO method	1899.50	21,020
LIFO method	2031.00	20,280

Which is the "right" value? The answer is that there is no one "right" value for inventory. Inventory is a relative value—how much a firm has in stock at the present time relative to how much it had in stock at some other time. Thus it is very important to be consistent. Otherwise, you'll wind up comparing apples and oranges and getting fruit salad.

Now sink your teeth into the practice problems in Section Test 15.1.

WORKSPACE

SECTION TEST 15.1 Inventory and Overhead

The following problems test your understanding of Section 15.1, Inventory Valuation.

A. Calculate the inventory value by the specific identification method.

1. Big Ben's Bolts made the following purchases of packing cartons:

January 14	975	@	$10.20
March 23	925	@	9.80
May 2	1000	@	10.25
July 1	875	@	10.00
September 23	725	@	10.20
November 14	500	@	9.70

   The year-end inventory counted a total of 1146 packing cartons in stock with the following costs:

Number	Cost
72	$10.20
133	9.80
116	10.25
132	10.00
352	10.20
341	9.70

   Determine the inventory value by the specific identification method.

2. The Rest-Not Restaurant Supply Company purchased the following numbers of cases of dinner plates:

April 3	30	@	$24.00
June 14	20	@	24.10
August 3	40	@	24.80
October 2	10	@	24.60

   The year-end inventory counted a total of 45 cases of plates in stock with the following costs:

Number	Cost
14	$24.00
12	24.10
13	24.80
6	24.60

   Determine the inventory value by the specific identification method.

3. The By the Byte Software House made the following purchases of $3\frac{1}{2}$ " high-density floppy diskettes:

January 25	1250 @	$0.59
March 23	1575 @	0.55
April 27	1420 @	0.63
June 25	1450 @	0.61
August 21	1625 @	0.58
October 25	1565 @	0.57
December 13	1115 @	0.60

The year-end inventory counted a total of 1676 of these high-density floppy diskettes in stock with the following costs:

Number	Cost
27	$0.59
102	0.55
113	0.63
97	0.61
186	0.58
512	0.57
639	0.60

Determine the inventory value by the specific identification method.

4. The Jolly Green Midget packages environmentally safe toys that are given away with children's meals in fast-food restaurants. During the last year, Jolly Green Midget purchased cases of toys in the following quantities:

January 14	580 @	$7.20
April 13	625 @	7.00
July 1	1060 @	7.10
September 23	750 @	7.12
November 27	985 @	7.20

The year-end inventory counted a total of 1145 cases of environmentally safe toys in stock with the following costs:

Number	Cost
33	$7.20
128	7.00
287	7.10
256	7.12
441	7.20

Determine the inventory value by the specific identification method.

B. Calculate the inventory value.

1. Big Ben's Bolts made the following packing carton purchases:

January 14	975 @ $10.20
March 23	925 @ 9.80
May 2	1000 @ 10.25
July 1	875 @ 10.00
September 23	725 @ 10.20
November 14	500 @ 9.70

    The year-end inventory counted a total of 1146 packing cartons in stock. Determine the inventory by the following methods:

    (a) Average cost   (b) FIFO   (c) LIFO

2. The Rest-Not Restaurant Supply Company made the following purchases of case of dinner plates:

April 3	30 @ $24.00
June 14	20 @ 24.10
August 3	40 @ 24.80
October 2	10 @ 24.60

    The year-end inventory counted a total of 45 cases of plates in stock. Determine the inventory by the following methods:

    (a) Average cost   (b) FIFO   (c) LIFO

3. The By the Byte Software House made the following purchases of $3\frac{1}{2}$" high-density floppy diskettes.

January 25	1250 @ $0.59
March 23	1575 @ 0.55
April 27	1420 @ 0.63
June 25	1450 @ 0.61
August 21	1625 @ 0.58
October 25	1565 @ 0.57
December 13	1115 @ 0.60

    The year-end inventory counted a total of 1676 of the high density floppy diskettes in stock. Determine the inventory by the following methods:

    (a) Average cost   (b) FIFO   (c) LIFO

4. The Jolly Green Midget packages environmentally safe toys that are given away with children's meals in fast-food restaurants. The following is a list of the toy purchases by the case.

January 14	580 @ $7.20
April 13	625 @ 7.00
July 1	1060 @ 7.10
September 23	750 @ 7.12
November 27	985 @ 7.20

The year-end inventory counted a total of 1145 cases of environmentally safe toys in stock. Determine the inventory by the following methods:

(a) Average cost   (b) FIFO   (c) LIFO

5. Claudia's Clip-Joint made the following purchases of shampoo during the year.

January 12	250 @ $4.97
March 3	320 @ $4.87
June 17	160 @ $5.19
September 9	540 @ $4.63
October 26	240 @ $5.02

The year-end inventory showed 115 bottles still on hand. Determine the ending inventory value by each of the following methods:

(a) average cost   (b) FIFO   (c) LIFO

Can You:

Estimate the inventory value using the retail method?
Allocate overhead by sales volume?
Allocate overhead by floor space?
. . . If not, you need this section.

You've probably noticed that all of the inventory methods we have covered so far call for a physical inventory. Physical inventories are expensive. They are time-consuming, requiring workers to stop producing or selling goods and count items. As a result, most companies do them only once or twice a year.

Ratio   A numeric comparison of two quantities.

Firms without computerized systems instead rely on the retail method of inventory valuation we will discuss in this section. Doing this type of valuation depends on ratios. A **ratio** is a numeric comparison of two quantities. For example, if 4 out of 20 members of your business mathematics class get an "A" in the course, the ratio of "A" students is

$$\text{Ratio} = \frac{\text{"A" students}}{\text{all students}} = \frac{4}{20} = 0.20$$

Ratios are also crucial in calculating the true costs of making or selling an item by allocating overhead.

Estimating Inventory Value Using the Retail Method

It's really quite simple to use the retail method and find inventory value without doing a physical inventory. Just follow these steps:

Steps: Valuing Inventory by the Retail Method

STEP 1.   *Calculate* the cost of goods available for sale at cost, that is, the cost of beginning inventory. Multiply the number of units times their cost and total.

STEP 2.   *Calculate* the cost of goods available for sale at retail, that is, the retail value of the beginning inventory.

$$\begin{array}{c}\text{Cost of goods available} \\ \text{for sale at retail}\end{array} = \left(\begin{array}{c}\text{total} \\ \text{goods available}\end{array}\right) \times \left(\begin{array}{c}\text{retail} \\ \text{price}\end{array}\right)$$

STEP 3.   *Calculate* the cost ratio and express as a decimal number. Do not round.

$$\text{Cost ratio} = \frac{\text{cost of goods available for sale at cost}}{\text{cost of goods available for sale at retail}}$$

STEP 4.   *Calculate* the ending inventory at retail.

$$\text{Ending inventory at retail} = \text{remaining inventory} \times \text{retail price}$$

STEP 5.   *Calculate* the ending inventory value (at cost).

$$\text{Ending inventory at cost} = \text{ending inventory at retail} \times \text{cost ratio}$$

Let's apply the retail method to the Count On Us example from Section 15.1, in which Count On Us wound up with 163 calculators in stock after making the following calculator purchases:

February 17	150 @ $12.50
April 25	120 @ 12.00
July 9	130 @ 11.50
September 7	100 @ 11.75

If Count On Us sells the calculators for $17.50, what is the inventory value using the retail method?

Step 1.

$$
\begin{aligned}
150 \times \$12.50 &= \$1875 \\
120 \times 12.00 &= 1440 \\
130 \times 11.50 &= 1495 \\
100 \times 11.75 &= \underline{1175}
\end{aligned}
$$

Cost of goods available for sale at cost = $5985

Step 2. Total number of calculators available for sale $= 150 + 120 + 130 + 100 = 500$

$$
\text{Cost of goods available for sale at retail} = \begin{pmatrix}\text{total no. of}\\ \text{calculators}\end{pmatrix} \times \begin{pmatrix}\text{retail}\\ \text{price}\end{pmatrix} = 500 \times \$17.50 = \$8750
$$

Step 3. Cost ratio $= \dfrac{\text{cost of goods available for sale at cost}}{\text{cost of goods available for sale at retail}} = \dfrac{\$5985}{\$8750} = 0.684$

Step 4. Ending inventory at retail = remaining inventory × retail price
$$= 163 \times \$17.50 = \$2852.50$$

Step 5. Ending inventory at cost = ending inventory at retail × cost ratio
$$= \$2852.50 \times 0.684 = \$1951.11$$

YOUR TURN

| WORK THIS PROBLEM |

The Question:

The Couch Potato made the following purchases of Multivision televisions.

January 7	35 @ $350
April 11	50 @ 365
August 15	65 @ 360
October 5	50 @ 370

The year-end inventory counted 57 Multivision televisions in stock. The Couch Potato sells the televisions for $452.50. What is the inventory value using the retail method?

| ✔ YOUR WORK |

The Solution:

Step 1.

$$
\begin{aligned}
35 \times \$350 &= \$12{,}250 \\
50 \times 365 &= 18{,}250 \\
65 \times 360 &= 23{,}400 \\
50 \times 370 &= \underline{18{,}500}
\end{aligned}
$$

Cost of goods available for sale at cost = $72,400

Step 2. Total number of TVs available for sale $= 35 + 50 + 65 + 50 = 200$

$$\text{Cost of goods available} \atop \text{for sale at retail} = \left(\text{Total number of} \atop \text{televisions available} \right) \times \text{retail price}$$

$$= 200 \times \$452.50 = \$90,500$$

Step 3. $\text{Cost ratio} = \dfrac{\text{Cost of goods available for sale at cost}}{\text{Cost of goods available for sale at retail}} = \dfrac{\$72,400}{\$90,500} = 0.8$

Step 4. Ending inventory at retail = remaining inventory × retail price
$$= 57 \times \$452.50 = \$25,792.50$$

Step 5. Ending inventory at cost = ending inventory at retail × cost ratio
$$= \$25,792.50 \times 0.8 = \$20,634$$

When calculating inventory valuation by the retail method, it is important *not* to round the ratio. Wait until the last step to round or you'll wind up with an inaccurate answer.

Using a calculator makes the process go more quickly.

Using a Calculator With Ratios:
Retail Method of Inventory Valuation

To see how the calculator can help you, let's return to our Count On Us example.

In the calculator box in Section 15.1, we discussed how to use a calculator to solve step 1 of our retail method for finding inventory value. Step 2 is a simple multiplication problem. After completing these steps, you would have the following information:

> Cost of goods available for sale at cost = $5985
> Cost of goods available for sale at retail = $8750
> Remaining inventory = 163
> Retail price = $17.50

You can now complete steps 3, 4, and 5 in one long calculation as follows:

5985 $\boxed{\div}$ 8750 $\boxed{\times}$ 163 $\boxed{\times}$ 17.5 $\boxed{=}$ $\longrightarrow$ 1951.11

Allocating Overhead by Sales Volume

Overhead All costs necessary in doing business but *not* directly accounted for in calculating costs per unit.

Overhead is all those costs necessary to doing business but not directly accounted for in calculating costs per unit. Overhead includes rent, salaries, taxes, and insurance on the business. *Allocating* means spreading these costs out over the number of units produced or sold so that the firm knows what its true costs are.

While retail valuations of inventory rely on a cost ratio, allocating overhead by sales volumes uses a department sales ratio and the following steps:

Steps: Allocating Overhead by Sales Volume

STEP 1.   *Calculate* the total sales by adding the sales from each department.

STEP 2.   *Calculate* the department sales ratio for each department and express as a decimal number. Do not round.

$$\text{Department sales ratio} = \frac{\text{department sales}}{\text{total sales}}$$

STEP 3.   *Calculate* the department overhead for each department.

$$\text{Department overhead} = \text{department sales ratio} \times \text{total overhead}$$

For example, Intense Tents had a weekly overhead of $4000. Its sales by department one week were as follows: recreational equipment, $12,000; backpacking equipment, $4,500; and climbing equipment, $3,500. How much overhead should Intense Tents allocate to each department using the sales volume method?

The following table is a convenient and organized method to perform the calculations.

Departments	Step 1 Weekly Sales	Step 2 Sales Ratio	Step 3 Calculating Department Overhead
Recreational	$12,000	$\frac{\$12,000}{\$20,000} = 0.6$	$0.6 \times \$4,000 = \quad \$2,400$
Backpacking	4,500	$\frac{\$4,500}{\$20,000} = 0.225$	$0.225 \times \$4,000 = \quad \$900$
Climbing	3,500	$\frac{\$3,500}{\$20,000} = 0.175$	$0.175 \times \$4,000 = \quad \$700$
Total	$20,000	1.000	$4,000

As a check, this total should be 1

As a check, this total should equal total overhead

YOUR TURN

WORK THIS PROBLEM

The Question:

The Stately Statue Company's weekly overhead was $1400. The cactus department had sales of $1500, the cow department $2100, the palm tree department $1200, and the pink flamingo department $3200. What is the overhead for each department using the sales volume method?

✔ YOUR WORK

The Solution:

Departments	Step 1 Weekly Sales	Step 2 Sales Ratio	Step 3 Calculating Department Overhead
Cactus	$1500	$\frac{\$1500}{\$8000} = 0.1875$	$0.1875 \times \$1400 = \$262.50$
Cow	2100	$\frac{\$2100}{\$8000} = 0.2625$	$0.2625 \times \$1400 = \$367.50$
Palm Tree	1200	$\frac{\$1200}{\$8000} = 0.15$	$0.15 \times \$1400 = \$210.00$
Pink Flamingo	3200	$\frac{\$3200}{\$8000} = 0.4$	$0.4 \times \$1400 = \$560.00$
Total	$8000	1.000 ⬆ Check	$1400.00 ⬆ Check

Allocating Overhead by Floor Space

Allocating overhead by floor space uses the department floor space ratio and the following steps:

Steps: Allocating Overhead by Floor Space

STEP 1. *Calculate* the total floor space by adding the floor space from each department.

STEP 2. *Calculate* the department floor space ratio for each department and express as a decimal number. Do not round.

$$\text{Department floor space ratio} = \frac{\text{department floor space}}{\text{total floor space}}$$

STEP 3. *Calculate* the department overhead for each department.

$$\text{Department overhead} = \text{department floor space ratio} \times \text{total overhead}$$

For example, Intense Tents had a weekly overhead of $4000. The following is the floor space for each department: recreational, 4500 square feet; backpacking, 1600 square feet; and climbing, 1900 square feet. What is the overhead for each department using the floor space method?

Again, using a table is a convenient and organized method to perform the calculations.

Departments	Step 1 Floor Space	Step 2 Floor Space Ratio	Step 3 Calculating Department Overhead
Recreational	4500 sq. ft.	$\dfrac{4500}{8000} = 0.5625$	$0.5625 \times \$4000 = \$2250$
Backpacking	1600 sq. ft.	$\dfrac{1600}{8000} = 0.2$	$0.2 \times \$4000 = \$\ 800$
Climbing	1900 sq. ft.	$\dfrac{1900}{8000} = 0.2375$	$0.2375 \times \$4000 = \$\ 950$
Total	8000 sq. ft.	1.000 ⬆ Check	$4000 ⬆ Check

YOUR TURN

| WORK THIS PROBLEM |

The Question:

The Stately Statue Company's weekly overhead was $1400. The following is the floor space for each department: cactus, 900 square feet; cow, 1200 square feet; palm tree, 700 square feet; and pink flamingo, 2200 square feet. What is the overhead for each department using the floor space method?

| ✔ YOUR WORK |

The Solution:

Departments	Step 1 Floor Space	Step 2 Floor Space Ratio	Step 3 Calculating Department Overhead
Cactus	900 sq. ft.	$\dfrac{900}{5000} = 0.18$	$0.18 \times \$1400 = \$252$
Cow	1200 sq. ft.	$\dfrac{1200}{5000} = 0.24$	$0.24 \times \$1400 = \$336$
Palm Tree	700 sq. ft.	$\dfrac{700}{5000} = 0.14$	$0.14 \times \$1400 = \$196$
Pink Flamingo	2200 sq. ft.	$\dfrac{2200}{5000} = 0.44$	$0.44 \times \$1400 = \$616$
Total	5000 sq. ft.	1.000 ⬆ Check	$1400 ⬆ Check

For a set of practice problems, turn to Section Test 15.2.

SECTION TEST 15.2   Inventory and Overhead

The following problems test your understanding of Section 15.2, Retail Method, Overhead.

A. Calculate the inventory value using the retail method.

1. Big Ben's Bolts made the following purchases of packing cartons:

January 14	975 @	$10.20
March 23	925 @	9.80
May 2	1000 @	10.25
July 1	875 @	10.00
September 23	725 @	10.20
November 14	500 @	9.70

   The year-end inventory counted a total of 1146 packing cartons in stock. The retail price is $14.95. What is the inventory value by the retail method?

2. The Rest-Not Restaurant Supply Company made the following purchases of cases of dinner plates:

April 3	30 @	$24.00
June 14	20 @	24.10
August 3	40 @	24.80
October 2	10 @	24.60

   The year-end inventory counted a total of 45 cases of plates in stock. The selling price is $29.95 per case. What is the inventory value by the retail method?

3. The By the Byte Software House had the following purchases of $3\frac{1}{2}''$ high-density floppy diskettes.

January 25	1250 @	$0.59
March 23	1575 @	0.55
April 27	1420 @	0.63
June 25	1450 @	0.61
August 21	1625 @	0.58
October 25	1565 @	0.57
December 13	1115 @	0.60

   The year-end inventory counted a total of 1676 of these high-density floppy diskettes in stock. The selling price is $0.75. What is the inventory value by the retail method?

4. The Jolly Green Midget packages environmentally safe toys that are given away with children's meals in fast-food restaurants. The following is a list of the toy purchases by case.

January 14	580 @	$7.20
April 13	625 @	7.00
July 1	1060 @	7.10
September 23	750 @	7.12
November 27	985 @	7.20

   The year-end inventory counted a total of 1145 cases of environmentally safe toys in stock. The selling price is $9.99. What is the inventory value by the retail method?

B. Calculate the overhead allocation.

1. Woody's Lumber Mill has a weekly overhead of $1950. The fence panel department has sales of $4200, the outdoor deck department $2640, indoor paneling $2100, and the outdoor shed department $3060. What is the overhead for each department using the sales volume method?

2. Gaylon's Boat Shop has a monthly overhead of $8500. The bass boat department had monthly sales of $21,175; the speed boat department $8,800; the jet ski department $12,375; and the pontoon craft department $12,650. What is the overhead for each department using the sales volume method?

3. The Chips and Bits Computer Store has a monthly overhead of $12,500. The laser printer department had $23,625 in monthly sales; the video/CRT department $16,500; the personal computer department $22,125; the accessory department $6750; and the memory chip department $6000. What is the overhead for each department using the sales volume method?

4. The Ko Minh Furniture Store has a monthly overhead of $14,500. The dining room furniture department had $33,250 in monthly sales; the bedroom furniture department $17,100; the outdoor furniture department $9975; the children's furniture department $7600; and the living room furniture $27,075. What is the overhead for each department using the sales volume method?

5. Woody's Lumber Mill has a weekly overhead of $1950. The following is the floor space for each department: fence panel department, 1900 square feet; the outdoor deck department, 1300 square feet; indoor paneling, 775 square feet; and the outdoor shed department, 1025 square feet. What is the overhead for each department using the floor space method?

6. Gaylon's Boat Shop has a monthly overhead of $8500. The following is the floor space for each department: bass boat department, 2880 square feet; the speed boat department, 1520 square feet; the jet ski department, 1920 square feet; and the pontoon craft department, 1680 square feet. What is the overhead for each department using the floor space method?

7. The Chips and Bits Computer Store has a monthly overhead of $12,500. The following is the floor space for each department: laser printer department, 2625 square feet; the video/CRT department, 1425 square feet; the personal computer department, 1950 square feet; the accessory department, 825 square feet; and the memory chip department, 675 square feet. What is the overhead for each department using the floor space method?

8. The Ko Minh Furniture Store has a monthly overhead of $14,500. The following is the floor space for each department: dining room furniture department, 2400 square feet; the bedroom furniture department, 1200 square feet; the outdoor furniture department, 750 square feet; the children's furniture department, 1050 square feet; and the living room furniture department, 2100 square feet. What is the overhead for each department using the floor space method?

9. Claudia has a weekly overhead of $2700. Her operators had the following sales figures for the first week in June: hair, $2345; nails, $1879; and tanning, $1568. Determine the allocation of overhead by sales volume for the given week.

WORKSPACE

Key Terms

average cost
FIFO (first in, first out)
inventory
inventory valuation methods

LIFO (last in, first out)
overhead
perpetual inventory system

physical inventories
ratio
specific identification

Chapter 15 at a Glance

Page	Topic	Key Point	Example
481	Specific identification method	Compile data on quantities and costs. Multiply each quantity by the appropriate cost and add the results.	A firm has in stock 6 radios it bought February 17 at $12 each and 9 radios it bought March 3 at $11 each. What is the total inventory value of radios at this store? $6 \times \$12 = \$\ 72$ $9 \times \$11 = \underline{\$\ 99}$ $\qquad\qquad\ \ \$171$
483	Average cost method	Total cost = (units × cost) + . . . Average cost = $\dfrac{\text{total cost}}{\text{total units}}$ Inventory value = no. of units in stock × average cost	Last year, a firm bought 100 radios on February 17 at $12 each and 200 radios March 3 at $11 each. What is the value of the 15 radios in stock? Total cost = $100 \times \$12 = \$1200$ $\qquad\qquad\ \ 200 \times \$11 = \underline{\$2200}$ $\qquad\qquad\qquad\qquad\qquad \$3400$ Average cost = $\$3400/300$ $\qquad\qquad\ = \$11.33$ rounded Inventory value = $15 \times \$11.33$ $\qquad\qquad\qquad = \$169.95$
484	FIFO method	Determine the no. of units in stock and the price paid for each, assuming they were the last items purchased. Multiply the units times their costs and add the results.	Solve the preceding problem assuming that 203 radios remain in stock and using the FIFO method. Last 200 radios in stock were bought March 3 for $11 each: $200 \times \$11 = \$2200$ Remaining 3 radios in stock were bought February 17 for $12 each: $3 \times \$12 = \$36$ $\$2200 + 36 = \$2236$

Page	Topic	Key Point	Example
485	LIFO method	Determine the no. of units in stock and the price paid for each, assuming they were the first items purchased. Multiply the units times their costs and add the results.	Solve the average cost method problem assuming that 203 radios remain in stock and using the LIFO method. First 100 radios in stock were bought February 17 for $12 each: $100 \times \$12 = \$1200$ Remaining 103 radios in stock were bought March 3 for $11 each: $103 \times \$11 = \$1133$ $\$1200 + 1133 = \$2333$
493	Retail method	Cost of good available for sale at cost = (units $\times$ cost) + (units $\times$ costs) + (units $\times$ costs)	A firm bought 100 radios on February 17 at $12 each and 200 radios March 3 at $11 each. If it sells these radios for $15 each and has 15 left in stock at the end of the year, what is the inventory value using the retail method? COG at cost $= 100 \times \$12 = \$1200$ $ 200 \times \$11 = \underline{\$2200}$ $ \$3400$
		Cost of goods available for sale at retail = total units $\times$ retail price Cost ratio $= \dfrac{\text{COG at cost}}{\text{COG at retail}}$ Ending inventory at retail = remaining inventory $\times$ retail price Ending inventory at cost = ending inventory at retail $\times$ cost ratio	COG at retail $= 300 \times \$15 = \$4500$ Cost ratio $= \dfrac{\$3400}{\$4500} = 0.755. . .$ Ending inventory at retail $= 15 \times \$15 = \$225$ Ending inventory at cost $= \$225 \times 0.7555555$ $= \$170.00$ rounded
495	Allocating overhead by sales volume	Find total sales by adding the sales of all departments.	A firm with a weekly overhead of $4000 had sales one week as follows: Dept. A, $12,000; Dept. B, $4,500; Dept. C, $3,500. Allocate overhead by sales volume. Total sales $= \$12,000 + \$4500 + \$3500$ $ = \$20,000$
		Calculate the department sales ratio for each dept and express as a decimal number. Sales ratio $= \dfrac{\text{department sales}}{\text{total sales}}$	Department ratios: Dept. A $= \dfrac{\$12,000}{\$20,000} = 0.6$ Dept. B $= \dfrac{\$4,500}{\$20,000} = 0.225$ Dept. C $= \dfrac{\$3,500}{\$20,000} = 0.175$
		Department overhead = department sales ratio $\times$ total overhead	Department overhead: Dept. A $= 0.6 \times \$4000 = \$2400$ Dept. B $= 0.225 \times \$4000 = \$900$ Dept. C $= 0.175 \times \$4000 = \$700$

Page	Topic	Key Point	Example
497	Allocating overhead by floor space	Determine total floor space.	A firm with a weekly overhead of $4000 has departments with the following floor space in square feet: Dept. A, 4500; Dept. B, 1600; Dept. C, 1900. Allocate overhead by floor space. Total floor space $= 4500 + 1600 + 1900$ $\qquad\qquad\qquad = 8000$ sq. feet
		Calculate the department floor space ratio and express it as a decimal number. Floor space ratio $=$ $\qquad \dfrac{\text{department floor space}}{\text{total floor space}}$	Floor space ratios: Dept. A $= \dfrac{4500}{8000} = 0.5625$ Dept. B $= \dfrac{1600}{8000} = 0.2$ Dept. C $= \dfrac{1900}{8000} = 0.2375$
		Department overhead $=$ $\qquad$ department floor space ratio $\qquad \times$ total overhead	Department overhead: Dept. A $= 0.5625 \times \$4000 = \$2250$ Dept. B $= 0.2 \times \$4000 = \$800$ Dept. C $= 0.2375 \times \$4000 = \$950$

WORKSPACE

Name

Date

Course/Section

The following problems test your understanding of inventory and overhead.

1. The OKC Community College Bookstore made the following purchases of cases of mechanical pencils:

January 5	125 @	$28.80
May 14	90 @	29.10
August 3	210 @	29.20
October 10	75 @	29.40

The year-end inventory counted a total of 163 cases of mechanical pencils in stock with the following costs:

Number	Cost
11	$28.80
45	29.10
48	29.20
59	29.40

Determine the inventory value by the specific identification method.

2. J.R.'s Construction Company made the following hard hat purchases:

March 23	225 @	$5.70
June 14	375 @	5.90
August 21	250 @	6.10
October 15	150 @	6.20

The year-end inventory counted a total of 253 hard hats in stock with the following costs:

Number	Cost
17	$5.70
21	5.90
123	6.10
92	6.20

Determine the inventory value by the specific identification method.

3. The OKC Community College Bookstore made the following purchases of cases of mechanical pencils:

January 5	125 @	$28.80
May 14	90 @	29.10
August 3	210 @	29.20
October 10	75 @	29.40

The year-end inventory counted a total of 163 cases of mechanical pencils in stock. Determine the inventory by the following methods:

(a) Average cost   (b) FIFO   (c) LIFO   (d) Retail with $39.95 as the retail price

4. J.R.'s Construction Company made the following hard hat purchases:

March 23	225 @ $5.70
June 14	375 @ 5.90
August 21	315 @ 6.10
October 15	180 @ 6.20

The year-end inventory counted a total of 253 hard hats in stock. Determine the inventory by the following methods:

(a) Average cost   (b) FIFO   (c) LIFO   (d) Retail with $10.95 as the retail price

5. Crystal's Diamonds has a total monthly overhead of $6800. The women's watch department had monthly sales of $8050; Austrian crystal sales $11,040; bridal set department $17,480; and men's watch department $9430. What is the overhead for each department using the sales volume method?

6. Two Wheels Are Enough, a bicycle shop, has a total monthly overhead of $11,500. The men's racer department had $11,205 in monthly sales; the tandem department $34,860; the touring department $8715; the children's department $12,450; and the women's racer department $15,770. What is the overhead for each department using the sales volume method?

7. Crystal's Diamonds has a total monthly overhead of $6800. The following is the floor space for each department: women's watch department had 1425 square feet; Austrian crystal sales, 1650 square feet; bridal set departments, 2700 square feet; and men's watch department, 1725 square feet. What is the overhead for each department using the floor space method?

8. Two Wheels Are Enough, a bicycle shop, has a total monthly overhead of $11,500. The following is the floor space for each department: men's racer department, 975 square feet; the tandem department, 2860 square feet; the touring department, 715 square feet; the children's department, 1040 square feet; and the women's racer department, 910 square feet. What is the overhead for each department using the floor space method?

Depreciation

CHAPTER OBJECTIVES

When you complete this chapter successfully, you will be able to:

1. Construct a depreciation schedule using the straight-line and units-of-production methods.

2. Develop a depreciation schedule using the declining-balance and sum-of-the-years'-digits methods.

3. Use the ACRS and MACRS depreciation methods.

■ I. M. Exquisite Architects purchased a new computer system for $5000. It has a useful life of five years with a salvage value of $800. Construct a depreciation schedule using the straight-line method.

■ Shiring Investments purchased a new paper shredder for $450. It has a useful life of four years with a salvage value of $50. Construct a depreciation schedule using the double-declining-balance method.

■ Petal Pushers bought a light truck for $18,750. Construct depreciation schedules using the ACRS and MACRS methods.

Look around you. Everything you see is wearing out to some degree—your clothes, your funiture, your car, your books. In business, an asset's value becomes less and less as time goes on, a pattern called *depreciation*. In this chapter, we will consider the different methods businesses use to depreciate assets for accounting and tax purposes.

Get ready to understand and "appreciate" depreciation.

SECTION 16.1: Straight-Line and Units-of-Production Methods

Can You:

Construct a depreciation schedule using the straight-line method, given the cost, useful life, and salvage value?

Construct a depreciation schedule using the units-of-production method, given the cost, salvage value, estimated units of production, and number of units produced per year?

. . . If not, you need this section.

Capital Expenditures Purchases of physical assets that have a useful life of over one year.

Depreciation The portion of a capital expenditure's cost that may be deducted for income tax purposes to compensate for the asset's wearing out.

Initial Cost The purchase price of an asset.

Useful Life The estimated length of time an asset can/will be used.

Salvage Value The estimated value of an asset at the end of its useful life.

Businesses have many "assets"—their reputation, their employees, their creativity. But these assets do not lose their value (depreciate). Physical assets such as machinery, building, vehicles, and office furniture do wear out, though. These items, which have a useful life of over one year, are called **capital expenditures.**

Under U.S. tax laws, not all, but only a portion of the cost of capital expenditures—the **depreciation**—may be deducted each year. The amount deducted each year depends on the item and the depreciation method used. For example, when a business purchases a sophisticated personal computer for $7000, it expects to get several years' use from the computer. Because of the time factor in the use of the computer, the business may not deduct the full $7000 as a business expense during the first year. The total loss of value of the computer must be spread out over at least a portion of the years the computer will be used.

To determine depreciation, you need to know the three characteristics of all assets: an initial cost, a useful life, and a salvage value. The **initial cost** is the purchase price of the asset. The **useful life** is the estimated length of time the asset is to be used. The **salvage value** of the asset is its estimated value at the end of its useful life. The salvage value may be thousands of dollars or zero or something in between.

Total Depreciation The total loss of an asset due to depreciation, it is the difference between the initial cost and the salvage value.

The **total depreciation** is the total loss of value of an asset due to depreciation. The total depreciation is the difference between the initial cost and the salvage value.

Total depreciation = initial cost − salvage value

Both the useful life and salvage value are only *estimates,* made by the business based on such factors as usage, repair policy, replacement policy, and obsolescence. The Internal Revenue Service (IRS) publishes guidelines to aid businesses in making these estimations for income tax purposes.

For example, a new deluxe steam cleaner for Sparkle Cleaning cost $4500. It has an estimated useful life of five years, and at the end of that time can be sold for about $360. What is the total depreciation?

$$\text{Total depreciation} = \text{cost} - \text{salvage value}$$
$$= \$4500 - \$360 = \$4140$$

The steam cleaner's total loss of value due to depreciation is $4140.

YOUR TURN

WORK THIS PROBLEM

The Question:

Big Ben's Bolts purchased a new packaging machine for $7946. It has an estimated useful life of five years with a salvage value of $150. Calculate the machine's total depreciation.

✔ YOUR WORK

The Solution:

$$\text{Total depreciation} = \text{cost} - \text{salvage value}$$
$$= \$7946 - 150 = \$7796$$

The packaging machine's total loss of value is $7796. During the five years of the machine's useful life, it will depreciate $7796.

How much should an asset be depreciated each year? The answer depends on the depreciation method used.

Using the Straight-Line Method to Calculate Depreciation

Straight-Line Method
A method of depreciating assets that spreads depreciation evenly over the useful life of the asset.

The easiest method of depreciation, the **straight-line method,** spreads the depreciation equally throughout the useful life of the asset. That is, the depreciation is the same for each full year.

$$\text{Annual depreciation} = \frac{\text{total depreciation}}{\text{useful life}}$$

For example, using the straight-line method, what is the annual depreciation of Sparkle Cleaning's new steam cleaner? (The original cost was $4500, useful life 5 years, and salvage value $360.)

$$\text{Total depreciation} = \text{cost} - \text{salvage value}$$
$$= \$4500 - 360 = \$4140$$

$$\text{Annual depreciation} = \frac{\text{total depreciation}}{\text{useful life}} = \frac{\$4140}{5} = \$828$$

The depreciation is $828 per year for five years for a total of $5 \times \$828 = \$4140$

YOUR TURN

WORK THIS PROBLEM

The Question:

In the previous "Your Turn," we determined that the $7946 packaging machine bought by Big Ben's Bolts had a useful life of five years, a salvage value of $150, and total depreciation of $7796. Determine the annual depreciation using the straight-line method.

✔ YOUR WORK

The Solution:

$$\text{Annual depreciation} = \frac{\$7796}{5} = \$1559.20$$

Due to changes in the tax laws, the straight-line depreciation method can only be used for income tax purposes for assets placed in service by December 31, 1980. Today, the straight-line method is widely used by businesses for internal accounting purposes, however. The straight-line method is also the basis of the MACRS method currently in use for tax purposes and discussed in Section 16.3.

Constructing a Depreciation Schedule

Depreciation Schedule
A complete list of all years, annual depreciations, accumulated depreciations, and book values for a capital expenditure over its projected useful life.

Accumulated Depreciation Total depreciation to date.

Book Value The current value of an asset from an accounting standpoint.

Market Value The current value of an asset were it to be sold today.

In order to keep track of an asset's depreciation, accountants usually summarize depreciation information in a depreciation schedule. The **depreciation schedule** contains the year, annual depreciation, accumulated depreciation, and the book value. The **accumulated depreciation** is the total depreciation to date. The **book value** is the current value of an asset. That is, the book value is the accounting value, but it is not necessarily the current market value. Don't confuse the book value and market value. The **market value** is the amount it can be sold for. The book value is simply the difference between the initial cost and the accumulated depreciation.

Book value = cost − accumulated depreciation

Let's construct a depreciation schedule for Sparkle Cleaning's new steam cleaner. As you saw in earlier examples, this machine has a useful life of 5 years and will depreciate $828 per year using the straight-line method. Thus,

Year	Annual Depreciation	Accumulated Depreciation	Book Value
			$4500
1	$828	$828	4500 − 828 = 3672
2	828	828 + 828 = 1656	4500 − 1656 = 2844
3	828	1656 + 828 = 2484	4500 − 2484 = 2016
4	828	2484 + 828 = 3312	4500 − 3312 = 1188
5	828	3312 + 828 = 4140	4500 − 4140 = 360 Salvage value

Note that with the straight-line method, the book value at the end of the useful life is *always* equal to the salvage value. Always verify this; it's a good way to help check the problem.

YOUR TURN

| WORK THIS PROBLEM |

The Garden of Earthly Delights purchased a pot washer for $1800. It has an estimated useful life of six years with a salvage value of $288. Construct a depreciation schedule using the straight-line method.

| ✔ YOUR WORK |

The Solution:

$$\text{Total depreciation} = \$1800 - 288 = \$1512$$

$$\text{Annual depreciation} = \frac{\$1512}{6} = \$252$$

Year	Annual Depreciation	Accumulated Depreciation	Book Value
			$1800
1	$252	$252	1800 − 252 = 1548
2	252	504	1800 − 504 = 1296
3	252	756	1800 − 756 = 1044
4	252	1008	1800 − 1008 = 792
5	252	1260	1800 − 1260 = 540
6	252	1512	1800 − 1512 = 288 Salvage value

Units-of-Production Method   A method of depreciating assets that is based on the projected number of units to be produced by the asset.

Estimated Units of Production   The total number of units an asset is expected to produce over its useful life.

Applying the Units-of-Production Method to Depreciation

Accountants often use the **units-of-production method** to depreciate assets that are only occasionally used, such as a machine used in seasonal production. This method is based on an estimate of the total number of units to be produced by the asset—the **estimated units of production.** The units can be actual objects, like a nut or bolt, or miles for an automobile or hours for an aircraft.

To apply this method, follow these steps:

Steps: Determining Depreciation by Units of Production

STEP 1. *Calculate* total depreciation.
Total depreciation = cost − salvage value

STEP 2. *Calculate* the depreciation per unit of production.

$$\text{Depreciation per unit of production} = \frac{\text{total depreciation}}{\text{estimated units of production}}$$

STEP 3. *Calculate* the annual depreciation.
Annual depreciation = no. of units per year × depreciation per unit of production

For example, a machine that originally cost $1250 has an estimated production of 5500 units, and a salvage value of $150. Calculate the depreciation per unit of production and the annual depreciation for a year in which 1573 units were produced.

Step 1.   Total depreciation = cost − salvage value
= $1250 − 150 = $1100

Step 2.   Depreciation per unit of production $= \dfrac{\text{total depreciation}}{\text{estimated units of production}}$

$= \dfrac{\$1100}{5500} = \$0.20$

Step 3.   Annual depreciation = number of units × depreciation per unit
= 1573 × $0.20 = $314.60

YOUR TURN

WORK THIS PROBLEM

The Question:

A cam construction machine for the Camplex Manufacturing Company cost $4500. The estimated number of cams this machine is expected to produce in its lifetime is 8280. The salvage value of the machine is $360. Calculate the depreciation per cam produced, and the annual depreciation for a year in which 1736 cams were produced.

✔ YOUR WORK

The Solution:

Step 1.   Total depreciation = $4500 − 360 = $4140

Step 2.   Depreciation per cam $= \dfrac{\$4140}{8280} = \$0.50$

Step 3.   Annual depreciation = 1736 × $0.50 = $868

We can easily incorporate the units-of-production method in a depreciation schedule by including an additional column for the number of units produced each year.

For example, The Zippy Delivery Service purchased a used delivery van for $5900. On the basis of past experience, the total estimated mileage is 60,000 with a salvage value of $500. The actual mileage driven over a four-year period is

Year	Miles
1	15,473
2	19,857
3	16,152
4	8,518

Let's construct a depreciation schedule using the units-of-production method.

Step 1. Total depreciation = cost − salvage value
$$= \$5900 - 500 = \$5400$$

Step 2. Depreciation per mile = $\dfrac{\text{total depreciation}}{\text{estimated total miles}} = \dfrac{\$5400}{60,000} = \$0.09$

Step 3.

Year	Miles	Annual Depreciation	Accumulated Depreciation	Book Value
				$5900.00
1	15,473	15,473 × $0.09 = $1392.57	$1392.57	4507.43
2	19,857	19,857 × $0.09 = 1787.13	3179.70	2720.30
3	16,152	16,152 × $0.09 = 1453.68	4633.38	1266.62
4	8518	8518 × $0.09 = 766.62	5400.00	500.00

The book value at the end of the van's useful life, $500, is equal to the salvage value.

WORK THIS PROBLEM

YOUR TURN

The Question:

Lady Beautiful Chocolates purchased a mold for gigantic Valentine's hearts for $1800. The total estimated number of hearts to be produced is 4200, and the mold has a salvage value of $288. The actual production for six years is as follows:

Year	Hearts Produced
1	400
2	550
3	700
4	800
5	925
6	825

Construct a depreciation schedule using the units-of-production method.

✔ YOUR WORK

The Solution:

Step 1.   Total depreciation = $1800 − 288 = $1512

Step 2.   Depreciation per heart = $\dfrac{\$1512}{4200} = \$0.36$

Step 3.

Year	Hearts Produced	Annual Depreciation	Accumulated Depreciation	Book Value
				$1800
1	400	400 × $0.36 = $144	$144	1656
2	550	550 × 0.36 = 198	342	1458
3	700	700 × 0.36 = 252	594	1206
4	800	800 × 0.36 = 288	882	918
5	925	925 × 0.36 = 333	1215	585
6	825	825 × 0.36 = 297	1512	288

Remember the book value at the end of the useful life must equal the salvage value.

In the real world, no one figures depreciation by hand—they use computers or calculators.

Using a Calculator to Determine Depreciation by Units of Production

Most calculators allow you to multiply repeatedly by the same (constant) number, a handy feature in multiplying the units each year by a constant depreciation per unit in the units-of-production method of depreciation.

In the Zippy Delivery Service example,

Step 1.   5900 $\boxed{-}$ 500 $\boxed{=}$ ⟶ ░░ 5400. ░░

Step 2. $\boxed{÷}$ 60000 $\boxed{=}$ ⟶ ░░ 0.09 ░░

Step 3. $\boxed{×}$ 15473 $\boxed{=}$ ⟶ ░░ 1392.57 ░░ 19857 $\boxed{=}$ ⟶ ░░ 1787.13 ░░

   16152 $\boxed{=}$ ⟶ ░░ 1453.68 ░░ 8518 $\boxed{=}$ ⟶ ░░ 766.62 ░░

Notice in step 3 that you do not keep hitting the $\boxed{×}$ key—just the number and the $\boxed{=}$ key. There is no need to key in the constant multiplier 0.09 again.

After you are sure you understand depreciation by the straight-line and units-of-production methods, work the problems in Section Test 16.1.

SECTION TEST 16.1 Depreciation

The following problems test your understanding of Section 16.1, Straight-Line and Units-of-Production Methods.

A. Complete a depreciation schedule using the straight-line method.

1. Woody's Lumber Mill purchased a new copy machine for $6000. It has a useful life of four years with a salvage value of $500. Construct a depreciation schedule using the straight-line method.

2. Shiring Investments purchased a new paper shredder for $450. It has a useful life of four years with a salvage value of $50. Construct a depreciation schedule using the straight-line method.

3. I. M. Exquisite Architects purchased a new computer system for $5000. It has a useful life of five years with a salvage value of $800. Construct a depreciation schedule using the straight-line method.

4. The Zippy Delivery Service purchased a new delivery truck for $22,500. It has a useful life of five years with a salvage value of $1500. Construct a depreciation schedule using the straight-line method.

5. The Spitting Image purchased a new color copier for $9000. It has a useful life of six years with a salvage value of $600. Construct a depreciation schedule using the straight-line method.

6. The Wild Child Recording Studio purchased new office furniture for $4000. It has a useful life of ten years with a salvage value of $700. Construct a depreciation schedule using the straight-line method.

B. Complete a depreciation schedule using the units-of-production method.

1. The Oiled Well Garage purchased a Super-Luber for $6000. The total estimated lubes for the machine are 80,000 lubes. The salvage value is $400. The actual performance was

Year	Lubes
1	15,800
2	23,200
3	24,600
4	16,400

Construct a depreciation schedule using the units-of-production method.

2. The Green Acres Lawn Service purchased a riding lawn mower for $1800. The mower has an estimated lifetime of 50,000 hours of mowing. The salvage value is $300. The actual use was

Year	Hours
1	9,245
2	14,350
3	13,225
4	13,180

Construct a depreciation schedule using the units-of-production method.

3. The Zippy Delivery Service purchased a new delivery truck for $19,500. The total estimated mileage is 100,000 miles. The salvage value is $1500. The actual mileage was

Year	Miles
1	15,285
2	23,162
3	21,475
4	22,317
5	17,761

Construct a depreciation schedule using the units-of-production method.

4. The Dependable Airport Transportation Service purchased a van for $17,800. The total estimated mileage is 80,000 miles. The salvage value is $2300. The actual mileage was

Year	Miles
1	12,480
2	20,880
3	22,400
4	16,960
5	7,280

Construct a depreciation schedule using the units-of-production method.

5. Ava's Air Tours purchased an aircraft generator unit for $15,400. The total estimated aircraft start capability is 50,000 starts. The salvage value is $1200. The actual performance was

Year	Starts
1	7,250
2	14,725
3	14,420
4	13,605

Construct a depreciation schedule using the units-of-production method.

Can You:

Construct a depreciation schedule using the declining balance?
Construct a depreciation schedule using sum-of-the-years'-digits methods?
. . . If not, you need this section.

The straight-line method has the same depreciation each year. But many assets, such as cars, depreciate more at the beginning of their useful life and less at the end. The declining-balance and sum-of-the-years'-digits methods have a larger initial depreciation and smaller depreciations each subsequent year.

Using the Declining-Balance Method of Depreciation

Declining-Balance Method A method of depreciating assets that is expressed as the book value times the depreciation rate; it may be used for taxes only on assets put into service by December 31, 1980.

With the **declining-balance method,** we always use the book value (*balance*) rather than the total depreciation.

> Annual depreciation = rate × book value

The rate in the declining-balance method may take one of three forms:

Double-Declining-Balance Method A form of declining-balance method depreciation in which the rate is equal to 2 divided by the useful life of the asset.

1. 2 ÷ useful life for the **double-declining-balance method** (also called the *200% declining-balance method*). This method, which is twice the straight-line rate, is commonly used only for new assets.
2. 1.5 ÷ useful life for the **150% declining-balance method,** which is commonly applied to used assets.
3. 1.25 ÷ useful life for the **125% declining-balance method.**

150% Declining-Balance Method A form of declining-balance method depreciation in which the rate is equal to 1.5 divided by the useful life of the asset.

For tax purposes, companies may use the declining-balance method only for assets placed in service by December 31, 1980. However, this system is still used for in-house accounting purposes in some firms. It is also a basis for the MACRS method used for tax purposes and discussed in Section 16.3.

For example, let's return to Sparkle Cleaning's purchase of a new deluxe steam cleaner for $4500. It has an estimated useful life of five years with a salvage value of $360.

125% Declining-Balance Method A form of declining-balance method depreciation in which the rate is equal to 1.25 divided by the useful life of the asset.

The first year's depreciation by the double-declining-balance method is

$$\text{First-year depreciation} = \frac{2}{\text{useful life}} \times \text{book value}$$

$$= \frac{2}{5} \times \$4500 = \$1800$$

This depreciation is much larger than the $828 we obtained with the straight-line method in the previous section. The first-year depreciation is *always* greatest with the declining-balance method.

Before calculating the depreciation for the second year, you must determine the new book value. The book value will change each year and must be calculated before computing the next year's depreciation.

> New book value = former book value − latest depreciation

Thus, in the above example,

$$\text{New book value} = \$4500 - 1800 = \$2700$$

We can then calculate the second-year depreciation as

$$\text{Second-year depreciation} = \frac{2}{\text{useful life}} \times \text{book value} = \frac{2}{5} \times \$2700 = \$1080$$

Using the same information as in the straight-line method, we can summarize this material in a depreciation schedule.

Year	Annual Depreciation	Accumulated Depreciation	Book Value
			$4500.00
1	$\frac{2}{5} \times \$4500 = \$1800$	$1800.00	2700.00
2	$\frac{2}{5} \times 2700 = 1080$	2880.00	1620.00

YOUR TURN

WORK THIS PROBLEM

The Question:

Complete the entries for the third and fourth years. Remember to use the new book value each year.

✔ YOUR WORK

The Solution:

$$\text{New book value} = \$2700 - 1080 = \$1620$$
$$\text{Third-year depreciation} = \tfrac{2}{5} \times 1620 = 648$$
$$\text{New book value} = \$1620 - 648 = \$972$$
$$\text{Fourth-year depreciation} = \tfrac{2}{5} \times 972 = 388.80$$
$$\text{New book value} = \$972.00 - 388.80 = \$583.20$$

Year	Annual Depreciation	Accumulated Depreciation	Book Value
			$4500.00
1	$\frac{2}{5} \times \$4500 = \$1800$	$1800.00	2700.00
2	$\frac{2}{5} \times 2700 = 1080$	2880.00	1620.00
3	$\frac{2}{5} \times 1620 = 648$	3528.00	972.00
4	$\frac{2}{5} \times 972 = 388.80$	3916.80	583.20

When using the declining-balance method, you must *never* depreciate below the salvage value. The IRS would frown on such a situation. (So would you when they come after you.)

In our problem, the fifth-year calculation is $\frac{2}{5} \times 583.20 = \$233.28$. But this calculation leads to trouble. Remember that at the start of this example we said that the salvage value was $360. A depreciation of $233.28 is too large because it results in a book value of $583.20 − 233.28 = $349.92, which is below the salvage value. In the fifth year, the asset can only be depreciated down to the salvage value. The depreciation will be the difference between the previous book value and the salvage value.

The completed depreciation schedule looks like this:

Year	Annual Depreciation	Accumulated Depreciation	Book Value
			$4500.00
1	$\frac{2}{5} \times \$4500 = \$1800$	$1800.00	2700.00
2	$\frac{2}{5} \times 2700 = 1080$	2880.00	1620.00
3	$\frac{2}{5} \times 1620 = 648$	3528.00	972.00
4	$\frac{2}{5} \times 972 = 388.80$	3916.80	583.20
5	$223.20 to salvage value	4140.00	360.00

YOUR TURN

WORK THIS PROBLEM

The Question:

Sew What purchased a new fabric-cutting machine for $14,000. It has an estimated life of six years with a salvage value of $1500. Construct a depreciation schedule using the double-declining-balance method.

✔ YOUR WORK

The Solution:

$$\text{First-year depreciation} = \tfrac{2}{6} \times \$14{,}000 = \$4666.67$$
$$\text{New book value} = \$14{,}000 - 4666.67 = \$9333.33$$
$$\text{Second-year depreciation} = \tfrac{2}{6} \times \$9333.33 = \$3111.11$$
$$\text{New book value} = \$9333.33 - 3111.11 = \$6222.22$$
$$\text{Third-year depreciation} = \tfrac{2}{6} \times \$6222.22 = \$2074.07$$
$$\text{New book value} = \$6222.22 - 2074.07 = \$4148.15$$
$$\text{Fourth-year depreciation} = \tfrac{2}{6} \times \$4148.15 = \$1382.72$$
$$\text{New book value} = \$4148.15 - 1382.72 = \$2765.43$$
$$\text{Fifth-year depreciation} = \tfrac{2}{6} \times \$2765.43 = \$921.81$$
$$\text{New book value} = \$2765.43 - 921.81 = \$1843.62$$
$$\text{Sixth-year depreciation} = \tfrac{2}{6} \times \$1843.62 = \$614.54$$
$$\text{New book value} = \$1843.62 - 614.54 = \$1229.08$$

Oops—this last new book value is below the salvage value of $1500. We need to recalculate the sixth year.

$$\text{Sixth-year depreciation} = \$1843.62 - 1500 = \$343.62$$

Year	Annual Depreciation	Accumulated Depreciation	Book Value
			$14,000.00
1	$\frac{2}{6} \times$ $14,000.00 = $4666.67	$ 4,666.67	9,333.33
2	$\frac{2}{6} \times$ 9,333.33 = 3111.11	7,777.78	6,222.22
3	$\frac{2}{6} \times$ 6,222.22 = 2074.07	9,851.85	4,148.15
4	$\frac{2}{6} \times$ 4,148.15 = 1382.72	11,234.57	2,765.43
5	$\frac{2}{6} \times$ 2,765.43 = 921.81	12,156.38	1,843.62
6	$343.62 to salvage value	12,500.00	1,500.00

You would use exactly the same procedure but a different rate to calculate the depreciation by the 150% or by the 125% declining-balance methods. For example a six-year depreciation by the 150% method would have a rate of $\frac{1.5}{6}$ and by the 125% method would have a rate of $\frac{1.25}{6}$.

Computing depreciation with the declining-balance method—regardless of rate—goes faster if you use a calculator.

Using a Calculator to Determine Depreciation By Declining Balance

Let's solve the Sparkle Cleaning problem we worked in text but use a calculator.

For the first year, key in the following:

2 [÷] 5 [×] 4500 [=] ⟶   **1800.**   (first-year depreciation)

4500 [−] 1800 [=] ⟶   **2700.**   (new balance)

Repeat the process for each year, writing down the answers as you go, to find the full depreciation schedule.

Applying the Sum-of-the-Years'-Digits Method of Depreciation

Sum-of-the-Years'-Digits Method   A method of depreciating assets that uses a large initial depreciation and smaller depreciations at the end of each subsequent year; it may be used for taxes only on assets put into service by December 31, 1980.

Like the declining-balance method, the **sum-of-the-years'-digits method** involves a large initial depreciation and a smaller depreciation each subsequent year. In the sum-of-the-years'-digits method, the book value at the end of the useful life is *always* equal to the salvage value.

The sum-of-the-years'-digits method uses the total depreciation. Each year, the depreciation is the product of the total depreciation and a fraction that varies each year.

For example, if the useful life is four years, the fractions have numerators of 4, 3, 2, and 1. The numerator is 4 for the first year, 3 for the second year, 2 for the third year, and 1 for the fourth year. The denominator is the same each year and is the sum of the numerators. For a useful life of four years, the denominator is $4 + 3 + 2 + 1 = 10$. This is where the sum-of-the-years'-digits method gets its name.

Thus for a four-year depreciation the fractions are $\frac{4}{10}$, $\frac{3}{10}$, $\frac{2}{10}$, and $\frac{1}{10}$. The fractions in a sum-of-the-years'-digits problem will *always* add up to 1. In this case, the fractions add up to 10/10 or 1.

YOUR TURN

[WORK THIS PROBLEM]

The Question:

What are the fractions for an asset with a useful life of seven years?

[✔ YOUR WORK]

The Solution:

There will be seven fractions:

1. The numerators will be 7, 6, 5, 4, 3, 2, and 1.
2. The denominator will be $7 + 6 + 5 + 4 + 3 + 2 + 1 = 28$.
3. The fractions, then, are $\frac{7}{28}$, $\frac{6}{28}$, $\frac{5}{28}$, $\frac{4}{28}$, $\frac{3}{28}$, $\frac{2}{28}$, and $\frac{1}{28}$.

Check: $\dfrac{7}{28} + \dfrac{6}{28} + \dfrac{5}{28} + \dfrac{4}{28} + \dfrac{3}{28} + \dfrac{2}{28} + \dfrac{1}{28} = \dfrac{28}{28} = 1$

It's really easy, but can be quite lengthy for assets with a long useful life. For an asset with a useful life of 50 years, the numerators will be 50, 49, 48, 47, . . . , 3, 2, 1. The denominator will be $50 + 49 + 48 + 47 + \cdots + 3 + 2 + 1$.

Luckily, there is a shortcut method to calculate the denominator using the following formula:

$$\text{Denominator} = \frac{N \times (N + 1)}{2}, \text{ where } N = \text{useful life}$$

For a useful life of 50 years.

Useful life

$$\text{Denominator} = \frac{50 \times (50 + 1)}{2} = \frac{50 \times 51}{2} = 1275$$

Isn't that easier than adding all those numbers together?

YOUR TURN

[WORK THIS PROBLEM]

The Question:

What are the first five fractions for an asset with a useful life of 35 years?

> ✔ YOUR WORK

The Solution:

Useful life

$$\text{Denominator} = \frac{35 \times (35 + 1)}{2} = \frac{35 \times 36}{2} = 630$$

The fractions for the first five years of depreciation are $\frac{35}{630}$, $\frac{34}{630}$, $\frac{33}{630}$, $\frac{32}{630}$, and $\frac{31}{630}$.

For tax purposes, the sum-of-the-years'-digits method may be used only for assets placed in service by December 31, 1980. However, this method is still used as an internal accounting device in some companies.

Calculating depreciation using the sum-of-the-years'-digits method is simple if you follow these steps:

Steps: Sum-of-the-Years'-Digits Method

STEP 1.   *Find* the fractions for the useful life of the asset.

STEP 2.   *Calculate* total depreciation.

$$\text{Total depreciation} = \text{cost} - \text{salvage value}$$

STEP 3.   *Calculate* depreciation for each year by multiplying total depreciation by the correct fraction for that year.

For example, the Sparkle Cleaning Co. purchased a new deluxe steam cleaner for $4500. It has an estimated useful life of five years with a salvage value of $360.

Step 1. $\text{Denominators} = \dfrac{5 \times (5 + 1)}{2} = \dfrac{5 \times 6}{2} = \dfrac{30}{2} = 15$ or

Denominator = $5 + 4 + 3 + 2 + 1 = 15$
Numerators = 5, 4, 3, 2, 1
Fractions = $\dfrac{5}{15}$, $\dfrac{4}{15}$, $\dfrac{3}{15}$, $\dfrac{2}{15}$, $\dfrac{1}{15}$

Step 2.   Total depreciation = cost − salvage value = $4500 − 360 = $4140

Step 3.   First-year depreciation $= \frac{5}{15} \times \$4140 = \$1380$
Second-year depreciation $= \frac{4}{15} \times \$4140 = \$1104$
Third-year depreciation $= \frac{3}{15} \times \$4140 = \$828$
Fourth-year depreciation $= \frac{2}{15} \times \$4140 = \$552$
Fifth-year depreciation $= \frac{1}{15} \times \$4140 = \$276$

We can then complete the depreciation schedule as follows:

Year	Annual Depreciation	Accumulated Depreciation	Book Value
			$4500
1	$\frac{5}{15} \times \$4140 = \$1380$	$1380	3120
2	$\frac{4}{15} \times 4140 = 1104$	2484	2016
3	$\frac{3}{15} \times 4140 = 828$	3312	1188
4	$\frac{2}{15} \times 4140 = 552$	3864	636
5	$\frac{1}{15} \times 4140 = 276$	4140	360 Salvage value

The book value at the end of the fifth year must be the same as the salvage value. The book value at the end of the useful life is *always* the same as the salvage value. This is a good way to help check your calculation.

(WORK THIS PROBLEM)———————————————————————

The Question:

The Garden of Earthly Delights purchased a pot scrubber for $1800. It has an estimated life of six years with a salvage value of $288. Construct a depreciation schedule using the sum-of-the-years'-digits method.

(✔ YOUR WORK)———————————————————————

The Solution:

Step 1. Denominator $= \dfrac{6 \times (6 + 1)}{2} = \dfrac{42}{2} = 21$

The numerators will be 6, 5, 4, 3, 2, and 1
The fractions are $\frac{6}{21}$, $\frac{5}{21}$, $\frac{4}{21}$, $\frac{3}{21}$, $\frac{2}{21}$, and $\frac{1}{21}$

Step 2. Total depreciation = cost − salvage value = $1800 − 288 = $1512

Step 3. First-year depreciation $= \frac{6}{21} \times \$1512 = \$432$
Second-year depreciation $= \frac{5}{21} \times 1512 = 360$
Third-year depreciation $= \frac{4}{21} \times 1512 = 288$
Fourth-year depreciation $= \frac{3}{21} \times 1512 = 216$
Fifth-year depreciation $= \frac{2}{21} \times 1512 = 144$
Sixth-year depreciation $= \frac{1}{21} \times 1512 = 72$

Year	Annual Depreciation	Accumulated Depreciation	Book Value
			$1800
1	$\frac{6}{21} \times \$1512 = \$432$	$432	1368
2	$\frac{5}{21} \times 1512 = 360$	792	1008
3	$\frac{4}{21} \times 1512 = 288$	1080	720
4	$\frac{3}{21} \times 1512 = 216$	1296	504
5	$\frac{2}{21} \times 1512 = 144$	1440	360
6	$\frac{1}{21} \times 1512 = 72$	1512	288 Salvage value

Now, turn to Section Test 16.2 for a set of practice problems involving depreciation by the declining-balance and sum-of-the-years'-digits methods.

SECTION TEST 16.2 Depreciation

Name _____

Date _____

Course/Section _____

The following problems test your understanding of Section 16.2, Declining-Balance and Sum-of-the-Years'-Digits Methods.

A. Complete a depreciation schedule using the double-declining-balance method.

1. Woody's Lumber Mill purchased a new copy machine for $6000. It has a useful life of four years with a salvage value of $500. Construct a depreciation schedule using the double-declining-balance method.

2. Shiring Investments purchased a new paper shredder for $450. It has a useful life of four years with a salvage value of $50. Construct a depreciation schedule using the double-declining-balance method.

3. I. M. Exquisite Architects purchased a new computer system for $5000. It has a useful life of five years with a salvage value of $800. Construct a depreciation schedule using the double-declining-balance method.

4. The Zippy Delivery Service purchased a new delivery truck for $22,500. It has a useful life of five years with a salvage value of $1500. Construct a depreciation schedule using the double-declining-balance method.

5. The Spitting Image purchased a new color copier for $9000. It has a useful life of six years with a salvage value of $600. Construct a depreciation schedule using the double-declining-balance method.

6. The Wild Child Recording Studio purchased new office furniture for $4000. It has a useful life of ten years with a salvage value of $700. Construct a depreciation schedule using the double-declining-balance method.

B. Complete a depreciation schedule using the sum-of-the-years'-digits method.

1. The Oiled Well Garage purchased a car lift for $6000. The salvage value is $400. The useful life is four years. Construct a depreciation schedule using the sum-of-the-years'-digits method.

2. The Green Acres Lawn Service purchased a riding lawn mower for $1800. The salvage value is $300. The useful life is four years. Construct a depreciation schedule using the sum-of-the-years'-digits method.

3. The Zippy Delivery Service purchased a new delivery truck for $19,500. The salvage value is $1500. The useful life is five years. Construct a depreciation schedule using the sum-of-the-years'-digits method.

4. The Dependable Airport Transportation Service purchased a van for $17,800. The salvage value is $2300. The useful life is five years. Construct a depreciation schedule using the sum-of-the-years'-digits method.

5. Ava's Air Tours purchased an aircraft generator unit for $15,400. The salvage value is $1200. The useful life is four years. Construct a depreciation schedule using the sum-of-the-years'-digits method.

6. J.R.'s Construction Company purchased a commercial paint sprayer for $3500. The salvage value is $300. The useful life is four years. Construct a depreciation schedule using the sum-of-the-years'-digits method.

7. Dynamite Deals paid $770 for an integrated filing system. The salvage value was estimated to be $150 after an expected life of 8 years. The accounting department is trying to decide which depreciation method to use for their internal records.
   (a) Make a depreciation schedule for the 8 years using the double-declining-balance method.
   (b) Make a depreciation schedule for the 8 years using the sum-of-the-years'-digits method.
   (c) Which method gives the most depreciation in the first year?
   (d) Which method completes the depreciation process fastest?
   (e) Write a sentence giving an overall comparison of the two methods.

SECTION 16.3: The ACRS and MACRS Methods

Can You:

Construct a depreciation schedule using the ACRS method?
Develop a depreciation schedule using the MACRS method?
. . . If not, you need this section.

Under U.S. federal tax laws, companies must depreciate all assets placed in service after December 31, 1980 using the ACRS or MACRS methods.

Using the ACRS Method of Depreciation

ACRS (Accelerated Cost Recovery System) Method A method of depreciating assets that accelerates depreciation and shortens the recovery period; it may be used only for assets put into service after December 31, 1980 and before January 1, 1987.

Assets placed in service after December 31, 1980 and before January 1, 1987 are depreciated using the **accelerated cost recovery system (ACRS) method.** This method has several advantages over the older techniques we have discussed.

1. The ACRS is an accelerated depreciation method. That is, it permits businesses to write off expenses quickly on their taxes.
2. The time period over which the asset is depreciated—the *recovery period*—is shorter than the time period in the older methods. There are only four different recovery periods.

ACRS Recovery Periods

Recovery Period	Type of Assets
3 years	Automobiles, light trucks, and machinery with a useful life of 7 years or less
5 years	Heavy-duty trucks, most machinery and equipment, most office equipment (copiers, computers), and furniture.
10 years	Manufactured homes
15 years	Most real estate

3. Depreciation calculations using the ACRS method are very easy. Each year, the depreciation is simply the product of a given depreciation rate times the cost.

ACRS Depreciation Rates

Recovery Year	Recovery Period			
	3-Year	5-Year	10-Year	15-Year
1	25%	15%	8%	5%
2	38%	22%	14%	10%
3	37%	21%	12%	9%
4		21%	10%	8%
5		21%	10%	7%
6			10%	7%
7			9%	6%
8			9%	6%
9			9%	6%
10			9%	6%
11				6%
12				6%
13				6%
14				6%
15				6%

To use the ACRS method, just use the preceding tables and the following steps:

Steps: Calculating ACRS Depreciation

STEP 1. *Use* the ACRS Recovery Periods table to determine the recovery period for the asset.

STEP 2. *Use* the recovery period and the ACRS Depreciation Rates table to obtain the depreciation rates and convert them to decimals.

STEP 3. *Find* the depreciation.

$$\text{Depreciation} = \text{rate} \times \text{cost}$$

For example, let's calculate the first-year depreciation on a new grinding machine with an original cost of $12,500.

Step 1. Three-year recovery

Step 2. First-year rate is 25% = 0.25
Second-year rate is 38% = 0.38
Third-year rate is 37% = 0.37

Step 3. First-year depreciation = first-year rate × cost = 0.25 × $12,500 = $3125

YOUR TURN

WORK THIS PROBLEM

The Question:

Calculate the second- and third-year depreciations for the example given and construct a depreciation schedule.

✔ YOUR WORK

The Solution:

We have already completed steps 1 and 2.

Step 3.   Second-year depreciation = $0.38 \times \$12,500 = \$4750$
Third-year depreciation = $0.37 \times \$12,500 = \$4625$

Year	Depreciation	Accumulated Depreciation	Book Value
			$12,500
1	$3125	$ 3,125	9,375
2	4750	7,875	4,625
3	4625	12,500	0

Applying the MACRS Method of Depreciation

MACRS (Modified Accelerated Cost Recovery System) Method   A method of depreciating assets that slows depreciation and lengthens the recovery period compared to the ACRS method; it may be used only for assets put into service after December 31, 1986.

The Tax Reform Act of 1986 modified the ACRS method and produced the **modified accelerated cost recovery system (MACRS) method**. MACRS is used for income tax purposes for all assets placed in service after December 31, 1986.

The MACRS method has longer recovery periods for most assets than the ACRS method. For example, the recovery period for automobiles jumped from three years in the ACRS method to five years in the MACRS method.

MACRS Recovery Periods

Recovery Period	Type of Assets
3 year	Certain types of horses
5 year	Automobiles, light trucks, typewriters, calculators, copiers, computers
7 year	Office furniture and fixtures
10 year	Vessels, barges, tugs, and similar water transportation equipment
15 year	Telephone distribution plant
20 year	Municipal sewers
27.5 year	Residential rental property
31.5 year	Nonresidential real property

Rates in the MACRS method also differ from those of ACRS method.

MACRS Depreciation Rates: Half-Year Convention

Recovery Year	Recovery Period					
	3-Year	**5-Year**	**7-Year**	**10-Year**	**15-Year**	**20-Year**
1	33.33%	20.00%	14.29%	10.00%	5.00%	3.750%
2	44.45%	32.00%	24.49%	18.00%	9.50%	7.219%
3	14.81%	19.20%	17.49%	14.40%	8.55%	6.677%
4	7.41%	11.52%	12.49%	11.52%	7.70%	6.177%
5		11.52%	8.93%	9.22%	6.93%	5.713%
6		5.76%	8.92%	7.37%	6.23%	5.285%
7			8.93%	6.55%	5.90%	4.888%
8			4.46%	6.55%	5.90%	4.522%
9				6.56%	5.91%	4.462%
10				6.55%	5.90%	4.461%
11				3.28%	5.91%	4.462%
12					5.90%	4.461%
13					5.91%	4.462%
14					5.90%	4.461%
15					5.91%	4.462%
16					2.95%	4.461%
17						4.462%
18						4.461%
19						4.462%
20						4.461%
21						2.231%

Half-Year Convention
In the MACRS depreciation method, the assumption that an asset is placed in service by midyear.

Midquarter Convention
In the MACRS depreciation method, a method of calculating depreciation when an asset is *not* placed in service by midyear.

Calculating depreciation is a bit more complicated because the MACRS depreciates most assets using a **half-year convention,** which assumes that the asset is placed in service by midyear. Assets placed in service in the last half of the year are subject to a **midquarter convention** and yet another set of tables, which we will not go into in this text.

Using the half-year convention results in the depreciation being spread over one year more than the recovery period. For example, if you look in the MACRS Depreciation Rates table, you will note that the three-year recovery period is spread over four years. The MACRS rates are based on a combination of the straight-line and the declining-balance methods.

To calculate depreciation with the MACRS method, use these steps:

Steps: Calculating MACRS Depreciation

STEP 1. *Use* the MACRS Recovery Periods table to determine the recovery period for the asset.

STEP 2. *Use* the recovery period and the MACRS Depreciation Rates table to find the depreciation rates and convert them to decimals.

STEP 3. *Find* the depreciation.

$$\text{Depreciation} = \text{rate} \times \text{cost}$$

For example, let's use the MACRS method to calculate the first-year depreciation on a new grinding machine with an original cost of $12,500.

Step 1. Five-year recovery

Step 2.
First-year rate = 20%   = 0.20
Second-year rate = 32%   = 0.32
Third-year rate = 19.2% = 0.192
Fourth-year rate = 11.52% = 0.1152
Fifth-year rate = 11.52% = 0.1152
Sixth-year rate = 5.76% = 0.0576

Step 3. First-year depreciation = first-year rate × cost = 0.20 × $12,500 = $2500

WORK THIS PROBLEM

The Question:

Calculate the remaining depreciation for the example given using the MACRS method and construct a depreciation schedule.

✔ YOUR WORK

The Solution:

Again, steps 1 and 2 have already been solved.

Step 3. Second-year depreciation = 0.32 × $12,500 = $4000

or 12500 ⨯ 32 % ⟶ 4000.

Third-year depreciation = 0.192 × $12,500 = $2400

or 12500 ⨯ 19.2 % ⟶ 2400.

Fourth-year depreciation = 0.1152 × $12,500 = $1440
Fifth-year depreciation = 0.1152 × $12,500 = $1440
Sixth-year depreciation = 0.0576 × $12,500 = $720

Year	Depreciation	Accumulated Depreciation	Book Value
			$12,500
1	$2500	$ 2,500	10,000
2	4000	6,500	6,000
3	2400	8,900	3,600
4	1440	10,340	2,160
5	1440	11,780	720
6	720	12,500	0

For a set of practice problems on the ACRS and MACRS methods, turn to Section Test 16.3.

SECTION TEST 16.3   Depreciation

The following problems test your understanding of Section 16.3, the ACRS and MACRS Methods.

1. Bob's Beeflike Burgers bought a company car for $17,500. Construct a depreciation schedule using the (a) ACRS method and (b) MACRS method.

2. Petal Pushers bought a light truck for $18,750. Construct a depreciation schedule using the (a) ACRS method and (b) MACRS method.

3. Shiring Investments bought new office furniture for $4200. Construct a depreciation schedule using the (a) ACRS method and (b) MACRS method.

4. Stauffer Research and Development purchased new desks for $2800. Construct a depreciation schedule using the (a) ACRS method and (b) MACRS method.

5. Bells-a-Ringing Telecommunications built a telephone distribution plant at a total cost of $5,550,000. This has an estimated salvage value of $400,000 after an expected life of 16 years. While trying to determine the appropriate form of depreciation to use internally, each of the following values was needed. Find each of them.
    (a) Book value for year 10, using straight line.
    (b) Depreciation for year 4 using sum-of-the-years'-digits.
    (c) Accumulated depreciation for year 3 using double-declining balance.
    (d) Book value for year 5 using ACRS (treat as a 15-year item).
    (e) Depreciation for year 7 using MACRS (treat as a 15-year item).

Key Terms

accumulated depreciation
ACRS (accelerated cost recovery system) method
book value
capital expenditures
declining-balance method
depreciation
depreciation schedule
double-declining-balance method

estimated units of production
half-year convention
initial cost
MACRS (modified accelerated cost recovery system) method
market value
midquarter convention
150% declining-balance method

125% declining-balance method
salvage value
straight-line method
sum-of-the-years'-digits method
total depreciation
units-of-production method
useful life

Chapter 16 at a Glance

Page	Topic	Key Point	Example
510	Total depreciation	Total depreciation = initial cost − salvage value	What is the total depreciation on a steam cleaner purchased for $4500 and having a salvage value of $360 with a useful life of five years? Total depreciation = $4500 − 360 = $4140
511	Straight-line method	Annual depreciation = $\dfrac{\text{total depreciation}}{\text{useful life}}$	What is the annual depreciation in the preceding example using the straight-line method? Annual depreciation = $\dfrac{\$4140}{5} = \$828$
512	Book value	Book value = cost − accumulated depreciation	What is the book value of the cleaner in the preceding examples after two years? Book value = $4500 − (2 × $828) = $4500 − 1656 = $2844
513	Units-of-production method	Depreciation per unit of production = $\dfrac{\text{total depreciation}}{\text{estimated units of production}}$ Annual depreciation = no. of units per year × depreciation per unit of production	What is the annual depreciation in the preceding example using the units-of-production method. Assume that the machine can clean a total of 6000 rugs and cleaned 1450 this year. Depreciation per unit of production = $\dfrac{\$4140}{6000}$ = $0.69 Annual depreciation = 1450 × $0.69 = $1000.50
521	Declining-balance method	Annual depreciation = rate × book value New book value = former book value − latest depreciation	Using the double declining-balance method, find annual depreciation for the cleaner in the preceding example for the first year and calculate the new book value after one year. Rate = $\dfrac{2 \text{ or } 1.5 \text{ or } 1.25}{\text{useful life}}$ Use the rate: $\dfrac{2}{\text{useful life}}$ Annual depreciation = $\dfrac{2}{5} \times \$4500 = \$1800$ New book value = $4500 − 1800 = $2700

Page	Topic	Key Point	Example
524	Sum-of-the-years'-digits methods	Find depreciation fraction denominator: $$\frac{N \times (N+1)}{2}$$ where N is useful life	What is the annual depreciation in year 4 of the useful life of the cleaner in the preceding example using the sum-of-the-years'-digits method? $$\text{Denominator} = \frac{5 \times (5+1)}{2} = 15$$
		Find depreciation fraction numerator: Assign total useful life number to year 1, useful life minus 1 to year 2, useful life minus 2 to year 3, and so on.	Numerators: Year 1 = 5 Year 2 = 4 Year 3 = 3 Year 4 = 2 Year 5 = 1 $$\text{Depreciation fraction} = \frac{2}{15}$$
		Annual depreciation = total depreciation × depreciation fraction	$$\text{Annual depreciation} = \$4140 \times \frac{2}{15} = \$552$$
533	ACRS method	Annual depreciation = cost × ACRS rate	What is the annual depreciation for the cleaner in the preceding example in year 3 of its useful life using the ACRS method? The cleaner has a useful life of less than seven years and so is depreciated over three years. The ACRS rate for the third year is 37%. Annual depreciation = $4500 × 0.37 = $1665
535	MACRS method	Annual depreciation = cost × MACRS rate	What is the annual depreciation for the cleaner in the preceding example in year 3 of its useful life using the MACRS method? Use a 5-year recovery period. The MACRS rate for the third year is 19.2% Annual depreciation = $4500 × 0.192 = $864

SELF-TEST Depreciation

The following problems test your understanding of depreciation.

1. Helen's House of Beauty purchased a tanning bed for $8000. It has a useful life of five years with a salvage value of $500. Construct a depreciation schedule using
   (a) Straight-line method
   (b) Double-declining-balance method
   (c) Sum-of-the-years'-digits method

2. The By the Byte Software House purchased a minicomputer for $12,000. It has a useful life of six years with a salvage value of $1500. Construct a depreciation schedule using
   (a) Straight-line method
   (b) Double-declining-balance method
   (c) Sum-of-the-years'-digits method

3. Petal Peddlers purchased a delivery truck for $18,000. The total estimated mileage is 100,000 miles and the truck has a salvage value of $1500. The actual mileage was

Year	Miles
1	15,780
2	32,910
3	29,520
4	21,790

   Construct a depreciation schedule using the units-of-production method.

4. Big Ben's Bolts purchased a bolt-producing machine for $13,500. The total estimated bolts produced by the machine are 60,000 bolts. The salvage value is $900. The actual production was

Year	Bolts
1	12,500
2	18,250
3	17,420
4	11,830

   Construct a depreciation schedule using the units-of-production method.

5. Over-and-Over Academy bought a minivan for $18,950. Construct a depreciation schedule using the (a) ACRS method and (b) MACRS method.

6. Crystal's Diamonds purchased new office equipment for $2700. Construct a depreciation schedule using the (a) ACRS method and (b) MACRS method.

WORKSPACE

Taxes

> Nothing is certain but death and taxes.

> Yes, but not in that order. You owe us money for 1995.

CHAPTER OBJECTIVES

When you complete this chapter successfully, you will be able to:

1. Determine employees' federal income tax and FICA withholdings.

2. Calculate employers' FUTA and SUTA unemployment taxes.

3. Find sales, excise, and property taxes.

■ Abe Beame's monthly gross pay is $3875. He is married with three exemptions. What is his FIT using the percentage method? What are his Social Security and Medicare deductions?

■ Mike is an employee of the Wild Child Recording Studio. He earned $5900 in the first quarter. The studio is not required to participate in a state unemployment plan. What is the studio's FUTA tax on Mike?

■ The OKC Community College Bookstore sold an English literature textbook for $41.55. If the sales tax rate is $6\frac{1}{4}\%$, what are the tax and the total price?

About 200 years ago, one of the founders of this country made an observation that has stood the test of time. As Benjamin Franklin noted, "Nothing is certain but death and taxes." In this chapter, we will consider some of the many forms of taxation that affect businesses in the United States. But as you read this chapter, keep in mind that while taxes may be certain, they are also ever changing. In fact, the U.S. tax codes have led to the rise in businesses who offer tax accounting or tax law services.

Fortunately, many of the underlying principles are straightforward. So don't worry. This chapter really is not very "taxing" at all.

SECTION 17.1: Federal Income Tax and FICA

Can You:

Determine federal income tax by the wage bracket method?
Calculate federal income tax by the percentage method?
Compute FICA?
. . . If not, you need this section.

Deduction Monies subtracted from gross pay to cover the costs of federal, state, city income tax withholding, FICA, health insurance, life insurance, annuities, union dues, and so on.

Look at any paycheck and you're sure to spot a large difference between wages or salary and what an employee actually gets to take home. The difference is the **deductions** removed.

Common deductions include federal income tax, FICA (Social Security and Medicare), state and city income taxes, health insurance, life insurance, annuities, and union dues. At times, the list seems endless and the total is staggering.

> Total deductions = sum of all deductions

Gross Pay Earnings before any deductions.

In Chapter 6, we calculated **gross pay**—earnings before any deductions. **Net pay**—"take-home pay"—is pay *after* deductions.

Net Pay Earnings after all deductions.

> Net pay = gross pay − total deductions

Understanding Federal Income Tax

Federal Income Tax Withholding (FIT) Deductions designed to cover most or all of an employee's tax debt to the government in a given year.

Usually, the largest deduction from a paycheck is the **federal income tax (FIT).** Employers are required to *withhold* a certain amount from each paycheck and to keep records of each employee's earnings and withholdings. To help employers compute the amount of federal income tax to withhold, the Internal Revenue Service (IRS) supplies businesses with tax tables in Circular E. Several commonly used tables from this circular appear in this section.

In addition, each quarter (every 13 weeks), employers must send a summary of employees' earnings and tax withholdings to the IRS, using Form 941, "Employer's Quarterly Federal Tax Return." Employers are required to make periodic transfers of taxes to the Federal Reserve Bank or to other authorized financial institutions. A Federal deposit coupon is sent along with the transfer. A sample from IRS Circular E, Form 8109, is shown.

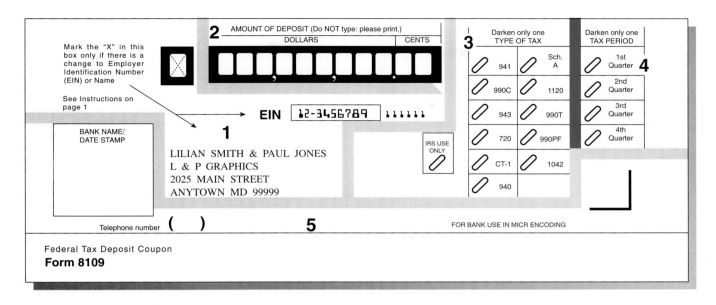

Finally, federal laws require all employees to file a "W4" form that declares the number of exemptions they wish to claim. An exemption is an allowance for a member of the household. A W4 form *must* be filed before income tax can be computed.

Using the Wage Bracket Method

There are two principal methods of calculating federal income tax: the wage bracket and percentage methods. The percentage method is generally used by computerized payrolls because it requires less storage for tables.

Wage Bracket Method
A method for calculating federal income tax withholding with the use of government-prepared tables.

The **wage bracket method** is the easiest to compute by hand. It only requires the use of a table for the appropriate period: weekly, biweekly, semimonthly, monthly, or daily payroll. For each payroll period, there is a separate table for single and married employees. This makes a total of ten tables: weekly single, weekly married, biweekly single, biweekly married, and so on. In this section on the next four pages, we have included the tables for Weekly Single and Weekly Married.

SINGLE Persons—WEEKLY Payroll Period
(For Wages Paid in 1995)

If the wages are—		And the number of withholding allowances claimed is—										
At least	But less than	0	1	2	3	4	5	6	7	8	9	10
		The amount of income tax to be withheld is—										
$0	$55	0	0	0	0	0	0	0	0	0	0	0
55	60	1	0	0	0	0	0	0	0	0	0	0
60	65	2	0	0	0	0	0	0	0	0	0	0
65	70	3	0	0	0	0	0	0	0	0	0	0
70	75	3	0	0	0	0	0	0	0	0	0	0
75	80	4	0	0	0	0	0	0	0	0	0	0
80	85	5	0	0	0	0	0	0	0	0	0	0
85	90	6	0	0	0	0	0	0	0	0	0	0
90	95	6	0	0	0	0	0	0	0	0	0	0
95	100	7	0	0	0	0	0	0	0	0	0	0
100	105	8	1	0	0	0	0	0	0	0	0	0
105	110	9	1	0	0	0	0	0	0	0	0	0
110	115	9	2	0	0	0	0	0	0	0	0	0
115	120	10	3	0	0	0	0	0	0	0	0	0
120	125	11	4	0	0	0	0	0	0	0	0	0
125	130	12	4	0	0	0	0	0	0	0	0	0
130	135	12	5	0	0	0	0	0	0	0	0	0
135	140	13	6	0	0	0	0	0	0	0	0	0
140	145	14	7	0	0	0	0	0	0	0	0	0
145	150	15	7	0	0	0	0	0	0	0	0	0
150	155	15	8	1	0	0	0	0	0	0	0	0
155	160	16	9	2	0	0	0	0	0	0	0	0
160	165	17	10	2	0	0	0	0	0	0	0	0
165	170	18	10	3	0	0	0	0	0	0	0	0
170	175	18	11	4	0	0	0	0	0	0	0	0
175	180	19	12	5	0	0	0	0	0	0	0	0
180	185	20	13	5	0	0	0	0	0	0	0	0
185	190	21	13	6	0	0	0	0	0	0	0	0
190	195	21	14	7	0	0	0	0	0	0	0	0
195	200	22	15	8	0	0	0	0	0	0	0	0
200	210	23	16	9	2	0	0	0	0	0	0	0
210	220	25	18	10	3	0	0	0	0	0	0	0
220	230	26	19	12	5	0	0	0	0	0	0	0
230	240	28	21	13	6	0	0	0	0	0	0	0
240	250	29	22	15	8	0	0	0	0	0	0	0
250	260	31	24	16	9	2	0	0	0	0	0	0
260	270	32	25	18	11	3	0	0	0	0	0	0
270	280	34	27	19	12	5	0	0	0	0	0	0
280	290	35	28	21	14	6	0	0	0	0	0	0
290	300	37	30	22	15	8	1	0	0	0	0	0
300	310	38	31	24	17	9	2	0	0	0	0	0
310	320	40	33	25	18	11	4	0	0	0	0	0
320	330	41	34	27	20	12	5	0	0	0	0	0
330	340	43	36	28	21	14	7	0	0	0	0	0
340	350	44	37	30	23	15	8	1	0	0	0	0
350	360	46	39	31	24	17	10	2	0	0	0	0
360	370	47	40	33	26	18	11	4	0	0	0	0
370	380	49	42	34	27	20	13	5	0	0	0	0
380	390	50	43	36	29	21	14	7	0	0	0	0
390	400	52	45	37	30	23	16	8	1	0	0	0
400	410	53	46	39	32	24	17	10	3	0	0	0
410	420	55	48	40	33	26	19	11	4	0	0	0
420	430	56	49	42	35	27	20	13	6	0	0	0
430	440	58	51	43	36	29	22	14	7	0	0	0
440	450	59	52	45	38	30	23	16	9	2	0	0
450	460	61	54	46	39	32	25	17	10	3	0	0
460	470	62	55	48	41	33	26	19	12	5	0	0
470	480	64	57	49	42	35	28	20	13	6	0	0
480	490	66	58	51	44	36	29	22	15	8	0	0
490	500	69	60	52	45	38	31	23	16	9	2	0
500	510	72	61	54	47	39	32	25	18	11	3	0
510	520	75	63	55	48	41	34	26	19	12	5	0
520	530	78	64	57	50	42	35	28	21	14	6	0
530	540	80	67	58	51	44	37	29	22	15	8	1
540	550	83	70	60	53	45	38	31	24	17	9	2
550	560	86	73	61	54	47	40	32	25	18	11	4
560	570	89	75	63	56	48	41	34	27	20	12	5
570	580	92	78	65	57	50	43	35	28	21	14	7
580	590	94	81	68	59	51	44	37	30	23	15	8
590	600	97	84	70	60	53	46	38	31	24	17	10

SINGLE Persons—WEEKLY Payroll Period (cont.)

(For Wages Paid in 1995)

If the wages are—		And the number of withholding allowances claimed is—										
At least	But less than	0	1	2	3	4	5	6	7	8	9	10
		The amount of income tax to be withheld is—										
$600	$610	100	87	73	62	54	47	40	33	26	18	11
610	620	103	89	76	63	56	49	41	34	27	20	13
620	630	106	92	79	65	57	50	43	36	29	21	14
630	640	108	95	82	68	59	52	44	37	30	23	16
640	650	111	98	84	71	60	53	46	39	32	24	17
650	660	114	101	87	74	62	55	47	40	33	26	19
660	670	117	103	90	76	63	56	49	42	35	27	20
670	680	120	106	93	79	66	58	50	43	36	29	22
680	690	122	109	96	82	69	59	52	45	38	30	23
690	700	125	112	98	85	71	61	53	46	39	32	25
700	710	128	115	101	88	74	62	55	48	41	33	26
710	720	131	117	104	90	77	64	56	49	42	35	28
720	730	134	120	107	93	80	66	58	51	44	36	29
730	740	136	123	110	96	83	69	59	52	45	38	31
740	750	139	126	112	99	85	72	61	54	47	39	32
750	760	142	129	115	102	88	75	62	55	48	41	34
760	770	145	131	118	104	91	78	64	57	50	42	35
770	780	148	134	121	107	94	80	67	58	51	44	37
780	790	150	137	124	110	97	83	70	60	53	45	38
790	800	153	140	126	113	99	86	72	61	54	47	40
800	810	156	143	129	116	102	89	75	63	56	48	41
810	820	159	145	132	118	105	92	78	65	57	50	43
820	830	162	148	135	121	108	94	81	67	59	51	44
830	840	164	151	138	124	111	97	84	70	60	53	46
840	850	167	154	140	127	113	100	86	73	62	54	47
850	860	170	157	143	130	116	103	89	76	63	56	49
860	870	173	159	146	132	119	106	92	79	65	57	50
870	880	176	162	149	135	122	108	95	81	68	59	52
880	890	178	165	152	138	125	111	98	84	71	60	53
890	900	181	168	154	141	127	114	100	87	74	62	55
900	910	184	171	157	144	130	117	103	90	76	63	56
910	920	187	173	160	146	133	120	106	93	79	66	58
920	930	190	176	163	149	136	122	109	95	82	68	59
930	940	192	179	166	152	139	125	112	98	85	71	61
940	950	195	182	168	155	141	128	114	101	88	74	62
950	960	198	185	171	158	144	131	117	104	90	77	64
960	970	201	187	174	160	147	134	120	107	93	80	66
970	980	204	190	177	163	150	136	123	109	96	82	69
980	990	206	193	180	166	153	139	126	112	99	85	72
990	1,000	209	196	182	169	155	142	128	115	102	88	75
1,000	1,010	212	199	185	172	158	145	131	118	104	91	77
1,010	1,020	215	201	188	174	161	148	134	121	107	94	80
1,020	1,030	218	204	191	177	164	150	137	123	110	96	83
1,030	1,040	222	207	194	180	167	153	140	126	113	99	86
1,040	1,050	225	210	196	183	169	156	142	129	116	102	89
1,050	1,060	228	213	199	186	172	159	145	132	118	105	91
1,060	1,070	231	216	202	188	175	162	148	135	121	108	94
1,070	1,080	234	219	205	191	178	164	151	137	124	110	97
1,080	1,090	237	222	208	194	181	167	154	140	127	113	100
1,090	1,100	240	225	210	197	183	170	156	143	130	116	103
1,100	1,110	243	228	213	200	186	173	159	146	132	119	105
1,110	1,120	246	231	216	202	189	176	162	149	135	122	108
1,120	1,130	249	235	220	205	192	178	165	151	138	124	111
1,130	1,140	253	238	223	208	195	181	168	154	141	127	114
1,140	1,150	256	241	226	211	197	184	170	157	144	130	117
1,150	1,160	259	244	229	214	200	187	173	160	146	133	119
1,160	1,170	262	247	232	217	203	190	176	163	149	136	122
1,170	1,180	265	250	235	220	206	192	179	165	152	138	125
1,180	1,190	268	253	238	223	209	195	182	168	155	141	128
1,190	1,200	271	256	241	226	211	198	184	171	158	144	131
1,200	1,210	274	259	244	229	215	201	187	174	160	147	133
1,210	1,220	277	262	247	233	218	204	190	177	163	150	136
1,220	1,230	280	266	251	236	221	206	193	179	166	152	139
1,230	1,240	284	269	254	239	224	209	196	182	169	155	142
1,240	1,250	287	272	257	242	227	212	198	185	172	158	145

$1,250 and over   Use Table 1(a) for a **SINGLE person** in "Tables for Percentage Method of Withholding."

MARRIED Persons—WEEKLY Payroll Period
(For Wages Paid in 1995)

If the wages are—		And the number of withholding allowances claimed is—										
At least	But less than	0	1	2	3	4	5	6	7	8	9	10
		The amount of income tax to be withheld is—										
$0	$125	0	0	0	0	0	0	0	0	0	0	0
125	130	1	0	0	0	0	0	0	0	0	0	0
130	135	1	0	0	0	0	0	0	0	0	0	0
135	140	2	0	0	0	0	0	0	0	0	0	0
140	145	3	0	0	0	0	0	0	0	0	0	0
145	150	4	0	0	0	0	0	0	0	0	0	0
150	155	4	0	0	0	0	0	0	0	0	0	0
155	160	5	0	0	0	0	0	0	0	0	0	0
160	165	6	0	0	0	0	0	0	0	0	0	0
165	170	7	0	0	0	0	0	0	0	0	0	0
170	175	7	0	0	0	0	0	0	0	0	0	0
175	180	8	1	0	0	0	0	0	0	0	0	0
180	185	9	2	0	0	0	0	0	0	0	0	0
185	190	10	2	0	0	0	0	0	0	0	0	0
190	195	10	3	0	0	0	0	0	0	0	0	0
195	200	11	4	0	0	0	0	0	0	0	0	0
200	210	12	5	0	0	0	0	0	0	0	0	0
210	220	14	7	0	0	0	0	0	0	0	0	0
220	230	15	8	1	0	0	0	0	0	0	0	0
230	240	17	10	2	0	0	0	0	0	0	0	0
240	250	18	11	4	0	0	0	0	0	0	0	0
250	260	20	13	5	0	0	0	0	0	0	0	0
260	270	21	14	7	0	0	0	0	0	0	0	0
270	280	23	16	8	1	0	0	0	0	0	0	0
280	290	24	17	10	3	0	0	0	0	0	0	0
290	300	26	19	11	4	0	0	0	0	0	0	0
300	310	27	20	13	6	0	0	0	0	0	0	0
310	320	29	22	14	7	0	0	0	0	0	0	0
320	330	30	23	16	9	1	0	0	0	0	0	0
330	340	32	25	17	10	3	0	0	0	0	0	0
340	350	33	26	19	12	4	0	0	0	0	0	0
350	360	35	28	20	13	6	0	0	0	0	0	0
360	370	36	29	22	15	7	0	0	0	0	0	0
370	380	38	31	23	16	9	2	0	0	0	0	0
380	390	39	32	25	18	10	3	0	0	0	0	0
390	400	41	34	26	19	12	5	0	0	0	0	0
400	410	42	35	28	21	13	6	0	0	0	0	0
410	420	44	37	29	22	15	8	1	0	0	0	0
420	430	45	38	31	24	16	9	2	0	0	0	0
430	440	47	40	32	25	18	11	4	0	0	0	0
440	450	48	41	34	27	19	12	5	0	0	0	0
450	460	50	43	35	28	21	14	7	0	0	0	0
460	470	51	44	37	30	22	15	8	1	0	0	0
470	480	53	46	38	31	24	17	10	2	0	0	0
480	490	54	47	40	33	25	18	11	4	0	0	0
490	500	56	49	41	34	27	20	13	5	0	0	0
500	510	57	50	43	36	28	21	14	7	0	0	0
510	520	59	52	44	37	30	23	16	8	1	0	0
520	530	60	53	46	39	31	24	17	10	3	0	0
530	540	62	55	47	40	33	26	19	11	4	0	0
540	550	63	56	49	42	34	27	20	13	6	0	0
550	560	65	58	50	43	36	29	22	14	7	0	0
560	570	66	59	52	45	37	30	23	16	9	1	0
570	580	68	61	53	46	39	32	25	17	10	3	0
580	590	69	62	55	48	40	33	26	19	12	4	0
590	600	71	64	56	49	42	35	28	20	13	6	0
600	610	72	65	58	51	43	36	29	22	15	7	0
610	620	74	67	59	52	45	38	31	23	16	9	2
620	630	75	68	61	54	46	39	32	25	18	10	3
630	640	77	70	62	55	48	41	34	26	19	12	5
640	650	78	71	64	57	49	42	35	28	21	13	6
650	660	80	73	65	58	51	44	37	29	22	15	8
660	670	81	74	67	60	52	45	38	31	24	16	9
670	680	83	76	68	61	54	47	40	32	25	18	11
680	690	84	77	70	63	55	48	41	34	27	19	12
690	700	86	79	71	64	57	50	43	35	28	21	14
700	710	87	80	73	66	58	51	44	37	30	22	15
710	720	89	82	74	67	60	53	46	38	31	24	17
720	730	90	83	76	69	61	54	47	40	33	25	18
730	740	92	85	77	70	63	56	49	41	34	27	20

MARRIED Persons—WEEKLY Payroll Period (cont.)
(For Wages Paid in 1995)

If the wages are—		And the number of withholding allowances claimed is—										
At least	But less than	0	1	2	3	4	5	6	7	8	9	10
		The amount of income tax to be withheld is—										
$740	$750	93	86	79	72	64	57	50	43	36	28	21
750	760	95	88	80	73	66	59	52	44	37	30	23
760	770	96	89	82	75	67	60	53	46	39	31	24
770	780	98	91	83	76	69	62	55	47	40	33	26
780	790	99	92	85	78	70	63	56	49	42	34	27
790	800	101	94	86	79	72	65	58	50	43	36	29
800	810	102	95	88	81	73	66	59	52	45	37	30
810	820	104	97	89	82	75	68	61	53	46	39	32
820	830	105	98	91	84	76	69	62	55	48	40	33
830	840	108	100	92	85	78	71	64	56	49	42	35
840	850	111	101	94	87	79	72	65	58	51	43	36
850	860	113	103	95	88	81	74	67	59	52	45	38
860	870	116	104	97	90	82	75	68	61	54	46	39
870	880	119	106	98	91	84	77	70	62	55	48	41
880	890	122	108	100	93	85	78	71	64	57	49	42
890	900	125	111	101	94	87	80	73	65	58	51	44
900	910	127	114	103	96	88	81	74	67	60	52	45
910	920	130	117	104	97	90	83	76	68	61	54	47
920	930	133	119	106	99	91	84	77	70	63	55	48
930	940	136	122	109	100	93	86	79	71	64	57	50
940	950	139	125	112	102	94	87	80	73	66	58	51
950	960	141	128	114	103	96	89	82	74	67	60	53
960	970	144	131	117	105	97	90	83	76	69	61	54
970	980	147	133	120	107	99	92	85	77	70	63	56
980	990	150	136	123	109	100	93	86	79	72	64	57
990	1,000	153	139	126	112	102	95	88	80	73	66	59
1,000	1,010	155	142	128	115	103	96	89	82	75	67	60
1,010	1,020	158	145	131	118	105	98	91	83	76	69	62
1,020	1,030	161	147	134	121	107	99	92	85	78	70	63
1,030	1,040	164	150	137	123	110	101	94	86	79	72	65
1,040	1,050	167	153	140	126	113	102	95	88	81	73	66
1,050	1,060	169	156	142	129	115	104	97	89	82	75	68
1,060	1,070	172	159	145	132	118	105	98	91	84	76	69
1,070	1,080	175	161	148	135	121	108	100	92	85	78	71
1,080	1,090	178	164	151	137	124	110	101	94	87	79	72
1,090	1,100	181	167	154	140	127	113	103	95	88	81	74
1,100	1,110	183	170	156	143	129	116	104	97	90	82	75
1,110	1,120	186	173	159	146	132	119	106	98	91	84	77
1,120	1,130	189	175	162	149	135	122	108	100	93	85	78
1,130	1,140	192	178	165	151	138	124	111	101	94	87	80
1,140	1,150	195	181	168	154	141	127	114	103	96	88	81
1,150	1,160	197	184	170	157	143	130	117	104	97	90	83
1,160	1,170	200	187	173	160	148	133	119	106	99	91	84
1,170	1,180	203	189	176	163	149	136	122	109	100	93	86
1,180	1,190	206	192	179	165	152	138	125	111	102	94	87
1,190	1,200	209	195	182	168	155	141	128	114	103	96	89
1,200	1,210	211	198	184	171	157	144	131	117	105	97	90
1,210	1,220	214	201	187	174	160	147	133	120	106	99	92
1,220	1,230	217	203	190	177	163	150	136	123	109	100	93
1,230	1,240	220	206	193	179	166	152	139	125	112	102	95
1,240	1,250	223	209	196	182	169	155	142	128	115	103	96
1,250	1,260	225	212	198	185	171	158	145	131	118	105	98
1,260	1,270	228	215	201	188	174	161	147	134	120	107	99
1,270	1,280	231	217	204	191	177	164	150	137	123	110	101
1,280	1,290	234	220	207	193	180	166	153	139	126	113	102
1,290	1,300	237	223	210	196	183	169	156	142	129	115	104
1,300	1,310	239	226	212	199	185	172	159	145	132	118	105
1,310	1,320	242	229	215	202	188	175	161	148	134	121	107
1,320	1,330	245	231	218	205	191	178	164	151	137	124	110
1,330	1,340	248	234	221	207	194	180	167	153	140	127	113
1,340	1,350	251	237	224	210	197	183	170	156	143	129	116
1,350	1,360	253	240	226	213	199	186	173	159	146	132	119
1,360	1,370	256	243	229	216	202	189	175	162	148	135	121
1,370	1,380	259	245	232	219	205	192	178	165	151	138	124
1,380	1,390	262	248	235	221	208	194	181	167	154	141	127

$1,390 and over   Use Table 1(b) for a **MARRIED** person in "Tables for Percentage Method of Withholding."

Steps: Calculating Federal Income Tax by the Wage Bracket Method

STEP 1.   *Locate* the correct payroll period and marital status table.

STEP 2.   *Find* the correct wage bracket in the two leftmost columns. This gives the correct row. When reading the column headings, be very careful. "But less than 280" means smaller than 280. That includes $279.99, but not $280.

STEP 3.   *Move* right in the row to the column with the correct number of exemptions. This figure is the federal income tax.

For example, Abby Adams is single and declares one exemption. Calculate her federal income tax (FIT) by the wage bracket method on her weekly wage of $379.36.

Step 1.   Use the SINGLE Persons—WEEKLY Payroll Period" table.

Step 2.   The current wage bracket is "at least 370, but less than 380."

Step 3.   The correct FIT amount in the "1 exemption" column is $42.

YOUR TURN

WORK THIS PROBLEM

The Question:

James Buchanan earned $1185.46 last week. He is married and declares four exemptions. Assume you are his employer, and compute his federal income tax (FIT) by the wage bracket method.

✔ YOUR WORK

The Solution:

Step 1.   Use the Married Weekly table.

Step 2.   The weekly wage is $1185.46. It's in the row "at least 1180, but less than 1190."

Step 3.   Move right in the row to 4 exemptions. The FIT is $152.

Isn't it easy? But remember, check each answer. Accuracy is very important in any business problem, especially payroll.

Applying the Percentage Method

Percentage Method A method for calculating federal income tax withholding; often used by computers, it takes into account tabular data and multiplies various stated rates by indicated income values.

Many employers prefer to use the **percentage method,** especially those using computer payrolls. This method has three steps:

Steps: Calculating Federal Income Tax by the Percentage Method

STEP 1. *Use* the "Percentage Method—Amount for One Withholding Allowance" table that follows to find the amount of withholding allowance for the payroll period. Multiply this amount by the number of exemptions and subtract from gross wages.

**PERCENTAGE METHOD—
AMOUNT FOR ONE
WITHHOLDING ALLOWANCE**

Payroll Period	One Withholding Allowance
Weekly	$48.08
Biweekly	96.15
Semimonthly	104.17
Monthly	208.33
Quarterly	625.00
Semiannually	1,250.00
Annually	2,500.00
Daily or miscellaneous (each day of the payroll period)	9.62

STEP 2. In the "Tables for Percentage Method of Withholding," *find* the correct table for the pay period and marital status.

STEP 3. *Locate* the correct wage bracket for the amount found in Step 1 (not the gross wage) in the two leftmost columns. Perform the calculation on the right.

For example, Mary Todd is married and declares two exemptions. Compute her federal income tax by the percentage method for a month in which she earned $4,107.49.

Step 1. Since she is paid monthly, the withholding allowance from the table is $208.33. For two exemptions the allowance is $2 \times \$208.33 = \$416.66$. Subtract $416.66 from her gross wages.

$$
\begin{array}{r}
\$4107.49 \\
-\ 416.66 \quad \boxed{2 \times \$208.33} \\
\hline
\$3690.83
\end{array}
$$

Step 2. Ms. Todd is paid monthly married, so we want Table 4(b), MONTHLY Payroll Period, MARRIED.

Step 3. Her wage is in the "over $3,588, but not over $7,213" bracket. Her federal income tax thus is

$$
\begin{aligned}
FIT &= \$458.25 \text{ plus } 28\% \text{ of excess over } \$3588 \\
&= 458.25 + [28\% \times (3690.83 - 3588)] \\
&= 458.25 + [0.28 \times 102.83] \\
&= 458.25 + 28.7924 = \$487.04 \text{ rounded}
\end{aligned}
$$

The phrase "of excess over $3,588" means her wage (from step 1, not her gross wage) minus $3,588. The remainder of the calculation—the "%-of" problem—is just like the ones we worked in Chapter 4.

This method is a little longer than the wage bracket method. But with a little practice, it's as easy. The IRS has mainframe computers and virtually every business today uses calculators or computers to find FIT.

Using a Calculator to Find FIT by the Percentage Method

When finding the FIT by the percentage method, you can do the calculations in one long step on your calculator and round the final answer.

In our example, we must calculate

$$FIT = 458.25 + (0.28 \times (3690.83 - 3588))$$

If you have parentheses on your calculator, just key in the preceding sequence, beginning with 458.25 and end with an equal sign $\boxed{=}$.

If you don't have parentheses, you must do the operation in the double parentheses first, then the operation in the single parentheses.

FIT = 3690.83 $\boxed{-}$ 3588 $\boxed{=}$ $\boxed{\times}$ 28 $\boxed{\%}$ $\boxed{+}$ 458.25 $\boxed{=}$ ▨▨▨ 487.0424

YOUR TURN

WORK THIS PROBLEM

The Question:

James Buchanan earned $1185.46 last week. He is married and declares four exemptions. Compute his federal income tax by the percentage method.

✔ YOUR WORK

The Solution:

Step 1. Since he is paid weekly, the withholding allowance is $48.08. For four exemptions, the allowance is $4 \times \$48.08 = \$192.32$.

$$\begin{array}{r} \$1185.46 \\ -192.32 \\ \hline \$\ 993.14 \end{array}$$ ◁ $4 \times \$48.08$

Step 2. Use the WEEKLY Payroll Period, MARRIED table (Table 1(b)).

Step 3. His wage bracket is "over $828, but not over $1,664." His federal income tax is

$$\begin{aligned} FIT &= \$105.75 \text{ plus } 28\% \text{ of excess over } \$828 \\ &= 105.75 + [28\% \times (993.14 - 828)] \\ &= 105.75 + [0.28 \times 165.14] \\ &= 105.75 + 46.2392 = \$151.99 \text{ rounded} \end{aligned}$$

Tables for Percentage Method of Withholding
(For Wages Paid in 1995)

TABLE 1—WEEKLY Payroll Period

(a) SINGLE person (including head of household)—

If the amount of wages (after subtracting withholding allowances) is:		The amount of income tax to withhold is:	
Not over $50		$0	
Over—	**But not over—**		**of excess over—**
$50	—$476	15%	—$50
$476	—$999 . . .	$63.90 plus 28%	—$476
$999	—$2,295 . . .	$210.34 plus 31%	—$999
$2,295	—$4,960 . . .	$612.10 plus 36%	—$2,295
$4,960	. . . .	$1,571.50 plus 39.6%	—$4,960

(b) MARRIED person—

If the amount of wages (after subtracting withholding allowances) is:		The amount of income tax to withhold is:	
Not over $123		$0	
Over—	**But not over—**		**of excess over—**
$123	—$828	15%	—$123
$828	—$1,664 . . .	$105.75 plus 28%	—$828
$1,664	—$2,839 . . .	$339.83 plus 31%	—$1,664
$2,839	—$5,011 . . .	$704.08 plus 36%	—$2,839
$5,011	. . . .	$1,486.00 plus 39.6%	—$5,011

TABLE 2—BIWEEKLY Payroll Period

(a) SINGLE person (including head of household)—

If the amount of wages (after subtracting withholding allowances) is:		The amount of income tax to withhold is:	
Not over $100		$0	
Over—	**But not over—**		**of excess over—**
$100	—$952 . . .	15%	—$100
$952	—$1,998 . . .	$127.80 plus 28%	—$952
$1,998	—$4,590 . . .	$420.68 plus 31%	—$1,998
$4,590	—$9,919 . . .	$1,224.20 plus 36%	—$4,590
$9,919	. . . .	$3,142.64 plus 39.6%	—$9,919

(b) MARRIED person—

If the amount of wages (after subtracting withholding allowances) is:		The amount of income tax to withhold is:	
Not over $246		$0	
Over—	**But not over—**		**of excess over—**
$246	—$1,656 . . .	15%	—$246
$1,656	—$3,329 . . .	$211.50 plus 28%	—$1,656
$3,329	—$5,679 . . .	$679.94 plus 31%	—$3,329
$5,679	—$10,021 . . .	$1,408.44 plus 36%	—$5,679
$10,021	. . . .	$2,971.56 plus 39.6%	—$10,021

TABLE 3—SEMIMONTHLY Payroll Period

(a) SINGLE person (including head of household)—

If the amount of wages (after subtracting withholding allowances) is:		The amount of income tax to withhold is:	
Not over $108		$0	
Over—	**But not over—**		**of excess over—**
$108	—$1,031 . . .	15%	—$108
$1,031	—$2,165 . . .	$138.45 plus 28%	—$1,031
$2,165	—$4,973 . . .	$455.97 plus 31%	—$2,165
$4,973	—$10,746 . .	$1,326.45 plus 36%	—$4,973
$10,746	. . . .	$3,404.73 plus 39.6%	—$10,746

(b) MARRIED person—

If the amount of wages (after subtracting withholding allowances) is:		The amount of income tax to withhold is:	
Not over $267		$0	
Over—	**But not over—**		**of excess over—**
$267	—$1,794	15%	—$267
$1,794	—$3,606	$229.05 plus 28%	—$1,794
$3,606	—$6,152	$736.41 plus 31%	—$3,606
$6,152	—$10,856 . . .	$1,525.67 plus 36%	—$6,152
$10,856	. . . .	$3,219.11 plus 39.6%	—$10,856

TABLE 4—MONTHLY Payroll Period

(a) SINGLE person (including head of household)—

If the amount of wages (after subtracting withholding allowances) is:		The amount of income tax to withhold is:	
Not over $217		$0	
Over—	**But not over—**		**of excess over—**
$217	—$2,063 . . .	15%	—$217
$2,063	—$4,329 . . .	$276.90 plus 28%	—$2,063
$4,329	—$9,946 . . .	$911.38 plus 31%	—$4,329
$9,946	—$21,492 . .	$2,652.65 plus 36%	—$9,946
$21,492	. . . .	$6,809.21 plus 39.6%	—$21,492

(b) MARRIED person—

If the amount of wages (after subtracting withholding allowances) is:		The amount of income tax to withhold is:	
Not over $533		$0	
Over—	**But not over—**		**of excess over—**
$533	—$3,588	15%	—$533
$3,588	—$7,213	$458.25 plus 28%	—$3,588
$7,213	—$12,304 . . .	$1,473.25 plus 31%	—$7,213
$12,304	—$21,713 . . .	$3,051.46 plus 36%	—$12,304
$21,713	. . . .	$6,438.70 plus 39.6%	—$21,713

Well, that's a little strange. In an earlier example, we worked the same problem by the wage bracket method and got a different answer, $152.

Yes, you will get different answers with each method. But if you use the amount in the middle of the wage bracket interval (the average), you will get approximately the same answer with both methods.

Try computing the income tax on a weekly wage of $1185.00, married with four exemptions. The tax is $152 with the wage bracket method and $151.86 with the percentage method. The wage $1185.00 is in the middle of the interval "at least $1180, but less than $1190." Rounding $151.86 to the nearest dollar produces $152 and gives the same answer with both methods.

YOUR TURN

| WORK THIS PROBLEM |

The Question:

Find the federal income tax to be deducted for each of the following employees by the percentage method.

Name	Marital Status	Exemptions	Semimonthly Wage	Income Tax
Bauer, Edie	Single	2	$1319.50	
Bean, L. L.	Married	4	1447.25	
Penney, J. C.	Single	1	1346.27	
Vernon, Lillian	Married	3	2145.65	

| ✔ YOUR WORK |

The Solution:

Bauer, Edie: 2 exemptions × $104.17 = $208.34

Step 1. $1319.50 − 208.34 = $1111.16

Step 2.   Use the SEMIMONTHLY Payroll Period, SINGLE table, Table 3(a).

Step 3.   FIT = $138.45 plus 28% of excess over $1031
    =   138.45 + [28% × (1111.16 − 1031)]
    =   138.45 + [0.28 × 80.16]
    =   138.45 + 22.4448 = $160.89 rounded

Bean, L. L.: 4 exemptions × $104.17 = $416.68

Step 1. $1447.25 − 416.68 = $1030.57

Step 2.   Use the SEMIMONTHLY Payroll Period, MARRIED table, Table 3(b).

Step 3.   FIT = 15% of excess over $267
    = 15% × (1030.57 − 267)
    = 0.15 × 763.57 = $114.54 rounded

Penney, J. C.: 1 exemption = $104.17

Step 1. $1346.27 − 104.17 = $1242.10

Step 2.   Use the SEMIMONTHLY Payroll Period, MARRIED table, Table 3(a).

Step 3. FIT = $138.45 plus 28% of excess over $1031
  = 138.45 + [28% × (1242.10 − 1031)]
  = 138.45 + [0.28 × 211.10]
  = 138.45 + 59.108 = $197.56 rounded

Vernon, Lillian: 3 exemptions × $104.17 = $312.51

Step 1. $2145.65 − 312.51 = $1833.14

Step 2. Use the SEMIMONTHLY Payroll Period, MARRIED table, Table 3(b).

Step 3. FIT = $229.05 plus 28% of excess over $1794
  = 229.05 + [28% × (1833.14 − 1794)]
  = 229.05 + [0.28 × 39.14]
  = 229.05 + 10.9592 = $240.01 rounded

Name	Marital Status	Exemptions	Semimonthly Wage	FIT
Bauer, Edie	Single	2	$1319.50	$160.89
Bean, L. L.	Married	4	1447.25	114.54
Penney, J. C.	Single	1	1346.27	197.56
Vernon, Lillian	Married	3	2145.65	240.01

Computing FICA

FICA (Federal Insurance Contributions Act)
Deductions for Social Security and Medicare.

Another required deduction is **FICA (Federal Insurance Contributions Act)**, commonly known as Social Security. When FICA was started in 1937, the original FICA rate was 1% on the first $3000 in wages. As the cost of living and Social Security expenses increased, the FICA rate and maximum wage base were increased. Prior to 1991, FICA included both Social Security and Medicare as a single rate. The 1990 rate was 7.65% on all wages up to $51,300.

In addition to the amount withheld from the employee's wages, the employer is required to match the deduction. That means that in 1990, the federal government actually received

$$2 \times 7.65\% = 15.3\%$$

In 1991, the rate remained 7.65%, but several changes were made. The IRS divided the FICA into two categories: Social Security and Medicare. The IRS now requires each to be reported separately. The Social Security rate is 6.2% with a wage base of $61,200. The Medicare rate is 1.45% on all wages. The total rate is 6.2% + 1.45% = 7.65%.

1995 FICA Rates

	Rate	Wage Base
Social Security	6.2%	$61,200
Medicare	1.45%	unlimited

There are two methods of calculating FICA: the table "lookup" method and the percentage method. Both methods give the same result. The table lookup method uses the relevant tables in IRS Circular E. The percentage method is the easiest method. You simply multiply gross pay by the Social Security and Medicare rates and round when necessary.

$$\text{Social Security} = 6.2\% \times \text{gross pay}$$
$$\text{Medicare} = 1.45\% \times \text{gross pay}$$

For example, let's calculate the Social Security and Medicare withholding on a monthly wage of $2865.

$$\text{Social Security} = 6.2\% \times \text{gross pay}$$
$$= 0.062 \times \$2865 = \$177.63$$
$$\text{Medicare} = 1.45\% \times \text{gross pay}$$
$$= 0.0145 \times \$2865 = \$41.54 \text{ rounded}$$

The employer in this case must withhold $177.63 for Social Security and $41.54 for Medicare from the employee's paycheck. The employer must also "contribute" an equal amount:

$$\$177.63 + 41.54 = \$219.17$$

WORK THIS PROBLEM

The Question:

Calculate the Social Security and Medicare withholding on a monthly wage of $1956.85.

✔ YOUR WORK

The Solution:

$$\text{Social Security} = 6.2\% \times \$1956.85$$
$$= 0.062 \times \$1956.85 = \$121.32 \text{ rounded}$$
$$\text{Medicare} = 1.45\% \times \$1956.85$$
$$= 0.0145 \times \$1956.85 = \$28.37 \text{ rounded}$$

In addition to federal income and FICA taxes, many other deductions are frequently made. Most states require a state income tax to be withheld. Although the state tax tables vary from state to state, the withholding methods are similar to those for the federal tax. Other possible deductions include health insurance, life insurance, union dues, annuities, bond purchases, retirement plans, and so on. The list seems endless.

A statement of deductions must be added to the payroll sheet. Starting with the gross pay, calculate the federal income tax and FICA. The total deductions will be the sum of federal income tax, FICA, and the other deductions. The net pay will be the gross pay minus the total deductions.

For example, if your gross pay is $2576.42, your FIT is $445.58, your Social Security is $159.74, your Medicare is $37.36, and your health insurance is $252.22. Your net pay would be

$$\text{Net pay} = \text{gross pay} - \text{total deductions}$$
$$= \$2576.42 - (445.58 + 159.74 + 37.36 + 252.22)$$
$$= \$2576.42 - 894.90 = \$1681.52$$

WORK THIS PROBLEM

The Question:

Complete the following weekly payroll sheet. Don't forget to total the gross pay, federal income tax, FICA, other deductions, total deductions, and net pay columns. Use the wage bracket method to calculate the federal income tax.

Name	Marital Status	No. of Exemptions	Gross Pay	FIT	Social Security	Medicare	Other Deductions	Total Deductions	Net Pay
Aquino, Corazon	S	1	$562				$52.70		
Gandhi, Indira	M	4	597				54.25		
Meir, Golda	S	1	553				51.95		
Thatcher, Margaret	M	3	584				53.20		
Totals									

✔ YOUR WORK

The Solution:

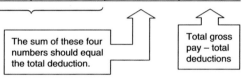

Name	Marital Status	No. of Exemptions	Gross Pay	FIT	Social Security	Medicare	Other Deductions	Total Deductions	Net Pay
Aquino, Corazon	S	1	$ 562	$ 75	$ 34.84	$ 8.15	$ 52.70	$170.69	$ 391.31
Gandhi, Indira	M	4	597	42	37.01	8.66	54.25	141.92	455.08
Meir, Golda	S	1	553	73	34.29	8.02	51.95	167.26	385.74
Thatcher, Margaret	M	3	584	48	36.21	8.47	53.20	145.88	438.12
Totals			2296	238	142.35	33.30	212.10	625.75	1670.25

The sum of these four numbers should equal the total deduction.

Total gross pay – total deductions

Be sure to check your work by totaling the columns.

For some practice problems calculating deductions, go to Section Test 17.1.

SECTION TEST 17.1 Taxes

The following problems test your understanding of Section 17.1, Federal Income Tax and FICA. Use the tables given in this chapter.

A. The following is a weekly payroll sheet for the Bridges Construction Company. For each individual, calculate the FIT, Social Security deduction, Medicare deduction, total deductions, and net pay. Use the wage bracket method to calculate the FIT.

Name	Mar. Stat.	No. Ex.	Gross Pay	FIT	Social Sec.	Medicare	Other Deduc.	Total Deduc.	Net Pay
1. G. Maddox	M	3	$492				$42.20		
2. M. Kaiser	M	4	517				43.25		
3. K. Fleck	S	1	485				41.95		
4. B. Scott	S	0	531				43.40		
5. S. Kamm	M	2	522				43.15		
6. L. Reed	S	1	543				43.35		
7. D. Bowie	S	0	497				42.45		
8. N. Daily	M	2	532				43.95		
9. E. Marlier	S	1	554				44.20		
10. F. Young	M	5	502				43.70		

B. The following is a semimonthly payroll sheet for the By the Byte Software House. For each individual, calculate the FIT, Social Security deduction, Medicare deduction, total deductions, and net pay. Use the percentage method to calculate the FIT.

Name	Mar. Stat.	No. Ex.	Gross Pay	FIT	Social Sec.	Medicare	Other Deduc.	Total Deduc.	Net Pay
1. B. Devo	S	1	$1455				$ 92		
2. Q. Trang	S	0	1425				89		
3. E. Nicklass	M	3	1512				98		
4. A. Carlton	M	4	1557				105		
5. K. Rose	S	0	1397				93		
6. R. Adams	M	2	1492				95		
7. B. Wang	M	5	1485				94		
8. Q. Jones	S	1	1388				87		
9. M. Todd	M	2	1525				102		
10. J. Pate	S	1	1492				91		

C. Solve these problems.

1. Abe Beame's monthly gross pay is $3875. He is married with three exemptions. (a) What is his FIT using the percentage method? What are his (b) Social Security and (c) Medicare deductions?

2. Colleen Dewhurst's monthly gross pay is $3920. She is married with four exemptions. (a) What is her FIT using the percentage method? What are her (b) Social Security and (c) Medicare deductions?

3. Enrico Fermi's semimonthly gross pay is $1775. He is married with four exemptions. (a) What is his FIT using the percentage method? What are his (b) Social Security and (c) Medicare deductions?

4. George Harrison has a semimonthly gross pay of $1850. He is married with two exemptions. (a) What is his FIT using the percentage method? What are his (b) Social Security and (c) Medicare deductions?

5. Jayne Kennedy has a biweekly gross pay of $1825. She is married with three exemptions. (a) What is her FIT using the percentage method? What are her (b) Social Security and (c) Medicare deductions?

6. Lee Marvin's biweekly gross pay is $2050. He is single and has one exemption. (a) What is his FIT using the percentage method? What are his (b) Social Security and (c) Medicare deductions?

7. Shirley Temple's weekly gross pay is $1005. She is single with one exemption. (a) What is her FIT using the percentage method? What are her (b) Social Security and (c) Medicare deductions?

8. Anne Boleyn's weekly gross pay is $850. She is married with two exemptions. (a) What is her FIT using the percentage method? What are her (b) Social Security and (c) Medicare deductions?

9. Charles Dickens's weekly gross pay is $472. He is married with two exemptions. (a) What is his FIT using the wage bracket method? What are his (b) Social Security and (c) Medicare deductions?

10. Eileen Farrell's weekly gross pay is $395. She is single with one exemption. (a) What is her FIT using the wage bracket method? What are her (b) Social Security and (c) Medicare deductions?

WORKSPACE

SECTION 17.2:   Unemployment Taxes

Can You:

Calculate federal unemployment taxes?
Determine state unemployment taxes?
. . . If not, you need this section.

In addition to withholding federal income taxes and to withholding and matching Social Security and Medicare deductions, employers are subject to unemployment taxes at both the federal and state levels. These taxes are designed to provide financial assistance to unemployed workers.

Calculating Federal Unemployment Tax

Federal Unemployment Tax Act (FUTA)   A federal government fund to which all employers must contribute and from which unemployed workers are paid.

By law, all U.S. employers must participate in the joint federal-state unemployment program established by the **Federal Unemployment Tax Act (FUTA).** Note that while unemployed workers receive the benefits, employers—not workers—pay all of this tax.

The 1991 FUTA rate is 6.2% on the first $7000 paid to each employee. Wages over $7000 per quarter are not taxed under FUTA. If the employer participates in a state unemployment program (discussed later in this section), the federal rate is reduced.

For example, the Fly-By-Night Aviary does not participate in a state unemployment program. Robin McCaw, an employee of the firm, earned $4500 in the first quarter cleaning bird cages. How much must Fly-By-Night "contribute" to FUTA?

$$\text{FUTA tax} = 6.2\% \times \$4500 = 0.062 \times \$4500 = \$279$$

Remember, FUTA tax must be paid only on the first $7000. Anything over $7000 is not taxed under FUTA. Use this fact to solve the next problem.

YOUR TURN

[WORK THIS PROBLEM]

The Question:

Big Ben's Bolt Company is not part of any state unemployment program. Hardy Weare, a supervisor for the firm, made $9250 in the first quarter of the year. What is the company's FUTA tax?

[✔ YOUR WORK]

The Solution:

Weare's earning exceeded the $7000 limit, so Big Ben's Bolts must pay tax on only the $7000.

$$\text{FUTA tax} = 6.2\% \times \$7000 = 0.062 \times \$7000 = \$434$$

At this time, $434 is the maximum FUTA tax per employee.

Determining State Unemployment Tax

State Unemployment Tax Act (SUTA) A state government fund in some states to which all employers of those states must contribute and from which unemployed workers in those states are paid.

Most states have their own **state unemployment tax act (SUTA).** Rates vary from state to state. In this text, we will use the most common rate—5.4% on the first $7000 quarterly earnings per employee.

As we noted earlier, when employers have to pay SUTA taxes, they pay a reduced FUTA tax. The FUTA rate is reduced by the SUTA rate up to a maximum of 5.4%. That is, when the SUTA tax rate is 5.4%, the reduced FUTA rate is 6.2% − 5.4% = 0.8%.

For example, The John Moose Tractor Company belongs to a SUTA program with a 5.4% rate. Eartha Tilly, a secretary for the firm, earned $5200 in the first quarter. How much FUTA tax does the firm owe? How much SUTA tax?

SUTA rate = 5.4%, so the reduced FUTA rate = 6.2% − 5.4% = 0.8%.

$$\text{FUTA tax} = 0.8\% \times \$5200 = 0.008 \times \$5200 = \$41.60$$
$$\text{SUTA tax} = 5.4\% \times \$5200 = 0.054 \times \$5200 = \$280.80$$

Remember to check whether the employee's earnings are over $7000.

YOUR TURN

| WORK THIS PROBLEM |

The Question:

The MausHaus, a cheese shop, belongs to a SUTA program with a 5.4% rate. Bree Limburger, head cheese buyer for the firm, earned $8750 in the first quarter. What are the MausHaus's FUTA and SUTA tax bills?

| ✔ YOUR WORK |

The Solution:

Ms. Limburger earned in excess of $7000, so the firm must pay taxes on $7000.

$$\text{SUTA rate} = 5.4\%$$
$$\text{Reduced FUTA rate} = 6.2\% − 5.4\% = 0.8\%$$
$$\text{FUTA tax} = 0.8\% \times \$7000 = 0.008 \times \$7000 = \$56$$
$$\text{SUTA tax} = 5.4\% \times \$7000 = 0.054 \times \$7000 = \$378$$

For a set of practice problems on calculating unemployment taxes, turn to Section Test 17.2.

SECTION TEST 17.2   Taxes

Name _____

Date _____

Course/Section _____

The following problems test your understanding of Section 17.2, Unemployment Taxes.

1. Bob's Beeflike Burgers is not required to participate in a state unemployment plan. If an employee earned $6700 in the first quarter, what is the company's FUTA tax?

2. Mike is an employee of the Wild Child Recording Studio. He earned $5900 in the first quarter. The studio is not required to participate in a state unemployment plan. What is the studio's FUTA tax on Mike?

3. Maria's Mannequins is not required to participate in a state unemployment plan. If an employee earned $7500 in the first quarter, what is the company's FUTA tax?

4. Bitten earned $8250 in the first quarter at the By the Byte Software House. By the Byte does not have to participate in a state unemployment plan. What is the company's FUTA tax on Bitten?

5. The Spitting Image employs a counter person who earned $7980 in the first quarter. The firm does not have to participate in the state unemployment plan. What is the Spitting Image's FUTA tax on the counter person?

6. Pia earned $9315 in the first quarter as manager of The Pizza Palace. The Pizza Palace is not required to participate in the state unemployment program. What does The Pizza Palace pay in FUTA tax on Pia?

7. The Zippy Delivery Service driver made $8752 in the first quarter. The company is not required to pay a state unemployment tax. What is the company's FUTA tax on the driver?

8. J.R.'s Construction Company is required to participate in FUTA and SUTA. The SUTA tax rate is 5.4% on the first $7000. If an employee earns $6300, what are the (a) FUTA tax and (b) SUTA tax?

9. Galyn made $6800 at Intense Tents. The company is required to participate in FUTA and SUTA. The SUTA tax rate is 5.4% on the first $7000. What are the (a) FUTA tax and (b) SUTA tax?

10. Mary Lou earned $7600 at Woody's Lumber Mill. The company is required to participate in FUTA and SUTA. The SUTA tax rate is 5.4% on the first $7000. What are the (a) FUTA tax and (b) SUTA tax?

11. The Fast Lube Garage is required to participate in FUTA and SUTA. The SUTA tax rate is 5.4% on the first $7000. If Betsy earns $8100, what are the (a) FUTA tax and (b) SUTA tax?

12. I. M. Exquisite Architects is required to participate in FUTA and SUTA. Grady earned $9350 from the company. The SUTA tax rate is 5.4% on the first $7000. What are the company's (a) FUTA tax and (b) SUTA tax?

13. King's Lawn Service paid Jerome $8975. The company is required to participate in FUTA and SUTA. The SUTA tax rate is 5.4% on the first $7000. What are the (a) FUTA tax and (b) SUTA tax?

14. The Dependable Airport Transportation Service is required to participate in FUTA and SUTA. The SUTA tax rate is 5.4% on the first $7000. If an employee earns $9629, what are the (a) FUTA tax and (b) SUTA tax?

15. The last week of March, Sally, an operator at Claudia's Clip-Joint, had a gross pay of $336.26. Sally is married, has 3 children, and claims all five exemptions. Claudia uses the percentage method to figure the federal withholding tax (FIT). FICA is split into the two categories: Social Security, 6.2%, and Medicare, 1.45%, of gross pay. State income tax is 3.8% of gross pay. Sally also has a health insurance deduction of $9.63 per week and a payroll savings deduction of $10.00 per week. Fill out the following pay stub for Sally. Claudia's Clip-Joint is required to participate in FUTA and SUTA. The SUTA tax rate is 5.4% on the first $7,000. What are the FUTA tax and SUTA tax on Sally's first quarter earnings of $4,371.38?

Gross Pay	$336.26
Deductions:	
FIT	
FICA—Social Security	
FICA—Medicare	
State Income Tax	
Other	
Net Pay	

FUTA _____

SUTA _____

In addition to payroll and unemployment taxes, businesses must collect and pay a wide range of taxes, including sales taxes, excise taxes, and property taxes.

Calculating Sales Tax

Sales Tax A tax on the price of a good sold at the time of its sale.

Most states and many cities and counties levy a sales tax. The **sales tax** is collected by the businesses from the buyer at the time of purchase. Manufacturers, wholesalers, and suppliers selling merchandise to other businesses for resale do not collect sales tax. The sales tax is only collected at the last sale—the sale to the consumer. This is usually done at the retail store.

Sales tax calculations may be made using a table or simply by multiplying by the tax rate. The answer is rounded to the nearest cent.

> Sales tax = rate × purchase price

Once you have calculated the sales tax, it's easy to calculate the total price:

> Total price = purchase price + sales tax

For example, Sy's Office Supplies (SOS) sold a file cabinet for $59.95. If the sales tax rate is 6%, what are the sales tax and the total price?

Rate   Purchase price

$$\text{Sales tax} = 6\% \times \$59.95 = 0.06 \times \$59.95$$
$$= \$3.597 = \$3.60 \text{ rounded}$$
$$\text{Total price} = \$59.95 + 3.60 = \$63.55$$

YOUR TURN

| WORK THIS PROBLEM |

The Question:

(a) Bits 'n' Bytes sells a computer modem for $357.84. If the sales tax rate is $5\frac{1}{2}\%$, what are the sales tax and the total price?

(b) Count On Us sells a "Struggling Student" calculator for $11.95. If the sales tax rate is 6%, what are the sales tax and the total price?

✔ YOUR WORK

The Solution:

(a)   Sales tax $= 5\frac{1}{2}\% \times \$357.84 = 5.5\% \times \$357.84$
            $= 0.055 \times \$357.84 = \$19.68$ rounded
    Total price $= \$357.84 + 19.68 = \$377.52$

(b)   Sales tax $= 6\% \times \$11.95 = 0.06 \times \$11.95 = \$0.72$ rounded
    Total price $= \$11.95 + 0.72 = \$12.67$

A calculator can make the simple task of finding sales tax and total price even easier.

Using a Calculator to Find the Total Purchase Price

When finding sales (or excise) tax and total purchase price, you can take advantage of the shortcut key sequence on most calculators. For example, to find the total price of a $59.95 file cabinet with a 6% sales tax rate, key in

59.95   ☒   6   ☒%   ⊞   ═   ⟶   **⌐⌐.547**

Note that you hit a plus sign just before the equal sign at the end of this sequence. The result rounds to $63.55, as we saw earlier in the text.

Determining Excise Tax

Excise Tax   A federal tax on the manufacture, sale, or consumption of designed items, usually items of a luxury or nonessential nature.

In addition to sales taxes, the federal government levies an **excise tax** on the manufacture, sale, or consumption of some items. Excise taxes are usually imposed on luxury or nonessential merchandise. Examples include alcohol, furs, tobacco products, vehicles, and telephone service.

Excise tax is calculated by the same method as sales tax.

> Excise tax = rate × price

For example, if the excise tax rate is 8%, a $10,000 fur coat would be subject to an excise tax of

$$\text{Excise tax} = \text{rate} \times \text{price}$$
$$= 8\% \times \$10,000 = 0.08 \times \$10,000 = \$800$$

YOUR TURN

WORK THIS PROBLEM

The Question:

Bits 'n' Bytes purchased a new delivery truck for $29,525. If the excise tax rate is 5%, what is the excise tax?

✔ YOUR WORK

The Solution:

$$\text{Excise tax} = 5\% \times \$29,525 = 0.05 \times \$29,525 = \$1476.25$$

Finding Property Tax

Real Property Taxes
State and local government taxes on property owned by businesses and individuals.

State and local governments levy **real property taxes** on property owned by businesses and individuals. In order to calculate property taxes, you must first determine the **assessed value** of the property, the value assigned to it by government assessors. The assessed value is not the same as the property's market value (what you could sell it for).

Steps: Calculating Assessed Value and Property Tax

STEP 1. *Calculate* the assessed value.

$$\text{Assessed value} = \text{assessment rate} \times \text{market value}$$

Assessed Value The value of a property as determined by government assessors; it is the basis for real property taxes on a property.

STEP 2. *Convert* the tax rate to a tax rate per $1. For example, if the rate is stated as so much per $100, divide by $100 to get the tax rate per $1. If the rate is stated per $1000, divide by $1000, etc.

STEP 3. *Calculate* the property tax.

$$\text{Property tax} = \text{tax rate per \$1} \times \text{assessed value}$$

For example, let's calculate the property tax for a building with a market value of $147,000, an assessment rate of 45%, and a property tax of $2.28 per $100 of assessed value.

Step 1. Assessed value = assessment rate × market value

$$= 45\% \times \$147{,}000 = 0.45 \times \$147{,}000 = \$66{,}150$$

Step 2. Tax rate per $1 $= \dfrac{\$2.28}{\$100} = 0.0228$

Step 3. Property tax = tax rate per $1 × assessed value
$$= 0.0228 \times \$66{,}150 = \$1508.22$$

Remember to look closely at how the property tax is stated. If it is stated "per $1,000" be sure to divide by 1000 to get the tax rate per $1.

YOUR TURN

WORK THIS PROBLEM

The Question:

The market value of a building is $172,500. If the assessment rate is 30% and the property tax is $9.58 per $1000, what is the property tax?

✔ YOUR WORK

The Solution:

Step 1.   Assessed value = 30% × $172,500 = 0.30 × $172,500 = $51,750

Step 2.   Tax rate per $1 = $\dfrac{\$9.58}{\$1000}$ = 0.00958

Step 3.   Property tax = 0.00958 × $51,750 = $495.77

Some areas of the United States express the property tax in "mills." One mill is one-thousandth of a dollar—one-tenth of a cent.

$$1 \text{ mill} = \$0.001 = 0.1$$
$$10 \text{ mills} = \$0.010 = 1¢$$

To change mills to a tax rate per $1, convert the mills to a decimal number by moving the decimal point three places to the *left*.

$$128 \text{ mills} = 0.128$$

$$39 \text{ mills} = 039 \text{ mills} = 0.039$$

$$42\tfrac{1}{2} \text{ mills} = 42.5 \text{ mills} = 042.5 \text{ mills} = 0.0425$$

YOUR TURN

WORK THIS PROBLEM

The Question:

The market value of the building that houses O'Hare's Yarn Shop is $87,000. If the assessment rate is 70% and the tax-rate is $20\tfrac{1}{2}$ mills, what is the property tax?

✔ YOUR WORK

The Solution:

Step 1.   Assessed value = 70% × $87,000 = 0.70 × $87,000 = $60,900

Step 2. $20\tfrac{1}{2}$ mills = 20.5 mills = 0.0205

Step 3.   Property tax = 0.0205 × $60,900 = $1248.45

When you "really" think you have mastered property taxes, turn to Section Test 17.3 and try the practice problems there.

SECTION TEST 17.3 Taxes ▬▬▬▬▬

The following problems test your understanding of Section 17.3, Sales Tax, Excise Tax, and Property Tax.

A. Calculate the tax and total purchase price.

	Selling Price	Tax Rate	Tax	Total Price
1.	$49.95	6%		
2.	21.15	7%		
3.	157.59	5%		
4.	1499.99	6%		
5.	17.67	$5\frac{1}{2}\%$		
6.	539.89	$6\frac{1}{2}\%$		
7.	323.99	$5\frac{1}{4}\%$		
8.	2007.12	$5\frac{3}{4}\%$		
9.	199.00	$6\frac{3}{4}\%$		
10.	25.77	$6\frac{1}{4}\%$		

B. Calculate the assessed value and property tax.

	Market Value	Assessment Rate	Assessed Value	Tax Rate	Property Tax
1.	$172,000	40%		$2.34 per $100	
2.	98,000	60%		$3.05 per $100	
3.	75,000	55%		$9.56 per $1000	
4.	112,000	62%		$8.92 per $1000	
5.	255,000	35%		$3.72 per $100	
6.	103,000	30%		$2.18 per $100	
7.	128,000	38%		28 mills	
8.	179,500	42%		19 mills	
9.	92,600	56%		$23\frac{1}{2}$ mills	
10.	85,500	47%		$25\frac{1}{2}$ mills	

C. Solve these problems.

1. Maria's Mannequins sold a mannequin for $137. If the sales tax rate is 6%, what are (a) the tax and (b) the total price?

2. Woody's Lumber Mill sold J.R.'s Construction ten 4-ft by 8-ft sheets of plywood. The price per sheet is $8.25. If the sales tax rate is 7%, what are (a) the tax and (b) the total price?

3. By the Byte Software House sold a word processing package for $199. If the sales tax rate is $5\frac{1}{2}$%, what are (a) the tax and (b) the total price?

4. The OKC Community College Bookstore sold an English literature textbook for $41.55. If the sales tax rate is $6\frac{1}{4}$%, what are (a) the tax and (b) the total price?

5. Crystal's Diamonds sold a tennis bracelet for $7200. If the excise tax rate is 12%, what is the excise tax?

6. Keep on Truckin' sold a truck for $18,699. If the excise tax rate is 4%, what is the excise tax?

7. Georgia Sand purchased a car for $21,345. If the excise tax rate is 5%, what is the excise tax?

8. Robert Stevenson bought new car tires for $235. If the excise tax rate is 5%, what is the excise tax?

9. Big Ben's Bolts owns a building with a market value of $145,000. If the assessment rate is 45% and the property tax rate is $2.19 per $100, what are (a) the assessed value and (b) the property tax?

10. The Ko Minh Furniture Store building has a market value of $215,000. If the assessment rate is 42% and the property tax rate is $2.19 per $100, what are (a) the assessed value and (b) the property tax?

11. Thomas Dylan's home has a market value of $115,000. If the assessment rate is 65% and the property tax rate is 27 mills, what are (a) the assessed value and (b) the property tax?

12. Julie Verne owns a house that has a market value of $82,000. If the assessment rate is 60% and the property tax rate is 24 mills, what are (a) the assessed value and (b) the property tax?

13. Dynamite Deals property has a market value of $598,072. The assessment rate is 72%, and the property tax rate is $5.32 per $1000 of assessed value. What is the property tax due on Dynamite Deals?

Key Terms

assessed value
deduction
excise tax
federal income tax withholding
Federal Unemployment Tax Act (FUTA)
FICA (Federal Insurance Contributions Act)
gross pay

net pay
percentage method
real property taxes
sales tax
State Unemployment Tax Act (SUTA)
wage bracket method

Chapter 17 at a Glance

Page	Topic	Key Point	Example
546	Net pay	Net pay = gross pay − total deductions	What is the net pay of a married worker with two exemptions who earns $650 per week and has total deductions of $203? Net pay = $650 − 203 = $447
547	Federal income tax withholding (wage bracket method, 1995 figures)	Use the appropriate wage bracket table to find tax.	What is the federal income tax withholding for the employee in the previous example using the wage bracket method? With two exemptions and a weekly salary of $650, this married employee's tax is $65.
553	Federal income tax withholding (percentage method, 1995 tax rates)	Taxable wages = gross wages − (no. exemptions × withholding allowance) per table in section 17.1. Tax depends on Tables for Percentage Method of Withholding.	Use the percentage method to solve the previous problem. Taxable wages = $650 − (2 × $48.08) = $553.84 Use table 1(b), Tax = 15% × ($553.84 − 123) = $64.63
557	FICA (1995 rates)	Social Security = 6.2% × gross pay up to $61,200 Medicare = 1.45% × gross pay FICA = Social Security + Medicare	What are the FICA deductions on the employee in the previous example? Social Security = 6.2% × $650 = $40.30 Medicare = 1.45% × $650 = $9.43 FICA = $40.30 + 9.43 = $49.73
565	FUTA (1995 rates)	FUTA = 6.2% × employee's gross pay up to $7000 per quarter	Find the FUTA the employer must pay on the employee in the previous example. Quarterly wages = 13 weeks × $650 = $8450 FUTA = 0.062 × $7000 = $434
566	SUTA	SUTA = rate × employee's gross pay up to state-set limit FUTA = (6.2% − SUTA rate) × employee's gross pay up to $7000 per quarter	Assume that the company for which the employee in the previous example works must pay a SUTA of 5.4% on the first $7000 quarterly earnings per employee. How much must the firm pay in SUTA and FUTA for this employee? Quarterly wages = 13 weeks × $650 = $8450 SUTA = 5.4% × $7000 = $378 FUTA = (6.2% − 5.4%) × $7000 = 0.8% × $7000 = $56

Page	Topic	Key Point	Example
569	Sales tax	Sales tax = rate × purchase price Total price = purchase price + sales tax	What is the total price on a $5500 copier with a 5% sales tax? Sales tax = 5% × $5500 = $275 Total price = $5500 + 275 = $5775
570	Excise tax	Excise tax = rate × price	What would be the excise tax on a $5500 copier with a 10% excise tax? Excise tax = 10% × $5500 = $550
571	Property tax	Assessed value = assessment rate × market value $\text{Tax rate per \$1} = \dfrac{\text{tax rate}}{\text{rate base}}$ Property tax = tax rate per $1 × assessed value	What is the property tax on a building with a market value of $147,000, an assessment rate of 45%, and a property tax of $2.28 per 100 of assessed value? Assessed value = $147,000 × 0.45 = $66,150 Tax rate per $1 = $2.28/$100 = 0.0228 Property tax = 0.0228 × $66,150 = $1508.22

SELF-TEST Taxes

The following problems test your understanding of taxes.

1. Janis Joplin's weekly gross pay is $423. She is married with three exemptions. (a) What is her FIT using the wage bracket method? What are her (b) Social Security and (c) Medicare deductions?

2. John Lennon's weekly gross pay is $382. He is married with two exemptions. (a) What is his FIT using the wage bracket method? What are his (b) Social Security and (c) Medicare deductions?

3. Patsy Cline's weekly gross pay is $415. She is single with one exemption. (a) What is her FIT using the wage bracket method? What are her (b) Social Security and (c) Medicare deductions?

4. Ricky Nelson's weekly gross pay is $295. He is single and declares zero exemptions. (a) What is his FIT using the wage bracket method? What are his (b) Social Security and (c) Medicare deductions?

5. Cass Elliott's monthly gross pay is $4025. She is married with two exemptions. (a) What is her FIT using the percentage method? What are her (b) Social Security and (c) Medicare deductions?

6. Karen Carpenter's monthly gross pay is $4175. She is married with four exemptions. (a) What is her FIT using the percentage method? What are her (b) Social Security and (c) Medicare deductions?

7. Buddy Holly's semimonthly gross pay is $1650. He is single with one exemption. (a) What is his FIT using the percentage method? What are his (b) Social Security and (c) Medicare deductions?

8. Jimi Hendrix has a semimonthly gross pay of $1785. He is single with one exemption. (a) What is his FIT using the percentage method? What are his (b) Social Security and (c) Medicare deductions?

9. Kate's Delicacies is not required to participate in a state unemployment plan. If an employee earned $6350 in the first quarter, what is the company's FUTA tax?

10. Mike, an employee of the Wild Child Recording Studio, earned $6125 in the first quarter. The studio is not required to participate in the state unemployment plan. What is the studio's FUTA tax on Mike?

11. Maria's Mannequins is not required to participate in a state unemployment plan. If an employee earned $8250 in the first quarter, what is the company's FUTA tax?

12. J.R.'s Construction Company is required to participate in FUTA and SUTA. The SUTA tax rate is 5.4% on the first $7000. If an employee earns $6775, what are (a) the FUTA tax and (b) the SUTA tax?

13. The tent darner at Intense Tents earned $5895. The company is required to participate in FUTA and SUTA. The SUTA tax is 5.4% on the first $7000. What are (a) the FUTA tax and (b) the SUTA tax?

14. Ashley earned $9650 at Woody's Lumber Mill. The company is required to participate in FUTA and SUTA. The SUTA tax rate is 5.4% on the first $7000. What are (a) the FUTA tax and (b) the SUTA tax?

15. The Nut Shop sold a set of tricams (climbing nuts) for $152. If the sales tax rate is 6%, what are (a) the tax and (b) the total price?

16. The Widget Company sold a microwidget for $12.95. If the sales tax rate is $6\frac{1}{2}$%, what are (a) the tax and (b) the total price?

17. Crystal's Diamonds sold an engagement ring for $950. If the excise tax rate is 4%, what is the excise tax?

18. Keep On Truckin' sold a light truck for $17,849. If the excise tax rate is 3%, what is the excise tax?

19. The Butter-em-Up Bakery owns a building with a market value of $95,000. If the assessment rate is 55% and the property tax rate is $2.35 per $100, what are (a) the assessed value and (b) the property tax?

20. The Ko Minh Furniture Store building has a market value of $175,500. If the assessment rate is 45% and the property tax rate is $9.79 per $1000, what are (a) the assessed value and (b) the property tax?

21. The Simpson's home has a market value of $98,500. If the assessment rate is 42% and the property tax rate is 26 mills, what are (a) the assessed value and (b) the property tax?

22. The Huxtables own a house that has a market value of $127,750. If the assessment rate is 65% and the property tax rate is $22\frac{1}{2}$ mills, what are (a) the assessed value and (b) the property tax?

WORKSPACE

Insurance

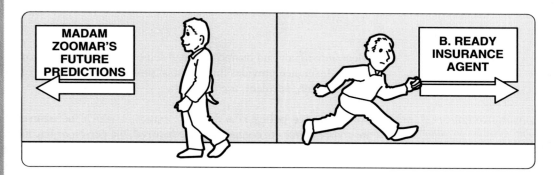

CHAPTER OBJECTIVES

When you complete this chapter successfully, you will be able to:

1. Calculate premiums, prorations, and coinsurance payouts for fire insurance.

2. Find premiums and payouts on liability, collision, comprehensive and medical payments for motor vehicle insurance policies.

3. Determine premiums and surrender options on various forms of life insurance policies.

■ The Wing and a Prayer Charter Flight Service carries two fire insurance policies: one with Company A for $60,000 and another with Company B for $40,000. Wing and a Prayer sustained a $42,000 loss in a recent fire. Calculate the amount of loss paid by each insurance company.

■ Mark Shiring, errand boy for Shiring Investments, uses his automobile for business. His insurance has 20/40/10 liability, $10,000 medical payments, and $100 deductible collision and comprehensive for an auto with price group E, age group 1. What is his premium?

■ Frances Woolworth, female, age 27, purchased a $75,000 ten-year term life insurance policy. What is her annual premium?

I t can happen in the twinkling of an eye—a fire, an automobile accident, a death. All these tragedies mean changes in the lives of the individuals and businesses struck by them.

Insurance can't prevent a loss. An insured building may still burn. An insured truck may still crash. An insured individual will die someday. What insurance does is to absorb some of the loss and to provide financial protection that can ease the path for survivors of a personal or business disaster of this sort. It will rebuild the building, replace the truck, and compensate a deceased person's family and business associates for the loss of income.

Ready? Put on your thinking cap and get ready to "absorb" this information-packed chapter. It's just good "insurance" for the future!

Can You:

Calculate fire insurance premiums?
Determine short-term policy premiums?
Find the prorations due in the event of cancellation by the insurer or insured?
Calculate the amount of loss paid by multiple insurers?
Determine losses paid on coinsurance policies?
. . . If not, you need this section.

Most individuals and businesses need some type of financial protection for personal or property losses. Insurance provides this financial protection against loss or damage by some contingency such as death, accident, fire, et cetera.

Insurance Policy   A written contract in which the insurer agrees to compensate the insured for specific losses or damages.

An **insurance policy** is a written contract in which the **insurer,** the company providing the insurance, agrees to compensate the **insured,** the person or business obtaining the insurance, for specific losses or damages. The **face value** of a policy is the maximum amount payable by the insurer for a loss to the insured. The insurance cost is called the **premium.**

Insurer   The company providing insurance.

Insurance cannot prevent a loss or damage. It only provides financial reimbursement in the case of a loss. For example, automobile insurance cannot prevent your auto from being stolen. But if it is stolen, the insurance company will compensate you for the financial loss.

Insured   The individual or organization covered by an insurance policy.

Calculating Fire Insurance Premiums

Face Value   The maximum amount payable by an insurer for a loss to the insured.

One of the most commonly purchased forms of insurance for both individuals and businesses is **fire insurance.**

Fire insurance provides protection against fire loss and related damage such as water damage caused by the process of extinguishing the fire. Most policies set a dollar limit on what the insurer will pay in the event of fire, though some policies guarantee to pay the **replacement value** of the building in the event of a total fire loss.

Premium   The cost of an insurance policy to the insured, sometimes expressed as a given amount per $100 of insurance coverage.

Fire insurance policies are generally written for either one or three years. A few are written for two years, and, under certain conditions, a four- or five-year policy may be issued.

Fire Insurance   A form of insurance in which the insurer agrees to recompense the insured in the event of fire damage to the property.

What will a policy cost you? The answer depends on many factors, beginning with the rate. Fire insurance rates vary widely, depending on the type of structure, location, maintenance, occupancy (owner or tenant), and proximity of fire hydrants.

Replacement Value   A clause in an insurance policy in which the insurer guarantees to pay the full price of replacing a property in the event of a loss.

Premium rates for fire insurance are usually expressed as some given amount per $100 of insurance coverage. For a one-year policy the premium is

Annual premium = rate × amount of insurance

For example, if a building is insured for $45,000 at a rate of $0.35 per $100, what is the annual premium?

The annual rate is $0.35 per $100, so the rate per $1 is 0.35/100 = 0.0035.

$$\text{Annual premium} = \text{rate} \times \text{amount of insurance}$$
$$= \frac{0.35}{100} \times \$45{,}000 = 0.0035 \times \$45{,}000 = \$157.50$$

This is the amount paid yearly to insure the building.

YOUR TURN

WORK THIS PROBLEM

The Question:

An accountant's office is insured for \$60,000. The fire insurance rate is \$0.27 per \$100. Calculate the annual premium.

✔ YOUR WORK

The Solution:

$$\text{Annual premium} = \frac{0.27}{100} \times \$60{,}000 = 0.0027 \times \$60{,}000 = \$162$$

When fire insurance is written for more than one year, most companies offer reduced premiums.

Number of Years	Premium
2	1.85 × annual premium
3	2.7 × annual premium

That is, for a two-year policy, you pay 1.85 × annual premium for both years combined rather than the 2 × annual premium for one-year policies.

For example, if a beauty parlor is insured for \$172,000 at a rate of \$0.42 per \$100, what are (a) the annual premium, (b) the two-year premium, and (c) the three-year premium?

(a) Annual premium = rate × amount of insurance

$$= \frac{0.42}{100} \times \$172{,}000 = 0.0042 \times \$172{,}000 = \$722.40$$

(b) Two-year premium = 1.85 × annual premium
$$= 1.85 \times \$722.40 = \$1336.44$$

(c) Three-year premium = 2.7 × annual premium
$$= 2.7 \times \$722.40 = \$1950.48$$

Without these discounts, two years' premiums for insurance would have been \$1444.80, and three years' premiums would have been \$2167.20.

YOUR TURN

WORK THIS PROBLEM

The Question:

Assume that premiums are based on the previous table.

(a) A store is insured for \$357,000 at a rate of \$0.31 per \$100. What is the two-year premium?

(b) A newsstand is insured for $73,000 at a rate of $0.37 per $100. What is the three-year premium?

✔ YOUR WORK

The Solution:

(a) Annual premium $= \dfrac{0.31}{100} \times \$357{,}000 = 0.0031 \times \$357{,}000 = \$1106.70$

Two-year premium $= 1.85 \times \$1106.70 = \$2047.395 = \$2047.40$ rounded

(b) Annual premium $= \dfrac{0.37}{100} \times \$73{,}000 = 0.0037 \times \$73{,}000 = \$270.10$

Three-year premium $= 2.7 \times \$270.10 = \$729.27$

Determining Short-Term Policy Premiums

Short-Term Policy An insurance written for less than one year.

Occasionally it is desirable to obtain an insurance policy for less than one year. These policies, called **short-term policies,** are written at a higher rate than an annual policy. The standard short-rate table is shown.

Just follow these steps:

Steps: Calculating Short-Term Rates

STEP 1.   First, *calculate* the annual premium.

STEP 2.   *Find* the number of days and the rate in the Short-Rate table.

STEP 3.   The short-term premium is the product of the short rate found in step 2 and the annual premium.

$$\text{Short-term premium} = \text{short rate} \times \text{annual premium}$$

For example, if a building is insured for $455,000 at an annual rate is $0.45 per $100, what is the premium on a 90-day policy?

Step 1.   Annual premium $=$ rate $\times$ amount of insurance

$$= \dfrac{0.45}{100} \times \$455{,}000 = 0.0045 \times \$455{,}000 = \$2047.50$$

Step 2.   Looking at the Short-Rate table, we find that for 90 days, the rate is 35%.

Step 3.   Short-term premium $=$ short rate $\times$ annual premium

$$= 35\% \times \$2047.50 = 0.35 \times \$2047.50$$
$$= \$716.625 = \$716.63$$

The insurance would cost $2047.50 for one year or $716.63 for 90 days. Note that 90 days is about one-fourth (25%) of a year, but the short rate is 35%. The short rate is *always* a higher rate.

SHORT-RATE TABLE

Days Policy in Force	Percent	Days Policy in Force	Percent	Days Policy in Force	Percent
1	5	95–98	37	219–223	69
2	6	99–102	38	224–228	70
3–4	7	103–105	39	229–232	71
5–6	8	106–109	40	233–237	72
7–8	9	110–113	41	238–241	73
9–10	10	114–116	42	242–246	74
11–12	11	117–120	43	247–250	75
13–14	12	121–124	44	251–255	76
15–16	13	125–127	45	256–260	77
17–18	14	128–131	46	261–264	78
19–20	15	132–135	47	265–269	79
21–22	16	136–138	48	270–273	80
23–25	17	139–142	49	274–278	81
26–29	18	143–146	50	279–282	82
30–32	19	147–149	51	283–287	83
33–36	20	150–153	52	288–291	84
37–40	21	154–156	53	292–296	85
41–43	22	157–160	54	297–301	86
44–47	23	161–164	55	302–305	87
48–51	24	165–167	56	306–310	88
52–54	25	168–171	57	311–314	89
55–58	26	172–175	58	315–319	90
59–62	27	176–178	59	320–323	91
63–65	28	179–182	60	324–328	92
66–69	29	183–187	61	329–332	93
70–73	30	188–191	62	333–337	94
74–76	31	192–196	63	338–342	95
77–80	32	197–200	64	343–346	96
81–83	33	201–205	65	347–351	97
84–87	34	206–209	66	352–355	98
88–91	35	210–214	67	356–360	99
92–94	36	215–218	68	361–365	100

YOUR TURN

WORK THIS PROBLEM

The Question:

A building is insured for $73,000 at an annual rate of $0.34 per $100. Calculate the premium on a 182–day policy.

✔ YOUR WORK

The Solution:

Step 1. Annual premium $= \dfrac{0.34}{100} \times \$73,000 = 0.0034 \times \$73,000 = \$248.20$

Step 2. For 182 days, the short rate is 60%.

Step 3. Short-term premium $= 60\% \times \$248.20 = 0.60 \times \$248.20$
$= \$148.92$

Canceling Property Insurance

Proration A partial refund to the insured in the event of cancellation of a policy by either insured or insurer before the expiration of the policy.

Either the insured or the insurer may cancel an insurance policy. In case of cancellation, the premium already paid must be **prorated**—divided between the two parties. The proration method depends on who cancels the policy.

When the *insured* cancels a policy, the cost of the insurance for the reduced time period is calculated using the Short-Rate table. This results in the insured paying for the policy at a rate higher than the annual rate. The exact number of days is used to calculate the time the policy was in force. For a review of exact time, return to Chapter 9, Section 9.1.

Thus, the steps in finding the proration are as follows:

Steps: Calculating Refunds When the Policyholder Cancels

STEP 1. *Calculate* the exact number of days the policy was in force.

STEP 2. *Find* the number of days and the corresponding rate in the Short-Rate table.

STEP 3. The amount of premium retained by the insurance company is the product of the short rate found in step 2 and the annual premium.

Premium retained by insurer = short rate × annual premium

STEP 4. *Calculate* the refund. The refund to the insured is the difference between the annual premium and the amount retained by the insurer.

Refund to insured = annual premium − premium retained by insurer

For example, Mr. Abraham insured his one-room law office for one year starting March 17 at an annual cost of $378. He canceled his policy on August 3. Calculate the amount of premium (a) retained by the insurance company and (b) the refund to Mr. Abraham.

Step 1.   March 17 to March 31   14 days ←——— 31 − 17 = 14 days
                            April   30
                             May   31
                            June   30
                            July   31
                  to August 3   3
             Total ——→ 139 days

Total days policy in force

Step 2. From the Short-Rate table, for 139 days the rate is 49%.

Step 3.

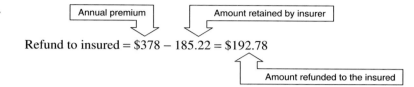

Premium retained by insurer = 49% × $378 = 0.49 × $378 = $185.22

Step 4.

Annual premium	Amount retained by insurer

Refund to insured = $378 − 185.22 = $192.78

Amount refunded to the insured

YOUR TURN

| WORK THIS PROBLEM |

The Question:

Figaro insured his barbershop for one year starting May 23 at an annual cost of $432. He canceled his policy on September 15. Calculate the amount of premium (a) retained by the insurer and (b) the refund to Figaro.

| ✔ YOUR WORK |

The Solution:

Step 1.

May 23 to May 31	8 days	← 31 − 23 = 8 days
June	30	
July	31	
August	31	
to September 15	15	
Total	115 days	

Step 2. From the Short-Rate table, for 115 days the short rate is 42%.

Step 3. Premium retained by insurer = 42% × $432 = 0.42 × $432 = $181.44

Step 4. Refund to insured = $432 − 181.44 = $250.56

It's easy if you follow the four steps.

When the *insurance company* cancels a policy, most states require a given number of days' notice in order to allow time for the insured to obtain new insurance. In this case, the amount of premium retained by the insurer is prorated using the *exact* number of days the policy is in force. The short-rate table is *not* used when the policy is canceled by the insurer.

Instead, follow these steps:

Steps: Calculating Refunds When the Insurer Cancels

STEP 1. *Calculate* the exact time the policy was in force.

STEP 2. *Calculate* the amount of premium retained by the insurance company as

$$\text{Premium retained by insurer} = \frac{\text{exact time}}{\text{total days in year}} \times \text{annual premium}$$

STEP 3.   *Calculate* the refund. The refund to the insured is the difference between the annual premium and the amount retained by the insurer.

Refund to insured = annual premium − premium retained by insurer

For example, Ms. Spode insured pottery shop for one year starting March 18 at an annual cost of $397. The policy was canceled by the insurance company on June 30. Calculate the amount of the premium (a) retained by the insurer and (b) the refund to Ms. Spode.

Step 1.   March 18 to March 31   13 days ←——— 31 − 18 = 13 days
                             April   30
                             May   31
                  to June 30   30
                  Total ———→   104 days

Step 2.

Days insurance in force   Annual premium

Premium retained by insurer = $\frac{104}{365}$ × $397 = $113.117 . . . = $113.12 rounded

Total days in year

Step 3.

Annual premium   Amount retained by insurer

Refund to insured = $397 − 113.12 = $283.88

Refund

WORK THIS PROBLEM

The Question:

YOUR TURN

Moe O'Hare insured his yarn shop for one year starting May 13 at a cost of $455. The policy was canceled by the insurance company on September 15. Calculate the amount of premium (a) retained by the insurer and (b) the refund to Mr. O'Hare.

✔ YOUR WORK

The Solution:

Step 1.   May 13 to May 31   18 days ←——— 31 − 13 = 18 days
                             June   30
                             July   31
                  August   31
                  to September 15   15
                  Total ———→   125 days

Step 2.   Premium retained by insurer = $\frac{125}{365}$ × $455
                  = $155.821 . . . = $155.82 rounded

Step 3.   Refund to insured = $455 − 155.82 = $299.18

When the insurance is canceled by the insured, *always* use the Short-Rate table. When insurance is canceled by the insurer, *always* use the exact number of days for the proration.

Determining Losses Paid by Multiple Insurers

Insurance on a very expensive piece of property is often divided among two or more insurance companies so that no one company is taking all the risk.

In the case of multiple insurers, each insurance company pays its pro-rata share of the total insurance for a claim. But in no case will any insurance company pay more than the face value of the policy.

$$\text{Each insurer's share} = \frac{\text{face value of insurer's policy}}{\text{total insurance}} \times \text{loss}$$

For example, Wreck-a-Mended Auto carries two fire insurance policies: Company A, $200,000, and Company B, $300,000. Wreck-a-Mended sustained a $36,000 fire loss. Calculate the amount of loss paid by each company.

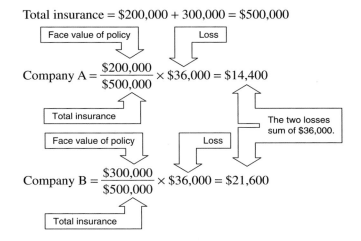

Total insurance = $200,000 + 300,000 = $500,000

Company A = $\dfrac{\$200,000}{\$500,000} \times \$36,000 = \$14,400$

Company B = $\dfrac{\$300,000}{\$500,000} \times \$36,000 = \$21,600$

You can check your answer by adding each insurer's share of the loss. The total should be the total loss. In this example, $14,400 + 21,600 = $36,000.

YOUR TURN

WORK THIS PROBLEM

The Question:

Maria's Mannequins carries two fire insurance policies: Company A, $25,000, and Company B, $15,000. Maria's sustained a $19,000 fire loss. Calculate the amount paid by each company.

✔ YOUR WORK

The Solution:

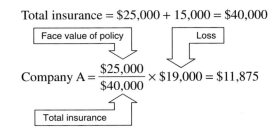

Total insurance = $25,000 + 15,000 = $40,000

Company A = $\dfrac{\$25,000}{\$40,000} \times \$19,000 = \$11,875$

$$\text{Company B} = \frac{\$15,000}{\$40,000} \times \$19,000 = \$7125$$

Check: $\$11,875 + 7125 = \$19,000$

What happens if the loss exceeds the face value of the policy or policies? The insurance company or companies pay the face value. The insured faces a financial loss on the remainder.

For example, Dorian Gray's Portrait Studio carries two fire insurance policies: Company A, $15,000, and Company B, $10,000. The studio sustained a $27,000 fire loss. Calculate the amount paid by each company.

$$\text{Total insurance} = \$15,000 + 10,000 = \$25,000$$

Since the total loss of $27,000 is larger than the total insurance, each company will pay the face value of their policy: Company A, $15,000, and Company B, $10,000. The studio will take a $2,000 loss.

YOUR TURN

| WORK THIS PROBLEM |

The Question:

The Fly-By-Night Aviary carries two fire insurance policies: Company X, $25,000, and Company Y, $15,000. Fly-By-Night had a $43,000 fire loss. Calculate the amount paid by each company.

| ✔ YOUR WORK |

The Solution:

$$\text{Total insurance} = \$25,000 + 15,000 = \$40,000$$

Since the loss is larger than the total insurance, each insurance company will pay the face value of its policy.

$$\text{Company X} = \$25,000$$
$$\text{Company Y} = \$15,000$$

Calculating Losses Paid by Coinsurance Policies

Coinsurance An agreement on the part of the insured to accept some of the risk on an insurance policy.

Since few fires result in a total loss, many businesses prefer to insure their property for only a portion of the total value. In order to compensate for underinsuring, most insurance companies require a **coinsurance** clause where the insured assumes some of the risk. The most common coinsurance rate is 80%. If the property is insured for at least 80% of its value, the insurer will pay for any loss up to the policy's face value. But if the business is insured for less than 80%, the insurance company will only pay a portion of the loss and the business must pay the rest. The maximum amount the insurance company will pay is

$$\text{Insurer pays} = \frac{\text{face value of policy}}{80\% \times \text{property value}} \times \text{loss}$$

For example, Shannel's Scent Shop carries $25,000 in fire insurance with an 80% coinsurance clause. The value of the property is $40,000. The company had a $12,000 fire loss. Calculate the amount paid by the insurer.

Face value of policy   Loss

$$\text{Insurer pays} = \frac{\$25,000}{80\% \times \$40,000} \times 12,000 = \frac{\$25,000}{\$32,000} \times 12,000 = \$9375$$

Property value

YOUR TURN

WORK THIS PROBLEM

The Question:

Len and Terry's Ice Cream stand carries $30,000 in fire insurance with an 80% coinsurance clause. The value of the property is $45,000. The corporation had a $20,000 fire loss. Calculate the amount paid by the insurer.

✔ YOUR WORK

The Solution:

$$\text{Insurer pays} = \frac{\$30,000}{80\% \times \$45,000} \times \$20,000$$

$$= \frac{\$30,000}{\$36,000} \times \$20,000 = \$16,666.666 \ldots = \$16,666.67 \text{ rounded}$$

A calculator makes it extra-easy to find coinsurance payoffs.

Using a Calculator with Coinsurance Policies

The simplest of calculators can speed this process by working the divisions and multiplications. Just start by finding the denominator. But if you have a calculator with parenthesis keys (and), it's even simpler.

For example, in the Shannel's Scent Shop example, punch in

25000 ÷ (.80 × 40000) × 12000 = ⟶   9375.

Now check your grasp of fire insurance in Section Test 18.1.

WORKSPACE

SECTION TEST 18.1 Insurance

The following problems test your understanding of Section 18.1, Fire Insurance.

A. Calculate the following premiums.

Amount of Insurance	Annual Rate per $100	Annual Premium	Two-Year Premium	Three-Year Premium
1. $147,000	$0.35			
2. 98,000	0.27			
3. 230,000	0.19			
4. 84,000	0.27			
5. 126,000	0.32			
6. 72,500	0.26			
7. 108,500	0.35			
8. 94,000	0.34			

B. Calculate the following short-term premiums. Use the Short-Rate table.

Amount of Insurance	Annual Rate per $100	Length of Policy in Days	Premium
1. $165,000	$0.32	60	
2. 120,000	0.42	120	
3. 96,000	0.36	90	
4. 140,000	0.29	30	
5. 87,000	0.33	180	
6. 150,000	0.39	125	
7. 225,000	0.27	98	
8. 179,000	0.25	64	

C. For the following canceled policies, calculate the time the policy was in force, the amount of premium retained by the insurer, and the refund to the insured. Use the Short-Rate table where necessary. (Assume it's not a leap year.) Note that the policies are sometimes canceled by the insurer and sometimes by the insured.

	Annual Premium	Canceled by	Policy Starting Date	Policy Cancellation Date	Time	Amount Retained by Insurer	Amount Refunded
1.	$375	Insurer	4/5	7/12			
2.	425	Insurer	6/2	10/5			
3.	375	Insured	3/7	5/12			
4.	423	Insured	6/15	10/2			
5.	505	Insured	7/12	11/15			
6.	415	Insurer	2/2	12/14			
7.	293	Insurer	3/17	11/2			
8.	342	Insured	1/5	5/17			

D. For the following multiple insurers, calculate the amount of the loss paid by each insurance company.

	Face Value of Policy Company A	Face Value of Policy Company B	Face Value of Policy Company C	Amount of Loss	Amount Paid Co. A	Amount Paid Co. B	Amount Paid Co. C
1.	$50,000	$25,000	$45,000	$ 75,000			
2.	40,000	35,000	25,000	55,000			
3.	50,000	40,000	50,000	70,000			
4.	60,000	50,000	30,000	84,000			
5.	50,000	60,000	70,000	144,000			
6.	30,000	40,000	50,000	96,000			
7.	60,000	50,000	70,000	90,000			
8.	75,000	70,000	55,000	112,000			

E. For the following problems, assume a coinsurance clause of 80%. Calculate the amount of the loss paid by the insurer.

	Value of Property	Face Value of Policy	Amount of Loss	Amount Paid by Insurer
1.	$120,000	$ 80,000	$25,000	
2.	96,000	60,000	14,000	
3.	250,000	150,000	37,000	
4.	95,000	75,000	23,000	
5.	155,000	100,000	17,500	
6.	80,000	65,000	22,000	
7.	135,000	110,000	27,500	
8.	175,000	125,000	35,000	

F. Solve these problems.

1. The By the Byte Software House building is insured for $125,000 at a rate of $0.27 per $100. What are (a) the annual premium, (b) the two-year premium, and (c) the three-year premium?

2. Bob's Beeflike Burgers' warehouse is insured for $95,000 at a rate of $0.32 per $100. What are (a) the annual premium, (b) the two-year premium, and (c) the three-year premium?

3. Shiring Investments' building is insured for $230,000 at a rate of $0.29 per $100. What are (a) the annual premium, (b) the two-year premium, and (c) the three-year premium?

4. The Butter-em-Up Bakery is insured for $190,000 at a rate of $0.24 per $100. What are (a) the annual premium, (b) the two-year premium, and (c) the three-year premium?

5. The Fin and Feather Pet Store is insured for $95,000 at an annual rate of $0.35 per $100. What is the premium on a 60-day policy?

6. The Robert and David Research and Development Company has insured its laboratory for $240,000 at an annual rate of $0.37 per $100. What is the premium on a 90-day policy?

7. The I. M. Exquisite Architects office building is insured for $150,000 at an annual rate of $0.28 per $100. What is the premium on a 120-day policy?

8. Maria's Mannequins has insured its mannequin warehouse for $98,000 at an annual rate of $0.27 per $100. What is the premium on a 150-day policy?

9. Della Street insured her paralegal office for one year starting May 12 at an annual cost of $475. She canceled the policy on September 7. What are (a) the amount of premium retained by the insurance company and (b) the refund to the insured?

10. Paul Drake insured his office for one year starting April 7 at an annual cost of $502. He canceled the policy on October 15. What are (a) the amount of premium retained by the insurance company and (b) the refund to the insured?

11. Petal Pushers insured its shop for one year starting at March 13 at an annual cost of $435. The policy was canceled on June 24 by the insurance company. What are (a) the amount of the premium retained by the insurance company and (b) the refund to the insured?

12. Ms. Spode insured her pottery shop on March 17 at a cost of $515. The policy was canceled by the insurance company on July 12. What are (a) the amount of the premium retained by the insurance company and (b) the refund to the insured?

13. The Zippy Delivery Service Company carries two fire insurance policies: one with Company A for $50,000 and another with Company B for $30,000. The Zippy Delivery Service Company sustained a $33,000 loss in a recent fire. Calculate the amount of loss paid by each insurance company.

14. The Wing and a Prayer Charter Flight Service carries two fire insurance policies: one with Company A for $60,000 and another with Company B for $40,000. The Wing and a Prayer Charter Flight Service sustained a $42,000 loss in a recent fire. Calculate the amount of loss paid by each insurance company.

15. J.R.'s Construction Company carries three fire insurance policies: one with Company A for $40,000, another with Company B for $30,000, and yet another with Company C for $50,000. The company sustained a $60,000 loss in a recent fire. Calculate the amount of loss paid by each insurance company.

16. The Wild Child Recording Studio carries three fire insurance policies: one with Company A for $30,000, another with Company B for $35,000, and yet another with Company C for $45,000. The studio sustained a $55,000 loss. Calculate the amount of loss paid by each insurance company.

17. Helen's House of Beauty carries $75,000 in fire insurance with an 80% coinsurance clause. The value of the property is $98,000. The boutique had an $18,000 loss. What is the amount paid by the insurer?

18. O'Hare's Yarn Shop carries $80,000 in fire insurance with an 80% coinsurance clause. The value of the property is $120,000. The shop had a $42,000 loss. What is the amount paid by the insurer?

19. Ava's Air Tours carries $160,000 in fire insurance with an 80% coinsurance clause. The value of the property is $250,000. The company had a $63,000 loss. What is the amount paid by the insurer?

20. Bits 'n' Bytes' main store carries $160,000 in fire insurance with an 80% coinsurance clause. The value of the property is $190,000. The store had a $23,000 loss. What is the amount paid by the insurer?

21. Dynamite Deals, with property valued at $598,072, carries $400,000 in fire insurance with an 80% coinsurance clause. A recent fire in the sporting goods department caused a $120,000 loss. What was the amount paid by the insurer on this claim?

WORKSPACE

SECTION 18.2: Motor Vehicle Insurance

Can You:

Determine driver classifications?
Calculate insurance payouts on auto liability coverage?
Determine premiums on liability policies?
Find premiums on collision and comprehensive coverage?
Calculate medical payments premiums?
Determine total premiums?
. . . If not, you need this section.

Because of the large number of automotive accidents, motor vehicle insurance is practically a necessity for individuals and businesses in the United States. Most states have legal requirements that vehicle owners and drivers carry insurance.

Motor vehicle insurance includes four main categories: liability, comprehensive, collision, and medical payments. We will explore each of these categories in this section.

Determining Driver Classification

Many factors help determine motor vehicle insurance premiums, including territory or region of residence, age and sex of drivers, driving record, whether the car is used in business or driven to work, and age, cost, and size of the auto.

The first step in calculating all kinds of motor vehicle insurance premiums is to determine the driver classification. The following is an abbreviated list of classes.

1A: No business use; no driver under 25; not driven to work
1B: No business use; no driver under 25; less than 10 miles to work one way
1C: No business use, no driver under 25; more than 10 miles to work one way
2A: Male under 25 not owner or principal driver; or married male under 25
2C: Male under 25 unmarried owner and/or principal driver
2F: Female driver under 25
3: Business use

For example, Karen Paisley, treasurer for Bits 'n' Bytes, uses her car for business. What is her driver classification? Business use is class 3.

YOUR TURN

| WORK THIS PROBLEM |

The Question:

Find the driver classification for the following individuals.

(a) Yuri Gagarin, age 32, drives his car to work 7 miles one way. He does not use his car in business.
(b) Krista McAullffe, age 21, drives her car to work 12 miles one way. She does not use her car in business.
(c) Stan Shepherd, age 52, uses his automobile in his business.
(d) Sally Ride, age 45, uses her car for pleasure only.

✔ YOUR WORK

The Solution:

(a) Class 1B
(b) Class 2F
(c) Class 3
(d) Class 1A

Calculating Liability Insurance Payouts

Liability Insurance   A form of motor vehicle insurance that covers medical and property damage to persons other than the insured.

One of the most important segments of motor vehicle insurance is **liability insurance.** Liability insurance covers medical and property damage to persons other than the insured. Although property damage is usually limited to the value of an automobile and its contents, it is very difficult to place a dollar value on an individual's injuries and resulting pain and suffering. Often the price of an injury is many times the property damage. Liability insurance is a virtual necessity if you don't want to lose your savings, home, and/or business!

Bodily injury and property damage are usually written together into an insurance specification, such as 50/100/25. All three amounts are thousands of dollars. The first two figures—50/100—denote the maximum amounts for bodily injury. The first number—50, or $50,000—is the maximum amount payable by the insurer to any one individual sustaining injuries in an accident. The second amount—100, or $100,000—is the maximum total amount payable to all individuals in a single accident for injuries. The third figure—25 or $25,000—is the maximum amount payable for property damage in an accident.

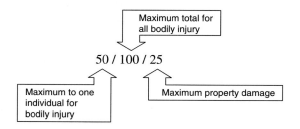

Common auto insurance specifications include 10/20, 20/40, 50/100, 100/300, and in even larger amounts. Property damage usually includes amounts of $5000, $10,000, $25,000, $50,000, or larger.

For example, Roger Reo's truck ran into Emma Edsel's car, causing injuries to Mrs. Edsel and her husband. Mr. Reo carries 10/20/10 liability insurance. Mrs. Edsel was awarded $8000 and her husband $12,500 for injuries. Damage to the Edsel's car was $8000. (a) What portion of the settlement is paid by Mr. Reo's insurance company? (b) What portion (if any) remains to be paid by Mr. Reo?

Since $10,000 is the maximum payment by the insurance company to an individual for injuries. Mr. Edsel is paid $10,000. Mrs. Edsel received $8000 for injuries. The insurer pays $8000 for the automobile.

Mr. Reo must pay the remaining $2500 for Mr. Edsel's injury.

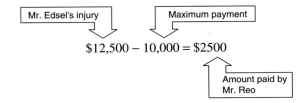

$$\$12,500 - 10,000 = \$2500$$

YOUR TURN

WORK THIS PROBLEM

The Question:

Denise Daimler's delivery van ran into Horace Hudson's car, causing injuries to Mr. Hudson and his wife. Ms. Daimler carries 10/15/10 liability insurance. The court awarded Horace Hudson $12,000 and his wife $4500 for injuries. Damage to the Hudson's car was $11,500. (a) What portion of the settlement is paid by Ms. Daimler's insurance company? (b) What portion must be paid by Ms. Daimler?

✔ YOUR WORK

The Solution:

(a) Paid by insurer:

Injury to Mr. Hudson	$10,000	($10,000 maximum)
Injury to Mrs. Hudson	4,500	
Auto damage	10,000	($10,000 maximum)
Total	$24,500	

(b) Paid by Ms. Daimler:

Injury to Mr. Hudson	$12,000 − 10,000 = $2000
Injury to Mrs. Hudson	0
Auto damage	$11,500 − 10,000 = 1500
Total	$3500

The moral of this problem is always carry adequate auto insurance.

Determining Liability Premiums on Motor Vehicle Insurance

Finding the premium for motor vehicle liability insurance requires you to apply the correct driver classification and the amount of coverage in a "motor vehicle liability insurance" table.

Premium tables vary from region to region, state to state, and even city to city, depending on state laws and local accident rates.

The following is a sample liability rate table. To use it, find the proper driver class in the left-hand column and read across the row to the appropriate bodily injury and property damage amounts.

MOTOR VEHICLE LIABILITY INSURANCE: SEMIANNUAL PREMIUM

Driver Class	Bodily Injury			Property Damage	
	25/50	50/100	100/300	$20,000	$50,000
1A	$168	188	212	151	156
1B	172	194	219	154	159
1C	177	201	227	159	165
2A	256	303	357	235	249
2C	282	337	401	248	263
2F	195	224	257	180	188
3	178	202	229	183	191

For example, let's calculate the semiannual liability insurance premium for 25/50/20 for driver class 1C.

Bodily injury 25/50	$177	
Property damage $20,000	159	
Total	$336	semiannual premium

As you can see, finding premiums for motor vehicle liability insurance is a simple two-step process:

Steps: Calculating Motor Vehicle Liability Premiums

STEP 1. *Determine* the driver classification.

STEP 2. *Use* the classification to find the premium on the premium table.

YOUR TURN

WORK THIS PROBLEM

The Question:

(a) Mr. and Mrs. N. Irv Usreck want to add their 19-year-old son as a driver on the family car. What would the semiannual premium be for 50/100/20 liability insurance?

(b) Zippy Delivery has decided to take a chance and take only 100/300/50 liability insurance on its new van, driven by 55-year-old Ima Snayle. What would be the semiannual premium?

✔ YOUR WORK

The Solution:

(a) *Step 1.*   Driver classification = 2A

Step 2.	Bodily injury	50/100	$303
	Property damage	$20,000	235
	Total		$538

(b) *Step 1.* Driver classification = 3

Step 2.	Bodily injury	100/300	$229
	Property damage	$50,000	191
	Total		$420

Finding Collision and Comprehensive Payouts and Premium

The property damage portion of motor vehicle liability insurance covers damage to property belonging to others, *not* the insured. If you wish to insure your car or truck and its contents, you must obtain *collision* and *comprehensive* coverage.

Collision Insurance A form of motor vehicle insurance that covers damages to the insured car if the insured is at fault or if the driver of the other car is at fault and unable to pay the damages.

Deductible An amount of damage that must be paid for by the insured before the insurer will cover any damages.

Collision covers damages to the insured car if the insured is at fault, or if the driver of the other car is at fault and unable to pay the damages. Most collision clauses carry a deductible provision. With a $100 **deductible** clause, the insured must pay the first $100 while the insurance company pays the rest of the damages. Collision coverage usually has $100, $250, $500, or higher deductibles.

For example, Jeri Ford was in an auto accident in which she was at fault. Her collision insurance coverage has a $150 deductible clause. The damage to her car was $2720. How much of the damages are paid by (a) Ms. Ford and (b) the insurance company?

(a) Ms. Ford must pay the deductible amount, $150.
(b) The insurance company pays $2720 − 150 = $2570.

YOUR TURN

WORK THIS PROBLEM

The Question:

Eddie Eagle was in an auto accident in which he was at fault. His collision insurance coverage has a $250 deductible clause. The damages to his car were $987. How much of the damages are paid by (a) Mr. Eagle and (b) the insurance company?

✔ YOUR WORK

The Solution:

(a) Mr. Eagle must pay the deductible amount—$250.
(b) The insurance company pays the damages less the deductible amount:
  $987 − 250 = $737.

Premiums for collision coverage depend on the driver classification, the auto's price and age, and the deductible.

The table on the following page is an abbreviated sample table. The auto's price group is determined by its original price. The auto's age group is determined by the age of the car. To calculate the collision premium, first find the proper row using the auto's price and age groups. Next, move right to the proper driver class.

For example, let's calculate the premium for a $100 deductible collision clause with price group E, age group 2, and driver class 1C.

**SEMIANNUAL MOTOR VEHICLE COLLISION
AND COMPREHENSIVE $100 DEDUCTIBLE**

Price Group	Age Group	Driver Class							Comprehensive
		1A	**1B**	**1C**	**2A**	**2C**	**2F**	**3**	
D	1	$109	$115	$119	$156	$212	$140	$132	$45
	2	92	98	103	148	201	132	125	39
	3	78	85	91	135	185	119	117	32
E	1	120	127	131	185	255	172	163	51
	2	103	109	104	168	231	161	155	45
	3	85	92	96	153	212	149	142	40
F	1	137	144	150	198	280	195	189	59
	2	121	128	133	182	263	180	176	52
	3	102	110	114	168	245	163	160	44

It's easy. Just find the row for price group E, age group 2. Then read across to class 1C. The premium is $104.

YOUR TURN

| WORK THIS PROBLEM |

The Question:

Calculate the premium for a $100 deductible collision clause with price group F, age group 3, and driver class 2A.

| ✔ YOUR WORK |

The Solution:

The collision premium is $168.

Comprehensive Insurance A form of motor vehicle insurance that pays for damages caused by fire, theft, vandalism, and so on.

At the right-hand end of the collision table is a column for comprehensive coverage. **Comprehensive** pays for damages caused by fire, theft, vandalism, and similar events. Comprehensive, like collision coverage, usually has a deductible amount. But only the auto's price and age groups are used to determine the premium.

For example, what is the premium for $100 deductible collision and comprehensive coverage with price group D, age group 1, and driver class 2F?

Collision	$140
Comprehensive	45
	$185

YOUR TURN

| WORK THIS PROBLEM |

The Question:

Calculate the premium for $100 deductible collision and comprehensive with price group E, age group 3, and driver classification 3.

✔ YOUR WORK

The Solution:

Collision	$142
Comprehensive	40
	$182

Calculating Medical Payments Premiums

Medical Payments A form of motor vehicle insurance that covers the insured in the event of injuries in an accident.

Just as you need collision and comprehensive insurance to protect your vehicle and its contents, so you need **medical payments** insurance to insure yourself against injuries. Driver classification may or may not affect the medical payments premium, depending on the state. A state without such a classification effect might have rates such as

AUTOMOBILE MEDICAL PAYMENTS

Amount	Semiannual Premium
$ 1,000	$16
2,000	18
5,000	23
10,000	33

For example, in such states the medical premium for $5,000 of coverage would be $23, regardless of driver classification.

YOUR TURN

WORK THIS PROBLEM

The Question:

Art Miller, age 54, is a traveling salesman who is "on the road" at least four days a week. What would his medical payments premium be for $10,000 of coverage using the preceding chart?

✔ YOUR WORK

The Solution:

$10,000 coverage = $33 semiannual premium

Total Motor Vehicle Insurance Premiums

In addition to liability, collision and comprehensive, and medical payments, you can obtain insurance to include uninsured motorists, towing, road service, and disability and death benefits.

Yet another factor that can play a significant role in your total insurance bill is your driving record—or the record of the firm's drivers. Having an accident in which you are at fault virtually always means your rates will go up. In fact, a single accident may double your rates—especially if you were drinking before the accident.

These additional considerations vary so much from policy to policy, state to state, and person to person that we can't calculate them here. What we can do is to calculate the cumulative effect of liability, collision and comprehensive, and medical payments insurance.

For example, what is the total semiannual premium for a motor vehicle policy with 50/100/50 liability, $5000 medical payments, and $100 deductible collision and comprehensive for a car with price group F, age group 1, and driver classification 1B?

Bodily injury 50/100	$194
Property damage $50,000	159
Medical payments $5,000	23
Collision	144
Comprehensive	59
Total	$579 Semiannual premium

It's easy if you work it carefully, step by step.

YOUR TURN

WORK THIS PROBLEM

The Question:

Betty Benz uses her car for business. Calculate her semiannual automobile insurance premium for 100/300/50 liability, $10,000 medical payments, and $100 deductible collision and comprehensive. Her car is in price group F, age group 1.

✔ YOUR WORK

The Solution:

Driver class 3.

Bodily injury 100/300	$229
Property damage $50,000	191
Medical payments $10,000	33
Collision	189
Comprehensive	59
Total	$701 Semiannual premium

For a set of practice problems on automobile insurance see Section Test 18.2.

SECTION TEST 18.2   Insurance

The following problems test your understanding of Section 18.2, Motor Vehicle Insurance.

A. Use the tables in the text to calculate the following automobile insurance premiums.

1. Constance Chung, age 23, has an automobile insurance policy with 25/50/50 liability, $5000 medical payments, and $100 deductible collision and comprehensive for an auto with price group E, age group 2. What is the semiannual premium?

2. Daniel Rathner, age 22, is unmarried and owns his automobile. His insurance policy has 25/50/50 liability, $5000 medical payments, and $100 deductible collision and comprehensive for an auto with price group F, age group 3. What is his semiannual premium?

3. Ed Bragley has an automobile used for business. His insurance policy has 100/300/50 liability, $10,000 medical payments, and $100 deductible collision and comprehensive for an auto with price group F, age group 1. What is his semiannual premium?

4. Crystal Yagle, owner of Crystal's Diamonds, has an automobile she uses for business. Her insurance policy has 50/100/50 liability, $10,000 medical payments, and $100 deductible collision and comprehensive for an auto with price group E, age group 1. What is her semiannual premium?

5. Peter Allnett, age 35, drives his car to work, 8 miles one way, and does not use his car for business. His insurance policy has 25/50/20 liability, $5000 medical payments, and $100 deductible collision and comprehensive for an auto with price group D, age group 2. What is his semiannual premium?

6. Gilda Retner, age 21, owns her automobile. Her insurance policy has 25/50/50 liability, $5000 medical payments, and $100 deductible collision and comprehensive for an auto with price group D, age group 1. What is her semiannual premium?

7. Mark Shiring, errand boy for Shiring Investments, uses his automobile for business. His insurance policy has 50/100/50 liability, $10,000 medical payments, and $100 deductible collision and comprehensive for an auto with price group E, age group 1. What is his semiannual premium?

8. Joan Stream has an automobile for business use. Her insurance policy has 100/300/50 liability, $10,000 medical payments, and $100 deductible collision and comprehensive for an auto with price group F, age group 2. What is her semiannual premium?

B. Work the following problems.

1. John Rutherford's auto ran into Nicki Landa's car, causing injuries to Ms. Landa. Mr. Rutherford carries 25/50/20 liability insurance. Ms. Landa was awarded $23,000 for injuries. Damage to Ms. Landa's car was $21,500. (a) What portion of the settlement is paid by Mr. Rutherford's insurance company? (b) What portion (if any) remains to be paid by Mr. Rutherford?

2. Jacqueline Stewart's auto ran into Rick Mears's car, causing injuries to Mr. Mears and his daughter. Ms. Steward carries 50/100/50 liability insurance. Mr. Mears was awarded $43,000 and his daughter $9500 for injuries. Damage to the Mears's car was $35,000. (a) What portion of the settlement is paid by Ms. Stewart's insurance company? (b) What portion (if any) remains to be paid by Ms. Stewart?

3. Angela Foyt's auto ran into Parnelli Jones's car, causing injuries to him and his wife. Ms. Foyt carries 25/50/20 liability insurance. Mr. Jones was awarded $27,000 and his wife $19,000 for injuries. Damage to the Jones car was $28,200. (a) What portion of the settlement is paid by Ms. Foyt's insurance company? (b) What portion (if any) remains to be paid by Ms. Foyt?

4. Russ Wallace's auto ran into Dick Petty's car, causing injuries to him, his wife and his son. Mr. Wallace carries 25/50/20 liability insurance. Mr. Petty was awarded $24,000, his wife $7500, and his son $8000. Damage to the Petty car was $23,500. (a) What portion of the settlement is paid by Mr. Wallace's insurance company? (b) What portion (if any) remains to be paid by Mr. Wallace?

5. Bob Allison was in an accident in which he was at fault. His collision coverage has a $250 deductible clause. The damages to his car were $2825. How much of the damages are paid by (a) Mr. Allison and (b) the insurance company?

6. Dale Earnhardt was in an automobile accident in which she was at fault. Her automobile collision coverage has a $300 deductible clause. The damages to her car were $6290. How much of the damages are paid by (a) Ms. Earnhardt and (b) the insurance company?

7. Ricki Rudd's auto ran into Darrell Waltrip's car, causing injuries to Mr. Waltrip, his wife, and his two daughters. Mr. Rudd carries 25/50/20 liability insurance. Mr. Waltrip was awarded $32,000, his wife $26,000, and each daughter $2,000. Damage to the Waltrip car was $27,000. (a) What portion of the settlement is paid by Mr. Rudd's insurance company? (b) What portion (if any) remains to be paid by Mr. Rudd?

8. Mary Andretti was in an automobile accident in which she was at fault. Her automobile collision coverage has a $150 deductible clause. The damages to her car were $5715. How much of the damages are paid by (a) Ms. Andretti and (b) the insurance company?

9. P&PIE Landscaping purchased a new truck in 1996 in price group F. This company, owned and run by 62-year-old Greenie Thumbkin, purchased insurance with liability coverage of 50/100/50, along with $100-deductible collision and comprehensive coverage and $10,000 coverage for medical payments. What is the total semiannual premium for this insurance policy?

WORKSPACE

SECTION 18.3: Life Insurance

Can You:

Calculate premiums for life insurance policies?
Determine your surrender option on life insurance policies?
. . . If not, you need this section.

Life Insurance Insurance that provides for a payment to a designated beneficiary upon the death of the insured.

Term Insurance Life insurance policies on which the insurer's obligation to pay benefits is limited to a stated number of years, while the insured is guaranteed a fixed premium.

Ordinary Life Insurance Life insurance policies that both pay death benefits and offer the option of cashing in the insurance at a later date for its "cash value"; payments must continue until the death of the insured or the cashing in of the policy.

Cash Value (Cash Surrender Value) The value of an ordinary life insurance policy; it depends on the size of the policy and the length of time it has been in force.

Limited-Payment Life Insurance Life insurance policies that provide death benefits and a cash value option; premiums are fixed and are completely paid up after a specified number of years.

Endowment A form of limited-payment life insurance policy whose high premiums are designed to build a large cash value.

The primary purposes of **life insurance** are to cover burial expenses and to provide continued support for the family of a deceased income earner. The insurance policy has a face value that is usually in multiples of $1000. The face value of the policy is usually paid to the designated beneficiary upon the death of the insured, though other options are available.

There are four basic types of life insurance: term, ordinary (straight), limited-payment, and endowment policies. **Term insurance** is simply life insurance for a certain term or period of time, usually one to ten years. If the insured dies during the term, the beneficiary receives the face value of the policy. Term insurance provides only death benefits; no other benefits are usually provided. The rates are relatively inexpensive for younger individuals. The rate increases with age and is very expensive for older people. Term insurance is very popular with employers who provide life insurance as a fringe benefit to employees.

Individuals may also take out **ordinary life insurance,** which provides both death benefits and builds a **cash value (cash surrender value)** that increases with the length of the policy. If, at some future date, the policy is discontinued, the insured may "cash in" the policy and receive the current cash value. The cash may provide retirement income, pay for a dependent's education, et cetera. The cash value may also be used as collateral for low-interest loans.

The premium on an ordinary life insurance policy depends on the insured's age when taking out the policy. That is, the younger a person is, the better (lower) the premium he or she will pay. The difference can really add up, because the premium remains the same throughout the length of the policy.

Like ordinary life insurance, **limited-payment life insurance** provides death benefits, builds a cash value, and has premiums that remain the same. But limited-payment is fully paid up after a limited number of years. The most common time period is 20 years, although other periods are available. Since the premiums are paid for only a limited time period, the rates are usually higher than ordinary life insurance.

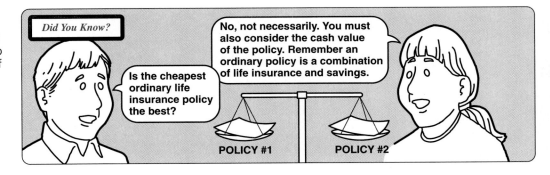

Finally, an **endowment** policy is a form of limited-payment life insurance, but the premiums are generally higher than the other types of insurance. An endowment policy is designed to build a large cash value after a certain number of years.

Calculating Life Insurance Premiums

Life insurance premiums depend on the age of the insured. The rate is usually stated per $1,000 of insurance. The following is a sample premium schedule for males. Because women live longer on the average, you must subtract three years from the ages of females to use the premium table for males.

ANNUAL LIFE INSURANCE PREMIUMS FOR MALES* PER $1,000

Age	10-Year Term	Ordinary Life	20-Payment Life	20-Year Endowment
18	$ 4.94	$13.07	$21.53	$41.17
20	5.02	13.72	22.60	41.23
22	5.10	14.42	23.75	41.36
24	5.16	15.18	24.94	41.57
26	5.21	16.09	25.91	41.72
28	5.29	17.02	27.36	41.89
30	5.42	17.93	29.17	42.06
35	5.98	21.52	32.28	43.23
40	7.72	25.61	36.42	44.42
45	11.26	30.46	41.50	46.95
50	17.02	37.48	48.92	50.18
55	26.12	49.47	57.43	57.42

*Subtract three years for female.

Steps: Calculating Life Insurance Premiums

STEP 1. *Determine* the basic rate in the table: Find the age in the left-hand column and read across the row to the desired life insurance type. For females subtract 3 years from the age.

STEP 2. *Multiply* the rate in the table by the number of $1000s of insurance.

For example, let's calculate the premium for a $5000 ordinary life insurance policy started by a man at age 24.

Step 1. From the table, the rate is $15.18 per $1000.

Step 2.

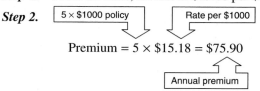

Premium = 5 × $15.18 = $75.90

YOUR TURN

WORK THIS PROBLEM

The Question:

(a) Calculate the annual premium for a $15,000 20-payment life insurance policy for Sam Malone, age 30.

(b) What is the premium for a $45,000 ten-year term insurance for Marge Simpson, age 29? (Hint: Remember to subtract three years from the age of a female.)

✔ YOUR WORK

The Solution:

(a) **Step 1.** The rate is $29.17 per $1000.
    Step 2. Premium = 15 × $29.17 = $437.55

(b) **Step 1.** Use age = 29 − 3 = 26, so the rate is $5.21 per $1000.
    Step 2. Premium = 45 × $5.21 = $234.45

Determining Surrender Options

Surrender (Nonforfeiture) Options The insured's options in the event of discontinuing an ordinary, limited-payment, or endowment life insurance policy.

If you discontinue an ordinary, limited-payment, or endowment life insurance policy, you may pick one of three common **surrender (nonforfeiture) options:** cash surrender value, paid-up insurance, and extended term insurance.

Cash Surrender Value A surrender option on a life insurance policy in which the insured receives money (the cash value of the policy at the time).

The **cash surrender value** is the amount of money the insured receives from the insurance company for "cashing in" or discontinuing a policy. The cash surrender value is usually stated per $1000 of insurance.

Another option, **paid-up insurance,** is a reduced amount of insurance on which no additional payments are due. The insurance is paid-up for the life of the insured; there are no more premiums. Paid-up insurance is usually stated per $1000 of insurance.

Paid-Up Insurance A surrender option on a life insurance policy in which the insured receives a reduced amount of insurance on which no additional premiums are due.

The third option, **extended term insurance,** provides insurance for the face value of the original policy for a given term or time period.

Extended Term Insurance A surrender option on a life insurance policy in which the insured receives insurance for the face value of the original policy for a given term or time period.

Each life insurance policy contains a table of surrender options for the age of the insured when the policy was started. The amount of each option depends on the length of time the policy was in force. The following is an abbreviated table for a policy issued at age 20 for a male or age 23 for a female.

SURRENDER OPTIONS FOR ORDINARY LIFE POLICY PER $1000

End of Year	Age 20 Male or Age 23 Female		
	Cash Surrender Value	Paid-Up Insurance	Extended Term Insurance
1	$ 0	$ 0	0y 0d
2	0	0	0y 0d
3	8	19	1y 50d
4	15	35	2y 232d
5	22	49	3y 159d
10	82	198	12y 341d
15	129	291	17y 118d
20	201	412	21y 47d

To find the surrender value, just follow these steps:

Steps: Calculating Surrender Value for Cash Surrender and Paid-up Insurance

STEP 1.   *Use* the table to find the value per $1000.

STEP 2.   *Multiply* the answer from step 1 by the number of $1000s in the policy.

For extended term insurance, the process is even easier—just look up in the chart the number of days and years the policy will remain in force.

For example, what are the three surrender options for a $15,000 ordinary life insurance policy issued at age 23 for a female and in force ten years?

Cash surrender value

Step 1.   Rate = $82 per $1000

Step 2.   Value = rate × policy thousands
             = $82 × 15
             = $1230

$15,000 policy

Paid-up insurance

Step 1.   Rate = $198 per $1000

Step 2.   Insurance = rate × policy thousands
              = $198 × 15
              = $2970

Extended term insurance = 12 years, 341 days

YOUR TURN

| WORK THIS PROBLEM |

The Question:

Dwayne Wayne is discontinuing his $12,000 ordinary life insurance policy issued at age 20. The policy has been in force five years. Calculate the (a) cash surrender value, (b) paid-up insurance, and (c) extended term insurance.

| ✔ YOUR WORK |

The Solution:

(a) *Step 1.*   Rate = $22
    Step 2.   Value = $22 × 12 = $264

(b) *Step 1.*   Rate = $49
    Step 2.   Insurance = $49 × 12 = $588

(c) Three years, 159 days

Selecting a Settlement Option

Insurers are not the only ones with decisions to make. When the insured dies, the beneficiary of a life insurance policy has several options for receiving the proceeds. One option is to receive the face value of the policy in a one lump-sum payment.

Another option is monthly (or annual, semiannual, or quarterly) payments to the beneficiary for a fixed number of years. The amount of each payment depends on the number of years selected by the beneficiary.

Finally, beneficiaries may choose to receive monthly payments for the rest of their lives. This option also contains a minimum guaranteed period of payments. The period is usually 10, 15, or 20 years. If the beneficiary dies during the guaranteed period, the beneficiary's heirs receive the payment for the remainder of the period. The amount of each payment depends on the age of the beneficiary and the length of the guaranteed period selected.

For a set of practice problems on life insurance, see Section Test 18.3.

WORKSPACE

SECTION TEST 18.3 Insurance

The following problems test your understanding of Section 18.3, Life Insurance.

A. Calculate the annual premium for the following life insurance policies. Use the Annual Life Insurance Premiums for Males per $1000 table.

Type of Insurance	Face Value	Age	Sex	Premium
1. Ordinary life	$25,000	26	M	
2. 20-payment life	20,000	21	F	
3. 10-year term	50,000	24	M	
4. Ordinary life	30,000	38	F	
5. 20-year endowment	15,000	27	F	
6. 20-year endowment	25,000	28	M	
7. 20-payment life	25,000	30	M	
8. 10-year term	60,000	25	F	
9. Ordinary	40,000	23	F	
10. 10-year term	75,000	35	M	

B. Calculate the three surrender options for the following ordinary life insurance policies. Use the Surrender Options for Ordinary Life Policy per $1000 table.

	Starting Age	Sex	Years Policy in Force	Face Value	Cash Surrender Value	Paid-up Insurance	Extended Term Insurance
1.	23	F	5	$20,000			
2.	20	M	20	10,000			
3.	20	M	15	25,000			
4.	23	F	10	30,000			
5.	23	F	20	15,000			
6.	23	F	15	50,000			
7.	20	M	10	40,000			
8.	20	M	4	12,000			
9.	23	F	4	21,000			
10.	20	M	5	35,000			

C. Solve these problems. Use the appropriate tables from Section 18.3.

1. Arthur Anderson, male, age 28, purchased a $20,000 ordinary life insurance policy. What is his annual premium?

2. Mary Kay, female, age 21, purchased a $50,000 ten-year term life insurance policy. What is her annual premium?

3. Liz Arden, female, age 38, purchased a $15,000 20-year endowment life insurance policy. What is her annual premium?

4. William Colgate, male, age 35, purchased a $25,000 20-payment life insurance policy. What is his annual premium?

5. Frances Woolworth, female, age 27, purchased a $75,000 ten-year term life insurance policy. What is her annual premium?

6. Walter Thompson, male, age 26, purchased a $30,000 ordinary life insurance policy. What is his annual premium?

7. Will Kellogg, male, age 24, purchased a $12,000 20-payment life insurance policy. What is his annual premium?

8. Connie Hilton, female, age 23, purchased a $10,000 20-year endowment life insurance policy. What is her annual premium?

9. Debbie Fields, female, purchased a $20,000 ordinary life insurance policy at age 23. What are the three surrender options after 15 years?

10. Tom Lipton, male, purchased a $15,000 ordinary life insurance policy at age 20. What are the three surrender options after ten years?

11. John Morgan, male, purchased a $25,000 ordinary life insurance policy at age 20. What are the three surrender options after five years?

12. At age 23, Elizabeth Clay Bourne, female, purchased a $40,000 ordinary life insurance policy. What are the three surrender options after 20 years?

Key Terms

cash surrender value
cash value (cash surrender value)
coinsurance
collision insurance
comprehensive insurance
deductible
endowment
extended term insurance
face value

fire insurance
insurance policy
insured
insurer
liability insurance
life insurance
limited-payment life insurance
medical payments
ordinary life insurance

paid-up insurance
premium
proration
replacement value
short-term policy
surrender (nonforfeiture) options
term insurance

Chapter 18 at a Glance

Page	Topic	Key Point	Example
582	Annual premium	Annual premium = rate × amount of insurance	What is the premium on a \$400,000 fire insurance policy when the rate is \$3.50 per \$1000? $\text{Annual premium} = \dfrac{\$3.50}{\$1000} \times \$400,000$ $= \$1400$
583	Multiyear policy premiums	Two-year policy premium = 1.85 × annual premium Three-year policy premium = 2.7 × annual premium	What is the three-year premium on the policy in the previous example? Premium = 2.7 × \$1400 = \$3780
584	Short-rate premiums	Short-rate premium = short rate from the Short-Rate table × annual premium	What is the short-rate premium for 180 days of coverage on the policy in the previous example? Short-rate premium = 60% × \$1400 = \$840
586	Refund when a policyholder cancels a policy	Premium retained by insurer = short rate from the short-rate table × annual premium Refund = annual premium − premium retained	What is the refund due the policyholder who cancels the policy in the above example after 120 days? Premium retained = 43% × \$1400 = \$602 Refund = \$1400 − 602 = \$798
587	Refund when an insurer cancels a policy	Premium retained by insurer = $\dfrac{\text{exact time}}{365} \times \dfrac{\text{annual}}{\text{premium}}$ Refund = annual premium − premium retained	What is the refund due the policyholder in the previous example if the insurer cancels after 120 days? Premium retained = $\dfrac{120}{365} \times \$1400$ $= \$460.27$ Refund = \$1400 − 460.27 = \$939.73

Page	Topic	Key Point	Example
589	Losses paid by multiple insurers	Each insurer's share = $\dfrac{\text{policy value}}{\text{total insurance}} \times \text{loss}$	If a company that carries a $200,000 fire insurance policy with Company A and a $300,000 policy with Company B suffers a $360,000 fire loss, how much does each insurer pay? Total insurance = $200,000 + 300,000 = \$500,000$ Company A $= \dfrac{\$200,000}{\$500,000} \times \$360,000$ $= \$144,000$ Company B $= \dfrac{\$300,000}{\$500,000} \times \$360,000$ $= \$216,000$
590	Losses paid by coinsurers	Insurer pays = loss $\times$ $\dfrac{\text{face value of property}}{\text{rate} \times \text{property value}}$	A company carrying a $250,000 fire insurance policy with an 80% coinsurance clause suffers a $120,000 loss on a $400,000 property. How much will the insurer pay? Insurer pays $= \$120,000 \times \dfrac{\$250,000}{80\% \times 400,000}$ $= \$93,750$
600	Liability insurance payouts	Insured's liability = injuries − maximum coverage	Roger Reo causes an accident to Emma Edsel's car. Roger carries 25/50/20 liability insurance. If Emma suffered $30,000 in injuries, her husband suffered $22,500 in injuries and her car had $14,500 in damage, what must Roger pay? On Emma's injuries: $30,000 − 25,000 = \$5000$ On her husband's injuries: $22,500 is less than $25,000. Pay nothing. On the car's damage: Coverage of $20,000 exceeds damage of $14,500, so Roger pays nothing.
601	Liability insurance premiums	Total liability premium = bodily injury premium + property damage premium	Using the table of sample rates in Section 18.2, what is the total liability premium for 50/100/20 coverage for business use? Driver class: 3 Bodily injury = $202 Property damage = $183 Total liability premium = $202 + 183 = \$385$

Page	Topic	Key Point	Example
603	Collision insurance coverage	Payment by insurer = covered damage − deductible	Roger Reo's car had $2720 in damages in an accident he caused. If he has $150 deductible on his collision insurance, how much must his insurer pay? Roger pays $150. Insurer pays = $2720 − 150 $\qquad\qquad\quad$ = $2570
604	Collision and comprehensive premiums	Use the table of sample rates in Section 18.2 to find premiums for the desired coverage level by auto price and age groups and driver class.	What is the premium for a $100 deductible collision and comprehensive clause with price E, age group 2, and driver class 1C? Collision = $104 Comprehensive = $45 Total = $149
605	Medical payments premiums	Use the sample rates in the Automobile Medical Payments table to find the premium.	What is the premium for $5000 medical payments coverage? $23 premium
612	Life insurance premiums	Premium = rate from Premiums for Males table (−3 years for females) × no. $1000s insurance	What is the premium for a $10,000 ordinary life policy for a 20-year-old male? Premium = $13.72 × 10 $\qquad\qquad$ = $137.20
613	Surrender values on an ordinary life policy	Cash surrender values = rate from the Surrender Options table × no. $1000s insurance Paid-up insurance value = rate from the Surrender Options table × no. $1000s insurance Extended term insurance period— see the Surrender Options table.	Find the three surrender values of a $10,000 ordinary life policy written on a 20-year-old male 15 years ago. Cash surrender value = $129 × 10 $\qquad\qquad\qquad\quad$ = $1290 Paid-up insurance value = $291 × 10 $\qquad\qquad\qquad\qquad$ = $2910 Extended term = 17 years, 118 days

WORKSPACE

SELF-TEST Insurance

The following problems test your understanding of insurance. Use the appropriate tables in the chapter.

1. The Wing and a Prayer Charter Flight Service hangar is insured for $175,000 at a rate of $0.23 per $100. What are (a) the annual premium, (b) the two-year premium, and (c) the three-year premium?

2. The Fly-by-Night Aviary is insured for $145,000 at a rate of $0.28 per $100. What are (a) the annual premium, (b) the two-year premium, and (c) the three-year premium?

3. The Tick Tock Clock factory is insured for $130,000 at an annual rate of $0.29 per $100. What is the premium on a 60-day policy?

4. Inge's German Bed and Breakfast is insured for $250,000 at an annual rate of $0.32 per $100. What is the premium on a 90-day policy?

5. Helen insured her House of Beauty for one year starting at September 15 at an annual cost of $482. The policy was canceled on December 2 by the insurance company. What are (a) the amount of the premium retained by the insurance company and (b) the refund to the insured?

6. Ms. Spode insured her pottery studio on August 7 at an annual cost of $375. The policy was canceled by the insurance company on November 12. What are (a) the amount of the premium retained by the insurance company and (b) the refund to the insured?

7. The By the Byte Software House carries three insurance policies: Company A, $25,000; Company B, $30,000; and Company C, $35,000. The firm sustained a $36,000 loss in a recent fire. Calculate the amount of loss paid by each insurance company.

8. Upward Bound Adventures carries three insurance policies: Company A, $75,000; Company B, $65,000; and Company C, $60,000. The company sustained a $72,000 loss in a recent fire. Calculate the amount of loss paid by each insurance company.

9. The Buy-a-Nose Horse Farm carries $60,000 in insurance with an 80% coinsurance clause. The value of the property is $92,000. The farm had a $37,500 loss. What is the amount paid by the insurer?

10. The Silver Gate Spa carries $150,000 in insurance with an 80% coinsurance clause. The value of the property is $225,000. The resort had an $18,000 loss. What is the amount paid by the insurer?

11. Hetty Green uses her car for business. Her automobile insurance policy has 100/300/50 liability, $10,000 medical payments, and $100 deductible collision and comprehensive for an auto with price group F, age group 1. What is the semiannual premium?

12. Frank Perdue, age 22, is unmarried and owns his automobile. His insurance policy has 25/50/50 liability, $5000 medical payments, and $100 deductible collision and comprehensive for an auto with price group E, age group 2. What is his semiannual premium?

13. Dan Kamm's auto ran into Wil Wright's vehicle, causing injuries to Mr. Wright. Mr. Kamm carries 15/30/10 liability insurance. Mr. Wright was awarded $17,000 for injuries. Damage to Mr. Wright's car was $11,300. (a) What portion of the settlement is paid by Mr. Kamm's insurance company? (b) What portion (if any) remains to be paid by Mr. Kamm?

14. Gary Geare's auto ran into Bob Unser's car, causing injuries to Mr. Unser and his daughter. Mr. Geare carries 10/20/10 liability insurance. Mr. Unser was awarded $12,500 and his daughter $9200 for injuries. Damage to the Unser's car was $4500. (a) What portion of the settlement is paid by Mr. Geare's insurance company? (b) What portion (if any) remains to be paid by Mr. Geare?

15. Anita Axle was in an accident in which she was at fault. Her collision coverage has a $250 deductible clause. The damages to her car were $3815. How much of the damages are paid by (a) Ms. Axle and (b) the insurance company?

16. Louise Braille was in an automobile accident in which she was at fault. Her automobile collision coverage has a $150 deductible clause. The damages to her car were $5845. How much of the damages are paid by (a) Ms. Braille and (b) the insurance company?

17. Phyllis Wheatley, female, age 38, purchased an $80,000 ten-year-term life insurance policy. What is her annual premium?

18. Crispus Attucks, male, age 24, purchased a $20,000 20-year endowment life insurance policy. What is his annual premium?

19. Sojourner Truth, female, purchased a $25,000 ordinary life insurance policy at age 23. What are the three surrender options after 15 years?

20. Marcus Garvey, male, purchased a $30,000 ordinary life insurance policy at age 20. What are the three surrender options after ten years?

WORKSPACE

CHAPTER 19

Financial Statement Analysis

That financial statement would look better if you took your glasses off!

$1 - 0$
$X - Y$
$Z^4 1$
$=$

CHAPTER OBJECTIVES

When you complete this chapter successfully, you will be able to:

1. Complete a horizontal and vertical analysis of two successive balance sheets.

2. Perform a horizontal and vertical analysis of two successive income statements.

3. Calculate current ratio, acid-test ratio, return on equity, and inventory turnover.

■ How do the 1996 and 1997 balance sheets for Bits 'n' Bytes compare when you perform a horizontal analysis?

■ How do the 1996 and 1997 income statements for Bob's Beeflike Burgers compare when you perform a vertical analysis?

■ A recent income statement for Intense Tents showed a beginning inventory of $37,000, ending inventory of $41,000, and cost of goods sold of $217,500. What is the inventory turnover ratio rounded to the nearest hundredth?

As many small businesspersons have found to their sorrow, failure to keep proper financial records can be disastrous. Well-kept financial summaries such as balance sheets and income statements can help owners and managers make sound financial decisions about the firm's future. In addition, a number of ratios provide information about day-to-day performance.

So pay attention and you may improve your own "financial state" in the future.

Can You:

Identify the elements of a balance sheet and the fundamental accounting equation?
Perform a horizontal analysis of two successive balance sheets?
Complete a vertical analysis of two successive balance sheets?
. . . If not, you need this section.

Reading Balance Sheets

Balance Sheet A form of financial statement that shows the firm's condition at a particular point in time in terms of its assets, liabilities, and owners' equity.

One of the most common forms of financial statement is the balance sheet. A firm's **balance sheet** shows its financial condition at a particular point in time. Thus, it is a sort of financial snapshot of a company's *assets* (what it owns of value), its *liabilities* (what it owes to others), and its *owners' equity* (the value of the company).

That is, a balance sheet presents a business's status according to this equation:

$$\text{Owners' equity} = \text{assets} - \text{liabilities}$$

Fundamental Accounting Equation In constructing and analyzing balance sheets, the rule that assets must always equal the combined value of liabilities and owners' equity.

We can also state this relationship as accountants do in what they call the **fundamental accounting equation:**

$$\text{Assets} = \text{liabilities} + \text{owners' equity}$$

We can summarize this information in an abbreviated balance sheet like that below. Notice that the balance sheet is broken into two main parts: (1) assets and (2) liabilities and owners' equity.

BALANCE SHEET
DATE: Today

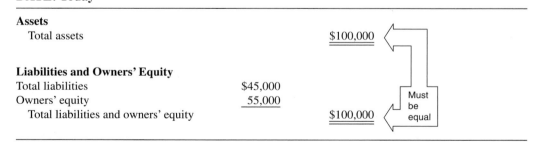

Assets
Total assets $100,000

Liabilities and Owners' Equity
Total liabilities $45,000
Owners' equity 55,000
   Total liabilities and owners' equity $100,000

Must
be
equal

The foregoing balance sheet satisfies the fundamental accounting equation:

$$\text{Assets} = \text{liabilities} + \text{owners' equity}$$
$$\$100,000 = \$45,000 + \$55,000$$

Managers use the information in balance sheets to determine where the business stands today. By comparing balance sheets from several years, managers can also see where the firm is headed and plan for the future.

Balance sheets used in business are much more detailed than the example, though. Consider the following balance sheet for Big Ben's Bolts:

BIG BEN'S BOLTS
BALANCE SHEET
December 31, 1995

(1) **Assets**
(2) *Current Assets*

Cash	$ 30,000	
Merchandise inventory	50,000	
Accounts receivable	70,000	
Total current assets		$150,000

(3) *Plant and Equipment*

Land	$ 80,000	
Buildings	170,000	
Equipment	100,000	
Total plant and equipment		$350,000
Total assets		$500,000

(4) **Liabilities and Owners' Equity**
(5) *Current Liabilities*

Accounts payable	$ 85,000	
Wages payable	30,000	
Total current liabilities		$115,000

(6) *Long-Term Liabilities*

Mortgage payable	$160,000	160,000
Total liabilities		$275,000

(7) *Owners' Equity*

Total owners' equity	$225,000	225,000
Total liabilities and owners' equity		$500,000

Assets → Liabilities and owners' equity

Assets All goods and property owned by a business.

Current Assets Assets that will be used or converted to cash in a short period of time—usually within one year; includes cash, merchandise on hand, and accounts receivable.

Fixed Assets Assets that will not be used or converted to cash within a year; includes land, buildings, and equipment.

Liabilities Amounts owed by the business to others.

Current Liabilities Amounts owed by a business that must be paid in a short period of time; includes accounts payable, wages payable, utilities payable, federal income tax payable, and notes payable.

Long-Term Liabilities Amounts owed by a business but not due for payment for more than a year.

Owners' Equity The difference between assets and liabilities.

(1) The **assets** of a business include the goods and property owned and claims that have not been collected. Assets are usually broken into two categories: (a) current assets and (b) fixed assets or plant and equipment.

(2) **Current assets** are assets that will be used or converted into cash in a short period of time—usually within one year. Current assets include cash, merchandise on hand, and accounts receivable (sales for which payments have not been received).

(3) **Fixed assets** or plant and equipment are items that will not be used or converted to cash within a year. Fixed assets include land, buildings, and equipment.

(4) **Liabilities** are the amounts owed by the business to others. Liabilities are also divided into two categories: current liabilities and fixed or long-term liabilities.

(5) **Current liabilities** are amount owed by the business that must be paid in a short period of time. They include accounts payable (money owed but not yet paid), wages payable, utilities payable, federal income tax payable, and notes payable that are due within one year.

(6) **Long-term (fixed) liabilities** are amounts that will be paid after a year. They include mortgages payable, bonds payable, and notes payable that are due later than one year.

(7) **Owners' equity** is the difference between the assets and liabilities. Owners' equity is also referred to as *net worth, stockholders' equity,* or *capital.*

Sound confusing? Just remember

$$\text{Assets} = \text{liabilities} + \text{owners' equity}$$

where
$$\text{Assets} = \text{values owned by the business}$$
$$\text{Liabilities} = \text{value owed to others}$$
$$\text{Owner's equity} = \text{net worth of the business}$$

Many businesses have an even more detailed balance sheet with many additional types of assets and liabilities. However, all balance sheets are presented in the prescribed format you just saw. The balance sheet is divided into two main sections: (1) assets, and (2) liabilities and owners' equity. The assets section is composed of current assets and fixed assets. The liabilities and the capital or owners' equity section is composed of current liabilities, long-term liabilities, and owners' equity.

Performing a Horizontal Analysis

Horizontal Analysis
Comparing financial statements from two different years or periods in order to determine the amount and percent of change in various categories.

A single balance sheet presents a static picture of a business at a given time. It tells us nothing about how the business will change over a certain period of time. To obtain a dynamic picture, one which includes changes in the business's financial picture during a given time period, you must compare two successive balance sheets. An analysis of two balance sheets to show the amount and percent of change in each item is called a **horizontal analysis.**

A horizontal analysis shows a manager how specific assets and liabilities have changed from year to year and thus hint at future trends. To perform a horizontal analysis, follow these steps:

Steps: *Performing a Horizontal Analysis*

STEP 1.   *Calculate* how much each item has increased or decreased on the second balance sheet compared to the first.

STEP 2.   *Determine* the percent increase or decrease and express the answer as a percent.

$$\text{Percent change} = \frac{\text{change}}{\text{first amount}}$$

If you need a review of %-type problems in order to complete step 2, see Chapter 4, Section 4.2.

Before we look at the analysis of an entire balance sheet, let's first consider the analysis of a single item. For example, two successive annual balance sheets of Big Ben's Bolts show that the total assets for 1995 are $500,000 and for 1996 are $600,000. Let's complete a horizontal analysis of this item.

Step 1. | 1996 Assets | | 1995 Assets |

$$\text{Increase} = \$600,000 - 500,000 = \$100,000$$

| Increase in total assets |

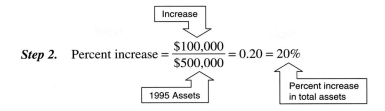

Step 2. Percent increase $= \dfrac{\$100,000}{\$500,000} = 0.20 = 20\%$

When calculating the percent of increase or decrease, always use the *first* balance sheet's amount (1995 in our example) as the denominator in step 2.

WORK THIS PROBLEM

YOUR
TURN

The Question:

Two balance sheets of Big Ben's Bolts show that the total liabilities for 1995 are $275,000 and for 1996 are $305,000. Complete a horizontal analysis. Round the percent to the nearest 0.1%.

✔ YOUR WORK

The Solution:

Step 1. Increase = $305,000 − 275,000 = $30,000

Step 2. Percent increase $= \dfrac{\$30,000}{\$275,000} = 0.1090 \ldots = 10.9\%$ rounded

Not all items on a balance sheet will be increases. Some items may decrease from one date to the next.

For example, two successive balance sheets of Big Ben's Bolts show that the mortgage payable for 1995 is $160,000 and for 1996 is $155,000. Let's complete a horizontal analysis of the mortgage payable, rounding the percent figure to the nearest 0.1%.

1995 mortgage payable	1996 mortgage payable

Step 1. Decrease = $160,000 − $155,000 = $5000

Decrease

Step 2. Percent decrease $= \dfrac{\$5000}{\$160,000} = 0.031 = 3.1\%$ rounded

1995 mortgage payable

On the balance sheet analysis, an asterisk (*) is used to denote decreases.

Notice that to calculate the change, we simply subtract the smaller amount from the larger. Since the amount in 1996 is smaller than the 1995 amount, there was a decrease.

Whether there is an increase or decrease, the percent change is always the amount of change divided by the amount from the first balance sheet.

YOUR
TURN

WORK THIS PROBLEM

The Question:

Big Ben's Bolts has balance sheets that show the mortgage payable for 1995 was $172,000 and for 1996 was $160,000. Complete a horizontal analysis of the mortgage payable and round the percent to the nearest 0.1%.

✔ YOUR WORK

The Solution:

Step 1.   Decrease = $172,000 − 160,000 = $12,000

Step 2.   Percent decrease = $\dfrac{\$12,000}{\$172,000} = 0.0697 \ldots = 7.0\%$ rounded

Now let's complete a horizontal analysis of two balance sheets.

BIG BEN'S BOLTS
COMPARATIVE BALANCE SHEETS   *Step 1*   *Step 2*
December 31, 1995 and 1996

	1995	1996	Increase or Decrease* Amount	Increase or Decrease* Percent
Assets				
Current Assets				
Cash	$ 30,000	$ 35,000	$ 5,000	16.7%
Merchandise inventory	50,000	70,000	20,000	40.0%
Accounts receivable	70,000	85,000	15,000	21.4%
Total current assets	$150,000	$190,000	$ 40,000	26.7%
Plant and Equipment				
Land	$ 80,000	$ 90,000	$ 10,000	12.5%
Buildings	170,000	190,000	20,000	11.8%
Equipment	100,000	130,000	30,000	30.0%
Total plant and equipment	$350,000	$410,000	$ 60,000	17.1%
Total assets	$500,000	$600,000	$100,000	20.0%
Liabilities and Owners' Equity				
Current Liabilities				
Accounts payable	$ 85,000	$110,000	$ 25,000	29.4%
Wages payable	30,000	40,000	10,000	33.3%
Total current liabilities	$115,000	$150,000	$ 35,000	30.4%
Long-Term Liabilities				
Mortgage payable	160,000	155,000	5,000*	3.1%*
Total liabilities	$275,000	$305,000	$ 30,000	10.9%
Owners' Equity				
Total owners' equity	$225,000	$295,000	$ 70,000	31.1%
Total liabilities and owners' equity	$500,000	$600,000	$100,000	20.0%

*Asterisk denotes a decrease.

It's easy if you take it step by step.

In a horizontal analysis, the percents cannot be added or subtracted. In this case, the individual items are compared "horizontally" or across, from one balance sheet to another. They are percent changes from one year to the next. The percent numbers *cannot* be added or subtracted *vertically* in a *horizontal* analysis.

WORK THIS PROBLEM

The Question:

Complete a horizontal analysis for Petal Pushers. Round all percent figures to the nearest 0.1%.

PETAL PUSHERS
COMPARATIVE BALANCE SHEETS
December 31, 1995 and 1996

	1995	1996	Increase or Decrease* Amount	Increase or Decrease* Percent
Assets				
Current Assets				
Cash	$ 36,000	$ 38,000		
Merchandise inventory	106,000	98,000		
Accounts receivable	58,000	52,000		
Total current assets	$200,000	$188,000		
Plant and Equipment				
Land	$ 25,000	$ 30,000		
Buildings	63,000	80,000		
Equipment	162,000	202,000		
Total plant and equipment	$250,000	$312,000		
Total assets	$450,000	$500,000		
Liabilities and Owners' Equity				
Current Liabilities				
Accounts payable	$ 78,000	$ 75,000		
Wages payable	29,000	30,000		
Total current liabilities	$107,000	$105,000		
Long-Term Liabilities				
Mortgage payable	100,000	90,000		
Total liabilities	$207,000	$195,000		
Owners' Equity				
Total owners' equity	243,000	305,000		
Total liabilities and owners' equity	$450,000	$500,000		

*Asterisk denotes a decrease.

✔ YOUR WORK

The Solution:

PETAL PUSHERS
COMPARATIVE BALANCE SHEETS   *Step 1*   *Step 2*
December 31, 1995 and 1996

	1995	1996	Increase or Decrease*	
			Amount	Percent
Assets				
Current Assets				
Cash	$ 36,000	$ 38,000	$ 2,000	5.6%
Merchandise inventory	106,000	98,000	8,000*	7.5%*
Accounts receivable	58,000	52,000	6,000*	10.3%*
Total current assets	$200,000	$188,000	$16,000*	6.0%*
Plant and Equipment				
Land	$ 25,000	$ 30,000	$ 5,000	20.0%
Buildings	63,000	80,000	17,000	27.0%
Equipment	162,000	202,000	40,000	24.7%
Total plant and equipment	$250,000	$312,000	$62,000	24.8%
Total assets	$450,000	$500,000	$50,000	11.1%
Liabilities and Owners' Equity				
Current Liabilities				
Accounts payable	$ 78,000	$ 75,000	$ 3,000*	3.8%*
Wages payable	29,000	30,000	1,000	3.4%
Total current liabilities	$107,000	$105,000	$ 2,000*	1.9%*
Long-Term Liabilities				
Mortgage payable	100,000	90,000	10,000*	10.0%*
Total liabilities	$207,000	$195,000	$12,000*	5.8%*
Owners' Equity				
Total owners' equity	243,000	305,000	62,000	25.5%
Total liabilities and owners' equity	$450,000	$500,000	$50,000	11.1%

*Asterisk denotes a decrease.

Whew! That's a lot of work! The analysis of two balance sheets is a long process, but if you work each item carefully, it's easy!

Performing a Vertical Analysis

Vertical Analysis Comparing financial statements from two different years or periods in order to determine changes in what percent of the total each represents.

As we noted earlier, every balance sheet is divided into two main categories: (1) assets and (2) liabilities and owners' equity. A **vertical analysis** determines what percent each item of a category is of the total. Managers look for changes in the percent versus the total. For example, did wages make up only 3.5% of liabilities last year but 5.9% this year?

Vertical analyses are also useful in comparing two businesses in the same field. The percent of land, buildings, and so on can be compared even if the companies are of different sizes.

To perform a vertical analysis, follow these steps:

Steps: Performing a Vertical Analysis

STEP 1. For each asset, *calculate* the percent of total assets. For each liability, calculate the percent of total liabilities and owners' equity.

$$\text{Percent} = \frac{\text{amount of individual item}}{\text{total}}$$

Be sure to convert the answer to a percent and round to the nearest 0.1%.

STEP 2. When completing an entire balance sheet, *check* your work by totaling the percent figures of each asset. The total should be approximately 100%, though it may not be exact due to rounding. The total percents of all liabilities and owners' equity should also be approximately 100%.

For example, on the December 31, 1995, balance sheet for Big Ben's Bolts, cash is $30,000 and total assets is $500,000. Let's perform part of a vertical analysis by determining what percent of the total assets is cash.

Step 1. Percent $= \dfrac{\overset{\boxed{\text{Cash}}}{\$30,000}}{\underset{\boxed{\text{Total assets}}}{\$500,000}} = 0.06 = 6\%$

**YOUR
TURN**

> ## WORK THIS PROBLEM

The Question:

On the December 31, 1995, balance sheet for Big Ben's Bolts the merchandise inventory is $50,000; the accounts receivable is $70,000; and the total assets is $500,000. Complete the vertical analysis for the merchandise inventory and accounts receivable. Find the merchandise inventory as a percent of total assets and the accounts receivable as a percent of total assets.

> ## ✔ YOUR WORK

The Solution:

Step 1. Merchandise inventory percent $= \dfrac{\$50,000}{\$500,000} = 0.10 = 10\%$

Accounts receivable percent $= \dfrac{\$70,000}{\$500,000} = 0.14 = 14\%$

The vertical analysis of a complete balance sheet is simply performed item by item. Each percent calculation is the same as the previous examples.

For example, the following balance sheet shows a complete vertical analysis of the December 31, 1995, balance sheet of Big Ben's Bolts.

Step 1.   **BIG BEN'S BOLTS**
BALANCE SHEET
December 31, 1995

	Amount	Percent
Assets		
Current Assets		
Cash	$ 30,000	6%
Merchandise inventory	50,000	10%
Accounts receivable	70,000	14%
Total current assets	$150,000	30%
Plant and Equipment		
Land	$ 80,000	16%
Buildings	170,000	34%
Equipment	100,000	20%
Total plant and equipment	$350,000	70%
Total assets	$500,000	100%
Liabilities and Owners' Equity		
Current Liabilities		
Accounts payable	$ 85,000	17%
Wages payable	30,000	6%
Total current liabilities	$115,000	23%
Long-Term Liabilities		
Mortgage payable	160,000	32%
Total liabilities	$275,000	55%
Owners' Equity		
Total owners' equity	225,000	45%
Total liabilities and owners' equity	$500,000	100%

Step 2.   Total assets percents = 6% + 10% + 14% + 16% + 34% + 20% = 100%
Total liabilities and owners' equity percents = 17% + 6% + 32% + 45%
= 100%

WORK THIS PROBLEM

The Question:

Complete the vertical analysis for the 1996 balance sheet of Big Ben's Bolts. (The 1995 analysis was completed in the previous example.) Round all percents to the nearest 0.1%.

BIG BEN'S BOLTS
COMPARATIVE BALANCE SHEETS
December 31, 1995 and 1996

	1995		1996	
	Amount	**Percent**	**Amount**	**Percent**
Assets				
Current assets				
Cash	$ 30,000	6.0%	$ 35,000	
Merchandise inventory	50,000	10.0%	70,000	
Accounts receivable	70,000	14.0%	85,000	
Total current assets	$150,000	30.0%	$190,000	
Plant and Equipment				
Land	$ 80,000	16.0%	$ 90,000	
Buildings	170,000	34.0%	190,000	
Equipment	100,000	20.0%	130,000	
Total plant and equipment	$350,000	70.0%	$410,000	
Total assets	$500,000	100.0%	$600,000	
Liabilities and Owners' Equity				
Current Liabilities				
Accounts payable	$ 85,000	17.0%	$110,000	
Wages payable	30,000	6.0%	40,000	
Total current liabilities	$115,000	23.0%	$150,000	
Long-Term Liabilities				
Mortgage payable	160,000	32.0%	155,000	
Total liabilities	$275,000	55.0%	$305,000	
Owners' Equity				
Total owners' equity	225,000	45.0%	295,000	
Total liabilities and owners' equity	$500,000	100.0%	$600,000	

✔ YOUR WORK

The Solution:

Step 1. **BIG BEN'S BOLTS**
COMPARATIVE BALANCE SHEETS
December 31, 1995 and 1996

	1995		1996	
	Amount	**Percent**	**Amount**	**Percent**
Assets				
Current Assets				
Cash	$ 30,000	6.0%	$ 35,000	5.8%
Merchandise inventory	50,000	10.0%	70,000	11.7%
Accounts receivable	70,000	14.0%	85,000	14.2%
Total current assets	$150,000	30.0%	$190,000	31.7%
Plant and Equipment				
Land	$ 80,000	16.0%	$ 90,000	15.0%
Buildings	170,000	34.0%	190,000	31.7%
Equipment	100,000	20.0%	130,000	21.7%
Total plant and equipment	$350,000	70.0%	$410,000	68.3%
Total assets	$500,000	100.0%	$600,000	100.0%
Liabilities and Owners' Equity				
Current Liabilities				
Accounts payable	$ 85,000	17.0%	$110,000	18.3%
Wages payable	30,000	6.0%	40,000	6.7%
Total current liabilities	$115,000	23.0%	$150,000	25.0%
Long-Term Liabilities				
Mortgage payable	160,000	32.0%	155,000	25.8%
Total liabilities	$275,000	55.0%	$305,000	50.8%
Owners' Equity				
Total owners' equity	225,000	45.0%	295,000	49.2%
Total liabilities and owners' equity	$500,000	100.0%	$600,000	100.0%

Step 2.   Total assets percents $= 5.8\% + 11.7\% + 14.2\% + 15.0\% + 31.7\% + 21.7\%$
$$= 100.1\%$$
Total liabilities and owners' equity percents $= 18.3\% + 6.7\% + 25.8\% + 49.2\%$
$$= 100.0\%$$

Notice that the long-term liability dropped from 32% to 25.8% and that owners' equity increased from 45% to 49.2%. These changes indicate that things are going well indeed at Big Ben's Bolts.

A calculator can speed the process of performing a vertical analysis.

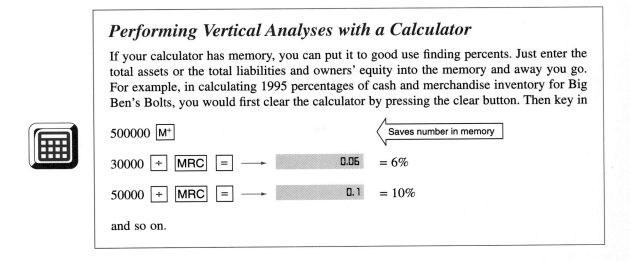

Performing Vertical Analyses with a Calculator

If your calculator has memory, you can put it to good use finding percents. Just enter the total assets or the total liabilities and owners' equity into the memory and away you go. For example, in calculating 1995 percentages of cash and merchandise inventory for Big Ben's Bolts, you would first clear the calculator by pressing the clear button. Then key in

500000 [M⁺]   ⟨ Saves number in memory

30000 [÷] [MRC] [=] ⟶ ░░░░░░ 0.06 = 6%

50000 [÷] [MRC] [=] ⟶ ░░░░░░ 0.1 = 10%

and so on.

For a set of practice problems on horizontal and vertical analysis of balance sheets, turn to Section Test 19.1.

WORKSPACE

SECTION TEST 19.1 Financial Statement Analysis

The following problems test your understanding of Section 19.1, Balance Sheets.

A. Complete the horizontal analysis of the following balance sheets. Round all percents to the nearest 0.1%. Use an * to denote any decreases.

1.

REDHAT TOOL AND DIE
COMPARATIVE BALANCE SHEETS
December 31, 1996 and 1997

	1996	1997	Increase or Decrease*	
			Amount	Percent
Assets				
Current Assets				
Cash	$ 33,000	$ 37,000		
Merchandise inventory	82,000	87,000		
Accounts receivable	79,000	78,000		
Total current assets	$194,000	$202,000		
Plant and Equipment				
Land	$ 32,000	$ 32,000		
Buildings	77,000	77,000		
Equipment	91,000	101,000		
Total plant and equipment	$200,000	$210,000		
Total assets	$394,000	$412,000		
Liabilities and Owners' Equity				
Current Liabilities				
Accounts payable	$ 74,000	$ 78,000		
Wages payable	31,000	35,000		
Total current liabilities	$105,000	$113,000		
Long-Term Liabilities				
Mortgage payable	83,000	80,000		
Total liabilities	$188,000	$193,000		
Owners' Equity				
Total owners' equity	206,000	219,000		
Total liabilities and owners' equity	$394,000	$412,000		

*Denotes a decrease.

2.

BITS 'N' BYTES
COMPARATIVE BALANCE SHEETS
December 31, 1996 and 1997

	1996	1997	Increase or Decrease*	
			Amount	Percent
Assets				
Current Assets				
Cash	$105,000	$ 97,000		
Merchandise inventory	126,000	136,000		
Accounts receivable	89,000	92,000		
Total current assets	$320,000	$325,000		
Plant and Equipment				
Land	$ 92,000	$105,000		
Buildings	123,000	145,000		
Equipment	51,000	63,000		
Total plant and equipment	$266,000	$313,000		
Total assets	$586,000	$638,000		
Liabilities and Owners' Equity				
Current Liabilities				
Accounts payable	$145,000	$162,000		
Wages payable	43,000	45,000		
Total current liabilities	$188,000	$207,000		
Long-Term Liabilities				
Mortgage payable	115,000	110,000		
Total liabilities	$303,000	$317,000		
Owners' Equity				
Total owners' equity	283,000	321,000		
Total liabilities and owners' equity	$586,000	$638,000		

*Denotes a decrease.

B. Complete the vertical analysis of the following balance sheets. Round all percents to the nearest 0.1%.

1.

REDHAT TOOL AND DIE
COMPARATIVE BALANCE SHEETS
December 31, 1996 and 1997

	1996		1997	
	Amount	Percent	Amount	Percent
Assets				
Current Assets				
Cash	$ 33,000		$ 37,000	
Merchandise inventory	82,000		87,000	
Accounts receivable	79,000		78,000	
Total current assets	$194,000		$202,000	
Plant and Equipment				
Land	$ 32,000		$ 32,000	
Buildings	77,000		77,000	
Equipment	91,000		101,000	
Total plant and equipment	$200,000		$210,000	
Total assets	$394,000		$412,000	
Liabilities and Owners' Equity				
Current Liabilities				
Accounts payable	$ 74,000		$ 78,000	
Wages payable	31,000		35,000	
Total current liabilities	$105,000		$113,000	
Long-Term Liabilities				
Mortgage payable	83,000		80,000	
Total liabilities	$188,000		$193,000	
Owners' Equity				
Total owners' equity	206,000		219,000	
Total liabilities and owners' equity	$394,000		$412,000	

2.

BITS 'N' BYTES
COMPARATIVE BALANCE SHEETS
December 31, 1996 and 1997

	1996		1997	
	Amount	**Percent**	**Amount**	**Percent**
Assets				
Current Assets				
Cash	$105,000		$ 97,000	
Merchandise inventory	126,000		136,000	
Accounts receivable	89,000		92,000	
Total current assets	$320,000		$325,000	
Plant and Equipment				
Land	$ 92,000		$105,000	
Buildings	123,000		145,000	
Equipment	51,000		63,000	
Total plant and equipment	$266,000		$313,000	
Total assets	$586,000		$638,000	
Liabilities and Owners' Equity				
Current Liabilities				
Accounts payable	$145,000		$162,000	
Wages payable	43,000		45,000	
Total current liabilities	$188,000		$207,000	
Long-Term Liabilities				
Mortgage payable	115,000		110,000	
Total liabilities	$303,000		$317,000	
Owners' Equity				
Total owners' equity	283,000		321,000	
Total liabilities and owners' equity	$586,000		$638,000	

3. Complete the balance sheet below and perform a complete vertical analysis.

P&PIE LANDSCAPING
BALANCE SHEET
December 31, 1995

	Amount	Percent
Assets		
Current Assets		
Cash	$10,000	
Merchandise inventory	5,000	
Accounts receivable	2,500	
Total current assets		
Plant and Equipment		
Land	$ 30,000	
Buildings	100,000	
Equipment	50,000	
Total plant and equipment		
Total assets		
Liabilities and Owners' Equity		
Current Liabilities		
Accounts payable	$1,500	
Wages payable	250	
Total current liabilities		
Long-Term Liabilities		
Mortgage payable	$65,000	
Total liabilities		
Owners' Equity		
Total owners' equity		
Total liabilities and owners' equity		

WORKSPACE

SECTION 19.2: Income Statements

Can You:

Identify the elements of an income statement?
Perform a horizontal analysis of two successive income statements?
Complete a vertical analysis of two successive income statements?
. . . If not, you need this section.

Reading Income Statements

Income Statement A type of financial statement that reflects the business's operation over a given period of time; it summarizes all income and expenses for that period.

Unlike a balance sheet, which represents a stationary financial picture of a firm, an **income statement** reflects the business's operation over a given period of time. An income statement is a summary of all income and expenses for a certain period of time, such as a month or a year.

The main purpose of an income statement is to show the profitability of a business over a certain time period. Thus, it is of interest not only to managers, but also to a firm's potential investors and lenders and to the government.

Income statements are sometimes called "profit and loss" statements because they result in the following equation:

Revenues − expenses = profit (or loss)

For example, if your business had revenues (sales) of $1000 and expenses of $600 this month, your profit would be

$$Profit = revenues − expenses$$
$$= \$1000 − 600 = \$400$$

But if your business had revenues (sales) of $600 and expenses of $1000 this month, you would have a loss:

$$Loss = revenues − expenses$$
$$= \$600 − 1000 = −\$400$$

A real-life income statement is a bit more complicated, as you might expect. It must also satisfy the following equation:

Net profit = gross profit − operating expenses

Let's look at an income statement for O'Hare's Yarns and what these terms mean.

O'HARE'S YARNS
INCOME STATEMENT
FOR THE YEAR ENDED DECEMBER 31, 1996

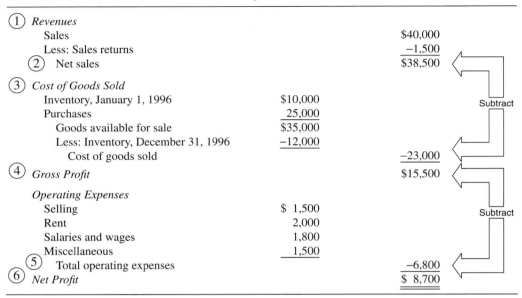

① *Revenues*		
Sales		$40,000
Less: Sales returns		−1,500
② Net sales		$38,500
③ *Cost of Goods Sold*		
Inventory, January 1, 1996	$10,000	
Purchases	25,000	
Goods available for sale	$35,000	
Less: Inventory, December 31, 1996	−12,000	
Cost of goods sold		−23,000
④ *Gross Profit*		$15,500
Operating Expenses		
Selling	$ 1,500	
Rent	2,000	
Salaries and wages	1,800	
Miscellaneous	1,500	
⑤ Total operating expenses		−6,800
⑥ *Net Profit*		$ 8,700

Subtract

Subtract

Revenues A firm's income from sales, after sales returns are deducted.

Net Sales Net income or revenues from the sale of goods and services; it is the difference between total sales and the amount of sales returns and allowances.

Cost of Goods Sold Beginning inventory value plus purchases minus ending inventory value.

Gross Profit (or Loss) The difference between net sales and cost of goods sold.

Total Operating Expenses The sum of all operating expenses for a business.

Net Profit (or Loss) The difference between gross profit and total operating expenses.

---

① **Revenues** are the company's income from sales after sales returns are deducted.

② The **net sales** is the net income or revenues from the sale of goods or services. It is the difference between the total sales and the amount of sales returns and allowances that have resulted from incorrect or damaged merchandise shipped to customers.

> Net sales = sales − sales returns

③ To find the **cost of goods sold,** add the beginning inventory to the purchases for the time period and subtract the value of the ending inventory from the total goods available for sale.

> Cost of goods sold = beginning inventory + purchases − ending inventory

④ The **gross profit** is the difference between the net sales and the cost of goods sold.

> Gross profit = net sales − cost of goods sold

⑤ The **total operating expenses** are the *sum* of all the operating expenses. Operating expenses may include selling expense, rent, utilities, advertising, taxes, wages, depreciation, and insurance.

⑥ Finally, the **net profit (or loss)** is calculated as the difference between the gross profit and the total operating expenses.

> Net profit = gross profit − total operating expenses

Performing a Horizontal Analysis

Just as we performed horizontal and vertical analyses on balance sheets, so we can similarly analyze income statements. The *horizontal* analysis of two income statements will show the increase or decrease in profit, along with changes in the various expense categories.

As with balance sheets, a horizontal analysis of income statements follows two steps:

Steps: Performing a Horizontal Analysis

STEP 1. *Calculate* how much each item has increased or decreased on the second income statement compared to the first.

STEP 2. *Determine* the percent increase or decrease and express the answer as a percent.

$$\text{Percent change} = \frac{\text{change}}{\text{first amount}}$$

For example, the income statements of O'Hare's Yarns show the gross profit for 1996 is $15,500 and for 1997 is $20,000. Let's complete a horizontal analysis of this item, rounding to the nearest 0.1%.

Step 1. | 1997 gross profit | | 1996 gross profit |

Increase = $20,000 − 15,500 = $4500

| Increase in profit |

Step 2.

| Increase |

$$\text{Percent increase} = \frac{\$4500}{\$15,500} = 0.2903 \ldots = 29.0\% \text{ rounded}$$

| 1996 gross profit | | Percent increase in profit |

WORK THIS PROBLEM

The Question:

The income statements of O'Hara's Yarns show the net sales for 1996 is $38,500 and for 1997 is $48,000. Complete a horizontal analysis of the net sales, rounding to the nearest 0.1%.

✔ YOUR WORK

The Solution:

Step 1.   Increase = $48,000 − 38,500 = $9500

Step 2.   Percent increase = $\dfrac{\$9500}{\$38,500} = 0.2467 \ldots = 24.7\%$ rounded

In business, a calculator or computer generally "runs the numbers" in analyses.

Using a Calculator to Perform a Horizontal Analysis of an Income Statement

A calculator not only makes it faster to do the calculations in a horizontal analysis, it also makes it immediately obvious whether an increase or a decrease has occurred.

For contrast, let's look at the gross profits of two companies in 1996 and 1997. Company A's income statement shows $20,000 in this category in 1996 and $15,500 in 1997. Company B's income statement shows $15,500 in this category in 1996 and $20,000 in 1997.

The process is the same for both situations.

Company A: 15500 $-$ 20000 $=$ $\div$ 20000 $=$ ⟶ 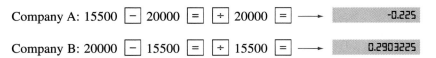   -0.225

Company B: 20000 $-$ 15500 $=$ $\div$ 15500 $=$ ⟶   0.2903225

The minus sign in front of Company A's answer tells you a *decrease* occurred to the tune of 22.5%, while Company B had an *increase* of 29.0%, rounded.

Always use the first income statement's amount as the base when calculating the percent change in a horizontal analysis.

The following table shows a completed horizontal analysis of two income statements for O'Hare's Yarns. An * is used to denote a decrease.

O'HARE'S YARNS
COMPARATIVE INCOME STATEMENTS   *Step 1*   *Step 2*
For the Years Ended December 31, 1996 and 1997

	1996	1997	Increase or Decrease*	
			Amount	Percent
Revenues				
Sales	$40,000	$50,000	$10,000	25.0%
Less: Sales returns	1,500	2,000	500	33.3%
Net sales	$38,500	$48,000	$ 9,500	24.7%
Cost of Goods Sold				
Inventory, January 1	$10,000	$12,000	$ 2,000	20.0%
Purchases	25,000	30,000	5,000	20.0%
Goods available for sale	$35,000	$42,000	$ 7,000	20.0%
Less: Inventory, December 31	12,000	14,000	2,000	16.7%
Cost of goods sold	$23,000	$28,000	$ 5,000	21.7%
Gross Profit	$15,500	$20,000	$ 4,500	29.0%
Operating Expenses				
Selling	$ 1,500	$ 2,500	$ 1,000	66.7%
Rent	2,000	2,500	500	25.0%
Salaries and wages	1,800	2,000	200	11.1%
Miscellaneous	1,500	1,400	100*	6.7%*
Total operating expenses	$ 6,800	$ 8,400	$ 1,600	23.5%
Net Profit	$ 8,700	$11,600	$ 2,900	33.3%

*Asterisk denotes a decrease.

As you can see, the analysis for income statements is just like that for balance sheets.

YOUR TURN

WORK THIS PROBLEM

The Question:

Complete the horizontal analysis of the following comparative income statements, rounding to the nearest 0.1%.

PETAL PUSHERS
COMPARATIVE INCOME STATEMENTS
For the Years Ended December 31, 1996 and 1997

	1996	1997	Increase or Decrease*	
			Amount	Percent
Revenues				
Sales	$150,000	$182,000		
Less: Sales returns	3,000	2,500		
Net sales	$147,000	$179,500		
Cost of Goods Sold				
Inventory, January 1	$ 42,000	$ 48,000		
Purchases	108,000	129,000		
Goods available for sale	$150,000	$177,000		
Less: Inventory, December 31	48,000	55,500		
Cost of goods sold	$102,000	$121,500		
Gross Profit	$ 45,000	$ 58,000		
Operating Expenses				
Selling	$ 18,000	$ 22,000		
Rent	5,000	5,500		
Salaries and wages	2,000	2,500		
Miscellaneous	3,000	2,500		
Total operating expenses	$ 28,000	$ 32,500		
Net Profit	$ 17,000	$ 25,500		

*Asterisk denotes a decrease.

✔ YOUR WORK

The Solution:

PETAL PUSHERS
COMPARATIVE INCOME STATEMENTS
For the Years Ended December 31, 1996 and 1997

Step 1   *Step 2*

	1996	1997	Increase or Decrease* Amount	Percent
Revenues				
Sales	$150,000	$182,000	$32,000	21.3%
Less: Sales returns	3,000	2,500	500*	16.7%*
Net sales	$147,000	$179,500	$32,500	22.1%
Cost of Goods Sold				
Inventory, January 1	$ 42,000	$ 48,000	$ 6,000	14.3%
Purchases	108,000	129,000	21,000	19.4%
Goods available for sale	$150,000	$177,000	$27,000	18.0%
Less: Inventory, December 31	48,000	55,500	7,500	15.6%
Cost of goods sold	$102,000	$121,500	$19,500	19.1%
Gross Profit	$ 45,000	$ 58,000	$13,000	28.9%
Operating Expenses				
Selling	$ 18,000	$ 22,000	$ 4,000	22.2%
Rent	5,000	5,500	500	10.0%
Salaries and wages	2,000	2,500	500	25.0%
Miscellaneous	3,000	2,500	500*	16.7%*
Total operating expenses	$ 28,000	$ 32,500	$ 4,500	16.1%
Net Profit	$ 17,000	$ 25,500	$ 8,500	50.0%

*Asterisk denotes a decrease.

Remember, you can *never* add or subtract percents in a horizontal analysis.

Performing a Vertical Analysis

As with balance sheets, a *vertical analysis* of income statements compares only items within a given year's statement. To perform this analysis, just calculate the percent of net sales as

$$\text{Percent} = \frac{\text{amount of individual item}}{\text{net sales}}$$

For example, the 1996 income statement for O'Hares Yarns lists total operating expenses as $6800 and net sales as $38,500. The total operating expenses can be expressed as a percent of net sales:

Total operating expenses

$$\text{Percent} = \frac{\$6800}{\$38,500} = 0.1766 = 17.7\% \text{ rounded}$$

Net sales

Total operating expenses were 17.7% of net sales in 1996.

YOUR TURN

WORK THIS PROBLEM

The Question:

On the 1996 income statement for O'Hare Yarns, the cost of goods sold is $23,000 and the net sales are $38,500. Complete a vertical analysis on this item. Round the percent to the nearest 0.1%.

✔ YOUR WORK

The Solution:

$$\text{Percent} = \frac{\$23,000}{\$38,500} = 0.5974 \ldots = 59.7\%$$

The vertical analysis of a complete income statement is simply performed item by item. Each percent calculation is performed as in the preceding problem.

As in analysis of balance sheets, the vertical analysis of income statements is usually performed on two successive sheets. Such an analysis enables managers to look for trends by analyzing any significant changes in percents from one income statement to the other.

The following is a complete vertical analysis of the income statements of O'Hare Yarns for 1996 and 1997.

O'HARE'S YARNS
COMPARATIVE INCOME STATEMENTS
For the Years Ended December 31, 1996 and 1997

	1996		1997	
	Amount	**Percent**	**Amount**	**Percent**
Revenues				
Sales	$40,000	103.9%	$50,000	104.2%
Less: Sales returns	1,500	3.9%	2,000	4.2%
Net sales	$38,500	100.0%	$48,000	100.0%
Cost of Goods Sold				
Inventory, January 1	$10,000	26.0%	$12,000	25.0%
Purchases	25,000	64.9%	30,000	62.5%
Goods available for sale	$35,000	90.9%	$42,000	87.5%
Less: Inventory, December 31	12,000	31.2%	14,000	29.2%
Cost of goods sold	$23,000	59.7%	$28,000	58.3%
Gross Profit	$15,500	40.3%	$20,000	41.7%
Operating Expenses				
Selling	$ 1,500	3.9%	$ 2,500	5.2%
Rent	2,000	5.2%	2,500	5.2%
Salaries and wages	1,800	4.7%	2,000	4.2%
Miscellaneous	1,500	3.9%	1,400	2.9%
Total operating expenses	$ 6,800	17.7%	$ 8,400	17.5%
Net Profit	$ 8,700	22.6%	$11,600	24.2%

Remember, all items on the 1996 income sheet are compared to the net sales for 1996. All items on the 1997 income sheet are compared to the net sales for 1997.

Notice that there was a decrease in the percent of purchases and amount of inventory. In other words, O'Hare's has had a slight percent decrease in costs causing a slight percent increase in gross profit. O'Hare's is doing well.

YOUR TURN

WORK THIS PROBLEM

The Question:

Complete a vertical analysis for the following 1996 and 1997 income statements for Petal Pushers.

PETAL PUSHERS
COMPARATIVE INCOME STATEMENTS
For the Years Ended December 31, 1996 and 1997

	1996		1997	
	Amount	**Percent**	**Amount**	**Percent**
Revenues				
Sales	$150,000		$182,000	
Less: Sales returns	3,000		2,500	
Net sales	$147,000		$179,500	
Cost of Goods Sold				
Inventory, January 1	$ 42,000		$ 48,000	
Purchases	108,000		129,000	
Goods available for sale	$150,000		$177,000	
Less: Inventory, December 31	48,000		55,500	
Cost of goods sold	$102,000		$121,500	
Gross Profit	$ 45,000		$ 58,000	
Operating Expenses				
Selling	$ 18,000		$ 22,000	
Rent	5,000		5,500	
Salaries and wages	2,000		2,500	
Miscellaneous	3,000		2,500	
Total operating expenses	$ 28,000		$ 32,500	
Net Profit	$ 17,000		$ 25,500	

✔ YOUR WORK

The Solution:

**PETAL PUSHERS
COMPARATIVE INCOME STATEMENTS
For the Years Ended December 31, 1996 and 1997**

	1996		1997	
	Amount	**Percent**	**Amount**	**Percent**
Revenues				
Sales	$150,000	102.0%	$182,000	101.4%
Less: Sales returns	3,000	2.0%	2,500	1.4%
Net sales	$147,000	100.0%	$179,500	100.0%
Cost of Goods Sold				
Inventory, January 1	$ 42,000	28.6%	$ 48,000	26.7%
Purchases	108,000	73.5%	129,000	71.9%
Goods available for sale	$150,000	102.0%	$177,000	98.6%
Less: Inventory, December 31	48,000	32.7%	55,500	30.9%
Cost of goods sold	$102,000	69.4%	$121,500	67.7%
Gross Profit	$ 45,000	30.6%	$ 58,000	32.3%
Operating Expenses				
Selling	$ 18,000	12.2%	$ 22,000	12.3%
Rent	5,000	3.4%	5,500	3.1%
Salaries and wages	2,000	1.4%	2,500	1.4%
Miscellaneous	3,000	2.0%	2,500	1.4%
Total operating expenses	$ 28,000	19.0%	$ 32,500	18.1%
Net Profit	$ 17,000	11.6%	$ 25,500	14.2%

Although the analysis of financial statements is a long process, don't panic. Simply break the problem down and work item by item.

For practice on horizontal and vertical analysis of income statements, turn to Section Test 19.2.

WORKSPACE

SECTION TEST 19.2 Financial Statement Analysis

The following problems test your understanding of Section 19.2, Income Statements.

A. Complete the horizontal analysis of the following income statements. Round all percents to the nearest 0.1%. Use an * to denote any decreases.

1.

BOB'S BEEFLIKE BURGERS
COMPARATIVE INCOME STATEMENTS
For the Years Ended December 31, 1996 and 1997

	1996	1997	Increase or Decrease* Amount	Percent
Revenues				
Sales	$102,000	$113,000		
Less: Sales returns	3,500	4,000		
Net sales	$ 98,500	$109,000		
Cost of Goods Sold				
Inventory, January 1	$ 17,000	$ 19,000		
Purchases	72,000	78,000		
Goods available for sale	89,000	97,000		
Less: Inventory, December 31	19,000	18,000		
Cost of goods sold	$ 70,000	$ 79,000		
Gross Profit	$ 28,500	$ 30,000		
Operating Expenses				
Selling	$ 17,000	$ 19,500		
Rent	4,000	4,500		
Salaries and wages	700	1,000		
Miscellaneous	600	900		
Total operating expenses	$ 22,300	$ 25,900		
Net Profit	$ 6,200	$ 4,100		

*Asterisk denotes decrease.

2.

ATONAL MUSIC STORE
COMPARATIVE INCOME STATEMENTS
For the Years Ended December 31, 1996 and 1997

	1996	1997	Increase or Decrease*	
			Amount	Percent
Revenues				
Sales	$223,000	$243,000		
Less: Sales returns	6,500	5,000		
Net sales	$216,500	$238,000		
Cost of Goods Sold				
Inventory, January 1	$ 37,000	$ 34,000		
Purchases	123,000	129,500		
Goods available for sale	160,000	163,500		
Less: Inventory, December 31	34,000	31,000		
Cost of goods sold	$126,000	$132,500		
Gross Profit	$ 90,500	$105,500		
Operating Expenses				
Selling	$ 36,000	$ 35,000		
Rent	21,000	23,500		
Salaries and wages	1,200	1,500		
Miscellaneous	1,300	1,500		
Total operating expenses	$ 59,500	$ 61,500		
Net Profit	$ 31,000	$ 44,000		

*Asterisk denotes decrease.

B. Complete the vertical analysis of the following income statements. Round all percents to the nearest 0.1%.

1.

BOB'S BEEFLIKE BURGERS
COMPARATIVE INCOME STATEMENTS
For the Years Ended December 31, 1996 and 1997

	1996		1997	
	Amount	Percent	Amount	Percent
Revenues				
Sales	$102,000		$113,000	
Less: Sales returns	3,500		4,000	
Net sales	$ 98,500		$109,000	
Cost of Goods Sold				
Inventory, January 1	$ 17,000		$ 19,000	
Purchases	72,000		78,000	
Goods available for sale	89,000		97,000	
Less: Inventory, December 31	19,000		18,000	
Cost of goods sold	$ 70,000		$ 79,000	
Gross Profit	$ 28,500		$ 30,000	
Operating Expenses				
Selling	$ 17,000		$ 19,500	
Rent	4,000		4,500	
Salaries and wages	700		1,000	
Miscellaneous	600		900	
Total operating expenses	$ 22,300		$ 25,900	
Net Profit	$ 6,200		$ 4,100	

2.

ATONAL MUSIC STORE
COMPARATIVE INCOME STATEMENTS
For the Years Ended December 31, 1996 and 1997

	1996		1997	
	Amount	**Percent**	**Amount**	**Percent**
Revenues				
Sales	$223,000		$243,000	
Less: Sales returns	6,500		5,000	
Net sales	$216,500		$238,000	
Cost of Goods Sold				
Inventory, January 1	$ 37,000		$ 34,000	
Purchases	123,000		129,500	
Goods available for sale	160,000		163,500	
Less: Inventory, December 31	34,000		31,000	
Cost of goods sold	$126,000		$132,500	
Gross Profit	$ 90,500		$105,500	
Operating Expenses				
Selling	$ 36,000		$ 35,000	
Rent	21,000		23,500	
Salaries and wages	1,200		1,500	
Miscellaneous	1,300		1,500	
Total operating expenses	$ 59,500		$ 61,500	
Net Profit	$ 31,000		$ 44,000	

Can You:

Calculate the current ratio?
Use the acid-test ratio?
Find the return on equity?
Determine inventory turnover?
. . . If not, you need this section.

Ratio A numeric comparison of two quantities found by dividing one quantity by the other.

Businesspeople often need to compare two quantities. In some cases, subtracting them to find their differences is a good comparison. But very often a better comparison can be made by finding their ratio. A **ratio** is a comparison of the sizes of two like quantities found by dividing one quantity by the other. It is a single number commonly written as a fraction.

For example, if your car weighs 2100 lb and you weigh 150 lb, the ratio of the car's weight to your weight is:

$$\text{Ratio of weights} = \frac{\text{car's weight}}{\text{your weight}} = \frac{2100 \text{ lb}}{150 \text{ lb}} = \frac{210}{15} = \frac{14}{1}$$

The ratio may be written as a fraction $\frac{210}{15}$ or $\frac{14}{1}$ or as 14:1 or sometimes simply written as 14. The ratio $\frac{14}{1}$ or 14:1 is read "14 to 1." Financial ratios are usually rounded to the nearest tenth.

YOUR TURN

⎡**WORK THIS PROBLEM**⎤

The Question:

Jack's height is 74 inches and Jill's height is 61 inches. Calculate the ratio of their heights.

⎡✔ **YOUR WORK**⎤

The Solution:

$$\text{Ratio of heights} = \frac{\text{Jack's height}}{\text{Jill's height}} = \frac{74 \text{ in.}}{61 \text{ in.}}$$

$$= 1.21 \text{ or } 1.2 \text{ to } 1, \text{ rounded}$$

The ratios most used in business finance are the current ratio, the acid-test ratio, the return on equity ratio, and the inventory turnover ratio.

Calculating the Current Ratio

Current Ratio The ratio of current assets to current liabilities; a measure of a firm's ability to meet its current debts.

One measure of the ability of a business to meet its current debts is the current ratio. The **current ratio** is the ratio of current assets to current liabilities. Both the current assets and liabilities may be obtained from the balance sheet.

663

$$\text{Current ratio} = \frac{\text{current assets}}{\text{current liabilities}}$$

The current ratio is a measure of the financial strength and short-term solvency (ability to pay bills) of the business. Bankers look closely at this ratio and generally prefer *not* to make loans to firms with a current ratio of less than 2.

For example, a recent balance sheet for Woody's Lumber shows current assets of $204,000 and current liabilities of $85,000. Calculate the current ratio.

$$\text{Current ratio} = \frac{\text{current assets}}{\text{current liabilities}} = \frac{\$204,000}{\$85,000} = 2.4$$

YOUR TURN

WORK THIS PROBLEM

The Question:

In a recent balance sheet for the Vitas' Vitamin Company, current assets are $35,150 and current liabilities are $18,500. Calculate the current ratio.

✔ YOUR WORK

The Solution:

$$\text{Current ratio} = \frac{\text{current assets}}{\text{current liabilities}} = \frac{\$35,150}{\$18,500} = 1.9$$

Using the Acid-Test Ratio

Liquid Assets Current assets that may be easily converted into cash; includes accounts receivable, short-term promissory notes receivable, and cash.

The current ratio is sometimes called the *working capital ratio* because it reflects how much of the firm's *capital*—its total funds—can be put to work immediately.

Not all current assets, such as inventories, can be readily converted into cash. **Liquid assets** are current assets that may be easily converted into cash. Liquid assets include accounts receivable, short-term promissory notes receivable, and cash.

Acid-Test Ratio (Quick Ratio) The ratio of liquid assets to current liabilities; a measure of a firm's ability to pay off its liabilities in a short time.

The **acid-test ratio (quick ratio)** is the ratio of liquid assets to current liabilities. Liquid assets and current liabilities are obtained from the balance sheet.

$$\text{Acid-test ratio} = \frac{\text{liquid assets}}{\text{current liabilities}}$$

The acid-test ratio measures the ability of a company to pay off its liabilities in a brief period of time. Thus, it is an indication of a firm's ability to meet unexpected demands for working capital quickly. The ratio uses the "quick" assets of cash and accounts receivable. It is called the acid-test ratio, however, because it is the ultimate test of a company's short-term solvency. A business in good condition usually has an acid-test ratio of at least 1.

For example, Wired for Sound's recent balance sheet showed cash $30,000, accounts receivable $51,000, and current liabilities $90,000. Calculate the liquid assets and acid-test ratio.

Cash · Accounts receivable

$$\text{Liquid assets} = \$30{,}000 + 51{,}000 = \$81{,}000$$

$$\text{Acid-test ratio} = \frac{\text{liquid assets}}{\text{current liabilities}} = \frac{\$81{,}000}{\$90{,}000} = 0.9$$

YOUR TURN

⟦WORK THIS PROBLEM⟧

The Question:

Wreck-a-Mended Auto's recent balance sheet showed cash $5000, accounts receivable $10,000, and current liabilities $12,500. Calculate the acid-test ratio.

✔ YOUR WORK

The Solution:

$$\text{Liquid assets} = \$5000 + 10{,}000 = \$15{,}000$$

$$\text{Acid-test ratio} = \frac{\text{liquid assets}}{\text{current liabilities}} = \frac{\$15{,}000}{\$12{,}500} = 1.2$$

Finding Return on Equity

Rate of Return on (Owners') Equity; also called Rate of Return on Investment (ROI) The ratio of net profit to owners' equity; a measure of the profitability of investing in a firm.

The **rate of return on (owner's) equity** shows the profitability of the funds invested in the firm—whether the money invested by a local beautician to set up a salon or the money invested by shareholders in a major corporation through the purchase of stock. The rate of return on equity is also called **return on investment (ROI)** because it gives the rate of return on the investment.

In the case of a stock company, the rate of return on equity is a major guide for potential investors. They usually seek a rate at least equal to the current savings interest rate.

The rate of return on equity is the ratio of net profit or income to the owner's equity. This figure is usually expressed as a percent rather than a ratio number. The net profit is obtained from the income statement and the owner's equity is from the balance sheet.

$$\text{Rate of return on equity} = \frac{\text{net profit}}{\text{owner's equity}}$$

For example, Crystal's Diamonds has owner's equity of $150,000 and a net profit of $21,000. Calculate the rate of return on equity.

$$\text{Rate of return on equity} = \frac{\text{net profit}}{\text{owner's equity}} = \frac{\$21{,}000}{\$150{,}000} = 0.14 = 14\%$$

The rate of return is 14%—a better rate of return than savings accounts or CDs.

YOUR TURN

The Question:

Lew's Restaurant has owners' equity of $76,000 and a net profit of $6000. Calculate the rate of return on equity. Round to the nearest 0.1%.

✔ YOUR WORK

The Solution:

$$\text{Rate of return on equity} = \frac{\text{net profit}}{\text{owners' equity}} = \frac{\$6000}{\$76,000}$$

$$= 0.0789 \ldots = 7.9\% \text{ rounded}$$

This is a much better rate of return than a passbook savings account or most money market funds.

Determining Inventory Turnover

Inventory Turnover The ratio of cost of goods sold to average inventory; a measure of how quickly a firm is selling its inventory.

A firm's **inventory turnover** is a measure of its volume of goods sold as compared to the amount of goods it has in inventory. It is the ratio of the cost of goods sold to the average inventory.

$$\text{Inventory turnover} = \frac{\text{cost of goods sold}}{\text{average inventory}}$$

The cost of goods sold is obtained from the income statement. The average inventory may be calculated using the beginning and ending inventories from the income statement.

$$\text{Average inventory} = \frac{\text{beginning inventory} + \text{ending inventory}}{2}$$

To calculate inventory turnover, then, follow these steps:

Steps: Determining Inventory Turnover

STEP 1.   *Calculate* the average inventory.

STEP 2.   *Calculate* the inventory turnover.

The resulting ratio will tell you how quickly the firm is selling its existing inventory. The higher the ratio, the more times the inventory cycles through or "turns over." The ratio varies depending on the type of business. In general, the higher the ratio, the greater the company's profit potential.

If the inventory turnover is equal to 1, the company sold its entire inventory during the period covered by the income statement. If the inventory turnover is 2, the firm sold twice its average inventory during this time. Inventory turnover ratios vary widely among different types of businesses.

For example, the recent income statement of Fast Track Athletic Shoes showed a beginning inventory of $8000, ending inventory of $10,000, and cost of goods sold $72,000. Let's calculate the inventory turnover.

Step 1. Average inventory $= \dfrac{\text{beginning inventory} + \text{ending inventory}}{2}$

$$= \dfrac{\$8000 + 10,000}{2} = \dfrac{\$18,000}{2} = \$9000$$

Step 2. Inventory turnover $= \dfrac{\text{cost of goods sold}}{\text{average inventory}} = \dfrac{\$72,000}{\$9000} = 8$

The inventory turnover is 8, which means that Fast Track "sold out" its inventory eight times during the time period covered by the income statement. It is a low inventory, rapid turnover business.

YOUR TURN

WORK THIS PROBLEM

The Question:

A recent income statement for the Fly-By-Night Aviary, showed a beginning inventory of $23,000, ending inventory $27,000, and cost of goods sold $180,000. Calculate the inventory turnover.

✔ YOUR WORK

The Solution:

Step 1. Average inventory $= \dfrac{\text{beginning inventory} + \text{ending inventory}}{2}$

$$= \dfrac{\$23,000 + 27,000}{2} = \dfrac{\$50,000}{2} = \$25,000$$

Step 2. Inventory turnover $= \dfrac{\text{cost of goods sold}}{\text{average inventory}} = \dfrac{\$180,000}{\$25,000} = 7.2$

Inventory turnover's a snap if you have a calculator.

Using a Calculator to Find Inventory Turnover

To see how the calculator can help, let's return to our example of Fast Track, which had a beginning inventory of $8,000, an ending inventory of $10,000, and a cost of goods sold equal to $72,000.

If you don't have parentheses on your calculator, key in

8000 $\boxed{+}$ 10000 $\boxed{=}$ $\boxed{\div}$ 2 $\boxed{=}$ ⟶ ▓▓▓▓▓▓ 9000.

Write the answer down or enter it in the calculator's memory. Then

72000 $\boxed{\div}$ 9000 $\boxed{=}$ ⟶ ▓▓▓▓▓ 8.

If your calculator uses parentheses, you can key in the sequence this way:

72000 $\boxed{\div}$ $\boxed{(}$ $\boxed{(}$ 8000 $\boxed{+}$ 10000 $\boxed{)}$ $\boxed{\div}$ 2 $\boxed{)}$ $\boxed{=}$ ⟶ ▓▓▓▓▓ 8.

Either way, inventory turnover is 8.

A set of practice problems on ratio analysis appears in Section Test 19.3.

Name

Date

Course/Section

The following problems test your understanding of Section 19.3, Financial Ratio Analysis.

A. Calculate the current ratio from the given information. Round to the nearest hundredth.

	Current Assets	Current Liabilities	Current Ratio
1.	$ 73,500	$ 32,000	
2.	152,300	85,500	
3.	243,000	162,000	
4.	89,500	65,400	
5.	375,000	158,000	
6.	192,000	103,000	

B. Calculate the acid-test ratio from the given information. Round to nearest hundredth.

	Liquid Assets	Current Liabilities	Acid-Test Ratio
1.	$128,000	$ 97,000	
2.	92,500	105,000	
3.	73,200	68,800	
4.	98,700	62,800	
5.	362,000	314,300	
6.	62,500	73,700	

C. Calculate the rate of return on equity from the given information. Round to the nearest 0.1%.

	Owner's Equity	Net Profit	Owner's Rate of Return
1.	$135,000	$14,200	
2.	235,000	9,250	
3.	275,000	37,500	
4.	82,600	5,350	
5.	456,000	38,700	
6.	247,000	32,400	

D. Calculate the average inventory and the inventory turnover from the given information. Round to the nearest hundredth.

	Cost of Goods Sold	Beginning Inventory	Ending Inventory	Average Inventory	Inventory Turnover
1.	$218,600	$21,500	$23,600		
2.	182,500	22,600	21,700		
3.	542,200	42,300	49,700		
4.	315,700	24,400	27,500		
5.	427,600	22,800	24,500		
6.	365,500	52,700	50,200		

E. Solve these problems.

1. Musselmann's Gym's recent balance sheet showed cash $42,000, accounts receivable $85,000, and current liabilities $92,500. What is the gym's acid-test ratio?

2. Helen's House of Beauty's balance sheet showed cash $27,500, accounts receivable $32,700, and current liabilities $42,800. What is the boutique's acid-test ratio?

3. In a recent balance sheet for J.R.'s Construction Company, current assets are $172,500 and current liabilities are $93,800. What is the company's current ratio?

4. In a recent balance sheet for Lew's Restaurant, current assets are $96,500 and current liabilities are $105,200. What is the company's current ratio?

5. The Fin and Feather Exotic Pet Store has owners' equity of $126,200 and net profit of $9650. What is the rate of return on equity? Round to the nearest 0.1%.

6. The Atonal Music Store has an owners' equity of $96,700 and a net profit of $11,200. What is the rate of return on equity? Round to the nearest 0.1%.

7. A recent income statement for Intense Tents showed a beginning inventory of $37,000, ending inventory of $41,000, and cost of goods sold of $217,500. What is the inventory turnover ratio rounded to the nearest hundredth?

8. A recent income statement for The Write Stuff showed a beginning inventory of $12,500, ending inventory of $15,500, and cost of goods sold of $123,700. What is the inventory turnover ratio? Round to the nearest hundredth.

9. Using the balance sheet for P&PIE Landscaping from problem B.3 in Section Test 19.1, find the current ratio and the acid-test ratio.

Key Terms

acid-test ratio (quick ratio)
assets
balance sheet
cost of goods sold
current assets
current liabilities
current ratio
fixed assets
fundamental accounting
   equation

gross profit (or loss)
horizontal analysis
income statement
inventory turnover
liabilities
liquid assets
long-term liabilities
net profit (or loss)
net sales

owners' equity
rate of return on (owners')
   equity
rate of return on investment
   (ROI)
ratio
revenues
total operating expenses
vertical analysis

Chapter 19 at a Glance

Page	Topic	Key Point	Example
628	Fundamental accounting equation for balance sheets	Assets = liabilities + owners' equity	What are the assets of a firm with total liabilities of $45,000 and owners' equity of $55,000? Assets = $45,000 + 55,000 = $100,000
630 and 649	Horizontal analysis of two balance sheets or two income statements	Calculate how much an item has increased or decreased between balance sheets. % increase or decrease = $\dfrac{\text{change}}{\text{first amount}}$	A firm's balance sheet showed total assets of $500,000 in 1995 and total assets of $600,000 in 1996. What is the % change? Change = $600,000 − $500,000 = $100,000 % change = $\dfrac{\$100,000}{\$500,000} = 0.2 = 20\%$
634 and 653	Vertical analysis of a balance sheet or an income statement	Calculate each asset as a percent of total assets and each liability as a percent of liabilities and owners' equity. Percent = $\dfrac{\text{amount of individual item}}{\text{total}}$ When completing an entire balance sheet, check by totaling percents to 100%.	A firm has $30,000 in cash as part of its total assets of $500,000. What percent of total assets is the cash? Percent = $\dfrac{\$30,000}{\$500,000} = 0.06 = 6\%$
647	Profit and loss equation	Profit (or loss) = revenues − expenses	If a firm has revenues of $10,000 and expenses of $6000 one month, what is its profit or loss for that month? Profit = $10,000 − 6000 = $4000
648	Net sales	Net sales = sales − sales returns	If a firm has sales of $40,000 and sales returns of $1500, what are its net sales? Net sales = $40,000 − 1500 = $38,500

Page	Topic	Key Point	Example
648	Cost of goods sold	Cost of goods sold = beginning inventory + purchases − ending inventory	If a firm begins with an inventory worth $15,000, makes purchases of $6000, and ends the year with an inventory valued at $5000, what is its cost of goods sold? Cost of goods sold $= \$15,000 + 6000 - 5000 = \$16,000$
648	Gross profit	Gross profit = net sales − cost of goods sold	Using the data from the previous two examples, find the firm's gross profit. Gross profit $= \$38,500 - 16,000$ $= \$22,500$
648	Net profit	Net profit = gross profit − total operating expenses	If the firm in the previous example had total operating expenses of $15,300, what was its net profit? Net profit $= \$22,500 - 15,300$ $= \$7,200$
663	Current ratio	Current ratio $= \dfrac{\text{current assets}}{\text{current liabilities}}$	What is the current ratio for a firm with current assets of $204,000 and current liabilities of $85,000? Current ratio $= \dfrac{\$204,000}{\$85,000} = 2.4$
664	Acid-test ratio	Acid-test ratio $= \dfrac{\text{liquid assets}}{\text{current liabilities}}$	What is the acid-test ratio for a firm with liquid assets of $81,000 and current liabilities of $90,000? Acid-test ratio $= \dfrac{\$81,000}{\$90,000} = 0.9$
665	Owners' rate of return ratio	Owners' rate of return $= \dfrac{\text{net profit}}{\text{owners' equity}}$	What is the owners' rate of return for a firm with a net profit of $21,000 and owners' equity of $150,000? Owners' rate of return $= \dfrac{\$21,000}{\$150,000}$ $= 0.14 = 14\%$
666	Inventory turnover ratio	Average inventory $= \dfrac{\text{beginning + ending inventory}}{2}$ Inventory turnover $= \dfrac{\text{cost of goods sold}}{\text{average inventory}}$	What is the inventory turnover ratio for a firm with a beginning inventory of $8000, an ending inventory of $10,000, and a cost of goods sold of $72,000? Average inventory $= \dfrac{\$8000 + 10,000}{2}$ $= \$9000$ Inventory turnover $= \dfrac{\$72,000}{\$9,000} = 8$

SELF-TEST   Financial Statement Analysis

The following problems test your understanding of financial statements.

1. Complete the horizontal analysis of the following balance sheets. Round all percents to the nearest 0.1%. Use an * to denote any decreases.

WOODY'S LUMBER YARD
COMPARATIVE BALANCE SHEETS
December 31, 1996 and 1997

	1996	1997	Increase or Decrease* Amount	Percent
Assets				
Current Assets				
Cash	$ 42,000	$ 39,000		
Merchandise inventory	96,000	117,000		
Accounts receivable	82,000	87,000		
Total current assets	$220,000	$243,000		
Plant and Equipment				
Land	$ 45,000	$ 61,000		
Buildings	83,000	105,000		
Equipment	89,000	92,000		
Total plant and equipment	$217,000	$258,000		
Total assets	$437,000	$501,000		
Liabilities and Owners' Equity				
Current Liabilities				
Accounts payable	$123,000	$119,000		
Wages payable	45,000	47,000		
Total current liabilities	$168,000	$166,000		
Long-Term Liabilities				
Mortgage payable	96,000	91,000		
Total liabilities	$264,000	$257,000		
Owners' Equity				
Total owners' equity	173,000	244,000		
Total liabilities and owners' equity	$437,000	$501,000		

*Denotes a decrease.

2. Complete the vertical analysis of the following balance sheets. Round all percents to the nearest 0.1%.

WOODY'S LUMBER YARD
COMPARATIVE BALANCE SHEETS
December 31, 1996 and 1997

	1996		1997	
	Amount	**Percent**	**Amount**	**Percent**
Assets				
Current Assets				
Cash	$ 42,000		$ 39,000	
Merchandise inventory	96,000		117,000	
Accounts receivable	82,000		87,000	
Total current assets	$220,000		$243,000	
Plant and Equipment				
Land	$ 45,000		$ 61,000	
Buildings	83,000		105,000	
Equipment	89,000		92,000	
Total plant and equipment	$217,000		$258,000	
Total assets	$437,000		$501,000	
Liabilities and Owners' Equity				
Current Liabilities				
Accounts payable	$123,000		$119,000	
Wages payable	45,000		47,000	
Total current liabilities	$168,000		$166,000	
Long-Term Liabilities				
Mortgage payable	96,000		91,000	
Total liabilities	$264,000		$257,000	
Owners' Equity				
Total owners' equity	173,000		244,000	
Total liabilities and owners' equity	$437,000		$501,000	

3. Complete the vertical analysis of the following income statements. Round all percents to the nearest 0.1%.

THE WRITE STUFF
COMPARATIVE INCOME STATEMENTS
For the Years Ended December 31, 1996 and 1997

	1996		1997	
	Amount	**Percent**	**Amount**	**Percent**
Revenues				
Sales	$122,000		$129,000	
Less: Sales returns	4,500		4,000	
Net sales	$117,500		$125,000	
Cost of Goods Sold				
Inventory, January 1	$ 26,000		$ 24,000	
Purchases	53,000		57,000	
Goods available for sale	$ 79,000		$ 81,000	
Less: Inventory, December 31	24,000		25,000	
Cost of goods sold	$ 55,000		$ 56,000	
Gross Profit	$ 62,500		$ 69,000	
Operating Expenses				
Selling	$ 14,000		$ 15,500	
Rent	24,000		26,000	
Salaries and wages	1,700		1,900	
Miscellaneous	1,600		1,800	
Total operating expenses	$ 41,300		$ 45,200	
Net Profit	$ 21,200		$ 23,800	

4. Complete the horizontal analysis of the following income statements. Round all percents to the nearest 0.1%. Use an * to denote any decreases.

THE WRITE STUFF
COMPARATIVE INCOME STATEMENTS
For the Years Ended December 31, 1996 and 1997

	1996	1997	Increase or Decrease*	
			Amount	Percent
Revenues				
Sales	$122,000	$129,000		
Less: Sales returns	4,500	4,000		
Net sales	$117,500	$125,000		
Cost of Goods Sold				
Inventory, January 1	$ 26,000	$ 24,000		
Purchases	53,000	57,000		
Goods available for sale	$ 79,000	$ 81,000		
Less: Inventory, December 31	24,000	25,000		
Cost of goods sold	$ 55,000	$ 56,000		
Gross Profit	$ 62,500	$ 69,000		
Operating Expenses				
Selling	$ 14,000	$ 15,500		
Rent	24,000	26,000		
Salaries and wages	1,700	1,900		
Miscellaneous	1,600	1,800		
Total operating expenses	$ 41,300	$ 45,200		
Net Profit	$ 21,200	$ 23,800		

*Denotes a decrease.

5. Big Boys' Toys RV Sales' recent balance sheet showed cash $89,200, accounts receivable $95,800, and current liabilities $145,600. What is the company's acid-test ratio?

6. The Garden of Earthly Delights' balance sheet showed cash $37,500, accounts receivable $45,300, and current liabilities $95,200. What is the nursery's acid-test ratio?

7. In a recent balance sheet for Crystal's Diamonds, current assets are $42,700 and current liabilities are $39,200. What is the company's current ratio?

8. In a recent balance sheet for the Frame-Up Shop, current assets are $87,300 and current liabilities are $98,700. What is the company's current ratio?

9. Bits 'n' Bytes has an owners' equity of $247,000 and a net profit of $23,800. What is the rate of return on equity? Round to the nearest 0.1%.

10. Dora's Boating Goods Store has an owners' equity of $152,800 and a net profit of $11,600. What is the rate of return on equity? Round to the nearest 0.1%.

11. A recent income statement for The Liki Valve Company showed a beginning inventory of $47,200, ending inventory of $51,800, and cost of goods sold of $317,600. What is the inventory turnover ratio? Round to the nearest hundredth.

12. A recent income statement for the Holistic Health Food Store showed a beginning inventory of $35,300, ending inventory of $37,700, and cost of goods sold $283,700. What is the inventory turnover ratio? Round to the nearest hundredth.

WORKSPACE

Statistics and Graphs

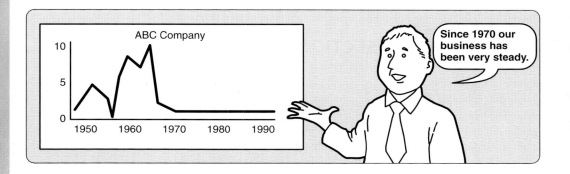

CHAPTER OBJECTIVES

When you complete this chapter successfully, you will be able to:

1. Calculate the mean, median, mode, and range for a set of data.

2. Interpret tables of numerical data.

3. Read and plot bar graphs, line graphs, and pie charts.

■ The following table shows the average salaries in Santa Lucia county for different occupational categories. Calculate the mean, median, and range of these salary averages.

Occupation	Average Salary
Services	$22,506
Tourism	9,803
Real estate	23,323
Medical/health	33,425
Hi-Tech	41,070
Oil/gas	38,518
Construction	28,482
Agriculture	13,522

■ A worker at Kate's Delicacies had the following croissant output: Monday, 125; Tuesday, 275; Wednesday, 340; Thursday, 310; Friday, 150. Show this data in a line graph and in a bar graph.

Information, mostly in the form of numbers, is the lifeblood of any modern business. Computers produce seemingly endless lists of numbers about costs, profits, income, sales, earnings, customer profiles, and the like. These numbers appear as lists, tables, graphs, and computer printouts in reports, newspapers, magazines, television news, and elsewhere.

If you don't know how to get the information you need from this flood of numbers, your business will suffer. In this chapter we will show you some simple skills used in handling business information.

Don't be a business "statistic." "Plot" your course for success by concentrating on this chapter.

SECTION 20.1: Business Statistics

Can You:

Find the mean, median, and mode of a set of numbers?
Calculate the range of a set of numbers?
Read tables of numerical information?
. . . If not, you need this section.

Finding the Mean

Mean The average of a group of numbers calculated as the sum of the numbers divided by the number of values.

Businesspeople often use numbers to describe important quantities, to communicate specific information about those quantities, and to help make predictions or decisions. One important type of numerical expression is the **mean** value of a set of numbers.

$$\text{Mean} = \frac{\text{sum of values}}{\text{number of values}}$$

The *mean* is the technical name given to what we commonly call the "average" value of a set of numbers. We use the name "mean" rather than "average" because there are other useful averages, as you will see in this section.

Let's use the mean to find the average (expected) dividend on stocks in a market such as the one represented by the following fast-growing stocks:

GROWTH STOCKS

Stock	Dividend, $ per share
Amal	1.72
Bost	1.16
Calgo	1.37
Dion	1.80
Egad	3.76
Fooy	0.40

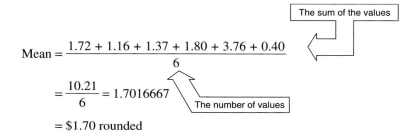

$$\text{Mean} = \frac{1.72 + 1.16 + 1.37 + 1.80 + 3.76 + 0.40}{6}$$

The sum of the values

$$= \frac{10.21}{6} = 1.7016667$$

The number of values

$$= \$1.70 \text{ rounded}$$

Notice that we usually round the mean to the same number of decimal places as the numbers in the set of data.

Also, be sure to check your calculation to be certain that it is reasonably representative of the set of numbers. It *must* be a number between the smallest and largest numbers in the data set.

YOUR TURN

> **WORK THIS PROBLEM**

The Question:

Find the mean number of sales of (a) radios and (b) televisions for the five appliance stores shown.

SALES AT FIVE APPLIANCE STORES

Store	Radios	Televisions
Amanda's	140	65
Eastinghome	172	130
Eddy's	185	200
Genial Electric	195	285
Kenless	190	375

✔ YOUR WORK

The Solution:

(a) Mean number of radios $= \dfrac{140 + 172 + 185 + 195 + 190}{5}$

$= \dfrac{882}{5} = 176$ rounded to the nearest unit

(b) Mean number of televisions $= \dfrac{65 + 130 + 200 + 285 + 375}{5}$

$= \dfrac{1055}{5} = 211$ units

Finding the Mean for Grouped Data

When some of the numerical values to be averaged appear more than once in the list, we can save time by grouping the values.

Steps: Finding the Mean for Grouped Data

STEP 1. *Multiply* each value by its frequency.

STEP 2. *Add* the results from step 1 to get the total value.

STEP 3. *Add* the frequencies to get the total frequency.

STEP 4. *Calculate* the mean by dividing the answer in step 2 by the answer in step 3.

For example, what is the mean price of a card at Mark Hall's Greeting Card Shop?

GREETING CARDS

Value (Price)	Frequency (number sold)
$0.90	86
$0.97	47
$1.29	132
$1.50	206
$1.79	54
$2.15	23

Applying step 1, step 2, and step 3, we get

Value	Frequency	Value × Frequency	
$0.90	86	$ 77.40	
0.97	47	45.59	
1.29	132	170.28	
1.50	206	309.00	*Step 1*
1.79	54	96.66	
2.15	23	49.45	
Totals	548	$748.38	

Step 3 *Step 2*

Step 4. Mean = $\dfrac{\text{Sum of (value} \times \text{frequency)}}{\text{Total frequency}} = \dfrac{\$748.38}{548}$

$= \$1.3656569 = \$1.37$ rounded

YOUR TURN

WORK THIS PROBLEM

The Question:

From the following data, find the mean hours worked last week at the Tru-Blu Manufacturing Company.

Value (hours)	Frequency (number of people)
35	6
40	28
45	11
50	4

✔ YOUR WORK

The Solution:

Value (hours)	Frequency (number of people)	Value × Frequency (total hours)	
35	6	210	
40	28	1120	*Step 1*
45	11	495	
50	4	200	
Totals	49	2025	

Step 3 *Step 2*

Step 4.   Mean $= \dfrac{2025}{49} = 41.326531 = 41$ hours rounded

---

Of course, businesspeople use calculators or computers to find the mean.

> ### *Finding the Mean with a Calculator*
>
> If your calculator does *not* have parentheses, you will have to perform each calculation in step 1 and step 3, write down the answers, and then key in the following to complete steps 2 and 4.
>
> 210 $\boxed{+}$ 1120 $\boxed{+}$ 495 $\boxed{+}$ 200 $\boxed{=}$ $\boxed{\div}$ 49 $\boxed{=}$ $\longrightarrow$ ▒ 41.326531 ▒
>
> If your calculator *does* have parentheses, you can perform the entire process in one step as follows:
>
> $\boxed{(}$ 35 $\boxed{\times}$ 6 $\boxed{)}$ $\boxed{+}$ $\boxed{(}$ 40 $\boxed{\times}$ 28 $\boxed{)}$ $\boxed{+}$ $\boxed{(}$ 45 $\boxed{\times}$ 11 $\boxed{)}$ $\boxed{+}$ $\boxed{(}$ 50 $\boxed{\times}$
>
> 4 $\boxed{)}$ $\boxed{=}$ $\boxed{\div}$ $\boxed{(}$ 6 $\boxed{+}$ 28 $\boxed{+}$ 11 $\boxed{+}$ 4 $\boxed{)}$ $\boxed{=}$ $\longrightarrow$ ▒ 41.326531 ▒

Determining the Median

The mean is a useful average in some situations, but it can be misleading in other cases. For example, consider the monthly wages paid to the nine employees of Greene's Landscape Company.

Name	Wages
Martin	$ 816
Fred	1100
Joyce	805
Dante	848
Jose	4680
Peter	1062
Eric	840
Norma	2150
Rico	814

The mean of these wages paid is $1457.22. But most of the employees earn much less than this average. It is not difficult to spot the high salaries of the boss and the bookkeeper. These high salaries distort the mean so that it is not really representative of the wages paid to the whole group.

Median The average of a group of numbers calculated as the number that falls in the middle of the values.

A better average to use in this case is the **median** or middle value. To calculate the median follow these steps:

Steps: Finding the Median

STEP 1. *Arrange* the numbers in increasing order from smallest to largest.

STEP 2. *Find* the middle value. If the set of numbers contains an *odd* number of values, the median is the middle value. If the set of numbers contains an *even* number of values, *the median is the mean of the two numbers in the middle.*

For example, to calculate the value of the median of the wages in the previous table,

Step 1. 805, 814, 816, 840, 848, 1062, 1100, 2150, 4680

the middle value

Step 2. In this case, there are nine numbers, so the middle value—the median—is the fifth one: $848.

WORK THIS PROBLEM

YOUR TURN

The Question:

Find the median value of the following set of numbers:

(a) 2, 2, 5, 6, 8, 14, 56, 3, 9, 11, 7
(b) 216, 315, 196, 452, 425, 108, 512, 627, 115, 460

✔ **YOUR WORK**

The Solution:

(a) *Step 1.* Arrange the numbers in increasing order:
2, 2, 3, 5, 6, 7, 8, 9, 11, 14, 56

The middle number is 7

Step 2. There are 11 numbers in this set, therefore the median is the middle or sixth number. The median is 7.

(b) *Step 1.* Arrange the numbers in increasing order:
108, 115, 196, 216, <u>315, 425</u>, 452, 460, 512, 627

The middle two numbers

Step 2. There are ten numbers, and the middle two are 315 and 425. Therefore the median is

$$\frac{315 + 425}{2} = \frac{740}{2} = 370$$

Calculating the Mode

Mode   The average of a group of numbers calculated as the number that occurs most frequently in a group of values.

A third average that is easy to calculate and often very useful in business is the mode. The **mode** is the value that occurs most frequently in a set of numbers. For example, the following numbers represent the daily output of shirts produced by one sewing machine operator over a two-week period:

$$36, 37, 39, 37, 41, 41, 45, 37, 32, 35, 37, 39$$

The number 37 appears four times and no other number appears more than twice. Therefore the mode of this set of numbers is 37.

The mode is an easy average to find, but it may not always provide the most representative number for the data.

Each of the three averages we have described provides a way of calculating a representative value for the set of data or group of numbers. If the numbers in the data set are nicely clustered about some central value, then the three averages may be roughly equal. In many business applications you may need to calculate all three averages.

Determining the Range

The mean, mode, and median are useful tools, but they tell you nothing about the spread of values in a set of data.

For example, Big Ben's Bolts has two machines producing megabolts. One of the machines has been acting up lately, producing bolts of very different weights. A sample of bolts from each machine produced the following data:

$$\text{Machine 1 (ounces): } 29, 20, 35, 27, 28, 53$$
$$\text{Machine 2 (ounces): } 32, 31, 32, 32, 33, 32$$

Range   The difference between the largest and smallest numbers in a set of values.

The mean weight in both cases is 32 ounces. But the first machine is producing megabolts in a wide variety of weights. This spread of values can be described by calculating the **range** of the set of numbers.

$$\boxed{\text{Range} = \text{largest value} - \text{smallest value}}$$

$$\text{Machine 1: range} = 53 - 20 = 33 \text{ ounces}$$
$$\text{Machine 2: range} = 33 - 31 = 2 \text{ ounces}$$

A large range tells us that the numbers in the data set vary greatly from one another, while a small range tells us that the numbers are very similar.

YOUR TURN

| WORK THIS PROBLEM |

The Question:

Find the mean, median, mode, and range for the following set of numbers:

$$8, 11, 3, 15, 12, 17, 26, 5, 12, 6, 9, 21 \text{ inches}$$

✔ YOUR WORK

The Solution:

$$\text{Range} = 26 - 3 = 23 \text{ in.}$$

$$\text{Mean} = \frac{8 + 11 + 3 + 15 + 12 + 17 + 26 + 5 + 12 + 6 + 9 + 21}{12}$$

$$= \frac{145}{12} = 12.083333 = 12 \text{ in., rounded}$$

There are 12 values—3, 5, 6, 8, 9, 11, 12, 12, 15, 17, 21, 26—so the median is the mean of the two middle values 11 and 12. The median is 11.5.

$$\text{Mode} = 12, \text{ the most frequently occurring number}$$

Reading Tables of Data

In addition to averages and ranges, businesspeople also look for other relationships between numbers in a set of data. To help you understand some (though by no means all) relationships of business interest, let's consider the following table summarizing the monthly sales for Bits 'n' Bytes:

Month	Sales, $
January	$ 85,000
February	92,000
March	89,000
April	95,000
May	104,000
June	99,000

From this table, we can answer many questions:

■ Which month had the largest sales? Which month had the smallest sales?
    May had the largest sales: $104,000
    January had the smallest sales: $85,000

■ What is the range of sales over the six-month period?
    Range = largest value − smallest value
           = $104,000 − 85,000 = $19,000

■ Which month had the biggest increase in sales?
    First, calculate the increase or decrease for each month by subtracting the previous month's sales:

        February increase = $92,000 −   85,000 = $7000
        March decrease   = $89,000 −   92,000 = − $3000
        April increase   = $95,000 −   89,000 = $6000
        May increase   = $104,000 − 95,000 = $9000
        June decrease   = $99,000 − 104,000 = − $5000

    The largest increase was in May.

■ What was the percent increase in sales from April to May?

The percent increase is easily calculated using the following formula:

$$\text{Percent increase} = \frac{\text{increase}}{\text{first amount}}$$

$$\text{Percent increase} = \frac{\$9000}{\$95,000} = 0.0947\ldots = 9.5\% \text{ rounded}$$

YOUR TURN

WORK THIS PROBLEM

The Question:

Consider the following chart of actual and projected stock values from Shiring Investments:

Company	Business	Price Earnings Ratio		Stock Price
		Latest 12 mos.	**Estimate**	
Chysler	Autos	3.1	5.4	$19\frac{1}{4}$
Amax	Aluminum	3.5	6.3	$24\frac{1}{4}$
Inco	Nickel	3.8	6.1	$26\frac{7}{8}$
Phlips	Copper	3.9	6.5	$62\frac{3}{4}$
MichNat	Banking	4.3	7.1	$49\frac{1}{2}$
Stone	Containers	4.5	6.2	$24\frac{1}{2}$

(a) Which stock has the highest price? the lowest?
(b) Which stock is estimated to have the highest jump in its price-earnings ratio for next year?
(c) What is the mean value for the stocks in this table?
(d) What is the percent increase between the price-earnings ratio for the last 12 months and the price-earning ratio as estimated for next year at Inco?

✔ YOUR WORK

The Solution:

(a) Phlips has the highest-priced stock: $62\frac{3}{4}$
Chysler has the lowest-priced stock: $19\frac{1}{4}$
(b) Chysler difference $= 5.4 - 3.1 = 2.3$
Amax difference $= 6.3 - 3.5 = 2.8$
Inco difference $= 6.1 - 3.8 = 2.3$
Phlips difference $= 6.5 - 3.9 = 2.6$
MichNat difference $= 7.1 - 4.3 = 2.8$
Stone difference $= 6.2 - 4.5 = 1.7$
Amax and MichNat, at 2.8, are estimated to have the highest price-earnings ratio jumps next year.

(c) Convert the stock prices to decimal numbers and add to get the numerator in the mean equation.

$$\text{Mean} = \frac{207.125}{6} = 34.520833 = 34.52 \text{ rounded}$$

(d) Percent increase $= \dfrac{6.1 - 3.8}{3.8} = 0.6052631 = 60.5\% \text{ rounded}$

If you solved these problems correctly, you're ready for the questions in Section Test 20.1. If you had difficulty, review this section.

WORKSPACE

SECTION TEST 20.1 Statistics and Graphs

Name

Date

Course/Section

The following problems test your understanding of Section 20.1, Business Statistics.

A. Calculate the mean, mode, median, and range for each of the following sets of numbers.

1. 60, 75, 70, 80, 65, 70, 85, 70, 75

2. $12, $15, $14, $10, $12, $17, $17, $16, $14, $17

3. 114, 125, 136, 123, 118, 125, 143, 131

4. 2.3, 2.4, 2.7, 3.1, 2.4, 2.5, 2.6, 2.9, 3.0

5. 0.08 in., 0.02 in., 0.04 in., 0.02 in., 0.03 in., 0.06 in., 0.05 in., 0.02 in.

6. $1.55, $0.85, $1.86, $1.37, $1.05, $1.37

7. 40, 37, 36, 40, 40, 35, 38, 40, 35

8. 32¢, 17¢, 25¢, 22¢, 25¢, 31¢, 28¢, 25¢, 20¢, 32¢, 28¢

9. 2720, 3125, 2540, 2670, 3005, 2540, 2475

10. 1.23, 1.27, 1.25, 1.24, 1.24, 1.26, 1.28

11. $56,427, $41,325, $64,271, $48,301, $27,008

12. 0.201, 0.198, 0.315, 0.201, 0.306, 0.254, 0.181

13. 16, 17, 18, 21, 23, 27, 17, 15, 14, 16, 17, 12, 14, 16, 22, 21

14. 225, 316, 204, 179, 355, 179, 307, 274, 211

15. $2.67, $2.43, $2,78, $2.51, $2.60, $2.43, $2.66

16. $12.98, $12.95, $13.50, $14.69, $15.10, $13.50, $15.65, $14.05

B. Solve these problems.

1. According to the latest U.S. Census, the population of the United States was 248,709,873 in 1990. The eight states in Fit to Print's Mid-Atlantic sales region had the following populations:

Delaware	666,168
District of Columbia	606,900
Kentucky	3,685,296
Maryland	4,781,468
North Carolina	6,628,637
Ohio	10,847,115
Virginia	6,187,358
West Virginia	1,793,477

   (a) Calculate the mean population of a state in this region.
   (b) Calculate the median population.
   (c) Calculate the range in population for these states.
   (d) What percent of the entire U.S. population was the population of North Carolina?
   (e) Virginia has a larger population than Maryland by what percent?

2. Four large retail sales chains in the United States had the following monthly sales for a selected month in 1997 (in billions of dollars).

Wal-Mart	$3.26
Sears	$2.54
Kmart	$2.53
J. C. Penney	$1.07

(a) Calculate the mean sales for this month for the "big four."

(b) Calculate the range of these sales figures.

(c) Calculate the median sales for this month.

3. The following table shows the average salaries in Santa Lucia county for different occupational categories.

Occupation	Average Salary
Services	$22,506
Tourism	9,803
Real estate	23,323
Medical/health	33,425
Hi-tech	41,070
Oil/gas	38,518
Construction	28,482
Agriculture	13,522

Calculate the mean, median, and range of these salary averages.

4. The following table shows the average total compensation for four classes of occupations in both 1960 and 1990.

Occupation	Average Total Compensation	
	1960	1990
Factory worker	$ 4,665	$ 22,998
School teacher	4,995	31,166
Engineer	9,828	49,365
Chief executive	190,383	1,952,806

(a) Calculate the percent increase for each of these four occupations.

(b) Calculate the range in 1960 and in 1990.

5. Audie Player, a salesman at Wired for Sound, made the following sales in a recent week: Monday, $368.50; Tuesday, $412.79; Wednesday, $351.60; Thursday, $450.16; Friday, $521.82; Saturday, $546.47. Calculate the mean daily sales and range of sales for this week.

6. Chessie Setter, the top salesperson in the furniture department of the Needless-Markup Department store had the following sales figures last year (rounded to the nearest $100):

Jan.	$19,100	July	$18,600
Feb.	$15,200	Aug.	$28,200
Mar.	$21,400	Sept.	$24,700
Apr.	$17,500	Oct.	$26,800
May	$21,300	Nov.	$28,200
June	$22,000	Dec.	$29,400

Calculate the mean, median, mode, and range for this set of sales figures.

7. Big Boys' Toys has sold the following numbers of RVs since the company began in 1989:

1989	216	1993	502
1990	321	1994	471
1991	346	1995	536
1992	412	1996	615

(a) Calculate the mean number of RVs sold per year.
(b) Calculate the median number of RVs sold per year.
(c) Calculate the range in RVs sold per year.
(d) By what percent did 1996 sales exceed the mean?
(e) By what percent did 1996 sales exceed those for 1995?

8. A survey of the seven computer stores found that the Lightning modem sold for the following prices: $199.95, $187.50, $212.25, $206.50, $199.95, $225.75, $209.99.
(a) Calculate the range of prices.
(b) Calculate the mean price.
(c) Calculate the median price.
(d) Calculate the mode price.
(e) By what percent did the highest price exceed the lowest?

9. Lots of Plots sold nine houses last month for the following prices:

$155,000 $326,400 $124,000
$115,600 $142,700 $211,300
$124,000 $ 91,800 $102,500

(a) Calculate the mean, median, and mode for these figures.
(b) Calculate the range of these sale prices.
(c) By what percent did the price of the most expensive house exceed the average?

10. The following table shows the results when temperatures were recorded every two hours at two different locations.

Pelican's Rest, Oregon

8 A.M.	42° F	4 P.M.	47° F	Midnight	41° F
10 A.M.	44° F	6 P.M.	44° F	2 A.M.	40° F
Noon	45° F	8 P.M.	43° F	4 A.M.	40° F
2 P.M.	48° F	10 P.M.	42° F	6 A.M.	41° F

Burning Bush, Arizona

8 A.M.	37° F	4 P.M.	82° F	Midnight	35° F
10 A.M.	41° F	6 P.M.	60° F	2 A.M.	33° F
Noon	72° F	8 P.M.	50° F	4 A.M.	33° F
2 P.M.	83° F	10 P.M.	42° F	6 A.M.	35° F

(a) Calculate the mean and median temperature at each location.
(b) Calculate the range of temperatures at each location.

11. The operators at Claudia's Clip-Joint had a competition for total services performed during June. The results are listed below. Find the mean, the median, the mode, and the range for the given data.

Sue	$4,325
Al	$3,250
Joan	$4,531
Sally	$3,250
Marianne	$3,874
Charles	$3,938

WORKSPACE

SECTION 20.2: Business Graphs*

> ## *Can You:*
>
> Read and plot bar graphs?
> Read and plot line graphs?
> Read and plot pie charts?
> . . . If not, you need this section.

Graph A pictorial display of information.

An old proverb notes that "a picture is worth a thousand words." This idea is at the heart of graphs. A **graph** is a pictorial display of information that allows us to take a jumble of confusing numbers and show trends or make comparisons. The most common type of graphs are bar graphs, line graphs, and pie charts.

Reading Bar Graphs

Bar Graph A method of visually comparing the sizes of different but related quantities by a series of bars plotted against two axes.

Axes (singular, *axis*) The vertical and horizontal scales on a bar or line graph.

To display and compare the sizes of different but related quantities, businesspeople use **bar graphs.** A bar graph has two scales or **axes** (singular, *axis*); a *horizontal* axis along the bottom and a *vertical* axis along the left side. Each axis measures a different facet of the data.

For example, consider the following bar graph, which compares revenues for the Big Bux Oil Company from 1992 to 1996.

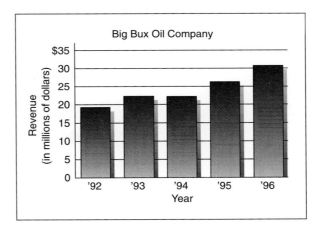

In this case, the vertical axis measures the firm's revenues. It goes from 0 to 35 in units of millions. In no year was the revenue above $35 million. The horizontal axis represents the year. It goes from 1992 to 1996.

Also note that the scale increases uniformly along both axes. Each vertical unit represents an increase of $5 million. Each horizontal unit represents an increase of one year.

To find the revenue in any given year, locate the top of the bar for that year. Move straight across to the vertical scale and read the amount. If the reading falls between two labeled units, estimate the value in decimal form. For example, the revenue earned in 1995 was about $27 million, since the top of the 1995 bar is just above the 25 line.

*This section is based on Robert Carman and Hal Saunders, *Mathematics for the Trades,* 4th edition (Englewood Cliffs, NJ: Prentice Hall, 1996).

YOUR TURN

WORK THIS PROBLEM

The Question:

Using the graph of revenues for the Big Bux Oil Company,

(a) What was the revenue earned in 1992?
(b) Calculate the percent increase revenue earned from 1993 to 1996.

✔ YOUR WORK

The Solution:

(a) The top of the 1992 bar falls just below the $20 million line. Revenues were about $19.5 million.
(b) The revenues earned in 1993 were about $22,000,000; those in 1996 were about $31,000,000.

$$\text{Increase} = \$31,000,000 - \$22,000,000 = \$9,000,000$$

$$\text{Percent increase} = \frac{9,000,000}{22,000,000} = 0.40909 \text{ or } 41\% \text{ rounded}$$

Double Bar Graph A form of bar graph in which two bars comparing related items are drawn side by side on a given axis.

We can also compare related quantities by using a **double bar graph.** For example, the double bar graph here compares the costs of two different types of insurance when purchased at various ages.

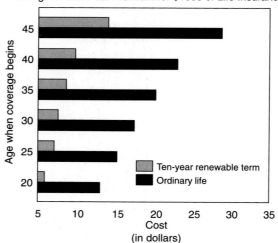

This graph illustrates another feature of bar graphs—they can be drawn with the bars extending either up or to the right. It all depends on what you put on which axis.

On this graph the light blue bars represent ten-year renewable term insurance, and the dark blue bars represent ordinary life insurance. To read the cost, follow the right edge of the bar down to the horizontal scale. For example, for coverage beginning at age 30, ten-year renewable term costs about $7.50 twice a year for $1000 coverage, while $1000 of ordinary life at the same age costs about $17.50 twice a year.

YOUR TURN

┌──────────────────────┐
│ WORK THIS PROBLEM │
└──────────────────────┘

The Question:

Using the double bar graph comparing two types of insurance,

(a) How much would $10,000 of ordinary life insurance cost at age 35?
(b) How much would $3000 of ten-year renewable term insurance cost at age 40?
(c) What is the semiannual difference in cost between ordinary life and ten-year renewable term for a $5000 policy at age 25?

┌──────────────────┐
│ ✔ YOUR WORK │
└──────────────────┘

The Solution:

(a) The end of the age 35 dark blue bar is at $20. The cost per $1000 of ordinary life is $20, or $20 × 10 = $200 for $10,000 of ordinary life insurance.
(b) The end of the age 40 light blue bar is at approximately 9.5. The cost per $1000 of renewable term insurance is $9.50, or $9.5 × 3 = $28.50 for $3000 of renewable term insurance.
(c) The end of the age 25 bars are at $15 and $6.5, so the costs for $5000 of coverage are $15 × 5 = $75 for ordinary life and $6.5 × 5 = $32.50 for renewable term insurance. The difference is $75.00 − 32.50 = $42.50.

Plotting Bar Graphs

Bar graphs are easy to read, and it is rather easy to draw bar graphs to illustrate numerical data. To draw a bar graph, follow these steps:

Steps: Drawing a Bar Graph

STEP 1. *Decide* what type of bar graph to use—single or double. The bars can be placed either horizontally or vertically—you may choose.

STEP 2. *Choose* suitable spacing for the axes and label each axis.

STEP 3. *Use* a straightedge to mark the length of each bar according to the data given. *Round* the numbers if necessary. Drawing your bar graph on graph paper will make the process easier.

For example, suppose you want to display the following data in a bar graph showing radio and television sales in five stores.

**SALES OF RADIOS AND TELEVISIONS
AT FIVE APPLIANCE STORES**

Store	Radios	Televisions
Amanda's	140	65
Eastinghome	172	130
Eddy's	185	200
Genial Electric	195	285
Kenless	190	375

Step 1.   In this case, we will use a double bar graph and place the bars horizontally.

Step 2.

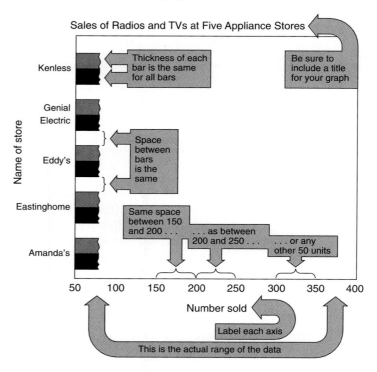

Notice that the numbers on the bottom axis are evenly spaced and cover the entire range of numbers in the data, in this case from 65 to 375.

Also notice that we arranged the stores in order of amount of sales. The biggest seller, Kenless, is at the top, and the smallest seller, Amanda's, is at the bottom. This is not necessary, but it makes your graph easier to read.

Step 3.

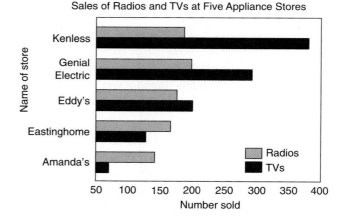

YOUR TURN

WORK THIS PROBLEM

The Question:

True-Built Construction Materials wants to provide the following technical information for its customers in a graphic way so that the customers can make easy comparisons. Make a bar graph of the following data.

AVERAGE ULTIMATE COMPRESSION STRENGTH OF COMMON MATERIALS

Material	Compression Strength (lb/in.^{2})
Hard bricks	12,000
Light red bricks	1,000
Portland cement	3,000
Portland concrete	1,000
Granite	19,000
Limestone and sandstone	9,000
Trap rock	20,000
Slate	14,000

✔ YOUR WORK

The Solution:

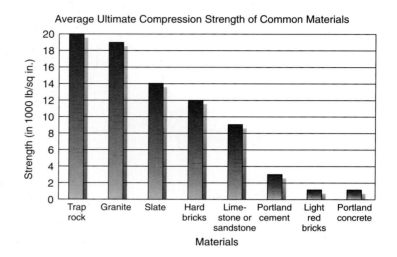

Average Ultimate Compression Strength of Common Materials

Notice that we have arranged the different materials from largest to smallest strength so that the graph will be easy to read. Also notice that the horizontal graph lines are marked every two units. We could have used markers every five units instead.

Reading Line Graphs

Line Graph   A method of visually displaying continuous changes in some quantity, usually over a period of time.

To show a continuous change in some quantity, usually over a period of time, businesspeople rely on **line graphs.** For example, the following graph shows changes in the population of Podunk, Pennsylvania, from 1700 to 1950.

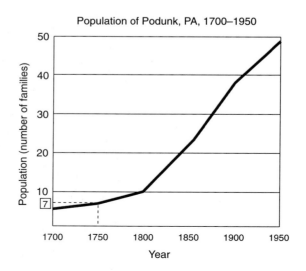

Population of Podunk, PA, 1700–1950

To read a line graph, find the desired time along the horizontal axis and follow the line on the grid upward until you reach the graph line. Then move horizontally to the left and read the value on the vertical axis. For example, the population in Podunk in 1750 was approximately 7 families.

WORK THIS PROBLEM

YOUR TURN

The Question:

What was the population of Podunk in 1850?

✔ YOUR WORK

The Solution:

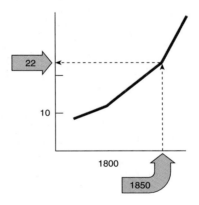

There were approximately 22 families.

Plotting Line Graphs

Plotting The process of locating the point on a line graph representing the values of both axes for a particular data item.

Constructing a line graph is similar to drawing a bar graph except that you must **plot** the points represented by the data.

Steps: Plotting Line Graphs

STEP 1. *Decide* on a suitable spacing or scale for each of the axes. The scale markings must be evenly spaced. In general, choose points spaced 1, 2, 5, 10—or some multiple of these units—apart. Time *must* be plotted along the horizontal axis.

STEP 2. *Plot* the graph points. For each pair of numbers, locate the year on the horizontal scale and the information for that year on the vertical scale. *Draw* imaginary lines up from the year and horizontally from the other characteristic. *Place* a dot where the lines intersect.

STEP 3. After all points are plotted on the graph, *connect* adjacent points.

For example, let's construct a line graph of the following data:

COMPARATIVE PRICES OF PAPER CLIPS

Year	Price per Box of 200
1990	$1.16
1991	1.33
1992	1.08
1993	1.57
1994	2.55
1995	2.95

Step 1.

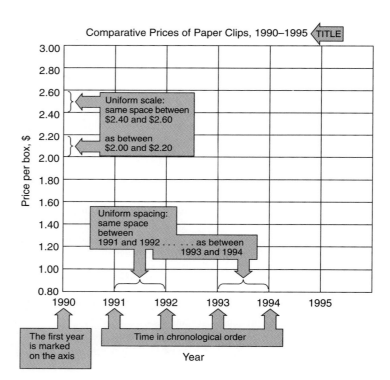

Step 2.

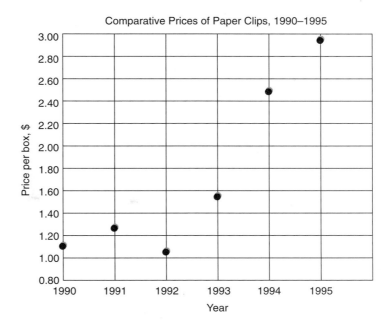

Step 3.

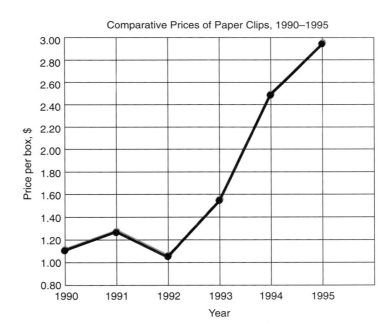

YOUR TURN

WORK THIS PROBLEM

The Question:

Plot the following data on a line graph.

**COMPARATIVE PRICES
OF RUBBER BANDS,
1990–1995**

Year	Price per Box of 1000
1990	$0.89
1991	0.97
1992	0.99
1993	1.21
1994	2.13
1995	2.72

✔ YOUR WORK

The Solution:

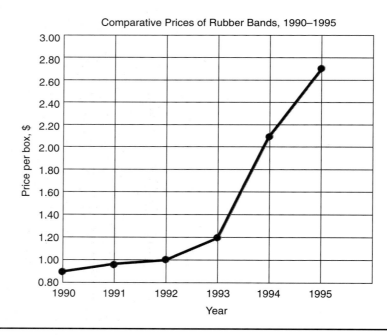

Reading Pie Charts

Pie Chart (circle graph)
A method of visually displaying the percentage of the whole represented by various data elements.

A *circle graph* or **pie chart** is used to show what fraction of the whole is represented by its different parts. For example, the circle graph shown below gives the distribution of expenses for a typical American family.

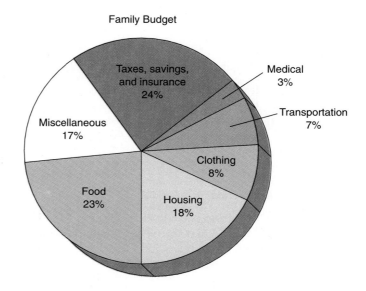

Family Budget

The area of the circle represents the total budget of the family, and the percents add up to 100%. Each category of expense takes a proportional share of the circle area. Since clothing represents 8% of the budget, it also represents 8% of the circle or a pie-shaped sector of 29°.

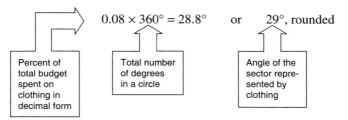

$$0.08 \times 360° = 28.8° \quad \text{or} \quad 29°, \text{ rounded}$$

Percent of total budget spent on clothing in decimal form

Total number of degrees in a circle

Angle of the sector represented by clothing

YOUR
TURN

WORK THIS PROBLEM

The Question:

Determine the angle of the sector represented by food in the pie chart shown earlier.

✔ **YOUR WORK**

The Solution:

$$0.23 \times 360° = 82.8° = 83° \text{ rounded}$$

Now suppose the family represented by this pie graph had a net income of $26,000. What actual dollar amount did they spend on housing?

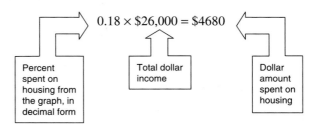

$$0.18 \times \$26,000 = \$4680$$

Percent spent on housing from the graph, in decimal form

Total dollar income

Dollar amount spent on housing

WORK THIS PROBLEM

The Question:

Use the preceding pie chart to answer the following questions:

(a) If the total income was $26,000, what dollar amount was spent on transportation?
(b) How much money went to taxes, savings, and insurance?
(c) If the family net income rose to $28,000 the following year, and if the percents remained the same, what would be the increase in their food expenditures?

✔ YOUR WORK

The Solution:

(a) 7% of $26,000 = 0.07 × $26,000 = $1820
(b) 24% of $26,000 = 0.24 × $26,000 = $6240
(c) 23% of $26,000 = $5980
    23% of $28,000 = $6440
    The increase is $6440 − 5980 = $460.

Constructing a Pie Chart

Constructing a pie chart often involves percent calculations (see Chapter 4). If the percents are already given, the work is simplified:

Steps: Constructing a Pie Chart

STEP 1.   *Convert* each percent figure to a decimal and multiply each by 360°. The result is the number of degrees in the circle each category will occupy.

STEP 2.   *Round* each angle to the nearest degree. (You really can't draw an angle more accurately than that.) Check to be certain that the angles add up to 360°. If the total is 359°, round up the angle with the largest fraction less than 0.5. If the total is 361°, round down the answer with the smallest fraction greater than 0.5.

STEP 3.   *Draw* a circle; then use a protractor to draw pie-shaped sectors corresponding to each angle. *Label* the areas by name and percent. Put a title on the graph.

For example, let's construct a pie chart based on the following data:

INCOME OF COLLEGE GRADUATES

Income Bracket	Percent of People
0 to $12,999	4%
$13,000 to $15,999	7%
$16,000 to $21,999	13%
$22,000 to $29,999	31%
$30,000 and up	45%

INCOME OF COLLEGE GRADUATES

Income Bracket	% of People	Pie Chart Segment
0 to $12,999	4%	$0.04 \times 360° = 14.4°$ or 14° rounded
$13,000 to $15,999	7%	$0.07 \times 360° = 25.2°$ or 25° rounded
$16,000 to $21,999	13%	$0.13 \times 360° = 46.8°$ or 47° rounded
$22,000 to $29,999	31%	$0.31 \times 360° = 111.6°$ or 112° rounded
$30,000 and up	45%	$0.45 \times 360° = 162.0°$ or 162°
Total		360°

Step 1   *Step 2*

Step 3.

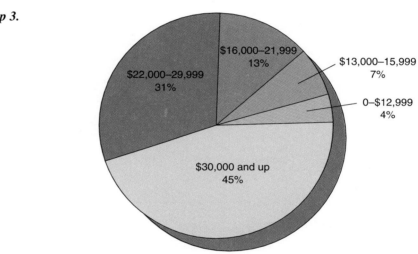

Income of College Grads

If you are an artist or have computer software to do the artwork, your pie chart might look like this:

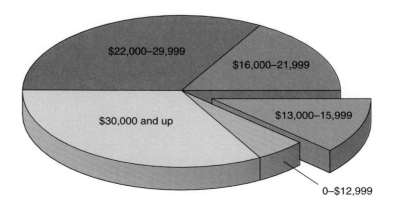

If the percent figures are not already given, you must first calculate them in order to draw a circle graph. For example, Mat Lien's 1997 property tax bill was $1955. This money was spent by the county as follows:

County indebtedness	$ 515
Fire protection/police	191
Schools	1030
Sanitation	109
Water	72
Flood control	30
Transportation	8
Total	$1955

In this case, rewrite each dollar amount as a percent of the total tax, then calculate the circle graph angle. For example, the first item, "county indebtedness," becomes

$$\frac{\$\ 515}{\$1955} = 0.2634 \qquad \text{or} \qquad 26.34\% \qquad \text{so} \qquad 0.2634 \times 360° = 94.8°$$

Divide each amount by the total

Multiply each decimal fraction by 360°

WORK THIS PROBLEM

YOUR
TURN

The Question:

Construct a pie chart for Mat Lien to show how his property tax money was spent.

✔ YOUR WORK

The Solution:

	Amount	Calculate percent — Percent	Step 1 — Angle	Step 2 — Angle, Rounded
County indebtedness	$ 515	26.3%	94.8°	95°
Fire protection/police	191	9.8%	35.3°	35°
Schools	1030	52.7%	189.7°	190°
Sanitation	109	5.6%	20.2°	20°
Water	72	3.7%	13.3°	13°
Flood control	30	1.5%	5.4°	5°
Transportation	8	0.4%	1.4°	2°
Total	$1955	100.0%		360°

(Notice that we rounded 1.4° to 2° in order to get a total of 360°.)

Step 3.

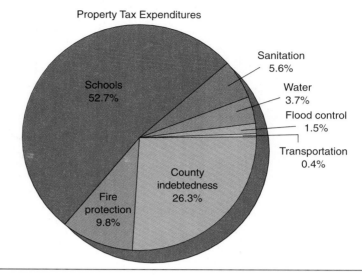

Property Tax Expenditures

You can use a calculator to find the angles in a pie chart.

Finding Angles in a Pie Chart Using a Calculator

Performing the calculations to construct a pie chart can be tedious. A calculator can shorten the process considerably.

For example, to find the angle for county indebtedness in the previous "Your Turn," key in the following:

515 $\boxed{\div}$ 1955 $\boxed{\times}$ 360 $\boxed{=}$ ⟶ 94.833756 or 95° rounded

If you have worked through all these problems correctly, you are ready for the questions in Section Test 20.2. If you had difficulty, review this section.

WORKSPACE

Name

Date

Course/Section

The following problems test your understanding of Section 20.2, Business Graphs.

A. Reading Graphs

1. Use these bar graphs to answer the questions that follow.

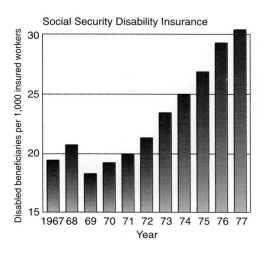

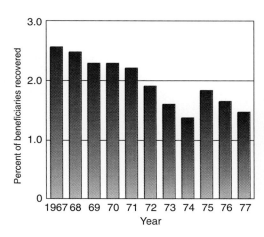

Social Security Disability Insurance

(a) In 1967, approximately how many disabled beneficiaries were there per 1000 insured workers?
(b) In 1976, approximately what percent of beneficiaries recovered?
(c) Did the number of disabled beneficiaries per 1000 ever decline? If so, when?
(d) If there were 2,500,000 disabled beneficiaries in 1976, approximately how many recovered?
(e) If there were 80 million insured workers in 1974, approximately how many of these were disabled?

2. Answer the following questions using the line graph shown.

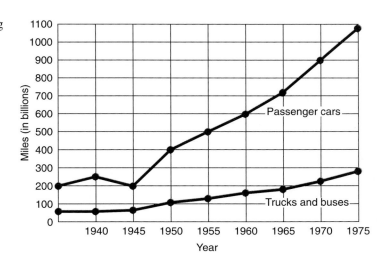

(a) How many miles were traveled by trucks and buses in 1950?
(b) How many miles were traveled by all vehicles in 1960?
(c) By how many miles did passenger car use increase from 1965 to 1975?

3. Answer the following questions using the circle graph.

Expenses of the Zzap Electrical Co.

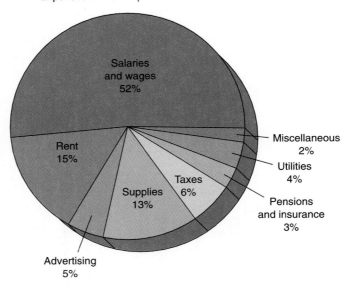

(a) What percent of Zzap's expenses were allotted to rent?
(b) What percent was spent on pensions and insurance?
(c) If the total expenses were $420,000, how much was spent on taxes? On rent?

4. Use the circle graph to answer the following questions.

Assets of the Allbrite Corp.

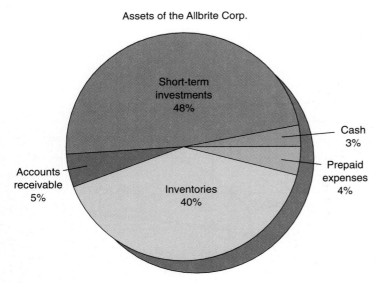

(a) What percent of the assets of the Allbrite Corporation is in the form of cash?
(b) If the total assets of the company are $90 million, how much is in inventory?
(c) If the total assets declined to $75 million and the company wished to maintain the same percent of short-term investments, how much will be in short-term investments?
(d) If the company had $2,500,000 in prepaid expenses, what would be its total assets?

5. Answer the following questions using the bar graph.

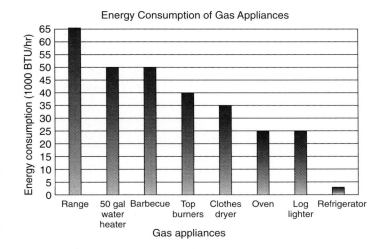

Energy Consumption of Gas Appliances

(a) Which appliance uses the most energy in an hour?
(b) Which appliance uses the least energy in an hour?
(c) How many BTU/hr does a gas barbecue use?
(d) How many BTU/*day* (24 hr) would a 50-gal water heater use?
(e) Is there any difference between the energy consumption of the range and that of the top burners plus oven?
(f) How many BTUs are used by a log lighter in 15 minutes?
(g) What is the difference in energy consumption between a 50-gal water heater and a clothes dryer?

6. Answer the following questions using the line graph.

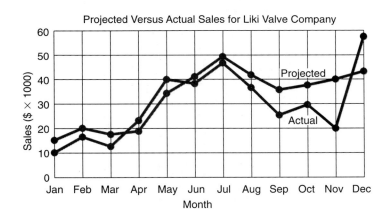

Projected Versus Actual Sales for Liki Valve Company

(a) What was the actual sales total in October?
(b) What was the projected sales total in May?
(c) During which month were actual sales highest?
(d) During which month were projected sales highest?
(e) During which month were actual sales lowest?
(f) During which month were projected sales lowest?
(g) During which months were actual sales lower than projected sales?
(h) During which month was the gap between actual and projected sales the largest? What was this difference?

B. Draw the indicated graphs.

Make bar graphs of the following sets of data (problems 1–5).

1. **SALES OF CONSTRUCTION
   MATERIAL: WOODY'S LUMBER
   in $1000**

Year	Wood	Masonry
1989	243	112
1990	289	131
1991	325	133
1992	296	106
1993	288	92
1994	307	94
1995	412	89

2. **NEW BUSINESS
   INCORPORATIONS, 1996**

Month	New Incorporations
January	30,100
February	29,500
March	29,000
April	30,800
May	28,000
June	31,600
July	29,400
August	32,200
September	32,000
October	32,300
November	33,500
December	33,900

3. **DISTRIBUTION OF PASSENGER CAR TRIPS**

Length of Trip	Percent of Total Trips	Percent of Vehicle Miles
Under 5 miles	59.6%	13.2%
5–9 miles	19.9%	15.4%
10–19 miles	12.3%	19.4%
20–49 miles	6.1%	21.2%
50 miles +	2.1%	30.8%

4. **AMOUNT OF STEEL USED BY BIG BEN'S BOLTS**
   (in thousands of pounds)

Month	Amount
January	85
February	103
March	132
April	138
May	121
June	100
July	75
August	112
September	106
October	135
November	158
December	127

5. **SALES OF RECORDED MUSIC**
   (in millions of dollars)

	1995	1996
Cassettes	2325.8	3472.4
CDs	2587.4	3451.6
Cassette singles	194.6	257.9
Music videos	166.0	172.3
Singles	116.4	94.4
LP/EPs	220.1	86.5

6. A worker at Kate's Delicacies had the following croissant output:

Work Output	Day
125	Mon.
275	Tue.
340	Wed.
310	Thu.
150	Fri.

(a) Draw a bar graph of this data.
(b) Draw a line graph of this data.

7. Construct a double line graph based on this data.

Year	Earnings per Share	Dividends per Share
1992	$1.12	$0.28
1993	1.27	0.32
1994	1.75	0.40
1995	1.92	0.40
1996	2.16	0.45
1997	2.85	0.55

8. Construct a double bar graph based on this data.

OUTPUT OF BRICKLAYERS FOR J.R.'S CONSTRUCTION

| Initials of Employee | Bricks Laid | |
	March	April
R.M.	1725	1550
H.H.	1350	1485
A.C.	890	1620
C.T.	1830	1950
W.F.	1175	1150
S.D.	2125	1875

9. Draw a line graph using the following set of data.

**AUTO FACTORY SALES
IN THE UNITED STATES–
BEFORE THE BOOM
(in thousands)**

Year	Sales
1900	4
1905	24
1910	181
1915	896
1920	1906
1925	3735
1930	3787
1935	3273
1940	3717
1945	70
1950	6665

Draw circle graphs illustrating the following sets of data (problems 10–14).

10. **CALIFORNIA'S
SOURCES OF
ELECTRICITY IN 1980**

Geothermal	1%
Nuclear	4%
Coal	7%
Gas	21%
Hydroelectric	24%
Oil	43%

11. **FAMILY BUDGET:**
Take-home Pay

Food	$660
Housing	850
Utilities	252
Transportation	378
Clothing	170
Entertainment	100
Savings	200
Other expenses	140

12. **HOUSING SITUATION:**
PINE VALLEY

Own home	16,500
Rent home	2,700
Own condominium	3,800
Rent condominium	1,600
Own apartment	650
Rent apartment	9,600
Other	1,200

13. **GOVERNMENT BUDGET:**
CENTER CITY

Education	$17,000,000
Police and fire	8,000,000
Health and sanitation	11,000,000
Pensions	6,000,000
Debt service	15,000,000
Miscellaneous	32,000,000

14. **THOREAU BOOKSTORE'S ANNUAL SALES (in $1000)**

Childrens	$40.2
Mysteries	46.9
Business and science	26.8
Romantic novels	80.4
Psychology and self-help	93.8
Miscellaneous	46.9

15. The following pie graph represents what happens to each part of the retail price of a gold and diamond pendant. The total selling price is $3640. Use the graph and your knowledge of percents to answer the following questions:
   a) What % goes to overhead?
   b) What is the jeweler's cost for the diamonds?
   c) How much profit does she make?

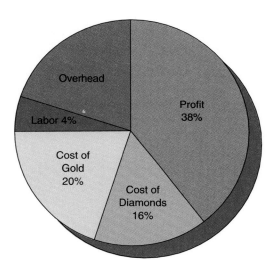

WORKSPACE

Key Terms

axes	line graph	pie chart (circle graph)
bar graph	mean	plotting
double bar graph	median	range
graph	mode	

Chapter 20 at a Glance

Page	Topic	Key Point	Example
680	Mean	$\text{Mean} = \dfrac{\text{sum of values}}{\text{no. of values}}$	What is the mean of the following sales figures? $2000, $2650, $2000, $3495, $7459, $2000, $9202 $\text{Mean} = \dfrac{\$28{,}806}{7} = \$4115$
682	Mean when multiples are involved	$\text{Mean} = \dfrac{\text{values} \times \text{frequency}}{\text{total frequency}}$	Find the mean for the following numbers: 32, 32, 33, 33, 33, 33, 34, 34, 34, 34. Values × frequency = $32 \times 2 = \ \ 64$ $33 \times 4 = 132$ $34 \times 4 = 136$ $64 + 132 + 136 = 332$ Total frequency = $2 + 4 + 4 = 10$ $\text{Mean} = \dfrac{332}{10} = 33$
684	Median	Median = middle number in set of numbers with odd number of values or mean of middle two numbers in set with even number of values.	Find the medians of the following series of numbers: (a) 30, 35, 37, 50, 80 (b) 30, 35, 37, 43, 60, 80 (a) Mode = 37 (b) Mode = $\dfrac{37 + 43}{2} = 40$
686	Mode	Mode = most frequently occurring number	Find the mode of the following series of numbers: 2, 3, 5, 3, 6, 5, 3, 3, 7, 9. Mode = 3 (occurs four times)
686	Range	Range = largest value − smallest value	Find the range of the following series of numbers: 29, 20, 35, 27, 28, 53. Range = 53 − 20 = 33
697	Reading bar graphs	Locate end of bar for a given year, move straight across to the appropriate axis, and read the amount, estimating when the end of the bar falls between two values.	See the beginning of Section 20.2.

Page	Topic	Key Point	Example
699	Constructing bar graphs	Decide on a single or double bar graph and on which axis which information will go. Choose spacing for information on axes and label axes. Draw lines to indicate ends of bars and complete bars.	See Section 20.2.
702	Reading line graphs	Find desired time on horizontal axis and follow the line on the grid until you reach the graphed line. Move horizontally to the left and read the value on the vertical axis.	See Section 20.2.
703	Plotting line graphs	Select spacing for each axis and plot time on horizontal axis. Plot points representing information and connect points.	See Section 20.2.
707	Finding pie chart sectors	Sector of pie chart = percent $\times$ 360°	If clothing is 8% of an average household budget, how big should the sector representing this information in a pie chart be? Sector = 0.08 $\times$ 360° = 28.8° = 29° rounded
707	Reading pie charts	Amount = percent $\times$ total	If a household has an annual income of $26,000 and a pie chart shows that it spends 18% of its income on housing, how much does it spend? Amount = 0.18 $\times$ $26,000 = $4860
708	Constructing pie charts	Find the sector of the pie chart represented by each element and round to the nearest degree, adjusting to reach 360° if necessary. Draw the circle and use a protractor to plot the angles. Label segments.	See Section 20.2.

SELF-TEST Statistics and Graphs

The following problems test your understanding of statistics and graphs.

1. Calculate the mean, median, mode, and range for the following set of numbers: 45, 50, 48, 50, 52, 42, 45, 50, 56, 51.

2. Calculate the mean, median, mode, and range for the following set of numbers: 163, 175, 172, 205, 172, 185, 175, 172.

3. The following table shows the monthly sales for Bits 'n' Bytes for the last half of the year. What are the mean, median, mode, and range of monthly sales?

Month	Sales
Jul.	$105,000
Aug.	102,000
Sep.	112,000
Oct.	98,000
Nov.	114,000
Dec.	123,000

4. The following table shows the daily sales of hexacentrics by the Nut Shop. What are the mean, median, mode, and range of daily sales?

Day	Hexacentrics
Mon.	34
Tues.	39
Wed.	29
Thurs.	35
Fri.	44
Sat.	53

5. Construct a bar graph for the following monthly sales.

Month	Sales
Jan.	$ 85,000
Feb.	91,000
Mar.	95,000
Apr.	88,000
May	93,000
Jun.	102,000
Jul.	108,000
Aug.	117,000

6. Construct a bar graph for the following yearly sales.

Year	Sales
1991	$355,000
1992	388,000
1993	402,000
1994	390,000
1995	425,000
1996	442,000

7. Construct a line graph for the following monthly sales.

Month	Sales
Jan.	$12,500
Feb.	15,800
Mar.	19,200
Apr.	23,900
May	21,000
Jun.	25,200
Jul.	28,100
Aug.	32,700
Sep.	32,000
Oct.	41,600
Nov.	42,000
Dec.	47,800

8. Construct a line graph for the following yearly sales.

Year	Sales
1987	$ 48,000
1988	58,000
1989	65,000
1990	82,000
1991	97,000
1992	89,000
1993	102,000
1994	123,000
1995	125,000
1996	143,000

9. Construct a circle graph for the following expenses.

Expenses	Amount
Salaries	$325,000
Rent	50,000
Utilities	25,000
Depreciation	40,000
Miscellaneous	60,000

10. Construct a circle graph for the following sales.

Store	Sales
#1	$495,000
2	135,000
3	72,000
4	108,000
5	90,000

11. Use the monthly sales table to answer the following questions.

Month	Sales
Jan.	$12,500
Feb.	15,800
Mar.	19,200
Apr.	23,900
May	21,000
Jun.	25,200
Jul.	28,100
Aug.	32,700
Sep.	32,000
Oct.	41,600
Nov.	42,000
Dec.	47,800

(a) Which month had the largest sales?
(b) Which month had the smallest sales?
(c) What is the range?
(d) Which month had the largest increase?
(e) Which month had the largest decrease?
(f) What was the percent increase from November to December?

12. Use the yearly sales table to answer the following questions.

Year	Sales
1987	$ 48,000
1988	58,000
1989	65,000
1990	82,000
1991	97,000
1992	89,000
1993	102,000
1994	123,000
1995	125,000
1996	143,000

(a) Which year had the largest sales?
(b) Which year had the smallest sales?
(c) What is the range?
(d) Which year had the largest increase?
(e) Which year had the largest decrease?
(f) Which was the percent from 1995 to 1996?

APPENDIX A

The Electronic Calculator

In a very short time, the calculator has become an essential tool for people in every occupation, from engineers to shoppers, from clerks to bank presidents. Just as computers have revolutionized society, the inexpensive pocket calculator has changed personal calculating. For millions of people, arithmetic will never be the same again.

It is reasonable to ask, If calculators are available, why study arithmetic at all? The answer is that while a calculator can be a very useful tool, it must not become a crutch. A calculator will help you to make mathematical computations more quickly, but it will not tell you *what* to do or *how* to do it. Because the calculator allows you to make very long and difficult calculations quickly, basic mathematical skills are more important than ever. The gadget itself can't make a mistake, but you can. If your basic mathematics skills are strong, you will recognize when an answer must be wrong.

To use a calculator intelligently and effectively in your work, you must be able to:

1. Multiply and add one-digit numbers quickly and correctly, almost like an instant reflex.
2. Read and write any number, very large or very small, decimal, whole number, or fraction.
3. Work with fractions and percents.
4. Estimate answers and check your calculations.

We have included a very careful review of these basic skills in the first four chapters of this textbook.

This appendix is designed to help you to understand the calculator and to use it correctly and effectively.

There are hundreds of calculators available. They differ in size, shape, color, cost, and, most important, in how they work and what they can do. In this brief introduction, we assume that the calculator you are using has the following characteristics:

■ It has at least an eight-digit display.

■ It performs at least the four basic arithmetic functions: addition (+), subtraction (−), multiplication (×), and division (÷).

■ It uses floating-point arithmetic, so that the position of the decimal point is given automatically in any answer.

■ It uses algebraic logic. (This is by far the most popular type of calculator in use today. If your calculator has a key marked $\boxed{+=}$, it is probably an "arithmetic logic" device, and some of this appendix will not be very helpful to you.)

Your calculator may have a great many additional features, of course, including square root $\boxed{\sqrt{}}$ or square $\boxed{x^2}$, trigonometric or more advanced functions, one or more memories, parentheses, and perhaps a programming capability. A simple business calculator would include a percent key $\boxed{\%}$, a change sign key, and some memory store and recall keys. We'll consider only the simple calculations here. In the following examples, the displays shown for our "average" calculator may differ slightly from what you find with your calculator, but the differences should not be confusing.

SECTION A.1: Simple Operations

First, check out the machine. You may find an on-off switch somewhere, perhaps a solar battery, a display, and a keyboard with both *numerical* (0, 1, 2, 3, . . . , 9) and basic *function* (+, −, ×, ÷, =, .) keys usually arranged like this:

$\boxed{7}\ \boxed{8}\ \boxed{9}\ \boxed{÷}$
$\boxed{4}\ \boxed{5}\ \boxed{6}\ \boxed{×}$
$\boxed{1}\ \boxed{2}\ \boxed{3}\ \boxed{−}$
$\boxed{0}\ \boxed{·}\ \boxed{=}\ \boxed{+}$

In addition, you will find a clear $\boxed{C}$ key somewhere on the keyboard, and perhaps others keys will appear such as $\boxed{\%}$, $\boxed{M+}$, $\boxed{MC}$, $\boxed{\sqrt{}}$, $\boxed{+/-}$, and so on.

Entering Numbers

Every number is entered into the calculator digit by digit, left to right. For example, to enter the number 37.4 the sequence looks like this:

Keyboard	Display
$\boxed{3}$	3.
$\boxed{7}$	37.
$\boxed{·}$	37.
$\boxed{4}$	37.4

Don't worry about leading zeros to the left of the decimal point (as in 0.5, 0.664, or 0.15627) or final zeros after the decimal point (as in 4.70, 32.500, or 8.25000). You need not enter these zeros; the calculator will automatically interpret the number correctly and display the zeros if they are needed. But no harm will be done if you *do* enter them.

	Keyboard	Display
	$\boxed{.}\ \boxed{5}$	0.5
or	$\boxed{0}\ \boxed{.}\ \boxed{5}$	0.5
or	$\boxed{.}\ \boxed{5}\ \boxed{0}$	0.50

For simple nonscientific calculators, fractions must be translated to decimal form before they can be entered. The fraction $4\frac{1}{4}$ is entered as 4.25. and $5\frac{1}{9}$ as 5.11111. . . .

Keyboard	Display
$\boxed{4}\ \boxed{.}\ \boxed{2}\ \boxed{5}$	4.25
$\boxed{5}\ \boxed{.}\ \boxed{1}\ \boxed{1}\ \boxed{1}\ \boxed{1}$	5.1111

Converting numbers from fraction to decimal form is covered in Chapter 3 of this textbook.

Clearing Numbers

If you make an error in entering the number, press either the clear $\boxed{C}$ or clear entry $\boxed{CE}$ key.

The *clear* key $\boxed{C}$ "clears" all entries and causes the number zero to appear in the display. Pressing the clear key means that you want to start the entire calculation over or begin a new calculation.

Keyboard	Display
$\boxed{C}$	0.

The *clear entry* key $\boxed{CE}$ "clears" the display only, leaves the previous number unchanged, and causes a zero to appear in the display. Pressing the clear entry key $\boxed{CE}$ means that you wish to remove the last entry only. For example, if you wanted to enter the number 46 and made an error, the following sequence might result:

Keyboard	Display	
$\boxed{4}$	4.	
$\boxed{7}$	47.	⇐ An error
$\boxed{CE}$	0.	⇐ Pressing the $\boxed{CE}$ key clears the entire number.
$\boxed{4}$	4.	⇐ *Start over,*
$\boxed{6}$	46.	⇐ The correct entry

Interpreting Results

When you solve a problem using a calculator, your result will be a number that appears in the display. To find the actual answer from this display may require a bit of interpretation on your part. For example, if you solve a business problem where the answer is in dollars, the display

| 37.6 | means | $37.60 |

| 0.07 | means | $ .07 or 7¢ |

Most calculators are capable of displaying an answer with at least eight numerical digits, so that in general you must round the displayed number to find the correct answer to your problem. For example, if the answer is in dollars, the display

| 6.7912 | means | $6.79 | rounding to the nearest cent, |

and

| 145.3086 | means | $145.31 | rounding to the nearest cent. |

Rounding is discussed in detail in Chapter 1 and Chapter 3.

Using Function Keys

Every numerical entry is built up digit by digit. The calculator does not know if the number is 3 or 37 or 37.4 or 37.492116 until you stop entering numerical digits and press a function key: $+$, $-$, $\times$, $\div$ or $=$.

For example, to add 48 + 37 = ? use this sequence:

Keyboard	Display	
4 8	48.	
+	48.	Pressing the $+$ key completes the entry of the first number and tells the calculator to add the next entry to the previous total.
3 7	37.	
=	85.	Pressing the $=$ key completes the entry of the second number, the two numbers are added, and the sum is displayed.

Another example: 7.42 × 3.5 = ?

Keyboard	Display	
7 . 4 2	7.42	
×	7.42	Pressing the $\times$ key completes the entry of the first number.
3 . 5	3.5	Pressing the $=$ key completes the entry of the second number. The two numbers are multiplied and the product is displayed.
=	25.97	

Estimating

In order to use a calculator most effectively and accurately, you need to develop the skill of *estimating*. Before beginning any calculation, you should first estimate the answer by doing simple arithmetic. This estimate gives you a quick and easy check of the calculator answer.

For example, before calculating 4820 + 1241 = ?, mentally round each number to the nearest thousand and add 5000 + 1000. Your final calculator answer should be roughly 6000. The actual sum is 6061 in this case.

Estimating the answer in this way greatly reduces the possibility that you will make an error and fail to detect it. The first law of effective calculating is

Never use the calculator until you have a good estimate of the answer.

Percent Key Calculations

A business calculator with a percent key $\boxed{\%}$ can greatly simplify percent calculations. For example, to calculate 16% of 40, first translate to a math equation: $16\% \times 40 = ?$.

Keyboard	Display	
$\boxed{4}\boxed{0}$	40.	Enter the number.
$\boxed{\times}$	40.	Press the arithmetic operation key.
$\boxed{1}\boxed{6}$	16.	Enter the percent number.
$\boxed{\%}$	6.4	Press the percent key.

Notice that the answer is displayed immediately after the percent key is pressed. There is no need to press the $\boxed{=}$ key to display the answer. In fact, in this calculation pressing the $\boxed{=}$ key after $\boxed{\%}$ multiplies by 40 again.

WORKSPACE

EXERCISES A.1: Simple Operations

Name _____

Date _____

Course/Section _____

Perform the following calculations. Round your answer to four decimal places.

1. $4726 + 378$

2. $8361 + 853$

3. $8645 - 836$

4. $9362 - 897$

5. 3648×647

6. 8464×529

7. $7355 \div 364$

8. $9476 \div 263$

9. 373.6×0.097

10. 90.63×0.074

11. $74.27 \div 4.86$

12. $0.0736 \div 0.0398$

13. What is the cost of seven bugaboos at $4.98 each?

14. Sam sold five hexacentrics for $30.75. What is the price of one hexacentric?

15. If 89.5 square feet of carpeting cost of $1517.92, what is the price per square yard?

16. If you earn $9.57 per hour, what is your pay for 25.25 hours?

WORKSPACE

SECTION A.2: Combined Operations

A calculator can only operate on two numbers at a time, but very often you need to use the result of one calculation in a second calculation or in a long string of calculations. Because the answer to any calculation remains in the calculator, you can carry out a very complex chain of calculations without ever stopping the calculator to write down an intermediate step. For example, to add a list of numbers simply continue the addition step.

$$4.1 + 0.72 + 12.68 + 3.2 = ? \qquad \textbf{Estimate: } 4 + 1 + 12 + 3 = 20$$

Keyboard	Display	
4 . 1	4.1	
+	4.1	
. 7 2	0.72	
+	4.82	⇐ The sum of the first two numbers
1 2 . 6 8	12.68	
+	17.5	⇐ The sum of the first three numbers
3 . 2	3.2	
=	20.7	⇐ The total sum of all four numbers

You can also combine addition and subtraction in the same sequence of operations. For example,

$$462 - 294 + 31 = ? \qquad \textbf{Estimate: } 460 - 300 + 30 = 190$$

Keyboard	Display	
4 6 2	462.	
−	462.	
2 9 4	294.	
+	168.	⇐ The difference of the first two numbers
3 1	31.	
=	199.	⇐ The answer

Multiplication of a list of factors is similar to addition and quite simple. For example,

$$3.2 \times 4.1 \times 0.53 \times 2.6 = ?\qquad \textbf{Estimate: } 3 \times 4 \times 0.5 \times 3 = 18$$

Keyboard	Display	
3 . 2	3.2	
×	3.2	
4 . 1	4.1	
×	13.12	⟸ Product of the first two numbers
. 5 3	0.53	
×	6.9536	⟸ Product of the first three factors
2 . 6	2.6	
=	18.07936	⟸ Final product of all four factors

You can easily combine multiplication and division operations. For example, calculate $2 \times 4 \div 5 = ?$.

Keyboard	Display
2	2.
×	2.
4	4.
÷	8.
5	5.
=	1.6

You can combine the operations of multiplication and division with the operations of addition and subtraction using a calculator, but you may have to write down the intermediate results. If your calculator has a memory, these intermediate results may be stored in the calculator and used in the rest of the calculation.

 Think before you combine unlike arithmetic operations.

For example, punch this problem into your calculator:

2 + 3 × 4 = ?

If you press the keys in the order given and your calculator displays the answer as 20, then your calculator does the operations sequentially, left to right. It has interpreted these instructions as

$$(2 + 3) \times 4 \quad \text{or} \quad 5 \times 4$$

A different model calculator may display the answer 14. This second kind of calculator interprets the instruction as

$$2 + (3 \times 4) \quad \text{or} \quad 2 + 12$$

This second calculator does multiplications or divisions first, then additions and subtractions.

To be certain there is no misinterpretation, we use parentheses to show which calculation should be done first. For example,

$$(3.1 \times 4.2) + 1.5 = ? \qquad \textbf{Estimate: } (3 \times 4) + 1 = 12 + 1 = 13$$

Keyboard   **Display**

Keyboard	Display	
[3][.][1]	3.1	
[×]	3.1	
[4][.][2]	4.2	
[=]	13.02	⟵ Pressing the [=] key gives the product of the multiplication in the parentheses.
[+]	13.02	
[1][.][5]	1.5	
[=]	14.52	⟵ The final answer

In a sequence of operations, a number to be subtracted may be displayed. To subtract this number use the *change sign* key [+/−]. For example, to calculate $8.45 - 4.5$ where 4.5 is already displayed on the calculator,

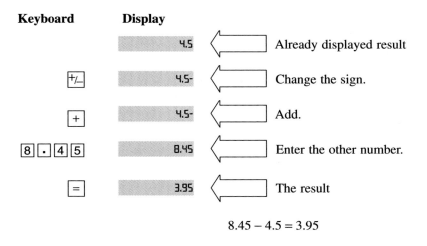

Keyboard   **Display**

Keyboard	Display	
	4.5	⟵ Already displayed result
[+/−]	4.5-	⟵ Change the sign.
[+]	4.5-	⟵ Add.
[8][.][4][5]	8.45	⟵ Enter the other number.
[=]	3.95	⟵ The result

$$8.45 - 4.5 = 3.95$$

WORKSPACE

EXERCISES A.2: Combined Operations

Name

Date

Course/Section

Perform the following calculations. Round your answer to four decimal places.

1. $63 + 852 + 73 + 173$

2. $836 + 84 + 95 + 2384$

3. $8.63 + 837.2 - 647.01$

4. $84.842 + 84 - 73.002$

5. $3.7 \times 6.27 \times 72$

6. $9.36 \times 0.036 \times 28.3$

7. $\dfrac{3.746 \times 0.37}{57.8 \times 23.83}$

8. $\dfrac{63.6 \times 28.04}{5.364 \times 0.036}$

9. $\dfrac{8.846 \times 0.036}{2.84 \times 23.87}$

10. $\dfrac{46.83 \times 34.7 \times 4.5}{23.74 \times 0.0036}$

11. 2150.35×1.025^2

12. 3541.95×1.015^4

13. $\dfrac{(5.83 \times 6.2) \times 7.4}{34.6 - 24.3}$

14. $\dfrac{(46.83 + 2.73) \times 2.73}{24.7 + 73.58}$

15. 462.46×1.05^5

16. 3851.84×1.035^6

WORKSPACE

APPENDIX B

Metric System

It's fourth down and centimeters to go!

Why worry about the metric system? Why use it? First, about 95% of the world's people already measure, think, buy, and sell in metric units.

Second, the United States already uses metric units in many ways: most electrical units such as volts, amperes, or watts are metric; cameras, film (8mm, 16mm, 35mm), lenses (50mm, 260mm), and other optical parts are measured in metric units; ball bearings, spark plugs, even skis and cigarettes come in metric sizes. Most important of all, the metric system is easier to use and simpler to learn. Working with metric units means there is much less use of fractions, fewer units to remember, less to memorize.

A pinch of salt, a teaspoonful of sugar, or a cup of flour will probably stay the same in the kitchen and inchworms won't become centimeter worms, but commercial, industrial, and government measurements may be metric. It is important that you start learning to read, use, and think metric.

The most important common units in the metric system are those for length or distance, speed, weight or mass, volume, and temperature. Other metric units are used in science but are not generally needed in day-to-day activities. Time units—year, day, hour, minute, second—are the same in the metric as in the English system.

In order to become familiar with the metric system it is important to *think metric*. Most people in the United States know about how much a foot is in length, how much a pound is in weight, and how much a quart is in volume. But what is a meter, a gram, and a liter?

The basic unit of length in the metric system is the meter (pronounced *meet-ur* and abbreviated m). One meter is 39.37 inches—a little longer than a yard.

The basic metric unit for weight or mass is the gram (abbreviated g). One ounce is equivalent to about 28 grams; one pound is equal to 454 grams.

The basic metric unit of volume is the liter (pronounced *leet-ur* and abbreviated ℓ). A liter is about 10% more than a quart.

Each of these metric measures can be broken down into smaller units or converted to larger units by the use of certain prefixes.

Prefix	Multiplier	Meters	Grams	Liters
kilo (k)	1000 ×	kilometer (km) = 1000 meters	kilogram (kg) = 1000 grams	kiloliter (kℓ) = 1000 liters
hecto (h)	100 ×	hectometer (hm) = 100 meters	hectogram (hg) = 100 grams	hectoliter (hℓ) = 100 liters
deka (da)	10 ×	dekameter (dam) = 10 meters	dekagram (dag) = 10 grams	dekaliter (daℓ) = 10 liters
deci (d)	0.1 ×	decimeter (dm) = $\frac{1}{10}$ meter	decigram (dg) = $\frac{1}{10}$ gram	deciliter (dℓ) = $\frac{1}{10}$ liter
centi (c)	0.01 ×	centimeter (cm) = $\frac{1}{100}$ meter	centigram (cg) = $\frac{1}{100}$ gram	centiliter (cℓ) = $\frac{1}{100}$ liter
milli (m)	0.001 ×	millimeter (mm) = $\frac{1}{1000}$ meter	milligram (mg) = $\frac{1}{1000}$ gram	milliliter (mℓ) = $\frac{1}{1000}$ liter

To put these measurements in perspective, consider the following relationships between metric and English measures:

Length
1. The thickness of a dime is approximately 1 millimeter.
2. A king-size cigarette is about 100 millimeters.
3. An unsharpened pencil is approximately 18 centimeters long.
4. A mile is about 1.6 km.

Weight
1. A paper clip weighs approximately 1 gram.
2. A nickel weighs 5 grams.
3. One kilogram is about 2.2 pounds.
4. A 100-pound barbell weighs about 45 kilograms.
5. A football player weighs about 120 kilograms.

Volume
1. One teaspoon is approximately 5 milliliters.
2. One fluid ounce is about 30 milliliters.
3. A gallon is approximately 3.8 liters.

Speed
1. Driving 55 miles per hour is about 90 kilometers per hour.
2. Driving 25 miles per hour is about 40 kilometers per hour.

SECTION B.1: Converting from Metric to Metric

The fact that all units in the metric system are multiples of ten times the basic unit makes it very easy to convert between metric measures.

For example, let's convert (a) 7.3 m to centimeters, (b) 775 dℓ to hectoliters, and (c) 8.72 hg to grams.

(a) 1 cm = 0.01 m   or   100 cm = 1 m

$$7.3 = 7.3 \; \cancel{m} \times \frac{100 \text{ cm}}{1 \; \cancel{m}} = 7.3 \times 100 \text{ cm} = 730 \text{ cm}$$

Notice that since 100 cm = 1 m, then 100 cm/1 m = 1.

When you multiply a number by 1, its value is unchanged. There are, however, two possible fractions: 100 cm/1 m and 1 m/100 cm. We used 100 cm/1 m in order to cancel the meter units, leaving the centimeter units.

(b) $775 \text{ dℓ} = 775 \; \cancel{dℓ} \times \dfrac{1 \text{ hℓ}}{1000 \; \cancel{dℓ}} = \dfrac{775}{1000} \text{ hℓ} = 0.775 \text{ hℓ}$

⬆ 1 hℓ = 1000 dℓ

(c) $8.72 \text{ hg} = 8.72 \; \cancel{hg} \times \dfrac{100 \text{ g}}{1 \; \cancel{hg}} = 8.72 \times 100 \text{ g} = 872 \text{ g}$

⬆ 1 hg = 100 g

WORKSPACE

EXERCISES B.1: Metric–Metric Conversions

Name _____

Date _____

Course/Section _____

Perform the following metric–metric conversions.

1. 12.3 kg = _____ g

2. 758 g = _____ kg

3. 375 mg = _____ dg

4. 1.5 hg = _____ cg

5. 64,300 cm = _____ km

6. 673 mℓ = _____ dℓ

7. 74 kg = _____ g

8. 82.4 dm = _____ mm

9. 7350 ℓ = _____ kℓ

10. 0.0075 hg = _____ mg

11. 3.7 km = _____ dam

12. 58 cℓ = _____ ℓ

13. 58 cℓ = _____ mℓ

14. 75 dm = _____ mm

15. 65 dag = _____ g

16. 5900 dℓ = _____ hℓ

17. 0.25 g = _____ mg

18. 125 dm = _____ dam

19. 50.4 dag = _____ dg

20. 8500 mℓ = _____ cℓ

SECTION B.2: Converting Between English and Metric Measures

It's much more difficult to "translate" English to metric or metric to English measures than it is to convert within the metric system because of the odd numbers in the English system. However, as U.S. companies do more and more business abroad, it is more and more crucial for American businesspeople to know how to translate between systems.

The following is a partial table of English measurements and their metric equivalents:

English-Metric Conversions

Length

1 inch = 2.54 cm exactly	0.0394 inch $\cong$ 1 mm	$\cong$ means "approximately equal"
1 foot $\cong$ 30.5 cm	0.394 inch $\cong$ 1 cm	
1 yard $\cong$ 91.4 cm	39.4 inches $\cong$ 1 m	
1 mile $\cong$ 1610 m	3.28 feet $\cong$ 1 m	
1 mile $\cong$ 1.61 km	1.09 yards $\cong$ 1 m	
	0.621 mile $\cong$ 1 km	

Weight/Mass

1 ounce $\cong$ 28.3 g	0.0353 ounce $\cong$ 1 g
1 pound $\cong$ 454 g	0.0022 pound $\cong$ 1 g
1 pound $\cong$ 0.454 kg	2.2 pounds $\cong$ 1 kg

Volume

1 gallon $\cong$ 3.79 ℓ	0.264 gallon $\cong$ 1 ℓ
1 quart $\cong$ 0.946 ℓ	1.06 quart $\cong$ 1 ℓ

For example, let's convert (a) 152 m to yards, (b) 3 lb to grams, and (c) 5 ℓ to gallons.

(a) $152 \text{ m} = 152 \text{ m} \times \dfrac{1.09 \text{ yd}}{1 \text{ m}} = 152 \times 1.09 \text{ yd} = 165.68 \text{ yd}$

$1 \text{ m} = 1.09 \text{ yd}$

(b) $3 \text{ lb} = 3 \text{ lb} \times \dfrac{454 \text{ g}}{1 \text{ lb}} = 3 \times 454 \text{ g} = 1362 \text{ g}$

$1 \text{ lb} \cong 454 \text{ g}$

(c) $5 \ell = 5 \ell \times \dfrac{0.264 \text{ gal}}{1 \ell} = 5 \times 0.264 \text{ gal} = 1.32 \text{ gal}$

$1 \ell \cong 0.264 \text{ gal}$

EXERCISES B.2:   English–Metric Conversions

Perform the following English-metric conversions. Round your answers to the nearest hundredth.

1. 12.3 kg = _____ lb

2. 25 oz = _____ g

3. 12 ft = _____ m

4. 15 cm = _____ in.

5. 5 gal = _____ ℓ

6. 3 m = _____ in.

7. 6 lb = _____ kg

8. 12 km = _____ mi

9. 7 ℓ = _____ gal

10. 7 in. = _____ cm

11. 15 km = _____ mi

12. 12 mi = _____ km

13. 5 m = _____ ft

14. 200 mm = _____ in.

15. 26 mi = _____ km

16. 20 gm = _____ oz

17. 3 qt = _____ ℓ

18. 3 gal = _____ ℓ

19. 10 in. = _____ cm.

20. 12 m = _____ yd

WORKSPACE

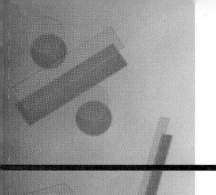

Answers for Students

CHAPTER 1: *Whole Numbers*

Section 1.1

A.   1. Twenty-eight
    3. Two hundred forty-three
    5. Seven hundred sixty-seven
    7. One thousand, nine hundred twenty-seven
    9. Seventeen thousand, nine hundred twenty-five
  11. Six hundred forty-five thousand, three
  13. Sixty-seven million, five hundred thirty-two
  15. Nine hundred seventy-three million, five hundred seventy-four thousand, eight hundred ten
  17. Seventy-two million, nine hundred one thousand, three hundred seventy-five
  19. One billion, two hundred fifty-five million, nine hundred eighty-seven thousand, one hundred twenty-three

B.   1. 27
    3. 387
    5. 59,765
    7. 100,100
    9. 32,416,075
  11. 57,212
  13. 16,989,300,052
  15. 10,000,000
  17. 16,434
  19. 600,000,227,005

C.   1. (a) 327,600   (c) 64,300   (e) 60,000
    2. (a) 875,000   (c) 83,000   (e) 42,000
    3. (a) 24,000,000   (c) 1,000,000   (e) 80,000,000

D.   1. $56,700; 62,900; 60,000; 67,300; 72,900; 62,400; 74,600; 64,800; 58,400; 61,400; 67,800; 85,000

Section 1.2

A.   1. 65   3. 81   5. 85
    7. 132   9. 121   11. 160
  13. 539 15. 1219 17. 829
  19. 1008 21. 1435 23. 1212

B.   1. 8231   3. 9346   5. 2665
    7. 10,912   9. 15,111 11. 10,051
  13. 15,354 15. 8132 17. 11,206
  19. 18,230

C. 1. 1724   3. 2053   5. 3243
   7. 8704   9. 3897   11. 8225
   13. 11,343   15. 19,233   17. 13,493
   19. 24,609

D. 1. 8866   3. 18,246   5. 17,136
   7. 24,893   9. 27,435

E. 1. 52
   3.

Branch	M	T	W	Th	F	S	Total
Downtown	$2574	$2148	$2014	$1957	$2683	$ 3,189	$14,565
Suburb	2956	2751	2515	2283	3108	3,573	17,186
Mall	3479	3216	3127	3065	3581	4,157	20,625
Totals	$9009	$8115	$7656	$7305	$9372	$10,919	$52,376

   5. 154,519   7. 99   9. $2467
   11. 105   13. $1373   15. 328

Section 1.3

A. 1. 11   3. 44   5. 25
   7. 29   9. 48   11. 17
   13. 129   15. 167   17. 85
   19. 377   21. 277   23. 236

B. 1. 659   3. 293   5. 823
   7. 2186   9. 3389   11. 1386
   13. 2091   15. 3365   17. 2889
   19. 864   21. 3891   23. 987
   25. 4508

C. 1. 18,820   3. 1596   5. 2775
   7. 26,296   9. 13,299   11. 50,378
   13. 29,176   15. 50,877   17. 12,036
   19. 29,584   21. 38,879   23. 28,081
   25. 2894

D. 1. 49   3. 49   5. $1216
   7. $470   9. $4200   11. 617
   13. $594   15. (a) 2345   (b) 11,725

Section 1.4

A. 1. 216   3. 246   5. 78
   7. 783   9. 553   11. 184

13. 468	15. 2660	17. 3384
19. 4028	21. 1400	23. 4346

B. 1. 23,086   3. 20,850   5. 65,340
    7. 20,720   9. 15,776   11. 38,361
    13. 38,637   15. 25,776   17. 195,750
    19. 124,960

C. 1. 177,230   3. 598,850   5. 529,254
    7. 492,954   9. 800,888   11. 3,330,768
    13. 7,231,015   15. 3,940,174   17. 24,317,654
    19. 59,918,608

D. 1. $7620   3. $53,196   5. $94,350
    7. 63,360   9. $2144   11. 180
    13. 504   15. $370

Section 1.5

A. 1. 12   3. 27   5. 37
    7. 52   9. 73   11. 85
    13. 61   15. 59   17. 79

B. 1. 25   3. 47   5. 19
    7. 19   9. 45   11. 92
    13. 37 r 13   15. 32 r 39   17. 28 r 45

C. 1. 93   3. 65   5. 83
    7. 83 r 21   9. 217   11. 272 r 43
    13. 235   15. 217 r 99   17. 319 r 277

D. 1. 29   3. 54   5. $1158
    7. $300   9. $4925   11. $465
    13. 104   15. 68   17. $185,246

Self-Test 1

1. Thirty-four million, seven hundred fifty-six thousand, one hundred five
3. 175,002,641
5. 455,506
7. 161,888
9. $1,782,264
11. 4691 r 201
13. $82,900; 87,500; 91,800; 83,500; 87,400; 105,700; 95,700; 99,500; 107,500; 96,400;
    113,600; 126,500
15.

Branch	M	T	W	Th	F	S	Weekly Total
Downtown	$ 8,462	$ 8,153	$ 7,826	$ 7,538	$ 9,163	$ 9,826	$ 50,968
Suburb	7,592	7,235	6,836	7,105	8,472	8,138	45,378
Mall	7,385	7,104	6,482	6,242	8,462	8,936	44,611
Totals	$23,439	$22,492	$21,144	$20,885	$26,097	$26,900	$140,957

17. $5177
19. $3540
21. $276

CHAPTER 2: *Fractions*

Section 2.1

A. 1. $\frac{14}{3}$ 3. $\frac{7}{2}$ 5. $\frac{47}{6}$ 7. $\frac{67}{9}$ 9. $\frac{95}{11}$ 11. $\frac{61}{4}$ 13. $\frac{207}{16}$ 15. $\frac{307}{15}$

B. 1. $2\frac{4}{5}$ 3. $11\frac{1}{2}$ 5. $3\frac{3}{8}$ 7. $4\frac{11}{16}$ 9. $3\frac{5}{12}$ 11. $12\frac{3}{4}$ 13. $7\frac{5}{6}$ 15. $12\frac{5}{8}$

C. 1. $\frac{2}{3}$ 3. $\frac{4}{5}$ 5. $\frac{3}{4}$ 7. $\frac{4}{7}$ 9. $\frac{3}{8}$ 11. $\frac{5}{7}$ 13. $\frac{5}{16}$ 15. $\frac{7}{9}$

D. 1. $\frac{12}{30}$ 3. $\frac{24}{36}$ 5. $\frac{24}{54}$ 7. $\frac{28}{32}$ 9. $\frac{55}{80}$ 11. $\frac{21}{28}$ 13. $\frac{215}{40}$ 15. $\frac{128}{36}$

Section 2.2

A. 1. $\frac{2}{3}$ 3. $\frac{2}{7}$ 5. $\frac{2}{7}$ 7. $\frac{1}{6}$ 9. $\frac{2}{7}$ 11. $\frac{1}{8}$ 13. 9   15. $3\frac{6}{7}$
    17. $\frac{9}{10}$ 19. $2\frac{1}{4}$ 21. $3\frac{1}{2}$

B. 1. 2   3. $3\frac{1}{3}$ 5. $1\frac{1}{3}$ 7. 24   9. $5\frac{1}{3}$ 11. 15   13. $3\frac{3}{4}$ 15. 20
    17. $12\frac{1}{2}$ 19. 10   21. 12

C. 1. $90\frac{1}{2}$ 3. $8\frac{3}{4}$ 5. $126\frac{1}{2}$ 7. $306\frac{2}{3}$ 9. $25\frac{1}{8}$; $83\frac{3}{4}$; $76\frac{4}{7}$; $251\frac{1}{4}$ 11. 234
    13. (a) $19\frac{1}{2}$ (b) $117   15. (a) $367\frac{15}{16}$ (b) $1839\frac{11}{16}$

Section 2.3

A. 1. $\frac{3}{4}$ 3. $2\frac{1}{4}$ 5. $\frac{4}{5}$ 7. $\frac{3}{4}$ 9. $\frac{4}{7}$ 11. $\frac{15}{28}$ 13. $16\frac{2}{3}$ 15. $\frac{3}{4}$
    17. $1\frac{1}{3}$ 19. 12   21. $4\frac{11}{16}$

B. 1. $\frac{4}{5}$ 3. $1\frac{1}{4}$ 5. $\frac{8}{9}$ 7. $\frac{5}{14}$ 9. $2\frac{2}{5}$ 11. $\frac{8}{15}$ 13. $\frac{3}{10}$ 15. $2\frac{5}{8}$
    17. $1\frac{1}{5}$ 19. 8   21. 16

C. 1. 400   3. (a) 30   (b) 24   5. 6   7. $1066\frac{2}{3}$ 9. 22   11. 80   13. 40
    15. 400

Section 2.4

A. 1. $1\frac{1}{7}$ 3. $1\frac{1}{2}$ 5. $\frac{1}{2}$ 7. $1\frac{9}{16}$ 9. $\frac{5}{8}$ 11. $\frac{7}{8}$ 13. $\frac{1}{6}$
    15. 1   17. $1\frac{1}{12}$ 19. $\frac{4}{15}$ 21. $\frac{23}{30}$

B. 1. $1\frac{7}{12}$ 3. $\frac{31}{36}$ 5. $\frac{17}{24}$ 7. $1\frac{3}{4}$ 9. $\frac{3}{8}$ 11. $21\frac{11}{24}$ 13. $13\frac{11}{18}$
    15. $56\frac{9}{10}$ 17. $7\frac{1}{4}$ 19. $16\frac{14}{45}$ 21. $25\frac{23}{36}$

C. 1. $\frac{1}{8}$ 3. $16\frac{2}{3}$ 5. $524\frac{1}{4}$ 7. $56\frac{5}{24}$ 9. $326\frac{1}{12}$ 11. $204\frac{5}{16}$ 13. $3\frac{3}{8}$
    15. $449\frac{3}{4}$ 17. $4299\frac{3}{4}$ sq ft

Self-Test 2

1. $\frac{89}{12}$ 3. $5\frac{3}{7}$ 5. $\frac{4}{5}$ 7. $\frac{45}{80}$ 9. 50

11. $27\frac{13}{18}$ 13. $74\frac{13}{30}$ 15. $\frac{9}{10}$ 17. (a) $364\frac{5}{6}$ (b) $85\frac{1}{6}$

19. (a) $368\frac{3}{4}$ (b) $271\frac{7}{8}$ (c) $640\frac{5}{8}$ 21. $512\frac{3}{4}$ 23. 160

CHAPTER 3: *Decimal Numbers*

Section 3.1

A. 1. 1.6   3. 6   5. 16
    7. 3.5   9. 1.8   11. 4.5
    13. 0.8   15. 20.2   17. 25.3
    19. 17.5   21. 12.38   23. 3.14

B. 1. 27.5   3. $41.51   5. 21.14
    7. $17.88   9. 54.85   11. 4.743
    13. 17.836   15. 2.969   17. 2.255
    19. 14.254   21. 743.204   23. 27.623

C. 1. 12.8872   3. $23.11   5. 43.57247
    7. 5.9635   9. 28.4723   11. 22.4638
    13. 0.47462   15. 321.72823   17. 882.2374
    19. 23.8077   21. 122.1704   23. 90.831783

D. 1. $147,378.28   3. $2444.09
    5. $162.77; $225.44; $367.69
    7. $439.94   9. $157.28
    11.

Month	North Division	South Division	East Division	West Division	Monthly Totals
April	$ 9,635.32	$ 7,345.19	$ 8,536.92	$10,534.83	$ 36,052.26
May	10,002.84	8,293.99	7,991.83	8,374.38	34,663.04
June	8,928.47	10,475.29	9,010.10	16,732.89	45,146.75
Quarterly Totals	$28,566.63	$26,114.47	$25,538.85	$35,642.10	$115,862.05

    13. 2.515   15. 1112.5

Section 3.2

A. 1. 0.001   3. 37.3   5. 2.88
    7. 1.898   9. 0.13838   11. 0.07106
    13. 0.00048546   15. 0.012103   17. 0.20944
    19. 0.0668419   21. 0.00440336   23. 0.00312664
    25. 4.353656   27. 29.987328   29. 0.377454
    31. 177.2820

B. 1. 230   3. 6.4   5. 680
    7. 26.5   9. 2.34   11. 67.5
    13. 348   15. 684

C. 1. 1.67   3. 4.29   5. 43.33
    7. 13.85   9. 15.33   11. 0.192
    13. 25.172   15. 0.083   17. 113.889
    19. 431.707

D. 1. $704.28   3. $132,269.28   5. 58.5
    7. $2153.11   9. (a) $24,680.25   (b) $25,271.25
    11. (a) 756   (b) $748.44   13. $8962.63   15. $660.49
    17. (a) $519.40   (b) $364.56   (c) $883.96
    19. (a) $99.90   (b) $139.65   (c) $4.95   (d) total $244.50

Section 3.3

A. 1. 0.75   3. 0.4   5. 0.5   7. 0.3125   9. 0.875
    11. 0.32   13. 2.125   15. 2.2   17. 1.5625   19. 3.34375

B. 1. 0.667   3. 0.143   5. 0.583   7. 0.889   9. 0.867
    11. 0.706   13. 2.833   15. 4.667   17. 5.222   19. 7.091

C. 1. $\frac{7}{10}$ 3. $\frac{2}{5}$ 5. $\frac{1}{5}$ 7. $\frac{27}{100}$ 9. $\frac{9}{20}$
    11. $\frac{12}{25}$ 13. $\frac{14}{25}$ 15. $3\frac{6}{25}$ 17. $\frac{1}{8}$ 19. $1\frac{7}{8}$
    21. $\frac{1}{400}$ 23. $7\frac{11}{16}$

D. 1. $65.61   3. $660.92   5. $17.08   7. $0.29   9. $114.91
    11. $287.14   13. $835.13   15. $0.55   17. (a) $24.48   (b) 11.71 saved

Self-Test 3

    1. 86.0277   3. 35.1135   5. 10.856
    7. 7.8   9. 6.83   11. $\frac{17}{25}$
    13. 2.375   15. $115,572.07
    17. (a) $10.70   (b) $3.26   (c) $6.27   19. (a) $1028.27   (b) $1285.15
    21. $0.33   23. $89.52

CHAPTER 4: *Percent*

Section 4.1

A. 1. 60%   3. 45%   5. 5%   7. 70%   9. 140%
    11. 2.5%   13. 0.5%   15. 25.8%   17. 200%   19. 605%   21. 337.5%

B. 1. 25%   3. 87.5%   5. 35%   7. 65%   9. 312.5%
    11. 237.5%   13. 481.25%

C. 1. 0.09   3. 0.29   5. 0.065   7. 2.56   9. 0.0005
    11. 0.85   13. 6.05

D. 1. $\frac{1}{20}$ 3. $\frac{6}{25}$ 5. $\frac{3}{5}$ 7. $2\frac{1}{2}$ 9. $\frac{1}{40}$ 11. $\frac{7}{8}$ 13. $\frac{2}{3}$

E. 1. 75%   3. $\frac{17}{25}$ 5. 0.75   7. 68.75%   9. $1\frac{3}{10}$ 11. 0.453
   13. 180%   15. $\frac{3}{8}$ 17. 0.051   19. 1234.375%   21. $\frac{5}{6}$

Section 4.2

A. 1. 7   3. 60   5. 150   7. 120   9. 95%
   11. 24%   13. 30   15. 170%   17. 280   19. 5500

B. 1. 8   3. 65   5. $66\frac{2}{3}\%$ 7. $13   9. $7.50
   11. 225%   13. 78%   15. 8.75   17. $4.48   19. 1.664

C. 1. $22,950   3. (a) $1.38   (b) 15%   5. (a) $15.95   (b) $56.55
   7. (a) $0.68   (b) $9.18   9. (a) $6.28   (b) 5%   11. (a) $206.25   (b) $2081.25
   13. (a) $62   (b) 4%   15. (a) 15   (b) 6.25%   17. (a) $7840   (b) 32%
   19. (a) 117   (b) 26%   21. (a) $4104   (b) $11,096   23. (a) $32.97   (b) 14%
   25. 37.5% rounded

Self-Test 4

1. 87.5%   3. 7.5%   5. 1.85   7. $\frac{17}{20}$ 9. 91
11. 7200   13. 65%   15. (a) $1.05   (b) $18.55   17. (a) $1.14   (b) 15%
19. (a) $98   (b) $82.32   21. (a) 1890   (b) 35%   23. (a) $6.37   (b) 26%
25. $60.72   27. $17,120.18

CHAPTER 5:   *Bank Records*

Section 5.1

1.

DEPOSIT TICKET		CASH	CURRENCY	472	00	
ACCOUNT NO. _123-4567_			COIN	27	83	
NAME _Bits 'n' Bytes_		LIST CHECKS SINGLY		473	95	
ADDRESS _514 Fox Ave., OKC_				236	27	**39-81**
				12	55	**1030**
DATE _September 3_ 19 _98_		TOTAL FROM OTHER SIDE				USE OTHER SIDE FOR ADDITIONAL LISTING
		TOTAL		1222	60	
		LESS CASH RECEIVED				
SIGN HERE FOR CASH RECEIVED (IF REQUIRED)		▶ NET DEPOSIT		1222	60	BE SURE EACH ITEM IS PROPERLY ENDORSED
"DEPOSITS MAY NOT BE AVAILABLE FOR IMMEDIATE WITHDRAWAL."						

Friendly Bank

I-240 & SOUTH PENN • I-240 & SOUTH WALKER
PO BOX 19208 • OKLAHOMA CITY, OK 73144 • 405-681-5221
MEMBER FDIC

5/90 RPJ CHECKS AND OTHER ITEMS ARE RECEIVED FOR DEPOSIT SUBJECT TO THE PROVISIONS OF THE UNIFORM COMMERCIAL CODE OR ANY APPLICABLE COLLECTION AGREEMENT.

3.

152	$ *96.82*

September 12 19 *98*
TO *Maples Electronics*
FOR *Surge suppressors*

	DOLLARS	CENTS
BALANCE FORWARD	2763	14
DEPOSIT		
TOTAL	2763	14
THIS CHECK	96	82
OTHER DEDUCTIONS		
BALANCE FORWARD	2666	32

DELUXE WALLET

Bits 'n' Bytes
514 Fox Ave.
Oklahoma City, OK 73159

152

39-81/1030

September 12 19 *98*

PAY TO THE
ORDER OF   *Maples Electronics* $ *96.82*

Ninety-six and *82*/100 _____ DOLLARS

Friendly Bank
I-240 & SOUTH PENN • I-240 & SOUTH WALKER
PO BOX 19208 • OKLAHOMA CITY, OK 73144 • 405-681-5221
MEMBER FDIC

MEMO *Surge suppressors*   *Karen M. Paisley*

⑆ 103000813 ⑆ ⑈ 123 ⑈ 4567 ⑈   152

SAFETY PAPER

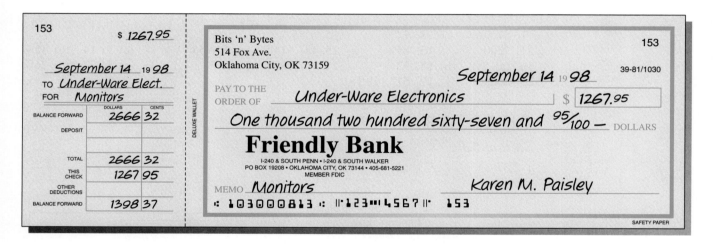

153	$ *1267.95*

September 14 19 *98*
TO *Under-Ware Elect.*
FOR *Monitors*

	DOLLARS	CENTS
BALANCE FORWARD	2666	32
DEPOSIT		
TOTAL	2666	32
THIS CHECK	1267	95
OTHER DEDUCTIONS		
BALANCE FORWARD	1398	37

DELUXE WALLET

Bits 'n' Bytes
514 Fox Ave.
Oklahoma City, OK 73159

153

39-81/1030

September 14 19 *98*

PAY TO THE
ORDER OF   *Under-Ware Electronics* $ *1267.95*

One thousand two hundred sixty-seven and *95*/100 — DOLLARS

Friendly Bank
I-240 & SOUTH PENN • I-240 & SOUTH WALKER
PO BOX 19208 • OKLAHOMA CITY, OK 73144 • 405-681-5221
MEMBER FDIC

MEMO *Monitors*   *Karen M. Paisley*

⑆ 103000813 ⑆ ⑈ 123 ⑈ 4567 ⑈   153

SAFETY PAPER

5.

RECORD ALL CHARGES OR CREDITS THAT AFFECT YOUR ACCOUNT

NUMBER	DATE	DESCRIPTION OF TRANSACTION	PAYMENT/DEBIT (–)		√ T	(IF ANY) (–) FEE	DEPOSIT/CREDIT (+)		BALANCE $ 2783 26	
836	11/2	Quang Van Tran Payroll	$ 1429	57		$	$		1353	69
837	11/5	P D Q Company Supplies	372	95					980	74
838	11/7	Sparkle Cleaning Cleaning services	85	50					895	24
	11/10	Deposit					2410	75	3305	99
839	11/10	Acme Widget Company Widgets	932	74					2373	25
840	11/15	Microsystems Computer equipment	1285	90					1087	35
841	11/16	DeTrop Surplus Misc.	207	28					880	07
842	11/19	Bargin Universe Office supplies	152	12					727	95
843	11/22	Instant Art Art work	78	32					649	63
844	11/25	Surety Insurance Insurance	42	50					607	13
845	11/25	Paper Cutters Computer Paper	217	21					389	92
	11/30	Deposit					562	07	951	99

7. (a) $110.41   (b) $53.27   (c) $968.61

Section 5.2

1.

RECORD ALL CHARGES OR CREDITS THAT AFFECT YOUR ACCOUNT

NUMBER	DATE	DESCRIPTION OF TRANSACTION	PAYMENT/DEBIT (−)		√ T	(IF ANY) (−) FEE	DEPOSIT/CREDIT (+)		BALANCE $
									952 06
512	2/6	Wreck-A-Mended Auto Truck repair	$ 517	89	√	$	$		434 17
513	2/6	Copy! Copy! Copy! Copy service	425	62	√				8 55
	2/7	Deposit			√		531	50	540 05
514	2/8	Midwest Labs Alignment	322	36	√				217 69
515	2/12	Aladdin Lock & Key Alarm system	208	95	√				8 74
	2/13	Deposit			√		1232	22	1240 96
516	2/15	Surety Insurance Insurance	57	91					1183 05
517	2/20	Maples Electronics Cabinets	128	85	√				1054 20
518	2/20	Bargin Universe Supplies	562	80					491 40
519	2/22	Microsystems Support service	419	74	√				71 66
520	2/24	Sparkle Cleaning Cleaning	62	50	√				9 16
	2/27	Deposit					1012	25	1021 41
		Service charge	6	75	√				1014 66

OUTSTANDING CHECKS WRITTEN TO WHOM	$ AMOUNT
# 516 Surety Insurance	57.91
# 518 Bargin Universe	562.80
TOTAL OF CHECKS OUTSTANDING $	620.71

BANK BALANCE (Last amount shown on this statement) $ 623.12

ADD + Deposits not shown on this statement (if any) $ 1012.25

TOTAL $ 1635.37

← SUBTRACT → $ 620.71

BALANCE $ 1014.66

3. Checkbook balance = $951.99 - 7.50 = $944.49

OUTSTANDING CHECKS WRITTEN TO WHOM	$ AMOUNT
# 841 De Trop Surplus	207.28
# 843 Instant Art	78.32
TOTAL OF CHECKS OUTSTANDING $	285.60

BANK BALANCE
(Last amount shown on this statement) $ 668.02

ADD +
Deposits not shown on this statement (if any) $ 562.07

TOTAL $ 1230.09

SUBTRACT $ 285.60

BALANCE $ 944.49

Self-Test 5

1.

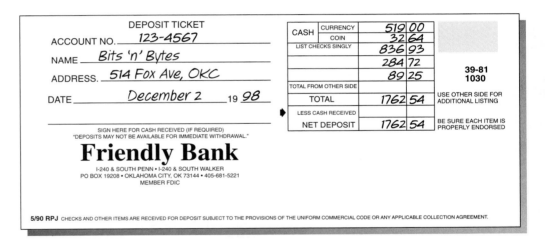

DEPOSIT TICKET

ACCOUNT NO. 123-4567

NAME Bits 'n' Bytes

ADDRESS. 514 Fox Ave, OKC

DATE December 2 19 98

SIGN HERE FOR CASH RECEIVED (IF REQUIRED)
"DEPOSITS MAY NOT BE AVAILABLE FOR IMMEDIATE WITHDRAWAL."

Friendly Bank

I-240 & SOUTH PENN • I-240 & SOUTH WALKER
PO BOX 19208 • OKLAHOMA CITY, OK 73144 • 405-681-5221
MEMBER FDIC

5/90 RPJ CHECKS AND OTHER ITEMS ARE RECEIVED FOR DEPOSIT SUBJECT TO THE PROVISIONS OF THE UNIFORM COMMERCIAL CODE OR ANY APPLICABLE COLLECTION AGREEMENT.

CASH	CURRENCY	519	00
	COIN	32	64
LIST CHECKS SINGLY		836	93
		284	72
		89	25
TOTAL FROM OTHER SIDE			
	TOTAL	1762	54
LESS CASH RECEIVED			
	NET DEPOSIT	1762	54

39-81
1030

USE OTHER SIDE FOR ADDITIONAL LISTING

BE SURE EACH ITEM IS PROPERLY ENDORSED

3.

	183	$ 265.74

December 16 19 98
TO Top Brand
FOR Paper

	DOLLARS	CENTS
BALANCE FORWARD	1236	24
DEPOSIT		
TOTAL	1236	24
THIS CHECK	265	74
OTHER DEDUCTIONS		
BALANCE FORWARD	970	50

DELUXE WALLET

Bits 'n' Bytes
514 Fox Ave.
Oklahoma City, OK 73159

183

December 16 19 98   39-81/1030

PAY TO THE ORDER OF Top Brand $ 265.74

Two hundred sixty-five and 74/100 DOLLARS

Friendly Bank

I-240 & SOUTH PENN • I-240 & SOUTH WALKER
PO BOX 19208 • OKLAHOMA CITY, OK 73144 • 405-681-5221
MEMBER FDIC

MEMO Paper   Karen M. Paisley

⑆ 103000813 ⑆ ⑈123⑈4567⑈ 183

SAFETY PAPER

5.

RECORD ALL CHARGES OR CREDITS THAT AFFECT YOUR ACCOUNT

NUMBER	DATE	DESCRIPTION OF TRANSACTION	PAYMENT/DEBIT (−)		√ T	(IF ANY) (−) FEE	DEPOSIT/CREDIT (+)		BALANCE 650 12	
592	2/2	Bargin Universe	$ 89	27	√	$	$		560	85
		Office supplies								
593	2/4	Under-Ware Electronics	42	95	√				517	90
		Connectors								
594	2/7	Paper Cutters	473	18	√				44	72
		Computer paper								
	2/8	Deposit			√		1650	25	1694	97
595	2/10	Spiffy Insurance Co.	119	55	√				1575	42
		Insurance								
596	2/10	Instant Art	238	21	√				1337	21
		Artwork								
597	2/19	J. R.'s Remodeling	251	79					1085	42
		Front office								
598	2/20	Microsystems	892	13	√				193	29
		Computer equipment								
599	2/25	Biff's Trash Service	18	57	√				174	72
		Trash pickup								
600	2/25	Kate's Delicacies	23	95					150	77
		Breakfast catering								
601	2/26	Sparkle Cleaning	119	85	√				30	92
		Cleaning								
	2/28	Deposit					532	24	563	16
		Service charge	5	75					557	41

OUTSTANDING CHECKS WRITTEN TO WHOM	$ AMOUNT
# 597 J.R.'s Remodeling	251.79
# 600 Kate's Delicacies	23.95
TOTAL OF CHECKS OUTSTANDING $	275.74

BANK BALANCE (Last amount shown on this statement) $ 300.91

ADD +
Deposits not shown on this statement (if any) $ 532.24

TOTAL $ 833.15

SUBTRACT $ 275.74

BALANCE $ 557.41

CHAPTER 6: *Payroll*

Section 6.1

A. 1. $1865   3. $2650   5. $272   7. $876   9. $3425   11. $895

B. 1. $2600; $1300; $1200; $600
    3. $1820; $910; $840; $420
    5. $2860; $1430; $1320; $660
    7. $3900; $1950; $1800; $900
    9. $3250; $1625; $1500; $750
    11. $4550; $2275; $2100; $1050
    13. $1430; $715; $660; $330
    15. $1950; $975; $900; $450

C. 1.

Name	Hours					Total Hours	Rate per Hour	Gross Pay
	M	T	W	T	F			
1. Bender, Keith	8	8	8	8	8	40	$ 8.12	$ 324.80
2. Chao, I-Na	8	7	6	5	4	30	7.55	226.50
3. Greenwood, Robert	6	8	5	8	7	34	6.38	216.92
4. Luliak, John	7	8	8	8	7	38	9.83	373.54
5. Pitt, David	8	6	0	8	6	28	10.27	287.56
6. Rhine, Alan	8	0	8	0	8	24	9.42	226.08
7. Tanner, Christie	8	7	7	7	8	37	6.83	252.71
8. Vasquez, Jose	8	6	5	6	0	25	7.98	199.50
Total								$2107.61

3.

Name	M	T	W	T	F	Total Hours	Regular Hours	Regular Rate	O.T. Hours	O.T. Rate	Regular Pay	O.T. Pay	Gross Pay
1. Cramer, Marvin	8	9	10	9	8	44	40	$9.27	4	$13.905	$ 370.80	$ 55.62	$ 426.42
2. Etheredge, Hugh	8	8	9	9	9	43	40	9.45	3	14.175	378.00	42.53	420.53
3. Lampron, George	9	8	7	6	5	35	35	8.52	0	—	298.20	0	298.20
4. McCabe, Lowell	10	10	10	10	9	49	40	8.99	9	13.485	359.60	121.37	480.97
5. Ngvyen, Tri	9	9	10	7	9	44	40	7.82	4	11.73	312.80	46.92	359.72
6. Reed, Harold	9	8	9	8	9	43	40	7.95	3	11.925	318.00	35.78	353.78
7. Tivis, Jerome	8	9	8	9	8	42	40	8.25	2	12.375	330.00	24.75	354.75
8. Wood, Regena	9	8	7	9	10	43	40	9.77	3	14.655	390.80	43.97	434.77
Totals											$2758.20	$370.94	$3129.14

Section 6.2

A. 1. 191; $429.75   3. 180; $383.40
    5. 174; $344.52   7. 207; $478.17

B. 1. 258; $503.10   3. 252; $491.40
    5. 306; $596.70   7. 258; $503.10

C. 1. $2295.65   3. $4756.50
    5. $2982.70   7. $3342.50

D. 1. $289.04   3. $235.37
    5. $254.60   7. $332.10

E. 1. $185.17; $365.17   3. $128.63; $258.63
    5. $172.46; $352.46   7. $139.55; $284.55

F. 1. $263.94

Self-Test 6

1.

Name	Hours					Total Hours	Regular Hours	Regular Rate	O.T. Hours	O.T. Rate	Regular Pay	O.T. Pay	Gross Pay
	M	T	W	T	F								
1. Burns, Tracy	8	9	9	9	9	44	40	$7.26	4	$10.89	$ 290.40	$ 43.56	$ 333.96
2. Grower, Charles	9	8	9	8	8	42	40	7.65	2	11.475	306.00	22.95	328.95
3. Jeter, Sandra	9	8	9	8	7	41	40	8.17	1	12.255	326.80	12.26	339.06
4. Lee, Russell	9	8	9	7	4	37	37	7.92	0	—	293.04	0	293.04
5. Mandrell, Carolyn	9	9	9	9	10	46	40	7.75	6	11.625	310.00	69.75	379.75
6. Nguyen, Emmanuelle	9	8	9	7	9	42	40	8.09	2	12.135	323.60	24.27	347.87
7. Palmer, Johnny	9	9	9	9	9	45	40	7.93	5	11.895	317.20	59.48	376.68
8. Roberts, Linda	10	9	8	7	9	43	40	8.11	3	12.165	324.40	36.50	360.90
Totals											$2491.44	$268.77	$2760.21

3. (a) $2925; (b) $1462.50; (c) $1350; (d) $675
5. 253; $493.35
7. 242; $447.70
9. 219; $324.12
11. 231; $358.05
13. $3404.25
15. $5692.50
17. $311.38
19. $239.04
21. $93.80; $298.80
23. $80.70; $275.70

CHAPTER 7: *The Mathematics of Buying*

Section 7.1

A. 1. $54; $396
   5. $66.24; $669.71

   3. $96.25; $1278.75
   7. $44.44; $801.95

B. 1. $403.56
   5. $398.50

   3. $2082.92
   7. $315.58

C. 1. 19.04%; $514.10
   5. 38.664%; $236.53

   3. 43.584%; $411.27
   7. 47.2%; $2059.49

D. 1. (a) $18.81 (b) $56.44
   5. $437.08
   9. (a) 52.75% (b) 53.6875%
   13. (a) 52.75% (b) 52.75% (c) yes

   3. $425.85
   7. $598.60
   11. (a) 38.8% (b) 39.28%
   15. $62.98

Section 7.2

A. 1. $22.50; $727.50
   5. $0; $632
   9. $14.16; $457.84
   13. $56.30; $2758.70

   3. $1.90; $93.10
   7. $8.79; $284.21
   11. $14.16; $693.84
   15. $7.32; $358.68

B. 1. $17.50; $857.50
   5. $29.36; $704.64

   3. $2.60; $127.40
   7. $0; $238

C. 1. $357.14; $492.86
   5. $120; $110

   3. $206.19; $263.81
   7. $515.46; $356.54

D. 1. $73.50
   5. $127.16
   9. $93.59
   13. (a) $510.20 (b) $376.80
   17. $443.78

   3. $1247.54
   7. $545.86
   11. $668.33
   15. (a) $257.73 (b) $214.27

Self-Test 7

   1. (a) $63.25 (b) $211.75
   5. $394.88
   9. (a) $111.11 (b) 35.4%
   13. $455.90
   17. $86.57
   21. $269.50
   25. (a) $257.73 (b) $337.27

   3. (a) $39.95 (b) $195.05
   7. $520.97
   11. (a) 35.875% (b) 35.328%
   15. $80.85
   19. $34.68
   23. $190.12

CHAPTER 8: *The Mathematics of Selling*

Section 8.1

A. 1.(a) $64.19 (b) $522.69
   5.(a) $495.60 (b) $718.62
   9.(a) $3388.83 (b) 42%
   13.(a) $632.75 (b) $303.72

   3. (a) $167.31 (b) 32%
   7. (a) $846.50 (b) $101.58
   11. (a) $283.50 (b) $459.27
   15. (a) $70.06 (b) 24%

B. 1. (a) $98.91   (b) 26%   3. (a) $335.40   (b) $452.79
    5. (a) $101.39   (b) $463.51   7. (a) $2437.50   (b) $292.50
    9. (a) $36.50   (b) $8.76   11. (a) $219.80   (b) $263.76
   13. (a) $75.44   (b) $311.19   15. (a) $98.75   (b) $39.50

Section 8.2

A. 1. (a) $176.49   (b) $62.01   3. (a) $1678.58   (b) 32%
    5. (a) $672.76   (b) $764.50   7. (a) $178.11   (b) $395.80
    9. (a) $164.32   (b) $92.43   11. (a) $220.17   (b) $268.50
   13. (a) $529.41   (b) 40%   15. (a) $192.14   (b) $295.60

B. 1. (a) $329.60   (b) 20%   3. (a) $20.37   (b) $48.50
    5. (a) $414.90   (b) $3457.50   7. (a) $137.87   (b) $24.33
    9. (a) $159.51   (b) $85.89   11. (a) $198.69   (b) $268.50
   13. (a) $5.07   (b) $19.50   15. (a) $117.18   (b) $68.82
   17. (a) $23.93   (b) 59.5% rounded

Section 8.3

A. 1. (a) $7.67   (b) $21.83   3. (a) 24%   (b) $62.89
    5. (a) $241.71   (b) $621.54   7. (a) 5%   (b) $499.32
    9. (a) 14%   (b) $13.79   11. (a) $572.40   (b) 15%

B. 1. (a) $12.54   (b) $63.46   3. (a) 25%   (b) $126.30
    5. (a) $44.70   (b) $327.80   7. (a) 13%   (b) $3697.50
    9. (a) $33.25   (b) $441.75   11. $49.98
   13. $228.90   15. $397.80

Self-Test 8

  1. (a) $525.45   (b) 24%   3. (a) $2325   (b) $3022.50
  5. (a) $295.40   (b) $822.90   7. (a) $732   (b) $212.28
  9. (a) $19.75   (b) 36%   11. (a) $678.42   (b) $942.25
 13. (a) $1597.50   (b) $3550   15. (a) $351.13   (b) $129.87
 17. (a) $7.15   (b) $20.35   19. (a) 12%   (b) $418
 21. $189

C H A P T E R 9 : *Simple Interest*

Section 9.1

A. 1. 134   3. 76
    5. 124   7. 275
    9. 131   11. 104

B. 1. 68   3. 72
    5. 69   7. 102
    9. 63   11. 313

C. 1. (a) 168   (b) 165   3. (a) 247   (b) 243
    5. (a) 72   (b) 72   7. (a) 94   (b) 94
    9. (a) 109   (b) 108   11. (a) 169   (b) 168

Section 9.2

A. 1. 195; $548.14; $6248.14   3. 95; $43.27; $918.27
    5. 146; $128; $1728   7. 128; $207.04; $3907.04
    9. 186; $247.75; $2747.75

B. 1. 93; $186; $4686   3. 110; $115.50; $1915.50
    5. 146; $570.62; $7270.62   7. 145; $155.07; $1905.07
    9. 156; $302.03; $3702.03

C. 1. 190; $180.50; $1980.50   3. 102; $163.20; $3043.20
    5. 132; $299.20; $5099.20   7. 137; $433.45; $7133.45
    9. 138; $589.38; $8089.38

D. 1. (a) $173.08 (b) $171.79 (c) $175.49   3. (a) $548.42 (b) $552.50 (c) $556.04
    5. (a) $485.90 (b) $494 (c) $494   7. (a) $171.34 (b) $170.78 (c) $173.72
    9. (a) $338.49 (b) $340.55 (c) $344.13   11. $5540.00

Section 9.3

A. 1. $7200   3. 17%
    5. 30   7. 18%
    9. 60   11. $3333.33
   13. 18%   15. $4500
   17. 144   19. 20%
   21. 72   23. $6000

B. 1. $7500   3. 17%
    5. 80   7. $4500
    9. $5000   11. 20%
   13. 146   15. $7008

Self-Test 9

 1. (a) 189 (b) 186   3. (a) 82 (b) 81
 5. (a) $275.82 (b) $274.95 (c) $279.65   7. (a) $382.36 (b) $383.44 (c) $388.73
 9. $2900   11. 16%
13. 40   15. $7900
17. 19%   19. 146

CHAPTER 10: *Bank Discount Loans*

Section 10.1

A. 1. $60; $2940   3. $56.25; $2443.75
    5. $13.54; $1236.46   7. $23.25; $1526.75
    9. $162.56; $3837.44

B. 1. $108.75; $4391.25   3. $49.87; $3150.13
    5. $22.05; $1827.95   7. $18.49; $1581.51
    9. $141.87; $4058.13

C. 1. (a) $66.50 (b) $2933.50   3. (a) $31.33 (b) $1468.67
    5. (a) $121.39 (b) $4878.61   7. (a) $40.80 (b) $1159.20
    9. (a) $138.13 (b) $6361.87

Section 10.2

A. 1. $4150; $83; $4067   3. $3296; $75.53; $3220.47
    5. $3843.75; $48.05; $3795.70   7. $1896.25; $17.91; $1878.34
    9. $4482.75; $191.64; $4291.11

B. 1. $2600; $58.64; $2541.36   3. $5125; $65.34; $5059.66
    5. $3609.38; $69.78; $3539.60   7. $3540.56; $26.55; $3514.01
    9. $3990; $111.72; $3878.28

C. 1. (a) $1248 (b) $38.83 (c) $1209.17   3. (a) $2776.50 (b) $55.53 (c) $2720.97
    5. (a) $2496 (b) $96.10 (c) $2399.90   7. (a) $2085 (b) $66.03 (c) $2018.97
    9. (a) $1947.50 (b) $38.95 (c) $1908.55 11. (a) $166.67 (b) $7333.33

Self-Test 10

 1. (a) $29.79 (b) $1070.21   3. (a) $63.19   (b) $3436.81
 5. (a) $45.60 (b) $1554.40   7. (a) $183.75 (b) $7316.25
 9. (a) $3352 (b) $107.26 (c) $3244.74   11. (a) $1642.67 (b) $38.33 (c) $1604.34
13. (a) $3848 (b) $157.77 (c) $3690.23

CHAPTER 11: *More Complex Loans*

Section 11.1

 1. (a) $7328   (b) $828   (c) $6000   (d) 13.2%   (e) 12.75%
 3. (a) $2632   (b) $132   (c) $1800   (d) 13.5%   (e) 13.25%
 5. (a) $1331.70 (b) $131.70 (c) $1050   (d) 15.8%   (e) 15.25%
 7. (a) $2701   (b) $251   (c) $1900   (d) 16.7%   (e) 16.00%
 9. (a) $4388   (b) $388   (c) $2600   (d) 14.3%   (e) 13.75%
11. (a) $22,708   (b) $5208   (c) $14,700   (d) 17.4%   (e) 15.75%

Section 11.2

 1. $27.60
 3. $10.15
 5. $16.17
 7. $22.02
 9. $58.20
11. $403

Section 11.3

1. (a) $792.87   (b) $11.89   (c) $1000.89
3. (a) $790.57   (b) $11.86   (c) $850.86
5. (a) $1028.77   (b) $12.86   (c) $1056.86
7. (a) $33,366.60   (b) $8366.60   (c) 12%   (d) $480.05

Self-Test 11

1. (a) $1266   (b) $66   (c) $900   (d) 13.5%   (e) 13.25%
3. (a) $18,838.80   (b) $4338.80   (c) $12,250   (d) 17.3%   (e) 15.75%
5. $12.69
7. $387.39
9. (a) $883.13   (b) $13.25   (c) $963.25

CHAPTER 12: *Compound Interest and Present Value*

Section 12.1

A. 1. $3714.87 3. $3675.62 5. $1307.93
    7. $5017.67 9. $2794.23

B. 1. $2810.28 3. $3931.04 5. $3840.75
    7. $823.03 9. $4181.85

C. 1. 8.160% 3. 8.328% 5. 12.360%
    7. 12.683% 9. 5.654%

D. 1. $4160.65 3. $5086.75 5. $4176.92
    7. $9960.23 9. 6.715% 11. 10.250%

Section 12.2

A. 1. $1748.86 3. $2146.12 5. $461.94
    7. $1267.92 9. $2199.33 11. $1853.43
    13. $2148.64 15. $485.04

B. 1. $2965.49 3. $1554.30 5. $1774.90
    7. $2224.12 9. $2389.77 11. $5189.61
    13. $1928.79 15. $7486.29 17. $11,982.78

Self-Test 12

1. $4713.99 3. $5314.34 5. $3102.66
7. $7420.17 9. $8303.94 11. 7.250%
13. 8.160% 15. 12.683% 17. $4710.80
19. $2168.59

CHAPTER 13: *Investments*

Section 13.1

A. 1. $2172.52 3. $3104.23 5. $2781.78 7. $8021.25
    9. $16,605.89

B. 1. $1907.40 3. $2319.03 5. $14,698.97 7. $16,524.71
    9. $5964.57

C. 1. $3856.54 3. $7203.66 5. $555.10 7. $3755.28
    9. $5516.84 11. $4295.03 13. (a) $9314.94 (b) $9177.28 rounded
    (c) annuity due

Section 13.2

A. 1. $78.64 3. $678.43 5. $407.78 7. $808.29
    9. $181.46 11. $215.03 13. $1130.71 15. $106.83

B. 1. $146.93 3. $880.39 5. $2332.15 7. $2665.31
    9. $1019.00 11. $2146.62 13. $563.70 15. $444.88

Section 13.3

1. $675	3. $1054	5. $119.60	7. $231
9. 14	11. 12	13. $87.50; $43.75	15. $93.75; $46.88
17. 9.02%	19. 9.10%	21. 1500	

Self-Test 13

1. $6195.83	3. $3973.77	5. $3309.21	7. $7523.70
9. $793.77	11. $2498.08	13. $444.88	15. $117
17. $87.50	19. 35	21. (a) $76.25 (b) $38.13	23. 15.27%

CHAPTER 14: *Real Estate Mathematics*

Section 14.1

1. $1447.53
3. $2254.26
5. (a) $761.86   (b) $822.89
7. (a) $999.94   (b) $254,978.40
9. (a) $2108.84   (b) $326,121.60
11. (a) 4   (b) $3400
13. (a) 2   (b) $1600
15. (a) 4   (b) $3120
17. (a) 2   (b) $1840
19. (a) 4   (b) $2920
21. (a) $74,800   (b) $656.42   (c) $161,511.20   (d) $255,011.20   (e) 4 points; $2992

Section 14.2

	Payment No.	Principal	P&I Payment	Interest Payment	Principal Payment
1.	1	$175,000.00	$1715.20	$1604.17	$111.03
	2	174,888.97	1715.20	1603.15	112.05
3.	1	195,000.00	2467.22	2112.50	354.72
	2	194,645.28	2467.22	2108.66	358.56
5.	1	135,000.00	1285.64	1237.50	48.14
	2	134,951.86	1285.64	1237.06	48.58
7.	1	73,000.00	750.89	730.00	20.89
	2	72,979.11	750.89	729.79	21.10
	3	72,958.01	750.89	729.58	21.31
9.	1	185,000.00	2086.50	2004.17	82.33
	2	184,917.67	2086.50	2003.27	83.23
	3	184,834.44	2086.50	2002.37	84.13

11. (a) $590.69   (b) $38.27   (c) $74,574.84   (d) $590.38   (e) $38.58   (f) $74,536.26

Self-Test 14

1. $1321.30
3. (a) $895.12   (b) $971.37
5. (a) $1190.40   (b) $303,544.00
7. (a) $2477.44   (b) $369,585.60
9. (a) 6   (b) $4680.00
11. (a) 2   (b) $1780.00

	Payment No.	Principal	P&I Payment	Interest Payment	Principal Payment
13.	1	$280,000.00	$2949.03	$2800.00	$149.03
	2	279,850.97	2949.03	2798.51	158.52
15.	1	155,000.00	1476.10	1420.83	55.27
	2	154,944.73	1476.10	1420.33	55.77
	3	154,888.96	1476.10	1419.82	56.28

CHAPTER 15: *Inventory and Overhead*

Section 15.1

A. 1. $11,444.90
   3. $985.51

B. 1. (a) $11,518.45   (b) $11,439.20   (c) $11,620.80
   3. (a) $986.56   (b) $988.77   (c) $971.80
   5. (a) $558.72   (b) $577.30   (c) $571.55

Section 15.2

A. 1. $11,518.45
   3. $986.56

B. 1. $682.50; $429.00; $341.25; $497.25
   3. $3937.50; $2750.00; $3687.50; $1125.00; $1000.00
   5. $741.00; $507.00; $302.25; $399.75
   7. $4375.00; $2375.00; $3250.00; $1375.00; $1125.00
   9. $1093.15; $875.91; $730.94

Self-Test 15

1. $4762.50
3. (a) $4745.26   (b) $4774.60   (c) $4705.80   (d) $4745.26
5. $1190.00; $1632.00; $2584.00; $1394.00
7. $1292.00; $1496.00; $2448.00; $1564.00

CHAPTER 16: *Depreciation*

Section 16.1

A.

Year	Annual Depreciation	Accumulated Depreciation	Book Value
1:			$6000.00
1	$1375.00	$1375.00	4625.00
2	1375.00	2750.00	3250.00
3	1375.00	4125.00	1875.00
4	1375.00	5500.00	500.00
3.			$5000.00
1	$840.00	$ 840.00	4160.00
2	840.00	1680.00	3320.00
3	840.00	2520.00	2480.00
4	840.00	3360.00	1640.00
5	840.00	4200.00	800.00
5.			$9000.00
1	$1400.00	$1400.00	7600.00
2	1400.00	2800.00	6200.00
3	1400.00	4200.00	4800.00
4	1400.00	5600.00	3400.00
5	1400.00	7000.00	2000.00
6	1400.00	8400.00	600.00

B.

Year	Units	Annual Depreciation	Accumulated Depreciation	Book Value
1.				$6000.00
1	15,800	$1106.00	$1106.00	4894.00
2	23,200	1624.00	2730.00	3270.00
3	24,600	1722.00	4452.00	1548.00
4	16,400	1148.00	5600.00	400.00
3.				$19,500.00
1	15,285	$2751.30	$2,751.30	16,748.70
2	23,162	4169.16	6,920.46	12,579.54
3	21,475	3865.50	10,785.96	8,714.04
4	22,317	4017.06	14,803.02	4,696.98
5	17,761	3196.98	18,000.00	1,500.00
5.				$15,400.00
1	7,250	$2059.00	$ 2,059.00	13,341.00
2	14,725	4181.90	6,240.90	9,159.10
3	14,420	4095.28	10,336.18	5,063.82
4	13,605	3863.82	14,200.00	1,200.00

Section 16.2

A.

Year	Annual Depreciation	Accumulated Depreciation	Book Value
1.			$6000.00
1	$3000.00	$3000.00	3000.00
2	1500.00	4500.00	1500.00
3	750.00	5250.00	750.00
4	250.00	5500.00	500.00
3.			$5000.00
1	$2000.00	$2000.00	3000.00
2	1200.00	3200.00	1800.00
3	720.00	3920.00	1080.00
4	280.00	4200.00	800.00
5	0	4200.00	800.00
5.			$9000.00
1	$3000.00	$3000.00	6000.00
2	2000.00	5000.00	4000.00
3	1333.33	6333.33	2666.67
4	888.89	7222.22	1777.78
5	592.59	7814.81	1185.19
6	395.06	8209.87	790.13

B.

Year	Annual Depreciation	Accumulated Depreciation	Book Value
1.			$6000.00
1	$2240.00	$2240.00	3760.00
2	1680.00	3920.00	2080.00
3	1120.00	5040.00	960.00
4	560.00	5600.00	400.00
3.			$19,500.00
1	$6000.00	$ 6,000.00	13,500.00
2	4800.00	10,800.00	8,700.00
3	3600.00	14,400.00	5,100.00
4	2400.00	16,800.00	2,700.00
5	1200.00	18,000.00	1,500.00
5.			$15,400.00
1	$5680.00	$ 5,680.00	9,720.00
2	4260.00	9,940.00	5,460.00
3	2840.00	12,780.00	2,620.00
4	1420.00	14,200.00	1,200.00

7(a). Double-declining balance method

Year	Annual Depreciation	Accumulated Depreciation	Book Value
			$770.00
1	$192.50	$192.50	577.50
2	144.38	336.88	433.12
3	108.28	445.16	324.84
4	81.21	526.37	243.63
5	60.91	587.28	182.72
6	32.72 to Salvage value	620.00	150.00

7(b). Sum-of-the-years'-digits method

Year	Annual Depreciation	Accumulated Depreciation	Book Value
			$770.00
1	$137.78	$137.78	632.22
2	120.56	258.34	511.66
3	103.33	361.67	408.33
4	86.11	447.78	322.22
5	68.89	516.67	253.33
6	51.67	568.34	201.66
7	34.44	602.78	167.22
8	17.22	620.00	150.00

(c) method (a); (d) method (a)

Section 16.3

	Year	Annual Depreciation	Accumulated Depreciation	Book Value
1(a).				$17,500.00
	1	$4375.00	$ 4,375.00	13,125.00
	2	6650.00	11,025.00	6,475.00
	3	6475.00	17,500.00	0
1(b).				$17,500.00
	1	$3500.00	$ 3,500.00	14,000.00
	2	5600.00	9,100.00	8,400.00
	3	3360.00	12,460.00	5,040.00
	4	2016.00	14,476.00	3,024.00
	5	2016.00	16,492.00	1,008.00
	6	1008.00	17,500.00	0
3(a).				$4200.00
	1	$630.00	$ 630.00	3570.00
	2	924.00	1554.00	2646.00
	3	882.00	2436.00	1764.00
	4	882.00	3318.00	882.00
	5	882.00	4200.00	0

Year	Annual Depreciation	Accumulated Depreciation	Book Value
3(b)			$4200.00
1	$ 600.18	$ 600.18	3599.82
2	1028.58	1628.76	2571.24
3	734.58	2363.34	1836.66
4	524.58	2887.92	1312.08
5	375.06	3262.98	937.02
6	374.64	3637.62	562.38
7	375.06	4012.68	187.32
8	187.32	4200.00	0

5. (a) $2,331.250   (b) $492,279.41   (c) $1,831,933.50
   (d) $3,385,500   (e) $327,450

Self-Test 16

Year	Annual Depreciation	Accumulated Depreciation	Book Value
1(a).			$8000.00
1	$1500.00	$1500.00	6500.00
2	1500.00	3000.00	5000.00
3	1500.00	4500.00	3500.00
4	1500.00	6000.00	2000.00
5	1500.00	7500.00	500.00
1(b).			$8000.00
1	$3200.00	$3200.00	4800.00
2	1920.00	5120.00	2880.00
3	1152.00	6272.00	1728.00
4	691.20	6963.20	1036.80
5	414.72	7377.92	622.08
1(c).			$8000.00
1	$2500.00	$2500.00	5500.00
2	2000.00	4500.00	3500.00
3	1500.00	6000.00	2000.00
4	1000.00	7000.00	1000.00
5	500.00	7500.00	500.00

Year	Miles	Annual Depreciation	Accumulated Depreciation	Book Value
3.				$18,000.00
1	15,780	$2603.70	$ 2,603.70	15,396.30
2	32,910	5430.15	8,033.85	9,966.15
3	29,520	4870.80	12,904.65	5,095.35
4	21,790	3595.35	16,500.00	1,500.00

Year	Annual Depreciation	Accumulated Depreciation	Book Value
5(a).			$18,950.00
1	$4737.50	$ 4,737.50	14,212.50
2	7201.00	11,938.50	7,011.50
3	7011.50	18,950.00	0
5(b).			$18,950.00
1	$3790.00	$ 3,790.00	15,160.00
2	6064.00	9,854.00	9,096.00
3	3638.40	13,492.40	5,457.60
4	2183.04	15,675.44	3,274.56
5	2183.04	17,858.48	1,091.52
6	1091.52	18,950.00	0

C H A P T E R 1 7 : *Taxes*

Section 17.1

A. 1. $34, $30.50, $7.13, $113.83, $378.17
    3. $58, $30.07, $7.03, $137.05, $347.95
    5. $46, $32.36, $7.57, $129.08, $392.92
    7. $69, $30.81, $7.21, $149.47, $347.53
    9. $73, $34.35, $8.03, $159.58, $394.42

B. 1. $228.00, $90.21, $21.10, $431.31, $1023.69
    3. $139.87, $93.74, $21.92, $353.53, $1158.47
    5. $240.93, $86.61, $20.26, $440.80, $956.20
    7. $104.57, $92.07, $21.53, $312.17, $1172.83
    9. $157.45, $94.55, $22.11, $376.11, $1148.89

C. 1. (a) $407.55   (b) $240.25   (c) $56.19
    3. (a) $163.70   (b) $110.05   (c) $25.74
    5. (a) $193.58   (b) $113.15   (c) $26.46
    7. (a) $198.56   (b) $62.31   (c) $14.57
    9. (a) $38.00   (b) $29.26   (c) $6.84

Section 17.2

 1. $415.40
 3. $434
 5. $434
 7. $434
 9. (a) $54.40   (b) $367.20
11. (a) $56   (b) $378
13. (a) $56   (b) $378
15. FIT $0
    FICA-SS $ 20.85
    FICA-MC $ 4.88
    STATE   12.78
    OTHER   19.63
    NET $278.12
    FUTA $ 34.97
    SUTA $236.05

Section 17.3

A. 1. $3.00; $52.95
    3. $7.88; $165.47
    5. $0.97; $18.64
    7. $17.01; $341.00
    9. $13.43; $212.43

B. 1. $68,800; $1609.92
    3. $41,250; $394.35
    5. $89,250; $3320.10
    7. $48,640; $1361.92
    9. $51,856; $1218.62

C. 1. (a) $8.22   (b) $145.22
    3. (a) $10.95   (b) $209.95
    5. $864
    7. $1067.25
    9. (a) $65,250   (b) $1428.98
   11. (a) $74,750   (b) $2018.25
   13. $2290.85

Self-Test 17

 1. (a) $24   (b) $26.23   (c) $6.13
 3. (a) $48   (b) $25.73   (c) $6.02
 5. (a) $463.95   (b) $249.55   (c) $58.36
 7. (a) $282.60   (b) $102.30   (c) $23.93
 9. $393.70
11. $434
13. (a) $47.16   (b) $318.33
15. (a) $9.12   (b) $161.12
17. $38
19. (a) $52,250   (b) $1227.88
21. (a) $41,370   (b) $1075.62

CHAPTER 18: *Insurance*

Section 18.1

A. 1. $514.50; $951.83; $1389.15
    3. $437; $808.45; $1179.90
    5. $403.20; $745.92; $1088.64
    7. $379.75; $702.54; $1025.33

B. 1. $142.56
    3. $120.96
    5. $172.26
    7. $224.78

C. 1. 98; $100.68; $274.32
    3. 66; $108.75; $266.25
    5. 126; $227.25; $277.75
    7. 230; $184.63; $108.37

D.   1. $31,250; $15,625; $28.125
     3. $25,000; $20,000; $25,000
     5. $40,000; $48,000; $56,000
     7. $30,000; $25,000; $35,000

E.   1. $20,833.33
     3. $27,750
     5. $14,112.90
     7. $27,500

F.   1. (a) $337.50   (b) $624.38   (c) $911.25
     3. (a) $667   (b) $1233.95   (c) $1800.90
     5. $89.78
     7. $180.60
     9. (a) $204.25   (b) $270.75
    11. (a) $122.75   (b) $312.25
    13. $20,625; $12,375
    15. $20,000; $15,000; $25,000
    17. $17,219.39
    19. $50,400
    21. $100,322.36

Section 18.2

A.   1. $612
     3. $701
     5. $486
     7. $640

B.   1. (a) $43,000   (b) $1500
     3. (a) $64,000   (b) $10,200
     5. (a) $250   (b) $2575
     7. (a) $70,000   (b) $12,000
     9. $630

Section 18.3

A.   1. $402.25
     3. $258
     5. $623.55
     7. $729.25
     9. $548.80

B.   1. $440; $980; 3y 159d
     3. $3225; $7275; 17y 118d
     5. $3015; $6180; 21y 47d
     7. $3280; $7920; 12y 341d
     9. $315; $735; 2y 232d

C.   1. $340.40
     3. $648.45
     5. $387
     7. $299.28
     9. $2580; $5820; 17y 118d
    11. $550; $1225; 3y 159d

Self-Test 18

1. (a) $402.50 (b) $744.63 (c) $1086.75
3. $101.79
5. (a) $103.00 (b) $379.00
7. (a) $10,000 (b) $12,000 (c) $14,000
9. $30,570.65
11. $701
13. (a) $25,000 (b) $3300
15. (a) $250 (b) $3565
17. $478.40
19. $3225; $7275; 17y 118d

CHAPTER 19: *Financial Statement Analysis*

Section 19.1

A. 1.

REDHAT TOOL AND DIE
COMPARATIVE BALANCE SHEETS
December 31, 1996 and 1997

	1996	1997	Increase or Decrease*	
			Amount	Percent
Assets				
Current Assets				
Cash	$ 33,000	$ 37,000	$ 4,000	12.1%
Merchandise inventory	82,000	87,000	5,000	6.1%
Accounts receivable	79,000	78,000	1,000*	1.3%*
Total current assets	$194,000	$202,000	8,000	4.1%
Plant and Equipment				
Land	$ 32,000	$ 32,000	0	0%
Buildings	77,000	77,000	0	0%
Equipment	91,000	101,000	10,000	11.0%
Total plant and equipment	$200,000	$210,000	10,000	5.0%
Total assets	$394,000	$412,000	18,000	4.6%
Liabilities and Owners' Equity				
Current Liabilities				
Accounts payable	$ 74,000	$ 78,000	4,000	5.4%
Wages payable	31,000	35,000	4,000	12.9%
Total current liabilities	$105,000	$113,000	8,000	7.6%
Long-Term Liabilities				
Mortgage payable	83,000	80,000	3,000*	3.6%*
Total liabilities	$188,000	$193,000	5,000	2.7%
Owners Equity				
Total owners' equity	206,000	219,000	13,000	6.3%
Total liabilities and owners' equity	$394,000	$412,000	18,000	4.6%

*Denotes a decrease

B.   1.

REDHAT TOOL AND DIE
COMPARATIVE BALANCE SHEETS
December 31, 1996 and 1997

	1996		1997	
	Amount	**Percent**	**Amount**	**Percent**
Assets				
Current Assets				
Cash	$ 33,000	8.4%	$ 37,000	9.0%
Merchandise inventory	82,000	20.8%	87,000	21.1%
Accounts receivable	79,000	20.0%	78,000	18.9%
Total current assets	$194,000	49.2%	$202,000	49.0%
Plant and Equipment				
Land	$ 32,000	8.1%	$ 32,000	7.8%
Buildings	77,000	19.5%	77,000	18.7%
Equipment	91,000	23.1%	101,000	24.5%
Total plant and equipment	$200,000	50.8%	$210,000	51.0%
Total assets	$394,000	100.0%	$412,000	100.0%
Liabilities and Owners' Equity				
Current Liabilities				
Accounts payable	$ 74,000	18.8%	$ 78,000	18.9%
Wages payable	31,000	7.9%	35,000	8.5%
Total current liabilities	$105,000	26.6%	$113,000	27.4%
Long-Term Liabilities				
Mortgage payable	83,000	21.1%	80,000	19.4%
Total liabilities	$188,000	47.7%	$193,000	46.8%
Owners' Equity				
Total owners' equity	206,000	52.3%	219,000	53.2%
Total liabilities and owners' equity	$394,000	100.0%	$412,000	100.0%

3. $10,000   5.1% $1,500   0.8%
        5,000   2.5%   250   0.1%
        2,500   1.3% $1,750   0.9%
      $17,500   8.9%
                                                     $65,000   32.9%
      $ 30,000   15.2% $66,750   33.8%
       100,000   50.6%
        50,000   25.3%   130,750   66.2%
      $180,000   91.1% $197,500   100.0%
      $197,500   100.0%

Section 19.2

A. 1.

BOB'S BEEFLIKE BURGERS
COMPARATIVE INCOME STATEMENTS
For the Years Ended December 31, 1996 and 1997

			Increase or Decrease*	
	1996	**1997**	**Amount**	**Percent**
Revenues				
Sales	$102,000	$113,000	$11,000	10.8%
Less: Sales returns	3,500	4,000	500	14.3%
Net sales	$ 98,500	$109,000	10,500	10.7%
Cost of Goods Sold				
Inventory, January 1	$ 17,000	$ 19,000	2,000	11.8%
Purchases	72,000	78,000	6,000	8.3%
Goods available for sale	89,000	97,000	8,000	9.0%
Less: Inventory, December 31	19,000	18,000	1,000*	5.3%*
Cost of goods sold	$ 70,000	$ 79,000	9,000	12.9%
Gross Profit	$ 28,500	$ 30,000	1,500	5.3%
Operating Expenses				
Selling	$ 17,000	$ 19,500	2,500	14.7%
Rent	4,000	4,500	500	12.5%
Salaries and wages	700	1,000	300	42.9%
Miscellaneous	600	900	300	50.0%
Total operating expenses	$ 22,300	$ 25,900	3,600	16.1%
Net Profit	$ 6,200	$ 4,100	2,100*	33.9%*

*Denotes a decrease

B. 1.

BOB'S BEEFLIKE BURGERS
COMPARATIVE INCOME STATEMENTS
For the Years Ended December 31, 1996 and 1997

	1996		1997	
	Amount	Percent	Amount	Percent
Revenues				
Sales	$102,000	103.6%	$113,000	103.7%
Less: Sales returns	3,500	3.6%	4,000	3.7%
Net sales	$ 98,500	100.0%	$109,000	100.0%
Cost of Goods Sold				
Inventory, January 1	$ 17,000	17.3%	$ 19,000	17.4%
Purchases	72,000	73.1%	78,000	71.6%
Goods available for sale	89,000	90.4%	97,000	89.0%
Less: Inventory, December 31	19,000	19.3%	18,000	16.5%
Cost of goods sold	$ 70,000	71.1%	$ 79,000	72.5%
Gross Profit	$ 28,500	28.9%	$ 30,000	27.5%
Operating Expenses				
Selling	$ 17,000	17.3%	$ 19,500	17.9%
Rent	4,000	4.1%	4,500	4.1%
Salaries and wages	700	0.7%	1,000	0.9%
Miscellaneous	600	0.6%	900	0.8%
Total operating expenses	$ 22,300	22.6%	$ 25,900	23.8%
Net Profit	$ 6,200	6.3%	$ 4,100	3.8%

Section 19.3

A. 1. 2.30
     3. 1.50
     5. 2.37

B. 1. 1.32
     3. 1.06
     5. 1.15

C. 1. 10.5%
     3. 13.6%
     5.   8.5%

D. 1. $22,550; 9.69
     3. $46,000; 11.79
     5. $23,650; 18.08

E. 1. 1.37
     3. 1.84
     5. 7.6%
     7. 5.58
     9. (a) 10.0   (b) 7.1

Self-Test 19

1.

WOODY'S LUMBER YARD
COMPARATIVE BALANCE SHEETS
December 31, 1996 and 1997

	1996	1997	Increase or Decrease*	
			Amount	Percent
Assets				
Currents Assets				
Cash	$ 42,000	$ 39,000	$ 3,000*	7.1%*
Merchandise inventory	96,000	117,000	21,000	21.9%
Accounts receivable	82,000	87,000	5,000	6.1%
Total current assets	$220,000	$243,000	23,000	10.5%
Plant and Equipment				
Land	$ 45,000	$ 61,000	16,000	35.6%
Buildings	83,000	105,000	22,000	26.5%
Equipment	89,000	92,000	3,000	3.4%
Total plant and equipment	$217,000	$258,000	41,000	18.9%
Total assets	$437,000	$501,000	64,000	14.6%
Liabilities and Owners' Equity				
Current Liabilities				
Accounts payable	$123,000	$119,000	4,000*	3.3%*
Wages payable	45,000	47,000	2,000	4.4%
Total current liabilities	$168,000	$166,000	2,000*	1.2%*
Long-Term Liabilities				
Mortgage payable	96,000	91,000	5,000*	5.2%*
Total liabilities	$264,000	$257,000	7,000*	2.7%*
Owners' Equity				
Total owners' equity	173,000	244,000	71,000	41.0%
Total liabilities and owners' equity	$437,000	$501,000	64,000	14.6%

*Denotes a decrease

3.

THE WRITE STUFF
COMPARATIVE INCOME STATEMENTS
For the Years Ended December 31, 1996 and 1997

	1996		1997	
	Amount	**Percent**	**Amount**	**Percent**
Revenues				
Sales	$122,000	103.8%	$129,000	103.2%
Less: Sales returns	4,500	3.8%	4,000	3.2%
Net sales	$117,500	100.0%	$125,000	100.0%
Cost of Goods Sold				
Inventory, January 1	$ 26,000	22.1%	$ 24,000	19.2%
Purchases	53,000	45.1%	57,000	45.6%
Goods available for sale	79,000	67.2%	81,000	64.8%
Less: Inventory, December 31	24,000	20.4%	25,000	20.0%
Cost of goods sold	$ 55,000	46.8%	$ 56,000	44.8%
Gross Profit	$ 62,500	53.2%	$ 69,000	55.2%
Operating Expenses				
Selling	$ 14,000	11.9%	$ 15,500	12.4%
Rent	24,000	20.4%	26,000	20.8%
Salaries and wages	1,700	1.4%	1,900	1.5%
Miscellaneous	1,600	1.4%	1,800	1.4%
Total operating expenses	$ 41,300	35.1%	$ 45,200	36.2%
Net Profit	$ 21,200	18.0%	$ 23,800	19.0%

5. 1.27
7. 1.09
9. 9.6%
11. 6.42

CHAPTER 20: *Statistics and Graphs*

Section 20.1

A. 1. Mean = 72;   Mode = 70;   Median = 70;   Range = 25
    3. Mean = 127 rounded;   Mode = 125;   Median = 125;   Range = 29
    5. Mean = 0.04 in.;   Mode = 0.02 in.;   Median = 0.035;   Range = 0.06 in.
    7. Mean = 38;   Mode = 40;   Median = 38;   Range = 5
    9. Mean = 2725;   Mode = 2540;   Median = 2670;   Range = 650
   11. Mean = $47,466;   Mode = none;   Median = $48,301;   Range = $37,263
   13. Mean = 18 rounded;   Mode = 16 and 17;   Median = 17;   Range = 15
   15. Mean = $2.58;   Mode = $2.43;   Median = $2.60;   Range = $0.35

B. 1. (a) Mean population = 4,399,552 rounded   (b) Median = 4,233,382
       (c) range = 10,240,215   (d) 2.7% rounded   (e) 29.4%
    3. Mean = $26,331;   Median = $25,903;   Range = $31,267

   5. Mean daily sales = $441.89;   Range = $194.87

   7. (a) Mean = 427 rounded   (b) Median = 442 rounded   (c) Range = 399   (d) 44.0%
      (e) 14.7%

   9. (a) Mean = $154,811 rounded;   Median = $124,000;   Mode = $124,000
      (b) Range = $234,600   (c) 110.8%

  11. (a) $3861.33   (b) $3906   (c) $3250   (d) $1281

Section 20.2

A. 1. (a) 19 disabled beneficiaries per 1000 workers   (b) 1.6%   (c) Decrease from 1968
       to 1969   (d) 40,000 recovered   (e) 2,000,000 disabled

   3. (a) 15%   (b) 3%   (c) Taxes = $25,200   Rent = $63,000

   5. (a) range   (b) refrigerator   (c) 50,000 BTU/hr   (d) 1,200,000 BTU/day
      (e) Range = 65,000 BTU/hr   (f) 6,250 BTUs   (g) 15,000 BTUs

B. 1.

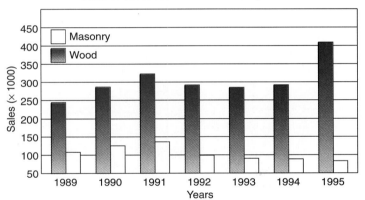

3.

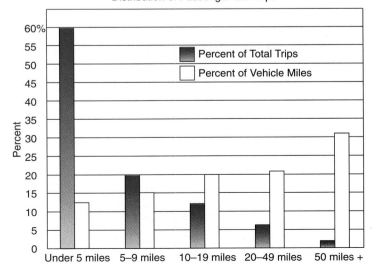

5.

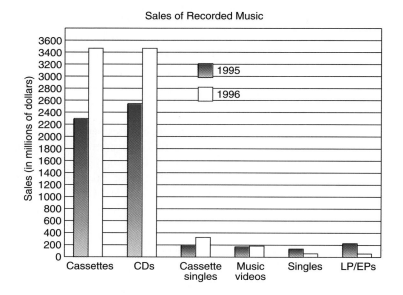

7.

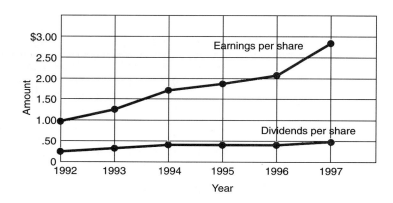

9.

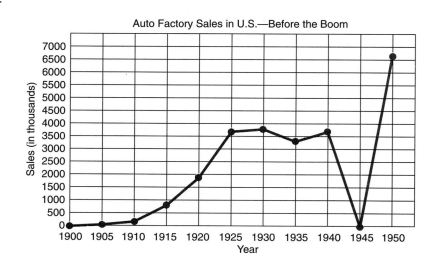

11.

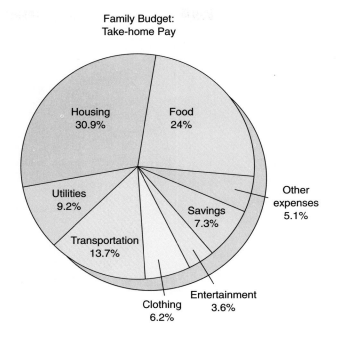

Family Budget:
Take-home Pay

TOTAL INCOME = $2750

	Amount	Percent	Angle	Angle Rounded
Food	$660	24%	86.4°	86°
Housing	850	30.9%	111.3°	111°
Utilities	252	9.2%	33.0°	33°
Transportation	378	13.7%	49.5°	50°
Clothing	170	6.2%	22.3°	22°
Entertainment	100	3.6%	13.1°	13°
Savings	200	7.3%	26.2°	26°
Other expenses	140	5.1%	18.3°	19°

(18.3° rounded to 19° to get 360° total)

13.

Government Budget:
Center City

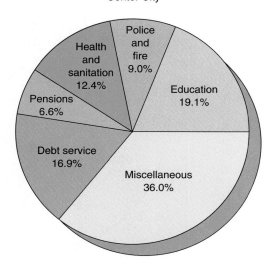

TOTAL BUDGET = $89,000,000

	Amount (in millions)	Percent	Angle	Angle Rounded
Education	$17	19.1%	68.8°	69°
Police • Fire	8	9.0%	32.4°	32°
Health • Sanitation	11	12.4%	44.5°	45°
Pensions	6	6.6%*	24.3°	24°
Debt Service	15	16.9%	60.7°	61°
Misc.	32	36.0%	129.4°	129°

*(we rounded to 6.6% to get 100%)

15. (a) 22% (b) $582.40 (c) $1383.20

Self-Test 20

1. Mean = 49; median = 50; mode = 50; range = 14
3. Mean = $109,000; median = $108,500; no mode; range = $25,000
5.

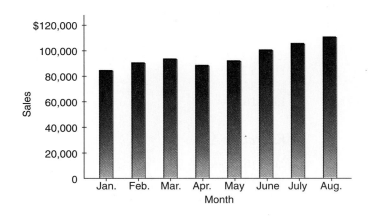

7.

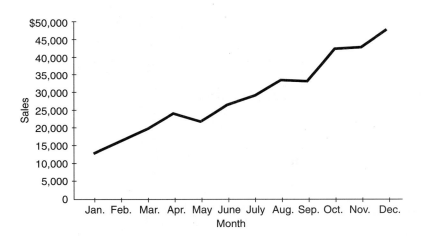

9.

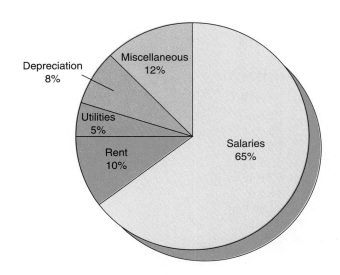

11. (a) December (b) January (c) $35,300 (d) October (e) May (f) 13.8%

APPENDIX A: *The Electronic Calculator*

Exercises A.1

1. 5104
3. 7809
5. 2,360,256
7. 20.2060
9. 36.2392
11. 15.2819
13. $34.86
15. $16.96

Exercises A.2

1. 1161
3. 198.82
5. 1670.328
7. 0.0010
9. 0.0047
11. 2259.2115
13. 8.6429
15. 590.2292

APPENDIX B: *Metric System*

Exercises B.1

1. 12,300 g
3. 3.75 dg
5. 0.643 km
7. 74,000 g
9. 7.35 kℓ
11. 370 dam
13. 580 mℓ
15. 650 g
17. 250 mg
19. 5040 dg

Exercises B.2

1. 27.06 lb
3. 3.66 m
5. 18.95 ℓ
7. 2.72 kg
9. 1.85 gal
11. 9.32 mi
13. 16.4 ft
15. 41.86 km
17. 2.84 ℓ
19. 25.4 cm

Index

accurate simple interest 321–23
addends 11
adding
    decimal numbers 109–110
    fractions 83–91
    whole numbers 11–22
accumulated depreciation 512
acid-test ratio 664–65
ACRS depreciation 533–35
aliquot parts 145
amortization 469–74
    interest payment 469
    principal payment 469
    schedule 471–72
    using a calculator 471
annuities 424–429
    due 424–27
    ordinary 424, 427–29
    table 426
    using a calculator 428–29
APR
    approximating 372–73
    converting from monthly rate 386
    converting to monthly rate 385
    installment loans by table 373–76
    table 374–75
assessed value 571
assets 629
average 680
average cost inventory method 483–84
average daily balance 388–90

balance 180
balancing 195
balance sheets 628–639
    assets 629
    current assets 629
    current liabilities 629
    fixed assets 629
    fundamental accounting equation 628
    horizontal analysis 630–34
    liabilities 629
    long-term liabilities 629
    owners' equity 629
    using a calculator 639
    vertical analysis 634–39
bank discount 351–363
    interest-bearing notes 357–360

noninterest-bearing notes 351–356
    using a calculator 353, 359
bank records 177–210
    completing check stubs 182–84
    making a deposit 179–80
    opening an account 178
    reconciling checking records 195–200
    using a check register 185–87
    writing a check 180–82
bankers' interest 326
bar graphs 697–702
base 140, 153
bearer bond 446
bonds 444–449
    bearer 446
    coupon 446
    current yield 447–48
    interest 445–46
    listing 446
    price 444
    registered 445
    using a calculator 448–49
book value 512
business graphs 697
buying 241–273
    cash discounts *See* cash discounts
    trade discounts *See* trade discounts

calculator A1–A13
    amortization 471
    annuities 428–29
    balance sheet 639
    bonds 448
    cash discounts 245, 246
    coinsurance 591
    compound interest 406
    depreciation 516, 524
    federal income tax 554
    finance charges 371
    financial statements 639, 650, 668
    horizontal analysis 650
    income statement 650
    insurance 591
    interest-bearing note 359
    inventory 482, 495, 668
    investments 428, 448
    markdowns 298, 299
    markup based on cost 281

markup based on selling price 290
mean 680–84
mortgages 465
noninterest-bearing note 353
payroll 216, 229
pie chart 706–11
present value 413–14
ratio analysis 668
real estate 465
sales tax 570
simple interest 322, 328
sinking fund 435–36
statistics and graphs 679–721
trade discounts 248
vertical analysis 639
canceled checks 195
capital expenditures 510
cash discounts 257–268
    due date 257–58
    EOM dating 260–61
    extra dating 262–63
    partial payments 263–65
    prox dating 260–61
    ROG dating 262
    using a calculator 245, 246, 248, 249, 264
cash surrender value 611
chain discount 245
check 178
check register 185
check stub 182
checking account *See* bank records
circle graphs 706–11
commission 226
common stock 439
compound interest 398–414
    annually 398
    daily interest 404–405
    daily interest table 405
    effective rate 402–403
    maturity value by table 399–402
    table 400
    using a calculator 406
comprehensive insurance 604–605
coinsurance 590–91
collision insurance 603–604
cost 276
cost of goods sold 648

coupon bond 446
cumulative preferred stock 440
current assets 629
current liabilities 629
current ratio 663
current yield, bond 447

days-in-year-calendar 313
decimal digits 108
decimal numbers 105–138
    adding 109–110
    converting from a fraction 127–29
    converting from a percent 146
    converting to a fraction 129–30
    converting to a percent 143
    dividing 118–126
    multiplying 117–18
    reading 106
    rounding 120
    subtracting 110–116
declining-balance depreciation 521–24
deductible, motor vehicle insurance
        603–604
deductions 546
denominator 57
deposit 179
depreciation 509–540
    accumulated depreciation 512
    ACRS method 533–35
    book value 512
    declining-balance method 521–24
    double-declining-balance method 521
    estimated units-of-production 513
    initial cost 510
    MACRS method 535–38
    salvage value 510
    schedule 512–13
    straight-line method 510–12
    sum-of-the-years'-digits method
        524–28
    total depreciation 510
    units-of-production method 513–16
    useful life 510
    using a calculator 516, 524
difference 24
differential piecework 224–26
digit 2
discount 242
dividend 41
    stock 439–42
dividing
    decimal numbers 118–126
    fractions 75–82
    whole numbers 41–50
divisor 41
double bar graph 698
double declining-balance method 521
driver classification 599
due date, calculation 257–58

effective rate 402–403
endowment, life insurance 611–12
EOM dating 260–61
exact time 310–314
    calendar method 313–14
excise tax 570
extended term insurance 613
extra dating 262–63

face value 582
face value of note 350
factors 31
federal income tax 546–60
    percentage method 553–57
    percentage method tables 555–56
    using a calculator 554
    wage bracket method 548–52
    wage bracket tables 549–52
federal tax deposit coupon 546–47
federal unemployment tax 565
Federal Unemployment Tax Act 565
FICA 557–60
    Medicare 557–58
    rates 557–58
    Social Security 557–58
FIFO inventory method 484–85
finance charges 370
    on average daily balance 388–90
    on unpaid balance 386–88
    using a calculator 371, 387
financial statements 627–70
    acid test 664
    balance sheets 628–39
    current ratio 663
    horizontal analysis 630–34, 649–50
    income statements 647–57
    inventory turnover 666–68
    rate of return on equity 666–67
    ratio analysis 663–68
    using a calculator 639, 650, 668
    vertical analysis 634–39, 653–57
fire insurance 582–591
    coinsurance 590–91
    multiple insurers 589–90
    premiums 582–85
    proration 586–88
    short-term rates 584–85
    short-term table 585
fixed assets 629
fractions 55–105
    adding 83–91
    changing an improper fraction to a
        mixed number 58–59
    converting a mixed number to an
        improper fraction 59
    converting from a percent 147
    converting to a decimal number 127–29
    converting to a percent 144
    division 75–82

improper 57
    least common denominator 86
    mixed number 58
    multiplication 67–74
    proper 57
    raising to higher terms 60–61
    reducing a fraction 61–63
    renaming fractions 55–66
    subtraction 92–99
fundamental accounting equation 628

graphs 697–711
    axes 697, 699
    bar 697–702
    double bar 698
    line 702–706
    pie chart 706–11
    See statistics and graphs
gross pay 213, 546
gross profit 648

horizontal analysis
    balance sheets 628–639
    income statements 649–53
    using a calculator 650
hourly pay 213

improper fractions 57
income statements 647–57
    cost of goods sold 648
    gross profit 648
    horizontal analysis 649–53
    net profit 648
    net sales 648
    reserves 648
    total operating expenses 648
    using a calculator 650
    vertical analysis 653–657
initial cost 510
installment loans 370–376
    APR, approximation 372–73
    APR, by table 373–76
    APR table 374–75
    finance charges 370
    rule of 78 379–82
insurance 581–618
    fire 582–591
    life insurance 611–15
    motor vehicle 599–606
    policy 582
    using a calculator 591
insured 582
insurer 582
interest 321
    compound 398–414
    present value 411–14
interest-bearing note 350
    bank discount 350–363
    using a calculator 359

interest payment 469
intermediate price 245
inventory 479–92
    average cost method 483–84
    FIFO method 484–85
    LIFO method 485–87
    perpetual 480
    physical 480
    retail method 493–95
    specific identification method 481–82
    turnover 666–68
    using a calculator 482, 495
investments 423–52
    annuities 424–29
    annuities due 424–27
    bonds 444–49
    ordinary annuities 424, 427–29
    sinking funds 433–36
    sinking fund table 434
    stock 439–444
    using a calculator 428, 448

knuckle method 257

least common denominator 86
liabilities 629
liability insurance 600
life insurance 611–15
    cash surrender value 611, 613
    endowment 611–12
    extended term 613
    limited-payment 611
    ordinary 611
    paid-up 613
    premiums 612–13
    settlement options 615
    surrender options 613–15
    surrender options table 613
    term 611
LIFO inventory method 485–87
like fractions 83
limited-payment life insurance 611
line graphs 702–706
liquid assets 664
list price 242
loans
    bank discount 351–363
    complex loans 369–390
    finance charge on average daily balance 388–90
    finance charges on unpaid balance 386–88
    installment loans 370–76
    open-end credit 385–90
    simple interest 321–333
long-term liabilities 629

MACRS depreciation 535–38
markdowns 297–301

rate 300–301
    series of markdowns 298–300
    shortcut 299
    using a calculator 298, 299
market rate 463
markup 276
markup based on cost 276–285
    finding cost 279–281
    finding markup 277
    finding rate 278
    using a calculator 281–82
markup based on selling price 287–291
    finding markup 287–88
    finding rate 288
    finding selling price 289
    using a calculator 290
mean 680–684
    from grouped data 682–84
median 684–85
medical payments 605
Medicare 557–58
metric system B1–B7
mills 572
minuend 23
mixed number 58, 84
mode 686
mortgages 460–68
    discount 464–65
    market rate 463
    P & I payments 460–62
    points 463–65
    total interest 462
    using a calculator 465
motor vehicle insurance 599–606
    collision 603–604
    comprehensive 604–605
    deductible 603
    driver classification 599
    liability insurance 600–602
    medical payments 605
multiplicand 31
multiplier 31
multiplying
    decimal numbers 117–18
    fractions 67–74
    whole numbers 31–40

net price 242
net proceeds 351
net profits 648
net sales 648
noninterest-bearing note 350
    bank discount 351–56
    using a calculator 353
numeral 2
numerator 57

odd lot 443
ordinary life insurance 611

ordinary simple interest at 30-day month time 325–26
ordinary simple interest at exact time 327–29
outstanding checks 197
overhead 493–98
    floor space method 497–98
    sales volume method 495–97
overtime pay 214
owner's equity 629

P & I payments 460–62
    table 461
paid-up insurance 613
partial payments, cash discounts 263–65
part of a percent problem 153
par value, stock 439
payroll 211–39
    commission 226
    differential piecework 224–26
    hourly pay 213–214
    overtime pay 214–17
    piecework 223–24
    salary 212
    salary plus commission 228–29
    sliding scale commission 227–28
    straight commission 226
    using a calculator 216, 229
percent 139–176
    %-type problems 159–61
    B-type problems 162–65
    converting from a decimal number 143
    converting from a fraction 144
    converting to a decimal 146
    converting to a fraction 147
    converting to and from 143–51
    decreases 161
    increases 161
    P-type problems 155–58
    solving percent problems 153–65
percentage method, federal income tax 553–57
perpetual inventory system 479
physical inventory 479
piecework 223–24
pie charts 706–11
place value system 2
preferred stock 439
premium 582
present value 411–14
    table 412
    using a calculator 413–14
price-earnings ratio 443
principal 321, 460
principal payment 460
product 31
promissory note 350
proper fractions 57
property tax 571–72

proration 586–88
prox dating 260–61

quotient 41

range 686–87
rate of return on equity 665–66
ratio 493
ratio analysis 663–68
    acid test ratio 664–65
    current ratio 663
    inventory turnover 666–68
    rate of return on equity 665–66
    using a calculator 668
real estate 459–474
    amortization 469–474
    discount 464–65
    mortgages 460–68
    P & I payments 460–62
    points 462–65
    using a calculator 465
real property tax 571–72
registered bond 445
remainder 43
repeating decimal 127
replacement value 582
retailer 244
retail method of inventory 493–95
revenues 648
ROG dating 262
rounding numbers 120
    whole numbers 5
round lot 443
rule of 78 379–82
    sum of digits 379–82

salary 212
salary plus commission 228–29
sales tax 569–70
    using a calculator 570
salvage value 510
selling 275–301
    markdowns *See* markdowns
    markup based on cost *See* markup
        based on cost
    markup based on selling price *See*
        markup based on selling price
selling price 276
service charge 180
settlement options 615
short-term policy 584
signature card 178
simple interest 321–348
    accurate interest 321–23
    ordinary interest at 30-day month time
        325–326
    ordinary simple interest at exact time
        327–329

solving for other interest variables
    335–341
solving for other interest variables
    using a calculator 340–41
solving for principal 336–37
solving for rate 337–38
solving for time 339–41
time 310–19
single equivalent discount rate 250–51
sinking funds 433–36
    table 434
    using a calculator 435–36
sliding scale commission 227–28
social security 557–58
specific identification inventory method
    481–82
statement, bank 195
state unemployment tax 566
State Unemployment Tax Act 566
statistics and graphs 679–721
    bar graphs 697–702
    double bar graphs 698
    line graphs 702–706
    mode 686
    mean 680–84
    median 684–85
    pie charts 706–11
    range 686–87
    table 687–89
    using a calculator A1–A13
straight commission 226
straight-line depreciation 510–12
stock 439–44
    common 439
    cumulative preferred 440
    dividend 439–44
    listing 443
    odd lot 443
    par value 439
    preferred 439
    price-earning ratio 443
    round lot 443
subtracting
    decimal numbers 110–16
    fractions 92–99
    whole numbers 23–31
subtrahend 23
sum 11
sum of digits 379–80
sum-of-the-years'-digits depreciation
    524–28
surrender options 613–14

tables of data 687–89
taxes 545–574
    excise tax 570

federal income tax 546–60
federal income tax by the percentage
    method 553–57
federal income tax by the wage bracket
    method 548–52
federal tax deposit coupon 546–47
federal unemployment tax 565
FICA 557–60
    percentage method tables 555–56
    property tax 571–72
    sales tax 569–70
    state unemployment tax 566
    unemployment taxes 565–66
    W4 547
    wage bracket tables 549–52
terminating decimal 127
term insurance 611
thirty-day-month time, 314–16
time
    calendar method 312–314
    days-in-year calendar 313
    exact time 310–314
    knuckle method 257
    thirty-day-month time 314–16
total depreciation 510
total operating expenses 648
trade discounts 244–55
    chain discounts 245–249
    shortcut 248
    shortcut using a calculator 249
    single equivalent discount rate 250–51
    using a calculator 248

unemployment taxes 565–66
    federal 565
    state 566
units-of-production depreciation 513–16
unpaid balance 385
useful life 510

vertical analysis
    balance sheets 634–639
    income statements 653–57
    using a calculator 639

W4 form 547
wage bracket method, federal income tax
    548–52
whole numbers 1–54
    adding 11–22
    dividing 41–50
    multiplying 31–40
    rounding 5
    subtracting 23–30
    writing numbers in words 3–9
wholesaler 244